Ecological Research Monographs

Series Editor

Yoh Iwasa, Department of Biology, Kyushu University, Fukuoka, Japan

The book series Ecological Research Monographs publishes refereed volumes on all aspects of ecology, including Animal ecology Population ecology Theoretical ecology Plant ecology Community ecology Statistical ecology Marine ecology Ecosystems Biodiversity Microbial ecology Landscape ecology Conservation Molecular ecology Behavioral ecology Urban ecology Physiological ecology Evolutionary ecology The series comprise books and edited collections by international experts in their fields.

More information about this series at https://link.springer.com/bookseries/8852

Futoshi Nakamura

Editor

Green Infrastructure and Climate Change Adaptation

Function, Implementation and Governance

 Springer

Editor
Futoshi Nakamura
Laboratory of Ecosystem Management,
Graduate School of Agriculture
Hokkaido University
Sapporo, Japan

ISSN 2191-0707   ISSN 2191-0715   (electronic)
Ecological Research Monographs
ISBN 978-981-16-6790-9   ISBN 978-981-16-6791-6   (eBook)
https://doi.org/10.1007/978-981-16-6791-6

Cover illustration: A flood control basin as a green infrastructure and a pair of Red-crowned cranes nesting there (Photos provided by Hokkaido Development Bureau and Naganuma Town)

This Springer imprint is published by the registered company Springer Nature Singapore Pte Ltd.
The registered company address is: 152 Beach Road, #21-01/04 Gateway East, Singapore 189721, Singapore

Preface

This volume consists mainly of studies conducted during two projects, titled "Green Infrastructure with a Declining Population and Changing Climate: Assessment of Biodiversity, Disaster Prevention, and Social Values (2015–2018)" and "Complementary Role of Green and Gray Infrastructures: Evaluation from Disaster Prevention, Environment, and Social and Economic Benefit (2018–2021)," that were supported by the Environment Research and Technology Development Funds (4-1504 and 4-1805) of the Ministry of the Environment of Japan. The leader of the two projects was Dr. Futoshi Nakamura. Before these projects, Dr. Nakamura and other members from various scientific fields organized the Japan Green Infrastructure Association, which aims to promote the establishment of a scientific basis of green infrastructure (GI), to apply GI for biodiversity conservation and disaster risk reduction under a climate change-affected environment, and to exchange scientific knowledge of GI with other countries. Thus, this volume includes articles provided by the members of the GI association and foreign scientists that exchange academic information with the association.

The aims of this volume are (1) to introduce the progress of the conception, evaluation, implementation, and governance of GI in Japan and other countries; (2) to provide basic information regarding the structure, function, and maintenance of GI for scientists, university students, government officers, and practitioners; and (3) to accelerate the transformative changes from gray-based strategies to green- or hybrid-based strategies to adapt to climate change.

The ecosystem functions of forests and wetlands have long been recognized and analyzed quantitatively; thus, the idea of GI is not new. However, developed countries such as Japan have lost natural and seminatural GI due to the historical overuse of natural resources and intensive land use development. To compensate for the loss of GI functions, many engineering gray infrastructures have been built. For example, Japan has been building continuous artificial levees and large dams to protect human lives and assets from floods in the last century. However, the construction and maintenance costs of gray infrastructure are enormous, and they have various negative impacts on biodiversity and ecosystem services. According to the WWF Living Planet Report 2014 Living Planet Index (LPI: a measure of the state of the world's biological diversity based on population trends of vertebrate species), the freshwater index has shown the greatest decline of any of the biome-based indices. The main threats are habitat loss and degradation, such as through direct impacts from dams and unsustainable water extractions, followed by overexploitation.

Moreover, climate change adds another constraint on the use of gray infrastructure because extraordinary events caused by global warming may frequently exceed the planning level of gray infrastructure, resulting in a great deal of damage to human lives and properties. Japan experienced frequent occurrences of megaflood disasters, such as torrential rain disasters in 2018 that caused landslides and flooding, which killed 200 people, and Typhoon Hagibis in 2019, which caused 140 breaks of artificial levees, 25,000 ha of inundation, and 64 deaths.

Thus, the strategy of dependence on gray infrastructure may not be acceptable, specifically in a depopulating society such as Japan. The drastic aging and depopulation in Japanese society, like other developed countries, will likely have impacts on social security (such as medical services and care), pensions, tax revenues, and the maintenance of existing infrastructures and will lead to farmland and forest abandonment. The Ministry of Agriculture, Forestry and Fisheries of Japan identified an increase in abandoned farmlands in Japan from 130,000 ha in the late 1980s to 400,000 ha in 2011. This outlook appears pessimistic, but it may provide other opportunities that we did not have in the past. If we can implement the best management of land use changes, these areas will become GIs that play simultaneous roles in disaster prevention and biodiversity conservation.

The purposes of implementing GI differ among countries. The United States strongly focuses on stormwater management in urban areas by GI, which reduces the amount of flooding and the polluted runoff that reaches sewers, streams, and lakes. In contrast, GI refers to the ecological network in the European Union, which delivers a wide range of ecosystem services and functions, such as biodiversity conservation, water purification, air quality, space for recreation, and climate mitigation and adaptation. The purpose of introducing GI in Japan is relatively closer to that of EU countries. In the EU, in May 2013, the Commission adopted an EU-wide strategy promoting investments in green infrastructure to restore the health of ecosystems, ensure that natural areas remain connected together, and allow species to thrive across their entire natural habitat. In Japan, GI was promoted in the National Spatial Planning and Priority Plan for Social Infrastructure Development in 2015. However, scientific research on GI in Japan is limited and is still in the beginning stage; there are many uncertainties and knowledge gaps in dealing with GI, especially the structure and function of individuals and networks of GI and management strategies, along with the stewardship of local communities.

The current major concern for people worldwide is COVID-19. As indicated by many scientists, the spread of viruses, loss of biodiversity, and climate change are interrelated with each other, and all are generated from the overuse of natural resources by humans. Many plants and animals have become extinct due to intensive exploitation of forest, river, wetland, and underground resources. Global warming proceeds with carbon dioxide emissions associated with the consumption of fossil fuels and the destruction of forests and wetlands. Once natural ecosystems are destroyed by human actions, opportunities for pathogens to pass between animals and people will increase. Green infrastructure will improve the situation by preserving natural and seminatural ecosystems and by restoring healthy ecosystems and will function as an adaptation strategy for climate change.

As a result of the pandemic, individuals may have to refrain from traveling and may have to telework full-time, isolated from coworkers, friends, and family, which is very stressful and makes it difficult to maintain physical and mental health. GI in the hometowns of these individuals may provide opportunities for walking and jogging and other recreation activities, which alleviate anxiety and stress resulting from the COVID-19 pandemic. I think the ultimate goal of the introduction of GI into society is the improvement of the quality of life, health, and well-being of individuals and communities.

I am grateful to the Ministry of Environment, Japan, for providing financial support for the two projects. Additionally, I would like to thank all the authors for submitting their chapters, all the reviewers who were willing to review the original manuscripts, and Ms. Fumiko Yamaguchi and other staff at the publisher, Springer Japan, for their encouragement and assistance.

Sapporo, Japan   Futoshi Nakamura

Contents

Part VIII Governance

Chapter 1
Introduction

Futoshi Nakamura

Abstract Worldwide, Green Infrastructure (GI) has mainly been discussed from an adaptation strategy perspective in cities and urban areas. However, we believe that GI can also function in rural and suburban areas where depopulation is prominent. From 2015 to 2021, my colleagues and I have launched two projects, titled "Green Infrastructure with a Declining Population and Changing Climate: Assessment of Biodiversity, Disaster Prevention, and Social Values" and "Complementary Role of Green and Gray Infrastructures: Evaluation from Disaster Prevention, Environment, and Social and Economic Benefit," which were supported by the Environment Research and Technology Development Funds (4-1504 and 4-1805) of the Ministry of the Environment of Japan. This volume introduces some of our achievements in the projects. Additionally, I invited active foreign scientists from the United Kingdom and the United States to contribute their experiences and knowledge to this volume. As suggested by the studies, one of the important characteristics of GI is multifunctionality, which maintains biodiversity and traditional landscapes. Using a natural and seminatural GI network in a watershed, we are able to adapt to elevated disaster risks in a changing climate while sustaining traditional land use and restoring natural ecosystems that provide a suite of ecosystem services and human welfare.

Keywords Hybrid infrastructure · Flood control · Climate change · Ecosystem services · Biodiversity · Multifunctionality

The global average air temperature has been increasing over the long term; since the 1890s, it has risen at a rate of 0.72 °C per 100 years (Ministry of the Environment et al. 2018). The IPCC (2013) showed that precipitation is different from air temperature, revealing an increasing trend across the Earth and increases in North

F. Nakamura (✉)
Laboratory of Ecosystem Management, Graduate School of Agriculture, Hokkaido University, Sapporo, Japan
e-mail: nakaf@for.agr.hokudai.ac.jp

© The Author(s) 2022
F. Nakamura (ed.), *Green Infrastructure and Climate Change Adaptation*,
Ecological Research Monographs, https://doi.org/10.1007/978-981-16-6791-6_1

America and Europe at the midlatitudes in the Northern Hemisphere since the 1900s. In Japan, the fluctuation of yearly precipitation has increased since the 1970s, and the frequency of hourly heavy rains of 50 mm or more has also increased (Ministry of the Environment et al. 2018). The climate projections for the twenty-first century in Japan indicate that the mean precipitation may increase by more than 10% (Kimoto et al. 2005), and other projections predict an increase in the frequency of high-magnitude floods and a reduced discharge from snowmelt floods.

The Japanese archipelago frequently suffers from not only climate disasters but also geological disasters, such as earthquakes and volcanic eruptions. In 2011, a large-magnitude tsunami hit Tohoku district following the Great East Japan Earthquake in 2011 and caused more than 30,000 people to die or go missing. In 2018, an earthquake struck the Iburi district in southern Hokkaido and triggered numerous landslides, reaching 44 km^2 in total area, killing 43 people and injuring 782 people. After the Great East Japan Earthquake, Japan enacted a basic law for National Resilience Contributing to Preventing and Mitigating Disasters for Developing Resilience in the Lives of the Citizenry in 2013. Since then, fundamental plans and action programs for national resilience have been established at the national, prefectural, and local government levels.

The world's population growth rate is slowly falling, and the population is projected to level off or decrease before the end of this century (United Nations 2019). In particular, in at least 55 out of 235 countries or areas, including Japan, populations are predicted to decline between 2019 and 2050. The United Nations (2014) has provided estimates showing that population decline has already been occurring in Germany since 2005 and in Italy and Japan since 2010. In a drastically depopulating society such as that in Japan, it will be more difficult to maintain engineered gray infrastructure with limited tax income.

Considering the above-described circumstantial background together with the importance of biodiversity conservation and the United Nations Sustainable Development Goals in Japanese society, national policies for disaster risk reduction have started to change from gray measures using artificially structured facilities such as dams and dikes to more environmentally friendly measures such as green infrastructure (GI) and nature-based solutions. In 2014, the former Prime Minister Shinzo Abe announced that the "application of the concept of green infrastructure using our rich natural environment is socially and economically effective and of great importance. We should preserve the natural environment and ecosystem services for future generations and use their functions for disaster risk reduction."

Japan has been developing traditional measures for disaster prevention since the sixteenth century. These are nature-friendly technologies and are recognized as GI from the present perspective (Nakamura Chap. 2). Unfortunately, the significance and necessity of these measures have been forgotten since modern technologies were introduced from European countries in the Meiji era (1868–1912). However, these modern technologies are vulnerable to extraordinary events such as mega-tsunamis and floods. Thus, we need to learn more about the wisdom and philosophy of traditional knowledge and technology. This volume includes those contents.

Worldwide, GI has mainly been discussed from an adaptation strategy perspective in cities and urban areas (e.g., Gill et al. 2007; Keeley et al. 2013; Netusil et al. 2014). However, we believe that GI can also function in rural and suburban areas where depopulation is prominent (Nakamura et al. 2020). Moreover, to protect cities, which are generally situated at lower, downstream elevations, we should explore the preservation and restoration of forest GI in headwater basins and floodplain wetland GI along rivers from a catchment perspective. Additionally, disaster risk reduction by a hybrid of green and gray infrastructure has been examined for stormwater, floods, and coastal flooding (Keeley et al. 2013; Sutton-Grier et al. 2015; Zelner et al. 2016), but very few studies have quantitatively examined flood risk, biodiversity, and socioeconomic benefit by defining existing GI (e.g., forest and wetland in a catchment) and additional layered GI (e.g., flood control basin along a river).

Moreover, farmlands, especially paddy fields, are one of the prevalent land uses in Southeast Asian countries, including Japan, which has an Asian monsoon climate, and play vital roles in providing various ecosystem services, such as biodiversity conservation and rain and floodwater retention, in addition to rice production (Natuhara 2013). We interpreted this type of seminatural environment as GI and evaluated its functions in this volume. Recently, some of these farmlands have been abandoned in the depopulating society. These abandoned farmlands may lose ecological and hydrological functions or may succeed in quasi-original natural environments after abandonment where various ecosystem services are provided (biodiversity, water retention, water quality, and recreation) (Queiroz et al. 2014, Koshida and Katayama 2018, Hanioka et al. 2018).

From 2015 to 2021, my colleagues and I have launched two projects, titled "Green Infrastructure with a Declining Population and Changing Climate: Assessment of Biodiversity, Disaster Prevention, and Social Values" and "Complementary Role of Green and Gray Infrastructures: Evaluation from Disaster Prevention, Environment, and Social and Economic Benefit," which were supported by the Environment Research and Technology Development Funds (4-1504 and 4-1805) of the Ministry of the Environment of Japan. This volume introduces some of our achievements in the projects. Additionally, I invited active foreign scientists from the United Kingdom and the United States to contribute their experiences and knowledge to this volume. The chapters are summarized below.

In Part I, the concept, history, theoretical approach, and practical model of green and hybrid infrastructure (i.e., the combination of green and gray infrastructure) are introduced. Nakamura (Chap. 2) presented a conceptual model of GI and gray infrastructure based on the model introduced by Onuma and Tsuge (2018) and then developed a hybrid model by combining these two models. He also introduced historical GI for floodwater management in Japan, which still provides important insights for current river management in a changing climate. Onuma (Chap. 3) developed an optimal hybrid model by economically maximizing social net benefits. Taki (Chap. 4) introduced one of the most advanced flood management policies in Shiga Prefecture, Japan, implementing GI at the watershed scale. Osawa and Nishida (Chap. 5) classified the types of GI (natural, seminatural, and artificial) and presented the principle to evaluate the implementation potential of GI.

In Part II, the forest ecosystem as a GI is the focus. Forests cover 67% of the land area in Japan, and 40% of the forests are plantations. There is a long history of forest studies examining the effects of forests on hydrological cycles, including rainwater storage and floodwater discharge attenuation. One of the current major concerns in Japan is the abandonment of artificial forest management, which results in high-density and thin-diameter trees. Unmanaged plantations are very vulnerable to windthrow and landslides; therefore, they may not be able to sustain forest GI functions. Nakamura (Chap. 6) focused on riparian forests as an interactive zone of green and blue infrastructure and discussed adaptation strategies to climate change using riparian forest GI. Tamura (Chap. 7) examined the effect of forest management on water discharge using a runoff model, focusing on the evaporation of intercepted rainfall and water storage in forest soil. Nisbet et al. (Chap. 8) provided guidance on designing appropriate and cost-effective forests for water payment schemes in the United Kingdom that support tree planting and forest management to protect and improve water quality.

In Part III, river and floodplain GI, including paddy fields and other farmlands, are the focus. Muto and Yokokawa (Chap. 9) built a hydraulic simulation model that is able to calculate both surface water flooding and river flooding, and functions for reducing flood risks by paddy fields with proper land-use management were evaluated. Imai et al. (Chap. 10) examined the negative impact of floodwater retention function after the abandonment of a paddy field. Osawa et al. (Chap. 11) reviewed the effects of the consolidation and abandonment of paddy fields in recent years on ecosystem services represented by habitat provision and regulating services (i.e., flood control) besides rice production.

In Part IV, the flood control basin (FCB) in an agricultural landscape that provides habitats for wetland flora and fauna was studied with special reference to biological conservation. In the past, flood control dams and diversion channels were dominant measures to prevent flood disasters. However, these engineered gray infrastructures have a detrimental influence on river and floodplain biota and have recently tended to be avoided by managers and practitioners. In contrast, FCB can be regarded as GI, functioning by attenuating peak discharge during a flood and providing wetland environment for a wide array of plants and animals and recreational opportunities for people during ordinary times. Ishiyama et al. (Chap. 12) introduced a case study in the Chitose River, Japan, which features a network of FCBs, regarding how FCBs and their networks contribute to the regional species biodiversity of various taxa. Morimoto et al. (Chap. 13) studied the succession of wetland vegetation in the FCB and proposed management practices to enhance the species diversity of wetland vegetation. Nishihiro et al. (Chap. 14) highlighted the need for human intervention and activities to maintain the FCB function of biodiversity.

In Part V, GI adaptation strategies in cities and urban areas are the focus. Cities and urban areas are vulnerable to heavy rains associated with climate change. Roads, buildings, and parking lots are paved and covered by impermeable surfaces, which increases surface runoff, leading to poor water quality and elevated peak discharge in urban streams. Additionally, most urban residents are eager to relax in green space after their hard official work. In this regard, Fukuoka (Chap. 15) introduced GI projects in various countries from the site scale to the urban land-use scale and discussed GI visions and frameworks that are needed to provide broader perspectives, ranging from urban heat mitigation and water disaster reduction to healthy and walkable cities. Interestingly, Ueno et al. (Chap. 16) paid attention to the use of GI before and during the COVID-19 pandemic in Japan and found that GI plays a role in maintaining health and refreshment during the pandemic. Watanabe and Ishida (Chap. 17) proposed a comprehensive land-use plan combined with GI, considering flood risk reduction in a depopulated local town on Shikoku Island, Japan. Finally, a case study of Portland, Oregon, which has a long history of implementing GI, was introduced by Shandas and Hellman (Chap. 18). They discussed the potential for activating a "green grid" in Portland that may help alleviate ongoing socioeconomic disparities.

In Part VI, GI and hybrid infrastructures in coastal and estuary ecosystems are studied. Since the Great East Japan Earthquake in 2011 and the subsequent tsunami disaster, disaster risk reduction in coastal zones has been a major nationwide concern for Japanese people (Suppasri et al. 2013). In addition to tsunamis triggered by earthquakes, high tidal waves associated with the rising sea level in the changing climate and their combinations are also anticipated disasters that Japan will certainly face in the future. Yamanaka and Nakagawa (Chap. 19) examined the effects of hybrid infrastructure consisting of a seashore, coastal embankment, coastal forest, and dunes on the spread of tsunamis and/or tidal waves in combination with the rising sea level. Matsushima and Zhong (Chap. 20) examined the effects of sand coverage on seawall slopes with the aid of local citizens on vegetation establishment. Kawata (Chap. 21) recognized mangrove forests in Jakarta as a GI mitigating flood damage and noticed floating garbage problems that may hinder the growth and regeneration of mangrove forests. Kuwae et al. (Chap. 22) summarized the current status of "blue carbon (carbon captured by marine organisms)" initiatives and reviewed three carbon offset projects in Japan.

In Part VII, public preference and willingness to pay (WTP) regarding GI and hybrid infrastructure are investigated. Shoji et al. (Chap. 23) found heterogeneous responses of the general public depending on their background knowledge of GI. Tsuge et al. (Chap. 24) conducted a survey in areas where giant seawalls were constructed after the Great East Japan Earthquake and found that citizens were strongly concerned about the negative impact of higher seawalls on the natural environment. Omori et al. (Chap. 25) quantified the economic value of coastal ecosystem services, including species richness, landscape, recreational services, and disaster risk reduction. They found that a hybrid infrastructure (seawalls + coastal forests) received higher positive responses. Valatin (Chap. 26) evaluated the climate change mitigation benefit of carbon storage in wood products and of carbon substitution associated with the use of wood instead of more fossil energy-intensive materials in the United Kingdom.

In Part VIII, governance systems to maintain and manage GI by various sectors and their collaborations are introduced. Asanami and Kamada (Chap. 27) studied the key role of the NPO group in maintaining collaborative activities for restoring and conserving pine forests in Kyushu, Japan. Kamada et al. (Chap. 28) introduced the Yolo Bypass Wildlife Area in the United States as a good example of GI governance through the collaboration of various sectors, such as federal, state, and local governments, landowners, NGOs, and citizens. Masuda et al. (Chap. 29) investigated the administrative plans of local governments across Japan to determine whether the plans contain the multifunctional features of farmland GI. They suggested effective strategies to implement GI, considering population, financial strength, extent of farmland and abandoned farmland, and flood risks.

As suggested by the above studies, one of the important characteristics of GI is multifunctionality, which maintains biodiversity and traditional landscapes. Using a natural and seminatural GI network in a watershed, we are able to adapt to elevated disaster risks in a changing climate while sustaining traditional land use and restoring natural ecosystems that provide a suite of ecosystem services and human welfare (Nakamura et al. 2020). In contrast, if we heavily depend on engineered gray infrastructure for disaster risk reduction, it may ruin biodiversity and traditional landscapes, as experienced with the seawall construction after the Great East Japan Earthquake. Even if the seawall guarantees safety against tsunamis in a return period of once every several decades to centuries, the local people who lost their houses to the tsunami would not return to live in a hometown where the original landscapes were destroyed by seawalls. The populations of Iwate, Miyagi, and Fukushima prefectures, where the tsunami disasters had prevailed in 2011, have rapidly decreased by approximately 15% between 2011 and 2020. Without considering an appropriate balance between gray and green infrastructures, we may create unacceptable land-use recovery plans for residents after such disasters. A land-use plan for adapting to climate change should be devised to allocate GI and gray infrastructure to function complementarily and comprehensively and to nurture local and regional landscapes in which GI contributes to an improved quality of life.

References

Gill SE, Handley JF, Ennos AR, Pauleit S (2007) Adapting cities for climate change: the role of the green infrastructure. Built Environ 33:115–133

Hanioka M, Yamaura Y, Senzaki M, Yamanaka S, Kawamura K, Nakamura F (2018) Assessing the landscape-dependent restoration potential of abandoned farmland using a hierarchical model of bird communities. Agric Ecosyst Environ 265:217–225

IPCC (2013) Climate change 2013: the physical science basis. In: Stocker TF, Qin D, Plattner GK, Tignor M, Allen SK, Boschung J et al (eds) Contribution of working group I to the fifth assessment report of the intergovernmental panel on climate change. Cambridge University Press, Cambridge, p 1535

Keeley M, Koburger A, Dolowitz DP, Medearis D, Nickel D, Shuster W (2013) Perspectives on the use of green infrastructure for stormwater management in Cleveland and Milwaukee. Environ Manag 51:1093–1108

Kimoto M, Yasutomi N, Yokoyama C, Emori S (2005) Projected changes in precipitation characteristics around Japan under the global warming. SOLA 1:085–088. https://doi.org/10.2151/sola.2005-023

Koshida C, Katayama N (2018) Meta-analysis of the effects of rice-field abandonment on biodiversity in Japan. Conserv Biol 32(6):1392–1402

Ministry of the Environment, Ministry of Education, Culture, Sports, Science and Technology, Ministry of Agriculture, Forestry and Fisheries, Ministry of Land, Infrastructure, Transport and Tourism, Meteorological Agency of Japan (2018) Kikou henndou no kannsoku yosoku eikyou hyouka nikannsuru tougou report 2018 nihonn no kikouhenndou to sono eikyou [Consolidated report on observations, projections and impact assessments of climate change: climate change and its impacts in Japan]. http://www.env.go.jp/earth/tekiou/report2018_full.pdf

Nakamura F, Ishiyama N, Yamanaka S, Higa M, Akasaka T, Kobayashi Y, Ono S, Fuke N, Kitazawa M, Morimoto J, Shoji Y (2020) Adaptation to climate change and conservation of biodiversity using green infrastructure. River Res Appl 36:921–933

Natuhara Y (2013) Ecosystem services by paddy fields as substitutes of natural wetlands in Japan. Ecol Eng 56:97–106

Netusil NR, Levin Z, Shandas V, Hart T (2014) Valuing green infrastructure in Portland, Oregon. Landsc Urban Plan 124:14–21

Onuma A, Tsuge T (2018) Comparing green infrastructure as ecosystem-based disaster risk reduction with gray infrastructure in terms of costs and benefits under uncertainty: a theoretical approach. Int J Disaster Risk Reduct 32:22–28

Queiroz C, Beilin R, Folke C, Lindborg R (2014) Farmland abandonment: threat or opportunity for biodiversity conservation? A global review. Front Ecol Environ 12(5):288–296

Suppasri A, Shuto N, Imamura F, Koshimura S, Mas E, Yalciner AC (2013) Lessons learned from the 2011 Great East Japan tsunami: performance of tsunami countermeasures, coastal buildings, and tsunami evacuation in Japan. Pure Appl Geophys 170(6):993–1018

Sutton-Grier AE, Wowk K, Bamford H (2015) Future of our coasts: the potential for natural and hybrid infrastructure to enhance the resilience of our coastal communities, economies and ecosystems. Environ Sci Pol 51:137–148

United Nations (2014) World population prospects: the 2012 revision. ESA/P/WP.235. United Nations, New York, NY

United Nations (2019) World population prospects 2019: data booket. ST/ESA/SER.A/424. United Nations, New York, NY

Zelner M, Massey D, Minor E, Gonzales-Meler M (2016) Exploring the effects of green infrastructure placement on neighborhood-level flooding via spatially explicit simulations. Comput Environ Urban Syst 59:116–128

Open Access This chapter is licensed under the terms of the Creative Commons Attribution 4.0 International License (http://creativecommons.org/licenses/by/4.0/), which permits use, sharing, adaptation, distribution and reproduction in any medium or format, as long as you give appropriate credit to the original author(s) and the source, provide a link to the Creative Commons license and indicate if changes were made.

The images or other third party material in this chapter are included in the chapter's Creative Commons license, unless indicated otherwise in a credit line to the material. If material is not included in the chapter's Creative Commons license and your intended use is not permitted by statutory regulation or exceeds the permitted use, you will need to obtain permission directly from the copyright holder.

Part I
Concept and Synthesis

Chapter 2
Concept and Application of Green and Hybrid Infrastructure

Futoshi Nakamura

Abstract Recently, Japan has suffered extraordinary damage from typhoons, heavy rains, and megafloods, each of which has exceeded the upper limit of control by managed infrastructure for flood mitigation. First, I present a conceptual framework of hybrid infrastructure at the watershed scale, combining (1) fundamental green infrastructure (GI), composed of forests and wetlands in the watershed; (2) additional multilevel GI, such as flood control basins; and (3) existing engineered disaster prevention infrastructure, such as dams and artificial levees. Second, I introduce the disaster risk reduction function of natural forests and wetlands and three representative traditional flood control measures: discontinuous levees, overflow embankments, and flood protection forests. This GI should be properly allocated and maintained at the watershed scale to reduce damage by megafloods. The multiple types and functions of GI may provide essential habitats for wildlife and recreational opportunities for local residents and others. Finally, I address key points for planning, implementation, and governance of GI at the watershed scale.

Keywords Hybrid infrastructure · Flood control · Climate change · Ecosystem services · Biodiversity · Ecological network

2.1 Introduction

Flood protection strategies using current gray infrastructure, such as dams and artificial levees, face limits and are not able to protect human lives and assets from megafloods (e.g., exceeding 100-year return interval) exacerbated under climate change. Here, I define gray infrastructure (GYI) as engineering infrastructure associated with water resources, such as water and wastewater treatment plants, pipelines, and reservoirs. Recently, Japan has been experiencing large flood disasters

F. Nakamura (✉)
Laboratory of Ecosystem Management, Graduate School of Agriculture, Hokkaido University, Sapporo, Japan
e-mail: nakaf@for.agr.hokudai.ac.jp

© The Author(s) 2022
F. Nakamura (ed.), *Green Infrastructure and Climate Change Adaptation*,
Ecological Research Monographs, https://doi.org/10.1007/978-981-16-6791-6_2

Fig. 2.1 Flooding of the Abukuma River by Typhoon Hagibis, Sukagawa city, central Japan (Ministry of Land, Infrastructure, Transport and Tourism, Japan)

every year associated with typhoons and seasonal rains, which have exceeded the river discharge that the infrastructure was designed to deal with (hereafter I call it "design flood discharge"). Therefore, floodwater spilled over the artificial levees and inundated farmlands and residential areas (Fig. 2.1).

There are two alternatives to respond to the increasing floodwater discharge: one is increasing the design flood discharge by conventional measures, such as construction of dams, excavation of channel sediment, setback of levees, and enhancement of levee height, and the other is keeping the design flood discharge at the current level while providing green infrastructure (GI) outside of the levees including forests on the hillside slopes and wetlands on floodplains to store rainwater and retain spilled floodwater there. Here, GI can be defined as "a strategically planned network of natural and semi-natural areas with other environmental features designed and managed to deliver a wide range of ecosystem services such as water purification, air quality, space for recreation and climate mitigation and adaptation" (European Commission 2016) (Fig. 2.2).

The revision of design flood discharge to a higher level is not easy. Reservoir dams have been built without considering future climate changes in design flood discharge. Floodwater control by levees and channelization started from downstream reaches (urban areas) where densely populated cities and towns are located and progressed to upper reaches (rural areas) where farmlands are developed. However, there are many rural areas in Japan where flood control measures have not yet been completed. Thus, if we raise the design flood discharge in response to climate change, we must restart flood control management again from the downstream reaches and leave upstream rural areas unprotected for a long time. Even if we can fulfill the flood control management corresponding to a certain higher level,

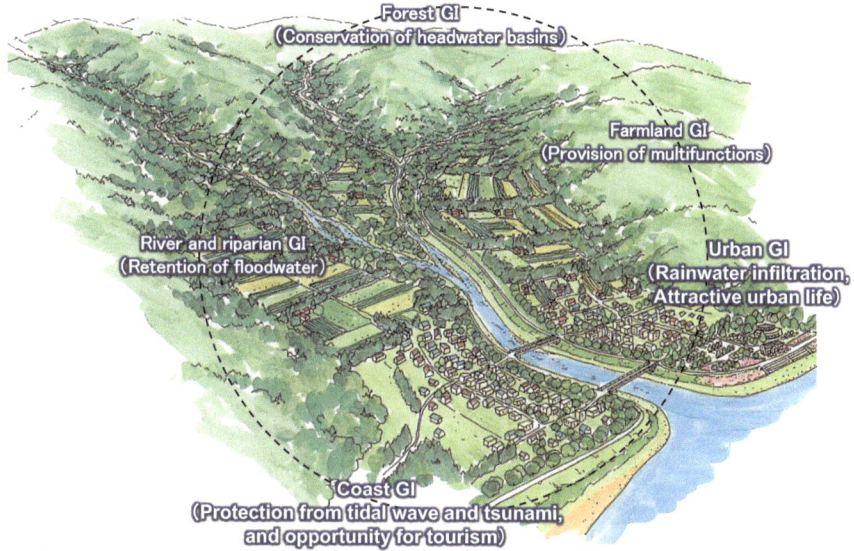

Fig. 2.2 Multi-functionality of GI and linkages at the watershed scale

intense flooding exceeding the revised management level is likely to occur under climate change. It is not realistic to revise the design flood discharge continuously and repeatedly.

When we explain green infrastructure (GI), we tend to situate GI as an opposing and conflicting concept against gray infrastructure (GYI). In some sense, this is true, but exaggerating the differences between GI and GYI may hinder productive discussions. Rather, under climate change, these two infrastructure types should complement each other and contribute to a comprehensive land-use plan that simultaneously provides amenities, convenience, and safety to local towns, cities, and communities (Green Infrastructure Association (Japan) 2017, 2020). For example, we consider that forest ecosystems are one of the most important types of GI and function to regulate water discharge from a basin. In Japan, someone who is opposed to dam construction may argue that forests are a surrogate for large dams (GYI); therefore, dam construction is not necessary if we leave forests in upper basins. This type of emotional argument is not productive; rather, we should manage plantation forests to keep appropriate tree densities and soil infiltration rates to reduce the dependence on water regulations by dams.

A suite of GI measures provide various ecosystem services across a watershed (Fig. 2.2). Our future land-use plan should be built from the viewpoint of how we can create a best mix of advantages provided by GI and GYI at the watershed scale. The objective of this chapter is to devise climate change adaptation strategies for disaster prevention and biodiversity conservation using GI in Japan. More specifically, I will (1) introduce a conceptual framework for hybrid infrastructure at the watershed level with concrete evidence of previous research results, (2) reevaluate Japanese traditional measures from the viewpoint of climate change adaptation, and (3) propose a best mix of GI and GYI at the watershed scale mainly focusing on flood protection and biodiversity conservation.

2.2   Conceptual Framework for Hybrid Infrastructure

First, I would like to define gray, green, and hybrid infrastructure at local and landscape (or watershed) scale. GYI at the local scale is represented by artificial elements, such as dams and channelization, whereas an individual natural ecosystem, such as a patch of wetland and forest, is at opposite end of the naturalness spectrum (Fig. 2.3). A mixture of natural and artificial elements, such as gardens and parks, is regarded as hybrid infrastructure. At the landscape (or watershed) scale, urban areas are built by various GYI, whereas wilderness areas consist of diverse natural ecosystems (Fig. 2.3). In this spectrum, a mosaic distribution of natural (or

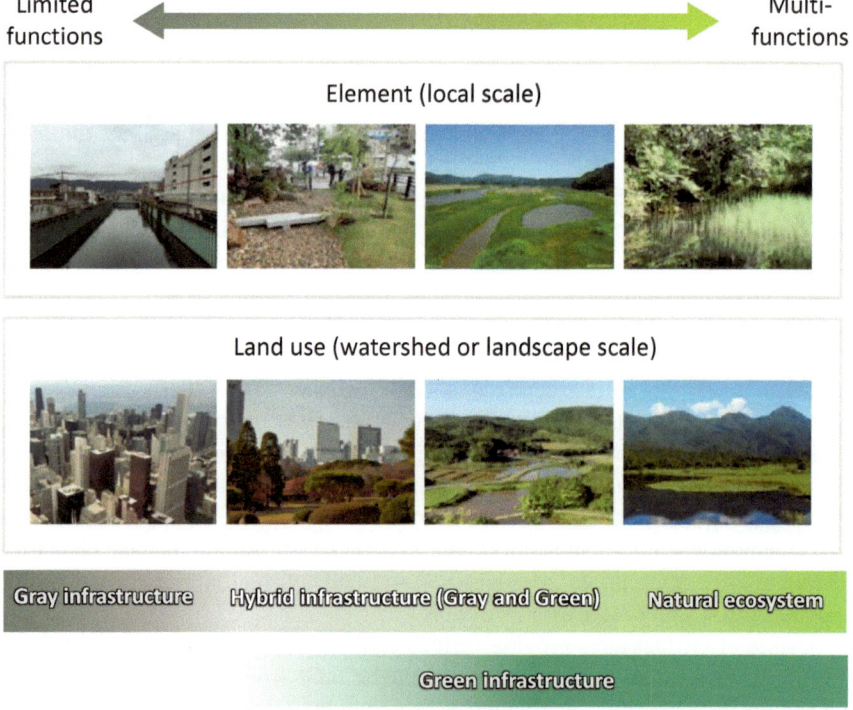

Fig. 2.3 Definition of green, gray, and hybrid infrastructure at local and watershed scale (adapted from Green Infrastructure Association (Japan) ed. (2020))

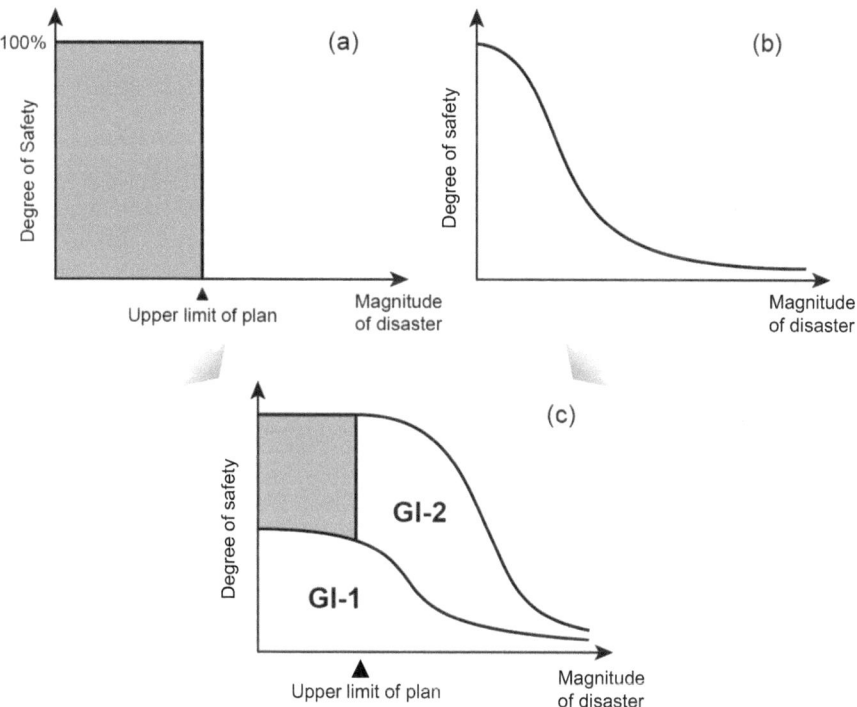

Fig. 2.4 Conceptual framework of gray (**a**), green (**b**), and hybrid infrastructure (**c**). Shaded and white areas denote the safety zones created by gray and green infrastructure (GI), respectively. The area denoted by GI-1 is the fundamental GI, while GI-2 is an additional multilevel GI. (**a**) and (**b**) are modified from Onuma and Tsuge (2018) (from Nakamura et al. 2020)

seminatural) and artificial ecosystems in a suburban and rural watershed can be referred to as hybrid infrastructure.

GYI elements, such as dams and artificial levees, usually assure 100% disaster protection until the magnitude of the disaster reaches the upper limit of the management plan, though unexpected risks associated with structural flaws and human errors still exist. However, once the magnitude exceeds the upper limit, the GYI will have no additional flood mitigation function; for example, floodwater will spill into residential areas where artificial levees are breached (see Fig. 2.1). Thus, the safety-magnitude curve for GYI has a rectangular shape (Fig. 2.4a). In contrast, I expect the response of GI to show a gradually decreasing trend. In addition, the disaster prevention function of GI may be sustained longer than that of GYI (Onuma and Tsuge 2018). However, the response curve of risk reduction against disaster magnitude by GI at the local scale is not well studied and can vary depending on the kind of GI (Fig. 2.4b). The uncertainty of the function is high for GI.

In the past, we have discussed the advantages and disadvantages of GYI and GI and compared them to determine which approach was better. Sometimes, such debates are not productive and promote the polarization of opinions between gray and green approaches. Here, I discuss a combination of GYI and GI with the aim of applying GI for disaster control in a society at high risk of various natural disasters, as is the case in Japan. I present a hybrid, combining the two types of infrastructure at watershed scale in Fig. 2.4c. In this conceptual diagram, GI-1 represents the fundamental GI composed of forests and wetlands in the watershed, while GI-2 is additional multilevel GI, such as flood control basins that function when floodwater exceeds the design flood discharge determined by artificial levees. In this diagram, an increase in the combined area of GYI and GI guarantees safety, even at a very high magnitude of flooding.

How much should we expand or reduce the areas of GYI or GI to meet present and future needs as we face climate change? We have to evaluate the effectiveness of hybrid infrastructure in terms of disaster prevention, biodiversity protection, and social and economic values to determine which combination is best in a given natural and social condition. If the area of GYI is expanded, high levels of disaster control may be achieved, but there may be losses in biodiversity and the hometown landscape and increases in maintenance costs. Historically, Japan has been losing forest and wetland GI-1 through overharvesting of forest resources and land conversion from wetland to agricultural fields (Nakamura et al. 2017). As a result, the peak flood discharge has gradually elevated, and Japanese government has had to compensate for the water retention ability that natural ecosystems provided in the past with GYI, such as dams and artificial levees. The combination of GYI and GI may change depending on future land use. We may preserve or rebuild GI using abandoned farmlands at lower cost in rural areas. In contrast, it may be difficult to restore natural ecosystems in highly populated urban areas, and GYI therefore still plays an important role in disaster risk reduction with a limited introduction of GI represented by gardens and city parks.

2.3   Attenuation of Flood Peaks by Forest and Wetland Ecosystems (Examples of GI-1 in Fig. 2.4c)

Natural forests and wetlands have the ability to reduce the disaster risk by attenuating and delaying flood peaks. The mechanism and detailed information are introduced in other chapters, and I will briefly explain their functions as GI-1 in Fig. 2.4c. These natural and seminatural ecosystems function to keep flood risks low by storing rainwater on hillslopes and floodplains at a watershed scale.

In one such example, Tsukamoto (1985) investigated how water discharge changed in association with urbanization and loss of forest cover (Fig. 2.5). From this figure, we understand that the peak discharge increased during floods, while the low-flow discharge decreased in the urbanized watershed. On the other hand, during the beginning of a series of rainfall events, the peak discharges from the forested watershed were greatly attenuated, and the low-flow discharge was slightly

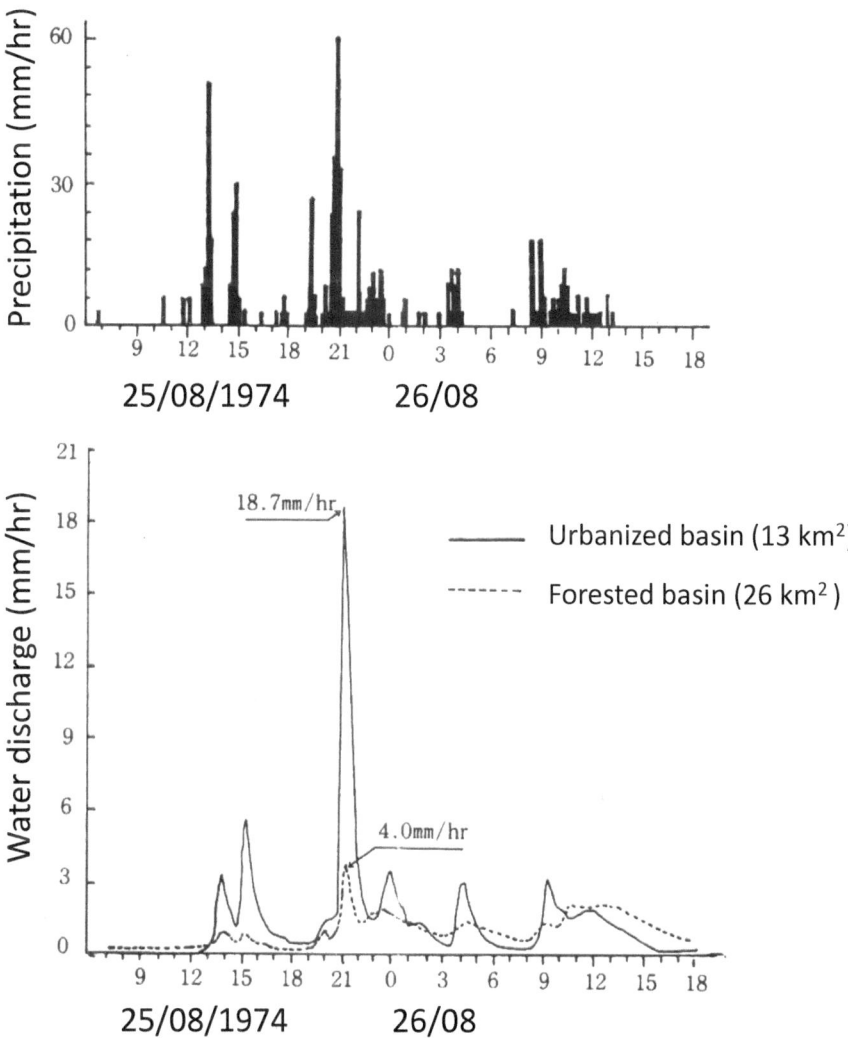

Fig. 2.5 Comparison of hydrographs between urbanized and forested basins (from Tsukamoto 1985)

increased. However, these effects were not clearly observed at the end of the sequence of rainfall events; the degree of peak discharge attenuation was reduced, and the low-flow discharge was greatly increased. The timing of flood peaks is almost simultaneous with rainfall peaks in the urbanized watershed throughout the investigated period in part because of their extensive paved surfaces. In contrast, delays in flood peaks can be observed in forested watersheds, especially at the end of the sequence of rainfall events. This is because most of the precipitation was absorbed by the forest soil and retained in the soil pores at the beginning of a series of rainfall events, and thereafter, precipitation added to the forest soils at the end of rainfall events was rapidly released from macropores and transported by saturated subsurface and surface flow. Thus, the flood risk reduction provided by forests may vary over time during a rainfall event. However, forest ecosystems certainly function as GI-1 and can stabilize variations in stream discharge.

Japan has traditionally recognized the functions of forest ecosystems, including the abovementioned hydrological effects, and developed forest protection laws starting in 1897. In contrast, in the past, wetlands have not been acknowledged as GI, but rather considered unproductive lands. Due to the lack of knowledge of wetland ecosystem services, wetlands have been converted to agricultural lands. Approximately 60% of the original wetlands in Hokkaido, a northern island in Japan, have disappeared over the last 50 years (Nakamura et al. 2005). However, people have gradually recognized the importance of the ecosystem services that wetlands provide, including biodiversity conservation, flood control, and improvement of water quality. In general, wetlands develop in relatively flat areas where streamwater and rainfall are retained, whereas forests become established not only in flat areas, but also on steep terrain. However, currently remaining forests are mainly on relatively steep land because most flat areas have been developed for other land uses. Thus, the remaining wetlands in flat areas are highly valued and should be retained and restored to capitalize on their capability of stormwater retention. In contrast, forests on steep slopes, where the hydraulic gradient is high, have lower water retention capability.

Nakamura et al. (2020) used the hydrological model GETFLOWS to simulate the effects of the Kushiro Mire on the water discharge of the Kushiro River in Hokkaido. The Kushiro Mire is the largest wetland complex in Japan, extending across approximately 26,000 ha. Three typhoons accompanied by heavy rains arrived in the Kushiro region from August to September 2016, producing three major flood peaks. The authors used the model to calculate water discharge for the case in which approximately 55% of wetlands were converted to residential lands. The simulated hydrograph of the modified wetlands showed a higher peak discharge, sharper rising and descending limbs, a 2-day earlier arrival of the peak, and a lower low-flow discharge than those of the present situation in which wetlands exist without 55% area-based land conversions (Fig. 2.6). The peak discharges of the simulation cases for the present and for the partial loss of wetlands were 390 and 580 m^3/s, respectively. The results of the model simulation clearly demonstrated that the Kushiro Mire acts as GI-1 (a large natural reservoir) that attenuates the peak discharge during floods.

As shown above, forests and wetlands in a watershed are fundamental GI that stabilize water discharge and substantially attenuate flood peaks. Thus, if we lose these GI-1 features, we will face a risk of flooding and have to compensate with more GYI. However, these GYI features lead to high maintenance costs and may not be economically feasible in depopulating societies, such as Japan and other countries.

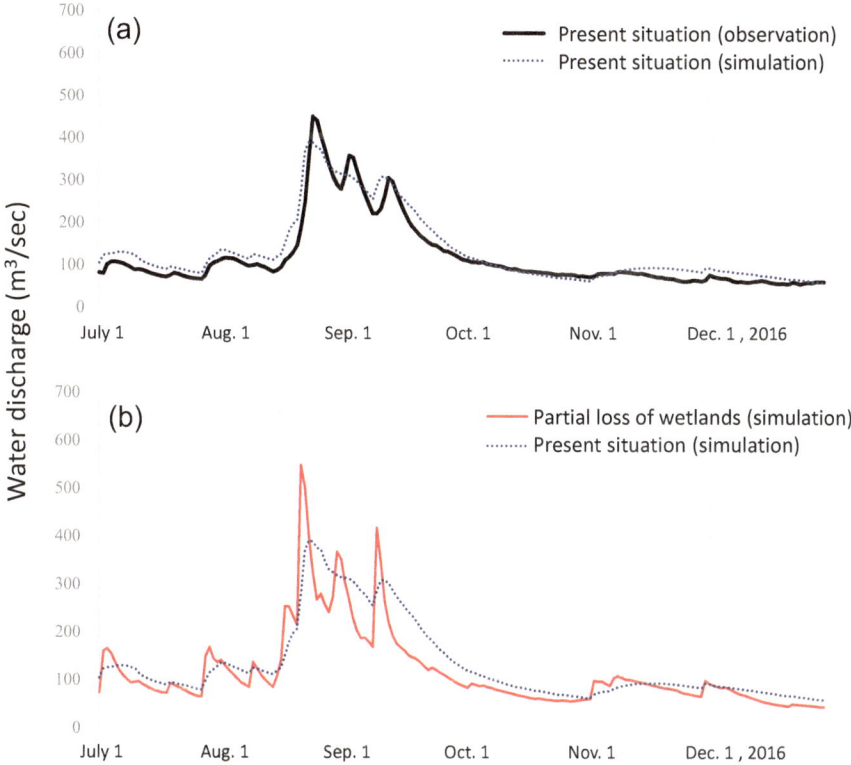

Fig. 2.6 Attenuation of flood peaks by the Kushiro Wetland. The fit of the simulation results to the observational hydrograph (**a**) and comparison of the simulation results with and without wetland green infrastructure (**b**) (from Nakamura et al. 2020)

2.4 Traditional Measures Against Large Floods (Examples of GI-2)

Japan has developed traditional flood control measures since the sixteenth century. These are nature-friendly technologies and are recognized as GI from the present perspective. Unfortunately, however, the significance and necessity of those measures for flood control have been forgotten since modern technologies were introduced from European countries in the Meiji era (1868–1912), which promoted channelization and artificial levee construction to convey floodwaters rapidly from headwater basins to river mouths. However, these modern technologies are vulnerable to megafloods, which easily exceed the design flood discharge. The Typhoon Hagibis disaster in 2019, for example, brought a huge amount of rainfall, and the floodwater level exceeded the upper limit of the artificial levees. As a result, artificial levees breached at 140 locations, and the floodwater broadly spilled into residential areas (see Fig. 2.1). The estimated flood damages were enormous.

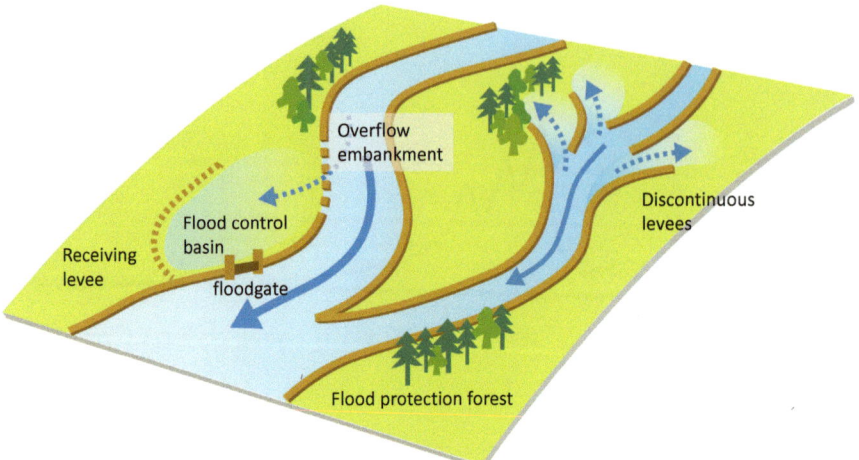

Fig. 2.7 Traditional flood control measures in Japan

Modern technologies for flood control tend to constrain floodwater in river channels by artificial levees, while traditional flood control measures do not confront flood power directly, but rather reduce it by the spread of the floodwaters over surrounding land-use areas, mainly agricultural lands. Thus, some of the floodwaters will be stored and retarded in land-use areas and slowly released when the floodwater level gradually recedes. This idea contrasts with modern techniques that focus on how rapidly floodwater can be discharged from basins. In the modern flood control framework, local and national governments tend to increase the heights of artificial levees to sustain safety in adjacent land-use areas once flood disasters occur. However, this policy encourages people to live closer to rivers and develop surrounding land-use areas more intensively. Under this situation, once artificial levees are breached by a large flood, the damage is enormous.

Here, I introduce three representative traditional flood control measures: discontinuous levees (*Kasumitei* in Japanese), overflow embankments (*Etsuryutei*), and flood protection forests (*Suiborin*) (Fig. 2.7). These measures still exist along Japanese rivers, but are greatly reduced through replacement with modern technologies. When the Japanese government now applies traditional measures to flood control, the intentional spillage of the floodwater into designated areas is required for some extraordinary floods exceeding the design flood discharge. The traditional measures including the designated areas should function as GI. In contrast, in modern techniques, we cannot predict where continuous artificial levees will break once the floodwater level exceeds the levee height. This may lead to serious disasters when levee breaches occur in intensively developed and populated areas. We should avoid this situation by learning the advantages of traditional measures and adopting GI.

2.4.1 Discontinuous Levees

The discontinuous levee was devised and developed by Shingen Takeda (Fig. 2.7), who was a military commander from 1521 to 1573. Discontinuous levees still existed along 54 rivers among the 109 major large rivers in 1987 in Japan (Okuma 1987). The roles of discontinuous levees differ among the locations where those measures were set. According to Okuma (1987), discontinuous levees built on high gradient (>1/500) alluvial, gravel-bed rivers aim to return floodwaters that spill over upstream levees and/or are generated from tributaries and inlands to the mainstem of the river. For example, many discontinuous, overlapping levees may be constructed over 1–2 km reaches, such as on the Tedori River, Ishikawa Prefecture (Fig. 2.8). In this case, even when the first levee closest to the river channel is broken by a flood, other levees situated behind the first levee will function to return the floodwater to the main channel repeatedly. This is a form of multiple protection measures.

In contrast, discontinuous levees built on low gradient (<1/1000) alluvial, sand-bed rivers aim to store floodwater by the backwater effect between discontinuous levees during periods of high discharge. Due to the gentle gradient of the riverbed, the storage capacity and retention effect of floodwater are expected to be large, and, therefore, substantially reduce the flood peak in the mainstem river.

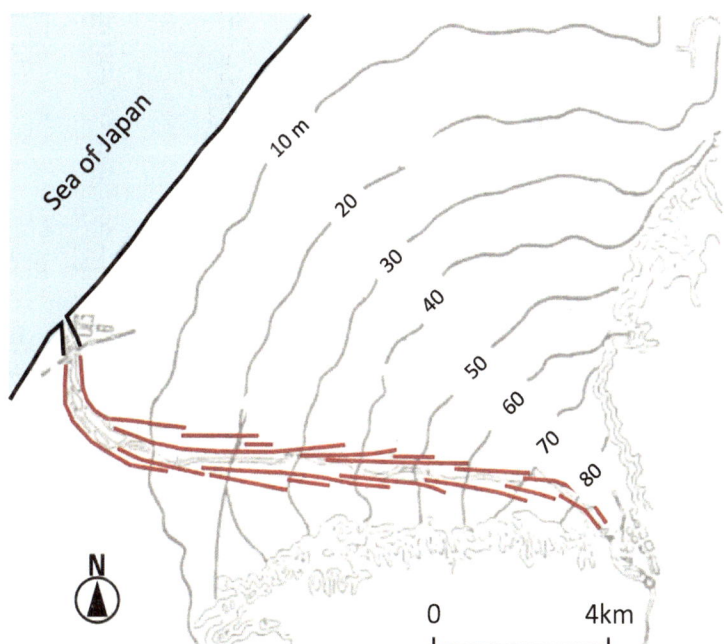

Fig. 2.8 Discontinuous levees (denoted by red lines) in the Tedori River, Ishikawa Prefecture

In both cases, natural confluences of mainstem and tributaries are maintained with discontinuous levees, and, thereby, the natural connectivity of the mainstem and tributaries can be sustained; otherwise, we have to build GYI, such as floodgate and sluice gate, to manage the discharge of tributaries into the mainstem. In general, there is an elevational gap between the water level of the mainstem and floodgate, and, therefore, it is difficult for fish to migrate into tributaries through the sluiceway. The open spaces created by the discontinuous levees function as GI and have been used for flood protection forests or agricultural lands. Moreover, the measure allows floodwater to inundate GI repeatedly, and, therefore, provides habitats for wetland plants and animals.

Currently, few local people and engineers understand the multiple functions of discontinuous levees; thus, many were changed to continuous levees following the approach of dam construction and flood control channels. In this way, functions originally provided by GI were replaced by those of GYI. However, Japan has recently experienced breaches of continuous levees, resulting in large flood disasters influenced by climate change.

The use of discontinuous levees represents the needed shift in Japan from a focus on constraining floodwater to the river channel to accepting more flow over floodplains. This is similar to a shift in thinking evident in Europe where the objectives are to cope with climate change and biodiversity conservation (Buijse et al. 2002). River managers in Japan and other developed countries should learn more about the advantages of traditional GI measures and keep them in our river landscape. Rather than the replacement of GI by GYI, we should consider the best mix of the two measures.

2.4.2 Overflow Embankment

The overflow embankment was devised and developed by Kiyomasa Kato, who was a military commander from 1562 to 1611. This measure determines in advance where the floodwater overflows the artificial levees (Fig. 2.7). Together with an overflow embankment, a secondary levee was constructed to receive the floodwaters, and the areas surrounded by the overflow embankment and receiving levees can function as GI-2.

Unfortunately, these traditional measures also tended to be removed after the introduction of modern technologies, such as strong embankment and road construction, and areas of GI have been developed as farmland and/or residential areas (Tanabe and Okuma 2001). However, most of the current artificial levees in Japan have not been equipped with overflow embankment system; therefore, we do not know where the floodwater will overflow the levees, which results in increased damage and delayed evacuation of residents. Thus, once the levees are partially breached by overbank flow, the exposure risk to flood damage increases. According to Nakajima et al. (2013), who conducted hydraulic simulations to evaluate the flood control effects of overflow embankments and discontinuous levees in the Jobarugawa River in Kyushu, the flood level in the mainstem did not change with increasing water discharge when floodwater overflowed these facilities.

2.4.3 Flood Protection Forest

Flood protection forests are one of 17 kinds of protection forests prescribed by the Japan Forest Law established in 1897. Flood protection forest is defined as riparian forests (including bamboo forest) located within, on, or outside of the levees (Fig. 2.7). Forests function to reduce damage to residential areas and farmlands behind levees during floods (Matsuura et al. 1988).

The root systems of trees and bamboo hold soils on levees and reduce flood velocity, thereby protecting riverbanks and levees from scouring and facilitating sediment deposition and driftwood retention. The floodwater running through protection forests is filtered and slowed, and, therefore, even if it inundates residential areas, the damage to houses and farmlands is greatly reduced. On the other hand, there is an argument that flood protection forests may add roughness elements to water flow, which may increase floodwater depth and increase the flooding risk. Moreover, they may not be able to withstand the flood power and may become a source of driftwood, which may increase the risks of wood-laden flooding. Also, floated large wood may damage riparian forest. In this situation, river managers may decide to cut flood protection forests. There are many uncertainties regarding the physical functions of flood protection forests and trade-offs to balance, and therefore further research is required.

To maintain the traditional GI, the understanding and cooperation of local communities and river managers are essential. Local residents who inhabit areas of GI may suffer flood damage repeatedly, and some watershed communities and government agencies should support them and compensate for this damage. In general, these GI features are situated in rural areas in the upstream and midstream reaches of the mainstream and tributaries. Thus, inundation by floodwater in upstream and midstream reaches certainly slows the arrival time of flood peaks and reduces the flood discharge in downstream reaches where cities and towns are generally situated. In this situation, people in urban areas should financially offset the damages in rural areas, and GI should be managed and maintained through the cooperative efforts of communities over the whole watershed. The education of river managers is also important. Young river managers may not know the functions and advantages of traditional GI in river management. We should publicly explain how these measures are applicable as an adaptation strategy for climate change.

2.5   Hybrid Infrastructure at Watershed Scale

The Japanese government has introduced new flood control management policies twice using hybrid (green/gray) infrastructure. There are some differences between these two management policies and the hybrid infrastructure concept in the present paper; the former strongly focuses on flood control measures by GYI, while the latter stresses GI having multi-functionalities including biodiversity conservation at a watershed scale. The first is called "Sogo-chisusui," which means comprehensive flood control measures. The measures focus on flood control in urbanized areas where the flood peaks were elevated as a result of land-use development in the 1970s, as shown in Fig. 2.5. During rapid economic growth in the 1960s and 1970s, asphalt pavement covered roads in urban areas and buildings were constructed. Streams in urban areas were channelized, and sewer systems were installed. Most of these facilities are impermeable and drained rainwater into sewer systems without allowing the water to penetrate into soils. The water flowing through drainage pipes reaches the straightened stream channel. Thus, the current water management policy aimed to flush out rainwater as fast as possible. As a result, flood peaks quickly increased, even if the amount of precipitation did not change (Fig. 2.5).

To control elevated streamflows in urban areas, the government had to change the policy from flushing rainwaters rapidly to retaining rainwaters in GI, such as woods, parks, gardens, and ponds, and in GYI, such as stormwater retention facilities underneath buildings. Then, the water is released slowly into the groundwater system. Additionally, the stream channels were partially widened, and sidewalk and parking pavement was changed to permeable materials. However, some of the recent stormwater retention facilities were built as GYI, which aims only at flood control without considering other functions, such as the provision of habitats for wildlife and recreational opportunities. Parks and schoolyards were built to function as stormwater retention facilities whose grounds were excavated and lowered relative to the surrounding areas.

After the introduction of the above flood control policy called "Sogo-chisusui," natural restoration or rehabilitation projects began in the 1990s, reflecting the world-wide movement to consider the protection of biodiversity and natural landscapes. These projects replaced concrete disaster prevention measures in streams, such as groundsill and revetment, with measures composed of stones which are more natural materials than concrete. The stream channel was partially broadened, and the vertical side slopes of the streambank were reformed to gentle slopes with some tree plantations. In the 1960s and 1970s, Japanese rivers and streams have been channelized and straightened for urban and agricultural land-use development and flood control. However, in the 1990s to 2000s, some of the straighten rivers are restored to original meandering rivers and wetlands by restoration projects. The river restoration projects function as an adaptation strategy to spread floodwater over floodplains and reduce the peak discharge of floods (Nakamura et al. 2014).

In recent years, Japan has experienced megaflood disasters due to climate change. A megaflood can easily exceed the design flood discharge of GYI (artificial levees and dams). We experienced serious flood damage in broad regions of eastern Japan during Typhoon Hagibis in 2019. In 2020, the Japanese government announced the need for "Ryuiki chisui," which is a new flood control concept at the watershed scale. This policy focuses on entire watershed areas from headwater basins covered by forests to midstream areas covered by farmlands to downstream areas where cities are situated. The policy encouraged active use of GI measures at the watershed scale. Natural forests in headwater basins (GI-1) should be preserved, and abandoned artificial forests should be managed by periodic thinning and pruning. Discontinuous levees, overflow embankments, and flood protection forests should be preserved, and land-use development in flood-prone areas should be restricted in midstream reaches (GI-2). Stormwater retention facilities and permeable pavement should be installed to compensate for the increase in water discharge volume caused by new developments.

This new policy matches the hybrid infrastructure concept presented in this chapter, but focuses on flood control with little emphasis on the multi-functionality and ecological network of GI as shown in Fig. 2.2. The Japanese government is still planning to build and/or reinforce GYI to adapt to climate change, and, therefore, it is unlikely that river managers will actively introduce GI at a watershed scale. One of the reasons why GI is not easily introduced is sectionalism between ministries and agencies. For example, the collaborative and coordinated actions between the Ministry of Land, Infrastructure, Transport and Tourism and the Ministry of Agriculture, Forestry and Fisheries of Japan is essential for flood control at a watershed scale, but it is hard to achieve due to strong divisions and cultural differences between administrative agencies. Another problem is the greater uncertainty and ambiguity of GI function for rainwater retention and floodwater control as compared to GYI. Thus, development of simulation models evaluating GI function quantitatively is required to find a best mix of GI and GYI for future river management.

2.6 Green Infrastructure Functioning as Ecological Networks

One of the important characteristics of GI is an interconnected network of green spaces (Maes et al. 2015). Dispersal and breeding success of red-crowned crane (*Grus japonensis*) using GI habitat provides an excellent example (Fig. 2.9). The red-crowned crane was distributed widely on the Japanese islands of Hokkaido, Honshu, and Kyushu at the beginning of the twentieth century. The population significantly declined to approximately 40 individuals with extensive farmland development of wetlands and overhunting (Masatomi and Masatomi 2018) and then recovered to approximately 1500 individuals with the help of artificial feeding during the winter season by local residents. The breeding and nesting sites of the cranes are concentrated in the Kushiro Mire (an example of GI-1 in Sect. 2.3), where

Fig. 2.9 Pair of red-crowned cranes that arrived in the Maizuru flood control basin in August 2012. The image in the upper right corner is an aerial photo of the entire basin

food and habitat limit the carrying capacity. The genetic diversity of the current population is low owing to the bottleneck effect (Masatomi and Masatomi 2018; Miura et al. 2013). Thus, the Ministry of the Environment plans to expand their nesting and breeding sites outside of the Kushiro region.

In the Chitose River watershed, six large-scale, flood control basins with areas of 150–280 ha were constructed to control extraordinary floods (acting as GI-2). These basins have also created a high-quality wetland landscape where many swans, greater white-fronted geese, and bean geese gather in the early spring. I, together with the Ecosystem Conservation Society of Japan, proposed restoring one of the flood control basins to wetlands dominated by a reed (*Phragmites australis*) that is preferred by the crane for nesting. A study that compared the abundance of different organisms (fish, aquatic insects, birds, and aquatic plants) among various water bodies (natural ponds, artificial channels, etc.) revealed that their abundance was highest in the flood control basins (see Chap. 12, Yamanaka et al. 2020). Thus, flood control basins have the potential to act as ecological networks for the conservation of the red-crowned crane on a large scale, employing the Kushiro Wetland as a hub wetland (GI-1) and flood control basins as satellite wetlands (GI-2). The GI networks promote migrations of cranes to extended areas, and add population stability in case of disaster, which may damage some habitats. In 2012, a pair of red-crowned cranes appeared in the Maizuru flood control basin, which is one of the six basins in the Chitose River watershed (Fig. 2.9). Since then, one or two pairs of

Fig. 2.10 Juvenile bird born in the Maizuru flood control basin in 2020

cranes have appeared in the Maizuru flood control basin and neighboring farmlands every year and finally overwintered, and one chick hatched in the Maizuru flood control basin in 2020 (Fig. 2.10).

2.7 Important Points for Planning, Implementation, and Governance of GI

Here are three important points for the planning and implementation of GI at the watershed scale.

1. Forests in headwater basins, the stream network (main channel and tributaries) and farmlands in midstream reaches, and green spaces in cities should be properly allocated to exploit the multiple functions of GI (e.g., flood control, biodiversity conservation, and recreation) together with existing GYI measures (Fig. 2.2).

   We should preserve natural and quasi-natural forests and wetlands still remaining in a watershed. The present natural ecosystems already perform ecosystem-based disaster risk reduction (Eco-DRR) (Figs. 2.5 and 2.6). Once we lose them, it takes enormous resources and time to restore them, and the restored ecosystems may never perform the multiple functions provided by the original ecosystems. The existing green parks and gardens should be preserved to maintain the retention capability for stormwater. In developed countries, abandoned tree plantations and farmlands will increase with depopulation in the future (Kobayashi et al. 2020). The passive and active restoration of natural forests, grasslands, and wetlands after abandonment is one of the challenging options to maintain the functions of GI (Morimoto et al. 2017).

The confluences of the mainstream and tributary rivers are key areas for the installation of GI. Approximately 80% of levee breaches by the 2019 flood disaster associated with Typhoon Hagibis occurred within 1 km of the confluences of the mainstream and tributaries. Levee breach sites at confluences accounted for 112 (62 rivers) out of 140 (71 rivers) breach sites in total (Asahi Newspaper, November 8 in 2019). Confluences are vulnerable to floodwaters and provide diverse habitats for various plants and animals (Benda et al. 2004). Thus, the riverbed should be wide, and traditional flood control measures, such as discontinuous levees, overflow embankments, and flood protection forests, should be used at appropriate locations.

2. One of the most important characteristics of GI is multi-functionality. When GI is used for Eco-DRR, the ecological functions of GI that create a network system that provides habitats and migration routes for wildlife should be considered, taking into account surrounding green areas, such as riparian forests, street trees, and windbreak forests. This ecological network will offer recreational opportunities for walking, jogging, and cycling and will contribute to creating attractive urban and rural landscapes. Moreover, consumers may selectively purchase agricultural goods produced from local communities with many GI measures contributing to the preservation of biodiversity, reduction of flood damage, and provision of recreational opportunities.

3. Collaborative management of GI.

Natural ecosystems functioning as GI can be maintained by letting ecosystems undergo natural succession. However, parks, farmlands, and other seminatural GI may require management activities. Currently, local and prefectural governments may conduct maintenance of those GI, but collaborative management by local residents with the help of the government is recommended to obtain long-term stewardship of GI.

Disaster risks are determined by hazard, exposure, vulnerability, and social capacity building (United Nations International Strategy for Disaster Reduction 2017). Collaborative management contributes to capacity building. GI may provide common lands (commons) for local people and can provide various ecosystem services, strengthen community solidarity through collaboration activities, and nurture the power of community resilience against natural disasters. Although local Japanese villages used to have commons in the form of forests for firewood and grasslands, those lands were subdivided into individual ownerships, and the commons system sustained by collaborative management was abolished. However, as indicated in Fig. 2.2, we should build a new commons system at the watershed scale that connects people living in up- and midstream forestry and/or agricultural villages with people living in downstream urban cities through the management of GI. The system will facilitate the exchange and cooperation of rural and urban people, which will promote social solidarity and economic benefits in local communities under normal circumstances and will provide evacuation shelters for urban people when they encounter serious disasters.

Acknowledgments The author sincerely appreciates all thoughtful comments and suggestions made by Dr. Frederick J. Swanson that significantly improve the quality of the original manuscript. This research was funded by the Environment Research and Technology Development Funds (4-1504 and 4-1805) of the Ministry of the Environment of Japan.

References

Benda L, Poff LR, Miller D, Dunne T, Reeves G, Pollock M, Pess G (2004) Network dynamics hypothesis: spatial and temporal organization of physical heterogeneity in rivers. Bioscience 54:413–427

Buijse AD, Coops H, Staras M, Jans LH, Van Geest GJ, Grift RE et al (2002) Restoration strategies for river floodplains along large lowland rivers in Europe. Freshw Biol 47(4):889–907

European Commission (2016) Supporting the Implementation of Green Infrastructure. European Commission, Rotterdam

Green Infrastructure Association (Japan) (ed) (2017) Authorized edition. (Ketteiban). Nikkei BP. 392p

Green Infrastructure Association (Japan) (ed) (2020) Practical edition. (Jissenban). Nikkei BP. 520p

Kobayashi Y, Higa M, Higashiyama K, Nakamura F (2020) Drivers of land-use changes in societies with decreasing populations: a comparison of the factors affecting farmland abandonment in a food production area in Japan. PLoS One 7:e0235846

Maes J, Barbosa A, Baranzelli C, Zulian G, e Silva FB, Vandecasteele I et al (2015) More green infrastructure is required to maintain ecosystem services under current trends in land use change in Europe. Landsc Ecol 30:517–534

Masatomi H, Masatomi Y (2018) Ecology of the red-crowned crane and conservation activities in Japan. In: Nakamura F (ed) Biodiversity conservation using umbrella species. Springer, Singapore, pp 83–105

Matsuura S, Yamamoto K, Hamaguchi T, Homma H (1988) Historical review of groves against flood hazards. In: Proceedings of the 8th historical study of Japanese civil engineering, pp 193–204 (in Japanese)

Miura Y, Shiraishi J, Shiomi A, Kitazawa T, Hiraga T, Matsumoto F et al (2013) Origin of three red-crowned cranes (*Grus japonensis*) found in Northeast Honshu and West Hokkaido, Japan, from 2008 to 2012. J Vet Med Sci 75:1241–1244

Morimoto J, Shibata M, Shida Y, Nakamura F (2017) Wetland restoration by natural succession in abandoned pastures with a degraded soil seed bank. Restor Ecol 25:1005–1014

Nakajima H, Ohgushi K, Hino T (2013) Numerical simulations of flood and inundation for evaluating the effects of Nokoshi and open levee in Jobaru River. Ann J Hydr Eng Japan Soc Civil Eng 69(4):I_1537–I_1542. (in Japanese with English abstract)

Nakamura F, Inahara S, Kaneko M (2005) A hierarchical approach to ecosystem assessment of restoration planning at regional, catchment and local scales in Japan. Landsc Ecol Eng 1:43–52

Nakamura F, Ishiyama N, Sueyoshi M, Negishi J, Akasaka T (2014) The significance of meander restoration for the hydrogeomorphology and recovery of wetland organisms in the Kushiro River, a lowland river in Japan. Restor Ecol 22:544–554

Nakamura F, Seo J Il, Akasaka T, Swanson FJ (2017) Large wood, sediment, and flow regimes: Their interactions and temporal changes caused by human impacts in Japan. Geomorphology 279:176–187

Nakamura F, Ishiyama N, Yamanaka S, Higa M, Akasaka T, Kobayashi Y, Ono S, Fuke N, Kitazawa M, Morimoto J, Shoji Y (2020) Adaptation to climate change and conservation of biodiversity using green infrastructure. River Res Appl 36:921–933

Okuma T (1987) A study on the function and etymology of open levee. In: Proceedings of the 7th historical study of Japanese civil engineering, pp 259–266. (in Japanese)

Onuma A, Tsuge T (2018) Comparing green infrastructure as ecosystem-based disaster risk reduction with gray infrastructure in terms of costs and benefits under uncertainty: a theoretical approach. Int J Disast Risk Reduct 32:22–28

Tanabe T, Okuma T (2001) A study on the roles and effects of "NOKOSHI" on the Jobaru-River Basin. Example of flood control that allows overflowing and its future possibility. Historic Stud Civil Eng 21(2):147–158. (in Japanese with English abstract)

Tsukamoto (1985) Effects of forests on flood and base flow discharge (*Shinrin no Suigen Kanyou Kinou*). Chisan 30-5:116–121. (in Japanese)

United Nations International Strategy for Disaster Reduction (2017) National disaster risk assessment: governance system, methodologies, and use of results. UNISDR. 77p

Yamanaka S, Ishiyama N, Senzaki M, Morimoto J, Kitazawa M, Fuke N, Nakamura F (2020) Role of flood-control basins as summer habitat for wetland species—a multiple-taxon approach. Ecol Eng 142:105617

Chapter 3
An Economic Analysis of Optimal Hybrid Infrastructure: A Theoretical Approach in a Hydro-Economic Model

Ayumi Onuma

Abstract This chapter features the functions of green infrastructure and gray infrastructure in a hydro-economic model and integrates them into a hybrid infrastructure. We show that adding green infrastructure to an existing gray infrastructure generates a new benefit, which is referred to as absorption effects. Using the model, we investigate the optimal hybrid infrastructure by maximizing the social net benefits, in which disaster risk reduction is defined. We then discuss how green infrastructure can be used to augment existing gray infrastructure from the perspective of economic optimality. We derive some conditions to determine whether both or either of gray and green infrastructures should be introduced to implement disaster risk reduction.

Keywords Green infrastructure · Gray infrastructure · Hybrid infrastructure · Disaster risk reduction · Hydro-economic model

3.1 Introduction

The purpose of this chapter is to model the functions of green infrastructure (GNI) and gray infrastructure (GYI) to integrate them into a hybrid infrastructure (HBI) and to express it in a simple hydro-economic model for deriving optimal HBI.

Although ecosystem-based disaster risk reduction has been widely attracting attention, there are few theoretical analyses in economics; Barbier (2012) and Barbier and Enchelmeyer (2014) provide theoretical frameworks in the context of GNI to prevent storm surges in coastal areas, but they only focus on GNI.

Meanwhile, Onuma and Tsuge (2018) present a hydro-economic framework to compare GYI and GNI in terms of cost–benefit by considering the size of the population and formulating the features of their respective disaster prevention

A. Onuma (✉)
Keio University, Minato-ku, Tokyo, Japan
e-mail: onuma@econ.keio.ac.jp

functions. In addition, disaster risk and disaster risk reduction are clearly defined in the above model. While this direction is suitable for discussing optimal disaster risk reduction, the model is based on a binary choice between GNI and GYI. In reality, even if a preventing flood infrastructure is called a GNI, it is often an HBI, i.e., a mixture of GNI and GYI. Examples include the "Slowing the Flow" project implemented in Pickering in the UK, which is known as a nature-based solution to flood management and uses one large storage bund to effectively reduce flooding (Nisbet et al., 2015). Therefore, providing a framework for the analysis of HBI is significant for the economic study of disaster prevention infrastructures, which are becoming increasingly important in the days of frequent natural disasters.

In this chapter, we formulate the features of HBI, which combine the features of GNI and GYI, and construct a hydro-economic model expressing HBI. We then show that adding GNI to an existing GYI to form an HBI generates a new benefit, which is referred to as absorption effects. We then discuss how GNI can be used to augment existing GYI from the perspective of economic optimality.

3.2 Green, Gray, and Hybrid Infrastructures

Let us consider a hazard or rainfall that leads to a flooding river. In this chapter, we use "hazard" and "rainfall" exchangeably and denote it by H. Residents directly face H if there is no preventing infrastructure, whereas H is mitigated if we introduce GNI and/or GYI.

3.2.1 Green Infrastructure

We consider a GNI, which is often exemplified by forests located in the upstream of the river. GNI has the ability to save water. In the case of forests, some of the water that flows in does not flow out because it penetrates the ground or stays on the leaves, while the capacity of water saving is limited.

Let X_f denote the direct outflow from the GNI or the forest. X_f is the level of H net of the level of flow absorbed by the GNI, which is denoted by A. Therefore, $X_f = H - A$. The level of rainfall at which the capacity of absorption is used up is expressed by $\hat{H}$. $\hat{H}$ is supposed to increase with the scale of the forest, K_f, so we define $\hat{H}$ as

$$\hat{H} = \hat{H}(K_f) \tag{3.1}$$

with $\hat{H}'(K_f) > 0$. The level of absorption depends on K_f and H. A larger forest flows out smaller X_f, that is, higher A. Thus, we write A as $A(H, K_f)$. We assume that A increases with H up to $\hat{H}$, whereas A is constant with H beyond $\hat{H}$. On the

Fig. 3.1 Outflow X_f and absorption $A(H, K_f)$ of GNI

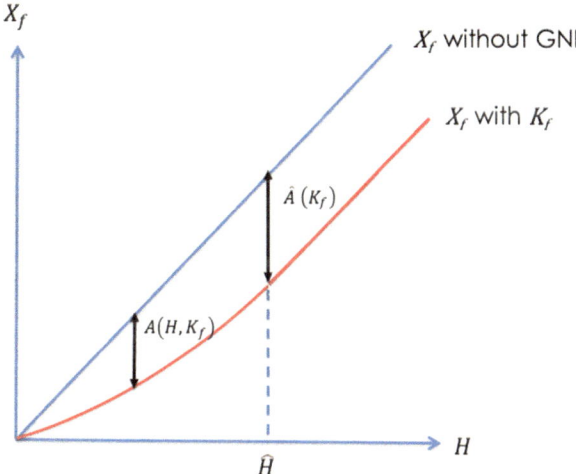

contrary, X_f also increases with H. That is, for $H < \hat{H}$, $0 < A'_H (\equiv \partial A / \partial H) < 1$, and $A''_H (\equiv \partial^2 A / \partial H^2) < 0$. Moreover, $A'_H = 0$ for $H \geq \hat{H}$ and $A'_{K_f} > 0$.

The maximum absorption of the forest, $\hat{A}$, is defined as

$$\hat{A} = A(\hat{H}, K_f) \equiv \hat{A}(K_f) \tag{3.2}$$

By definition, we have $\hat{A}' > 0$ and $A(H, K_f) < \hat{A}(K_f)$ for $H < \hat{H}$. In summary, the resulting outflow from GNI, X_f, is determined as

$$X_f = \begin{cases} H - A(H, K_f) \text{ for } H < \hat{H} \\ H - \hat{A}(K_f) \text{ for } H \geq \hat{H} \end{cases} \tag{3.3}$$

X_f is described in Fig. 3.1.

3.2.2 Gray Infrastructure

Let K_c represent the GYI. Let H_c be the hazard or flow arriving to GYI. Following Onuma and Tsuge (2018), GYI protects the flow completely up to a threshold $\hat{H}_c$, but it does not protect beyond $\hat{H}_c$. We call $\hat{H}_c$ *threshold hazard*, which increases with K_c; that is, the threshold hazard increases as the scale of the GYI increases. We assume that the threshold hazard is proportional to the scale of K_c:

$$\hat{H}_c = \theta K_c \tag{3.4}$$

Therefore, the resulting hazard that the residents face, X_c, is expressed as

$$X_c = \begin{cases} 0 & (H_c \leq \hat{H}_c) \\ H_c & (H_c > \hat{H}_c) \end{cases} \tag{3.5}$$

As noted by Onuma and Tsuge (2018), the most striking difference between GNI and GYI is their hazard prevention rates: in GYI, the rate is 1 up to the threshold but zero beyond that, while the rate never becomes 1 but is always positive in the case of GNI.

3.2.3 Hybrid Infrastructure

In our model, an HBI is defined as a mixture of GNI and GYI. A GNI is located upstream of the river, while a GYI is built downstream to protect nearby residents. Therefore, X_f in (3.3) flows to the GYI. That is,

$$H_c = X_f = H - A(H, K_f) < H \tag{3.6}$$

In this way, GNI mitigates the hazards coming to GYI. X_h denotes the hazard resulting from the HBI. GYI itself protects against the hazard completely up to a threshold of $\hat{H}_c$, but it fails beyond $\hat{H}_c$; that is,

$$X_h = \begin{cases} 0 & (X_f \leq \hat{H}_c) \\ X_f & (X_f > \hat{H}_c) \end{cases} \tag{3.7}$$

This is a function of the HBI (K_c, K_f). Under the HBI, we derive the level of hazard H^* such that

$$X_f^* \equiv H^* - A(H^*, K_f) = \hat{H}_c \Rightarrow H^* = \hat{H}_c + A(H^*, K_f) \tag{3.8}$$

That is, H^* is the level of rainfall in a forest where flow to GYI attains the threshold hazard $\hat{H}_c$. This shows that the threshold hazard increases from $\hat{H}_c$ under $(K_c, 0)$ to $A(H^*, K_f) + \hat{H}_c$ under (K_c, K_f). Let H^* be $\hat{H}_c + \omega$, which is the new threshold under HBI. Then, it holds that $\hat{H}_c + \omega = \hat{H}_c + A(\hat{H}_c + \omega, K_f)$, so

$$\omega = A(\hat{H}_c + \omega, K_f) \tag{3.9}$$

ω is an important variable when considering the HBI, in that the threshold hazard of GYI increases as a result of setting the GNI. We may call ω a *substantial increment in threshold hazard*. In the following lemma, we prove that ω is determined uniquely, given (K_c, K_f).

Lemma 1 *For any $\hat{H}_c > 0$, ω satisfying (3.9) exists uniquely.*

Proof From the definition of A, it holds that $0 < A(\hat{H}_c + \omega, K_f) \leq \hat{A}(K_f)$. This means that there always exists a ω belonging to $(0, \hat{H})$ that satisfies (3.9). Next, suppose that there exist ω_1 and ω_2 with $\omega_1 > \omega_2$ that satisfy (3.9) for some $\hat{H}_c > 0$. However, this implies

$$\omega_1 - \omega_2 = A(\hat{H}_c + \omega_1, K_f) - A(\hat{H}_c + \omega_2, K_f) \tag{3.10}$$

However, this is not feasible because we assume that $A'_H < 1$. Thus, ω that satisfies (3.9) must be unique. □

The next lemma shows that ω depends on GYI as well as GNI when GYI is not large enough.

Lemma 2 *For $\hat{H}_c > 0$ and ω satisfying (3.9),*

$$\frac{d\omega}{d\hat{H}_c} = \frac{A'_H}{1 - A'_H} \geq 0$$

where equality holds only if $A(H, K_f) = \hat{A}(K_f)$ or $\hat{H}_c + \omega \geq \hat{H}$.

Proof Differentiating (3.9) leads to

$$d\omega = A'_H(d\hat{H}_c + d\omega) \tag{3.11}$$

From here, we obtain $\frac{d\omega}{d\hat{H}_c} = \frac{A'_H}{1-A'_H}$, where $A'_H = 0$ if $\hat{H}_c + \omega \geq \hat{H}$. □

Then, we get the following result, which shows the magnitude of the substantial increment in the threshold hazard.

Proposition 1 *The HBI increases the threshold $\hat{H}_c$ by ω where (i) $\omega = \hat{A}(K_f)$ if $\hat{H}_c \geq \hat{H} - \hat{A}(K_f)$. (ii)$\omega < \hat{A}(K_f)$ if $\hat{H}_c < \hat{H} - \hat{A}(K_f)$.*

Proof For $\hat{H}_c = \hat{H} - \hat{A}(K_f)$,

$$A(\hat{H}_c + \omega, K_f) = A(\hat{H} - \hat{A}(K_f) + \omega, K_f) \tag{3.12}$$

It is obvious that if $\omega = \hat{A}(K_f)$, (3.9) is satisfied as

$$\omega = A(\hat{H}_c + \omega, K_f) = A(\hat{H}, K_f) = \hat{A}(K_f) \tag{3.13}$$

Then, from Lemma 1, $\omega = \hat{A}(K_f)$ is the solution of (3.9). Moreover, from Lemma 2, $A'_H = 0$ at $\hat{H}_c$, so $\omega = \hat{A}(K_f)$ is also the solution for $\hat{H}_c > \hat{H} - \hat{A}(K_f)$. Furthermore, for $\hat{H}_c < \hat{H} - \hat{A}(K_f)$, Lemma 2 says $\omega < \hat{A}(K_f)$. □

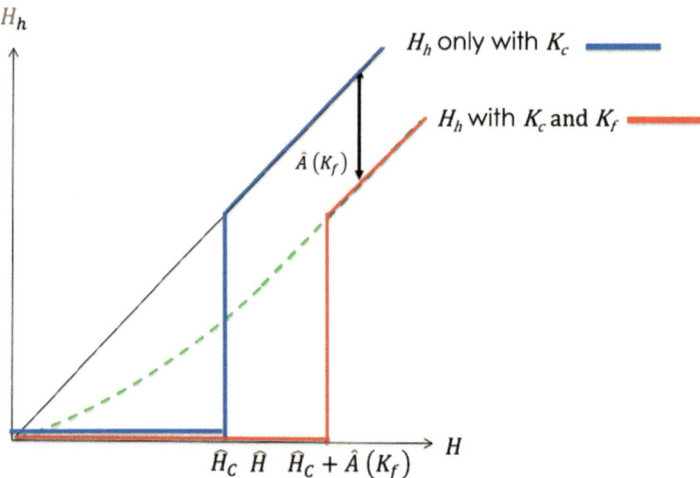

Fig. 3.2 Prevention of hazard by HBI and GYI

In summary, by introducing (K_c, K_f), the resulting hazard H_h is

$$H_h = \begin{cases} 0 & (H \le \hat{H}_c + \omega) \\ H - \omega & (H > \hat{H}_c + \omega) \end{cases} \tag{3.14}$$

Equation (3.14) is depicted in Fig. 3.2, where $\hat{H}_c \ge \hat{H} - \hat{A}(K_f)$ so that $\omega = \hat{A}(K_f)$. In the figure, the benefits of HBI can be graphically compared with GYI. HBI, in which GNI is added to the existing GYI, provides two benefits. First, even though the hazard is higher than $\hat{H}_c + \omega$, which means that the hazard cannot be prevented, it is always mitigated. It is worth noting that this benefit cannot be obtained by setting GYI solely; any hazard beyond the new threshold hazard is mitigated in HBI, due to the function of GNI.

Second, the threshold $\hat{H}_c$ of GYI increases by ω, despite the fact that GYI is kept the same as before. It might be interesting that the magnitude of ω sometimes depends on K_c as well as K_f, as demonstrated by Lemma 2. This occurs when K_c is not sufficiently large enough so that $\hat{H}_c < \hat{H} - \hat{A}(K_f)$ by proposition 1. This property might be important when we consider the replacement of a part of the preventing function of GYI with GNI. In the replacement, suppose that we keep the threshold hazard as before. It might be intuitive that ω depends only on K_f and is determined independently of K_c. However, if K_c is not large enough, if there is a decrease in K_c, it will reduce ω. Hence, without careful planning, such replacement could result in the reduction of the threshold hazard.

These features are depicted in Fig. 3.3, which expresses $A(H, K_f)$ as a green line. Suppose that the initial threshold hazard is $\hat{H}_c^0$, in which case the substantial increment in the threshold hazard ω, created by K_f, is ω_0. The figure shows that if

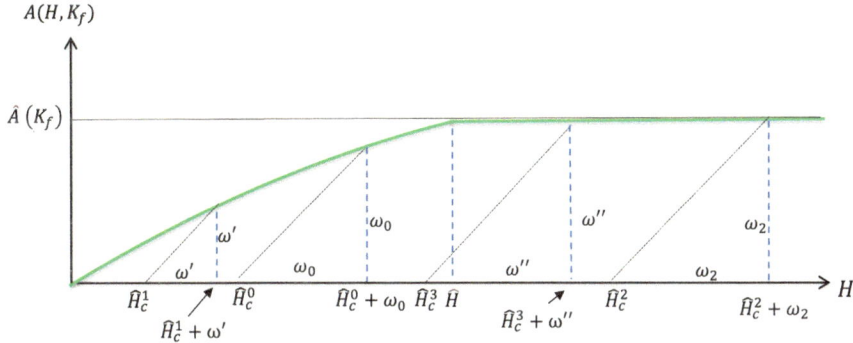

Fig. 3.3 Change of ω as a result of reducing $\hat{H}_c$

$\hat{H}_c^0$ is reduced to $\hat{H}_c^1$, ω_0 decreases to ω'. On the contrary, if we suppose that the initial threshold hazard is sufficiently large, $\hat{H}_c^2$, where $\omega = \omega_2$. In this case, even if $\hat{H}_c^2$ is reduced to $\hat{H}_c^3$, which changes ω_2 to ω'', the level of ω is equivalent.

3.2.4   Ratios of Prevented Hazard by GNI and GYI

As shown by (3.14), the HBI completely prevents any hazard H below $\hat{H}_c + \omega$ and partly H beyond it. Now that we have a mixture of two preventing hazard infrastructures, GNI and GYI, in HBI, seek each ratio of hazard that is prevented in each infrastructure. Given (K_c, K_f), we define $r_f = (H - X_f)/H = A(H, K_f)/H$ and $r_c = (X_f - X_h)/H$, which formally shows the ratio of hazard prevention. Thus, in terms of (3.14), both r_f and r_c are positive with $r_f + r_c = 1$ for $H(< \hat{H}_c + \omega)$, whereas $r_f > 0$, but $r_c = 0$ for $H(> \hat{H} + \omega)$, where $r_f + r_c < 1$.

Note that $r_f(H)$ decreases with H because $A''_H > 0$ for $H < \hat{H}$ and $A'_H = 0$ for $H \geq \hat{H}$. Note also that

$$r_f(0) = \lim_{H \downarrow 0} \frac{A(H, K_f)}{H} = A'_H(0, K_f) \tag{3.15}$$

by l'Hôpital's rule. Thus, we can describe the behavior of r_f and r_c as follows.

Proposition 2 *In HBI expressed by (3.14), the ratio of prevention by GNI, $r_f = A(H, K_f)/H$, decreases with H from $r_f(0) = A'_H(0, K_f)$. Meanwhile, the ratio of GYI, $r_c = (X_h - X_f)/H$, is equal to $1 - r_f$ for $H(< \hat{H}_c + \omega)$ and to zero for $H(\geq \hat{H}_c + \omega)$.*

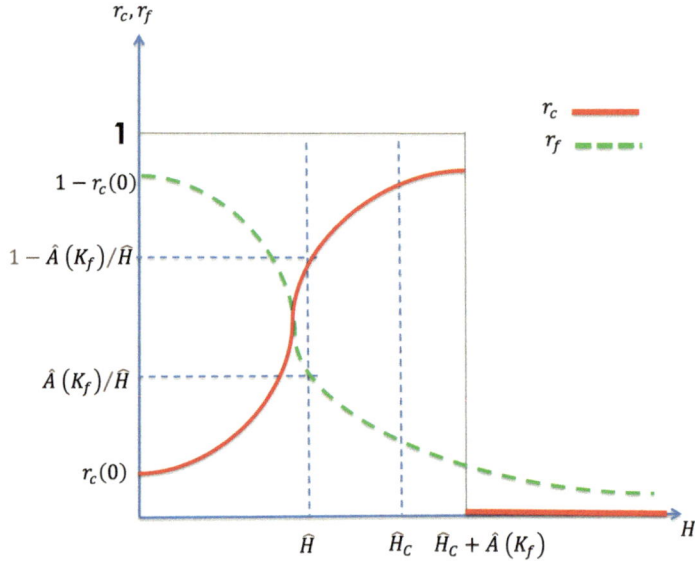

Fig. 3.4 r_c and r_f

Figure 3.4 depicts r_c and r_f.[1] GNI reduces the flow considerably when the rainfall is small.

3.3 Optimal Hybrid Infrastructure

In this section, we incorporate the function of HBI into a simple economic model, that is, a hydro-economic model, and analyze some properties of the optimal HBI. We define disaster risk reduction (DRR) in this model. Denoting the considerable maximum hazard with H^m, let $f(H) : [0, H^m] \rightarrow [0, 1]$ be the function expressing the probability that a hazard H occurs, with $\int_0^{H^m} f(H)dH = 1$. The hazard risk is defined as $HR = \int_0^{H^m} f(H)HdH$.

In contrast, $D(H)$ shows the disaster caused by hazard H, where $D(0) = 0, D' > 0, D'' > 0$. That is, the marginal disaster of hazard is positive and

---

[1] Although $A''_H < 0$, the signs of r''_f and r''_c are ambiguous. Figure 3.4 shows that A''_H is small enough ($|A''_H|$ is large enough), so that $r''_f < 0$ and $r''_c > 0$ for $H(< \hat{H})$. It is possible that the opposite case occurs, depending on the value of A''_H. However, $r''_f > 0$ for $H(\geq \hat{H})$.

increases with H. Disaster risk (DR) without preventing infrastructure, that is, $(K_c, K_f) = (0, 0)$, is defined as

$$DR(0, 0) = n \int_0^{H^m} f(H)D(H)dH \qquad (3.16)$$

where n is the population exposed to hazard H. On the contrary, DR under GYI, that is, $(K_c, K_f) = (K_c, 0)$ from (3.5) with its threshold hazard, $\hat{H}_c$, is expressed as

$$DR(K_c, 0) = n \int_{\hat{H}_c}^{H^m} f(H)D(H)dH \qquad (3.17)$$

Disaster risk reduction by installing GYI at the initial state without preventing infrastructure, DRR from $(0, 0)$ to $(K_c, 0)$ is, therefore, defined as

$$DRR((K_c, 0)|(0, 0)) = DR(0, 0) - DR(K_c, 0) = n \int_0^{\hat{H}_c} f(H)D(H)dH \qquad (3.18)$$

$DRR((K_c, 0)|(0, 0))$ is considered to be the benefit of setting GYI to the site without preventing infrastructures.

Now, let us suppose that the government strengthens an existing GYI to increase the threshold hazard from $\hat{H}_c$ by $\triangle H$ to $\hat{H}_c^*(= \hat{H}_c + \triangle H)$ by investing I_c into GYI and/or by introducing GNI. That is, we introduce (I_c, K_f) to satisfy

$$\theta I_c + \omega = \triangle H \qquad (3.19)$$

Here, we assume that $\hat{H}_c$ is large enough in view of proposition 1 to satisfy $\omega = \hat{A}(K_f)$. This increment can be implemented only with $I_c = \theta^{-1}\triangle H$, where GYI increases to $K_c + \theta^{-1}\triangle H$; or only with $K_f = \hat{A}^{-1}(\triangle H)$; or with a mixture of I_c and K_f satisfying (3.19).

Disaster risk under this improvement is

$$DR(K_c + I_c, K_f) = n \int_{\hat{H}_c^*}^{H^m} f(H)D(H - \hat{A}(K_f))dH \qquad (3.20)$$

Thus, DRR gained by this improvement is defined as

$$DRR((K_c + I_c, K_f)|(K_c, 0)) = n \int_{\hat{H}_c}^{\hat{H}_c^*} f(H)D(H)dH$$

$$+ n \int_{\hat{H}_c^*}^{H^m} f(H)(D(H) - D(H - \hat{A}(K_f)))dH \qquad (3.21)$$

The first term is the benefit from enhancing the threshold hazard, which we call the *threshold effect*. The second term, however, is the benefit from absorption by GNI, which we call the *absorption effect*.

Let us define the costs of implementing $\hat{H}_c + \triangle H$. We denote the cost of expanding GYI as $C_c(I_c)$, assuming $C'_c > 0$ and $C''_c \geq 0$. On the contrary, the cost of GNI is expressed as $\mu C_f(K_f)$, where $\mu (> 0)$ reflects the opportunity costs of setting GNI or the prices of land acquired for setting GNI because GNI requires a wider range of geographic spaces. We assume $C'_f > 0$ and $C''_f > 0$.

Meanwhile, environmental benefits and costs are included as $E_f = \alpha_f K_f$ ($\alpha_f > 0$) for GNI and $E_c = -\alpha_c K_c$ ($\alpha_c > 0$) for GYI, where $\alpha_f > 0$ and $\alpha_c > 0$. Then, social net benefit, W, is expressed by

$$W = DRR((K_c + I_c, K_f)|(K_c, 0))$$
$$- C_c(I_c) - \mu C_f(K_f) + \alpha_f K_f - \alpha_c I_c \qquad (3.22)$$

The optimal HBI is defined as maximizing (3.22) with (I_c, K_f) subject to (3.19). Assume that

$$\hat{A}(K_f) = \gamma K_f, \gamma > 0 \qquad (3.23)$$

The optimum must satisfy if it is interior (both I_c and K_f are positive)

$$n \int_{\hat{H}_c^*}^{H^m} f(H) D'(H - \gamma K_f) \gamma dH = (\mu C'_f - \alpha_f) - \frac{\gamma}{\theta}(C'_c + \alpha_c). \qquad (3.24)$$

γ/θ shows how much I_c is replaced by one unit, say, one hectare increase in K_f. Here, the left-hand side is the marginal absorption effect of this (marginal) replacement, while the right-hand side is the marginal net cost of replacement, including environmental benefits and costs. Equation (3.24) states that the marginal absorption effect must be equal to the net cost of replacement.

To make the argument clearer, we specify the function C_c as $C_c(I_c) = q I_c(q > 0)$, where q denotes the unit cost of investing in GYI. Moreover, the l.h.s. of (3.24) is denoted by $V(K_f)$ for convenience, where it holds that $V' < 0$. Let us denote the maximum feasible GNI as $\bar{K}_f \equiv \triangle H/\gamma$. Note that $V(0) > V(K_f) > V(\bar{K}_f)$ for $K_f \in (0, \bar{K}_f)$. Then, $K_f = 0$ becomes optimal if and only if

$$V(0) + \alpha_f + \frac{\gamma}{\theta}(q + \alpha_c) \leq \mu C'_f(0) \qquad (3.25)$$

where $I_c = \triangle H/\theta$. The l.h.s. of (3.25) includes the marginal benefit from launching GNI, marginal absorption benefit, and marginal environmental benefit, added by the saved marginal costs of GYI. Roughly, this occurs when $\mu C'_f(0)$ is sufficiently high.

Developing even the first hectare of GNI is very costly, relative to the unit investment cost q. However, $I_c = 0$ is optimal if and only if

$$V(\bar{K}_f) + \alpha_f + \frac{\gamma}{\theta}(q + \alpha_c) \geq \mu C'_f(\bar{K}_f) \tag{3.26}$$

This condition is satisfied when $\mu C'_f(\bar{K}_f)$ is still low enough, even though the targeted increase in threshold hazard is achieved only by GNI. We state these properties in the following proposition.

Proposition 3 *To enhance the threshold hazard from $\hat{H}_c$ by $\triangle H$, no use of GNI (i.e., $K_f = 0$) is optimal if $\mu C'_f(0)$ is sufficiently high to satisfy (3.25). On the contrary, zero investment in GYI, that is, $I_c = 0$, is optimal if $\mu C'_f(\bar{K}_f)$ is sufficiently low to satisfy (3.26). Otherwise, it holds that $(I_c, K_f) >> (0, 0)$, where (3.24) is satisfied.*

Let us suppose that the optimum is interior; that is, (3.24) is satisfied at the optimum. Now, we assume that μ depends on n, which can be justified by the fact that the opportunity cost of setting GNI increases with the size of the population. Thus, $\mu = \mu(n)$, with $\mu' > 0$. Let ϵ_μ express the population elasticity of opportunity cost, that is, $\epsilon_\mu = n\mu'/\mu$, which expresses how much the percentage of opportunity cost increases when population increases by 1%. We observe the sign of dK_f/dn to see how the use of GNI changes with population.

Proposition 4 *Suppose that both I_c and K_f are positive under the optimum HBI. Then, it holds*

$$\frac{dK_f}{dn} \underset{>}{\overset{<}{\lessgtr}} 0 \Leftrightarrow \epsilon_\mu \underset{<}{\overset{>}{\gtrless}} 1 - \frac{\alpha_f + \gamma\theta^{-1}(q + \alpha_c)}{\mu C'_f}$$

Proof Differentiating K_f with n under (3.24) leads to

$$\frac{dK_f}{dn} = \frac{\int_{\hat{H}_c^*}^{H^m} f(H)D'(H - \gamma K_f)\gamma dH - \mu'C'_f}{\mu C''_f + n\int_{\hat{H}_c^*}^{H^m} f(H)D''\gamma^2 dH}$$

$$= \frac{\mu C'_f(1 - \epsilon_\mu) - (\alpha_f + \gamma\theta^{-1}(q + \alpha_c))}{n(\mu C''_f + n\int_{\hat{H}_c^*}^{H^m} f(H)D''\gamma^2 dH)} \tag{3.27}$$

using $\int_{\hat{H}_c^*}^{H^m} f(H)D'(H - \gamma K_f)\gamma dH = ((\mu C'_f - \alpha_f) - \frac{\gamma}{\theta}(C'_c + \alpha_c))/n$ from (3.24). Then, since the denominator is positive, the claim is straightforward from the numerator. □

It should be noted that $0 < (\alpha_f + \gamma\theta^{-1}(q + \alpha_c))/\mu C'_f < 1$ holds because the r.h.s. of (3.24) must be positive. Therefore, a paradoxical case $dK_f/dn > 0$ occurs if ϵ_μ is sufficiently small, as in the claim. In other cases, the share of GNI use decreases when implementing $\triangle H$ as population increases.

3.4 Concluding Remarks

This chapter formalizes HBI and incorporates it into economics. Based on the model, we discuss the optimal mixture of GNI and GYI, that is, the optimal HBI.

To augment the disaster prevention function of the existing GYI, we show the nature of the optimal HBI. In particular, the logic for not using GNI at all is justified only when the marginal cost, including the opportunity cost when deployment is still zero, is sufficiently high, i.e., (3.25) holds. As the opportunity costs are expected to be lower for smaller populations, it is likely that the augmentation of disaster risk reduction infrastructure without the use of GNI would be inefficient, especially in sparsely populated areas.

In this chapter, GNI is assumed to be located upstream of GYI, but it is also possible that GYI is the first to receive the hazard and GNI is located behind it, as in the case of coastal seawalls. An HBI with a different arrangement of GNI and GYI could also be characterized in the same way as described in this chapter. This attempt is left for future research.

References

Barbier EB (2012) A spatial model of coastal ecosystem services. Ecol Econ 78:70–79

Barbier EB, Enchelmeyer BS (2014) Valuing the storm surge protection service of US Gulf Coast wetlands. J Environ Econ Policy 3(2):167–185

Nisbet T, Roe P, Marrington S, Thomas H, Broadmeadow S, Valatin G (2015) Project RMP5455: Slowing the Flow at Pickering Final Report: Phase II, Department for environment, food and rural affairs, UK

Onuma A, Tsuge T (2018) Comparing green infrastructure as ecosystem-based disaster risk reduction with gray infrastructure in terms of costs and benefits under uncertainty: A theoretical approach. Int J Disaster Risk Reduct 32:22–28

Chapter 4
Flood Management Policy in Shiga Prefecture, Japan: Implementation Approach of a Risk-Based Flood Management System at Catchment Scale

Kentaro Taki

Abstract Shiga Prefectural Government uses "site safety level" to assess the flood risk of major floodplains in the prefecture in order to implement an integrated floodplain management system in society. Site safety level is determined based on a risk matrix expressing the relationship between the frequency of inundation and the degree of damage at sites surrounding a river channel complex. A department has been set up for floodplain management separate from river management. The Department of Floodplain Management promotes disaster mitigation measures, such as land use and construction regulations, and conservation of traditional flood control facilities, including flood prevention forests, open levees, and ring levees. Such traditional facilities are highly functional as green infrastructure because they are focused not only on disasters but also on a daily basis in order to wisely utilize local ecosystem service.

Keywords Shiga Prefecture · Risk-based flood management · Risk assessment · Land use · Building regulation

4.1 Changes in Flood Risk Management in Japan

"Those who control the water, control the country." Flood control has been considered to be the basis of national administration since long ago, and the system for doing so has gradually changed over time to support the socio-economy of Japan.

Recently, the situation surrounding flood control has changed dramatically, and the risk of flooding is higher than ever before. Several challenges are emerging, including a decline in investment capacity owing to the decline in population,

K. Taki (✉)

School of Environmental Science, Lake Governance Research Center, University of Shiga Prefecture, Hikon, Shiga, Japan

e-mail: taki.k@ses.usp.ac.jp

© The Author(s) 2022   43

F. Nakamura (ed.), *Green Infrastructure and Climate Change Adaptation*,
Ecological Research Monographs, https://doi.org/10.1007/978-981-16-6791-6_4

changes in social structure and lifestyles, decline in literacy regarding water-related disaster mechanisms, and an increase in extreme events associated with climate change. Against this background, it is important to understand how we can eliminate human casualties and avoid catastrophic damage that would make it difficult to rebuild livelihoods. We are currently confronted with the issues that contribute to the initiation of flood control.

The general measures that have been taken to minimize the risk of flood inundation so far, including basin and in-channel measures such as rainwater storage and infiltration facilities, river improvement, and flood control facilities, are not adequate. In addition to these, there is a need for a multilayered approach for flood disaster risk reduction measures in floodplains including the maintenance and development of flood flow control facilities such as flood prevention forests, open levee systems (combination of discontinuous levees and secondary levees), and ring levees as shown in Fig. 4.1, land use and building regulations, and further enhancement of flood prevention activities and evacuation guidance. Floodplain in this section refers to the area of a plain that is inundated by the overflow of water from a river channel or other source during a flood. In other words, floodplain refers to the entire area of a valley floor plain, alluvial fan, alluvial plain, or delta that is inundated by floodwater.

Measures in the floodplain are not a new approach to disaster mitigation. Flood control during the Sengoku period (Warring States period, late fifteenth–late sixteenth century) was based on defensive measures to prevent flooding as exemplified by open levees; flood mitigation measures based on the premise of inundation rather than continuous embankments were the mainstream measures. In the Edo period (1603–1867), as the development of new rice fields and the intensification of land use progressed, flood control by continuous levees gradually became the mainstream owing to improvements in civil engineering techniques. The River Law was enacted in 1898 against the background of the modernization of civil

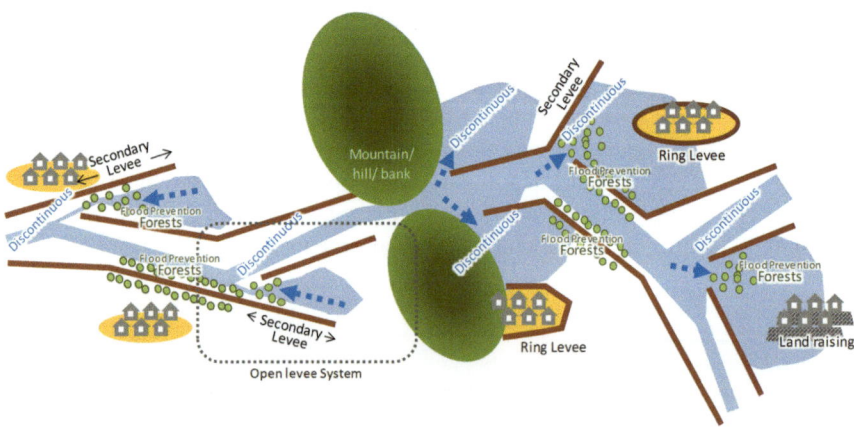

Fig. 4.1 Flood prevention forests, open levee system, and ring levees

Table 4.1 Changes in modern flood control planning (based on Hori et al. 2008)

First stage (largest recorded flooding) (Late 1890s)	The largest floods to date can be handled by the channels and dam reservoirs without causing flooding
Second stage (probable flooding) (Late 1950s)	The designed external force for flood control facilities is assessed according to the probability of exceeding the annual maximum rainfall; rainfall of a certain probability scale is used as the planned rainfall amount, and the main type of hydrographs produced by this rainfall is handled by the river channels and dam reservoirs without causing inundation
Third stage (reduce inflow from the watershed) (Late 1970s)	In addition to the measure of handling rainwater after it enters the channels, alternative plans include measures in the watershed to reduce the amount of rainwater that enters the channels itself
Fourth stage (flood risk management) (2000s)	Flood inundation is considered a premise, and alternative plans take into account floodplain damage reduction measures as well as watershed and channel facilities

engineering technology, and flood control systems using continuous levees became the basis for flood control in Japan. Currently, with the exception of a few nonurban rivers (e.g., Kita River in the Gokase River System (Miyazaki Prefecture), Kita River in the Kita River System (Fukui Prefecture), etc.), there are few flood control riparian forests that explicitly incorporate floodplain disaster mitigation measures.

Here, I would like to summarize the changes in flood control systems since the Meiji era. Hori et al. (2008) organized the changes in modern flood control planning in Japan into four stages (Table 4.1) and argued for the need to transition to the fourth stage. Flood control in the fourth stage is an expansion of flood control theory from the planning of watersheds (catchment areas) and river channels to encompass the entire basin including the floodplain, which is the affected area. This stage incorporates the concept of base defense into the continuous levee system.

Flood control in the fourth stage is based on the premise that flood inundation is a natural phenomenon; mitigation measures such as floodplain land use and housing practices can be considered as approaches to reduce damage by avoiding exposure and reducing vulnerability (Fig. 4.2).

4.2 Flood Management Policy in Shiga Prefecture: Policy Formulation

4.2.1 The History of Wise Land Use and Urban Development: Hints from History

The Amano River, which flows through Maibara City, Shiga Prefecture, was severely damaged by the Isewan Typhoon in 1959. Extensive riverbed excavation

Fig. 4.2 Approaches to damage factors and mitigation, and mitigation measures in the floodplain reduce damage by avoiding exposure and reducing vulnerability. (Sources: Ministry of the Environment, Natural Environment Bureau 2016; ADRC 2015)

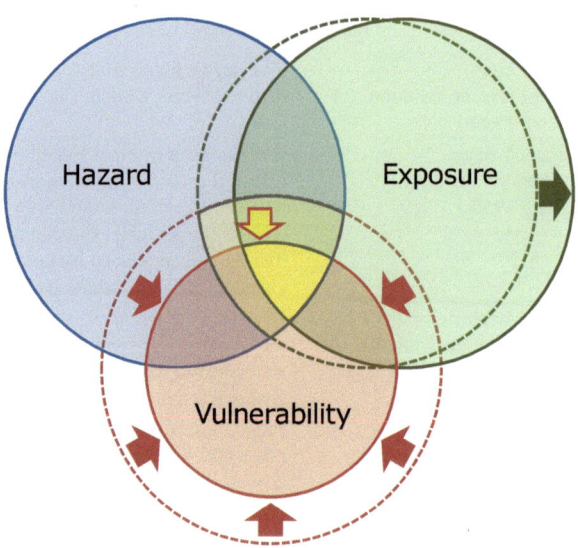

and revetment work was conducted in the recovery, but the prefectural public works office decided to keep the open levees. River administrators at that time well understood inundation characteristics and were also aware of excess flooding. The bullet train, Tokaido Shinkansen, crosses the Amano River. A railroad bridge was built across the river. Furthermore, in the sections of farmland, such as paddy fields, the viaduct extends instead of embankments, as shown in Fig. 4.3. Continuous embankments can interfere with the flow of floodwater and cause inundation. Therefore, bridges were built over non-riverine sections to allow floodwater to flow down through the section. The original track was planned with an embankment structure, but there was opposition from local residents against the embankment system owing to concerns about the retention of floodwater and inland water (Photo 4.1). In addition, the town of Kazuramaki in Higashiomi City, located near the confluence of the Hino River and the Hokyoji River, has autonomously raised the level of its residential land in conjunction with the height of its levees after restoration and repair based on its past experience with typhoons such as the Ise Bay typhoon. In areas where retention-type flooding occurs, such as at the confluence of a banked river, the frequency of outwash flooding is reduced by the embankments, but in the unlikely event that flooding does occur, floodwater and inland water may raise the depth of inundation to the height of the levee.

Inundation analysis throughout the prefecture showed that only approximately 1800 households would be submerged (inundated by more than 3.0 m) even with a rainfall probability of once in 200 years (Taki et al. 2010). In addition, many of these houses were developed after the late 1940s. In contrast, many old settlements

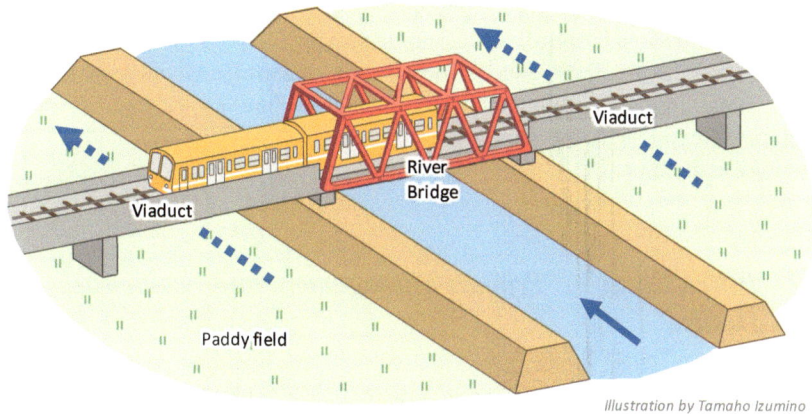

Illustration by Tamaho Izumino

Fig. 4.3 Extension of the viaduct onto the paddy field so as to not block the flow of floods

Photo 4.1 A sign opposing the embankment-type Shinkansen. (Source: Shiga Prefectural Government)

are located in slightly elevated areas such as natural levee belts (avoidance of exposure in Fig. 4.2) or have had measures implemented such as raising housing land (reduction of vulnerability in Fig. 4.2).

Several approaches described here were adopted in various parts of the prefecture and are still implemented in certain areas. However, with passage of time, many of these innovative measures have lost their importance. Open levees have also been gradually disappearing without their disaster mitigation effects being fully evaluated.

Until a few decades ago, river modification and urban development were linked. In the case of Lake Biwa, paddy fields are located on the lowlands around the river, and villages are located on very highlands such as natural levees. The result is a beautiful and rational landscape, harmonized with the providence of nature, in which the flow of the rivers and the positioning of farmlands and towns are all planned integrally. Such designs certainly made the best use of local resources or watershed ecosystem services given the size of the population and level of technology at the time.

4.2.2   Policy Features: Process Management and Framework Design

Shiga Prefecture's flood management policy began with the establishment of the Floodplain Management Office within the Public Works and Transportation Department in July 2006. This is a separate organization from the River Works Division.

In March 2012, approximately 6 years after its conception, the "Flood Risk Management Basic Policy in Shiga Prefecture" (hereinafter, "Basic Policy") was approved, and 2 years later, in March 2014, the "Flood Management Ordinance in Shiga Prefecture" was approved after rigorous discussions. The Flood Management Ordinance provides a legal basis for the Basic Policy and ensures the effectiveness of flood risk management measures conducted by Shiga Prefectural Government. The Basic Policy defines the objectives and measures of flood management in Shiga Prefecture as follows.

> Flood risk management integrates self-help, mutual aid, and public assistance to comprehensively implement both in-river and off-river measures for flood risk management to avoid (1) loss of life (the highest priority) and (2) damage that makes it difficult to rebuild daily life after any kind of flood.

The responsibility for river management is stipulated in the River Law. The obligatory responsibility for flood control is understood to be "setting up design floods and ensuring that design floods flow safely through river channels." The obligatory responsibility for flood control is understood to be "setting up design floods and ensuring that design floods flow safely in the river channels." The Basic Policy includes such river management systems according to the River Law. The Basic Policy positions conventional river management as a core project and specifies that "basin retention measures," "floodplain disaster mitigation measures," and "measures to improve local disaster preparedness" are to be implemented to achieve the flood control goals set forth (Table 4.1). Beyond the framework of the River Law, the City Planning Act, Building Standards Act, and Flood Fighting Act were included in the scope of protection. The framework is to achieve the goal by covering the areas that cannot be addressed by river maintenance with the other three measures. Of the three measures, the "floodplain disaster mitigation measures," in particular, are intended to "avoid exposure" and "reduce vulnerability" to flood inundation.

4.2.3   Site Safety Level: Basic Information for Policy Decisions

When considering flood control measures, the scope of defensive measures should be expanded to include floodplain land use and housing practices in order to move into the fourth stage of flood control. In addition, it is necessary to directly weigh the risks of the floodplains newly subject to planning. In other words, floodplain risk is the frequency at which a certain kind of damage occurs at each point surrounded by a river/channel complex. For example, if you buy the house shown in Fig. 4.4, you may be more concerned about how often and to what degree your home would be flooded rather than the individual safety of the rivers and waterways that surround it. The risk at each site surrounded by a river/channel complex in this way is referred to as the "site safety level" by SPG.

Site safety level can be calculated from hydraulic parameters (inundation depth and speed and fluid force) obtained from flood hydraulic analysis. For example, it is possible to express the site safety level as a risk matrix, as shown in Fig. 4.5, according to the frequency of damage and degree of damage at each point. The magnitude of damage can be classified into five levels: level 1 ($h \leq 0.1$ m) indicates no damage; level 2 ($0.1 \leq h \leq 0.5$ m) indicates damage from inundation below the ground floor of a building; level 3 ($0.5 \leq h < 3.0$ m) indicates severe damage from inundation above the ground floor; level 4 ($3.0 \leq h$) indicates fully submerged; and level 5 (2.5 m^3/s$^2 \leq u^2 h$) indicates completely destroyed. The probability of occurrence of each flood event was evaluated by means of its return period, namely, 2, 10, 30, 50, 100, 200, 500, and 1000 years, where h is depth (m) of the inundation and u is velocity (m/s). The term $u^2 h$ means the "fluid force."

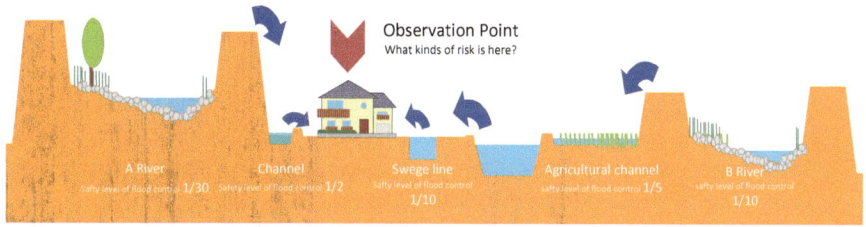

Fig 4.4 Site safety level. This is the level of safety of a point surrounded by a river/channel complex and for the setting of daily life (Source: Taki et al. 2010)

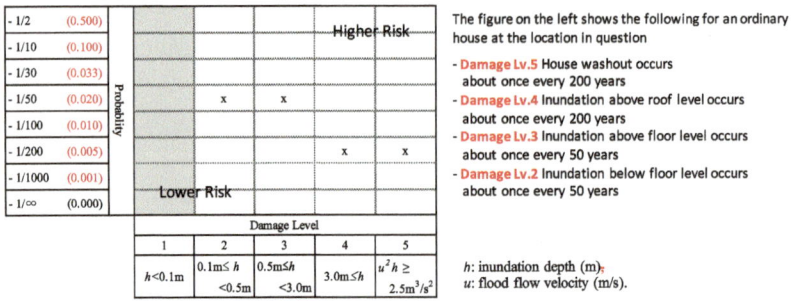

The figure on the left shows the following for an ordinary house at the location in question

- Damage Lv.5 House washout occurs about once every 200 years
- Damage Lv.4 Inundation above roof level occurs about once every 200 years
- Damage Lv.3 Inundation above floor level occurs about once every 50 years
- Damage Lv.2 Inundation below floor level occurs about once every 50 years

h: inundation depth (m);
u: flood flow velocity (m/s).

Fig 4.5 Site safety level at a point expressed using a risk matrix (Taki et al. 2010; Shiga Prefectural Government 2012)

SPG has developed and published a numerical analysis model for evaluating site safety level (Taki et al. 2013). The basic requirements of this numerical analysis model are as follows:

Requirement 1: Consider combined flooding from the river/channel complex surrounding a site on the floodplain (living sphere).

Requirement 2: Consider high- to low-frequency inundations (multiple occurrence probabilities).

Requirement 3: Be able to evaluate risk/safety level equally by policy units (in this case the entire prefecture).

Requirement 1 indicates that the model evaluates the risk/safety level at each location on the floodplain, rather than performance of each river levee or other disaster prevention facility. The model evaluates the performance of the system as a whole group of disaster prevention facilities rather than individual facilities. SPG incorporates (rainwater) sewage systems and agricultural drainage channels into the model besides the 240 major rivers and analyzes them in an integrated manner. Requirement 2 is to assess damage based on the frequency (or magnitude) of floods. Since administrative measures must be impartial within a jurisdiction, requirement 3 allows assessments by administrative units, such as prefectures and municipalities, in the same way.

The structure of the numerical analysis model and the flow of calculations are shown in Fig. 4.6 (Taki et al. 2019) and Fig. 4.7 (Taki et al. 2010). Although the calculation algorithm is orthodox, the most distinctive feature of the model is that it integrates data from not only major rivers managed by the national and prefectural governments but also from small rivers, (rainwater) sewage systems with different management categories, and agricultural drainage.

Modeled rainfall events with a probability of occurrence once in 2, 10, 30, 50, 100, 200, 500, and 1000 years estimated from the Shiga Prefecture rainfall intensity equation are used as the external force for the calculation.

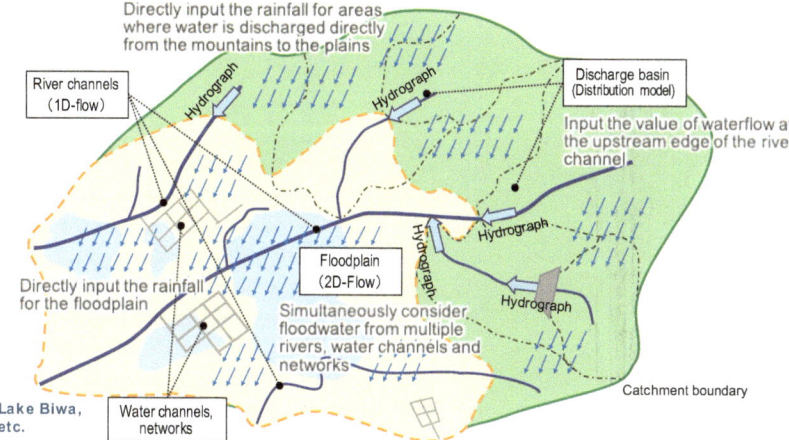

- River channels (approx. 240 rivers) are single dimensional and flood plain is two-dimensional
- Small rivers and large water channels are treated as channels of uniform flow.
- The scope of area for agricultural land readjustment and rainwater drainage development is obtained by deducting rainfall equivalent to discharge capacity and totaling in the downstream area.

Fig 4.6 Composition of the numerical analysis model (Source: Taki et al. 2019)

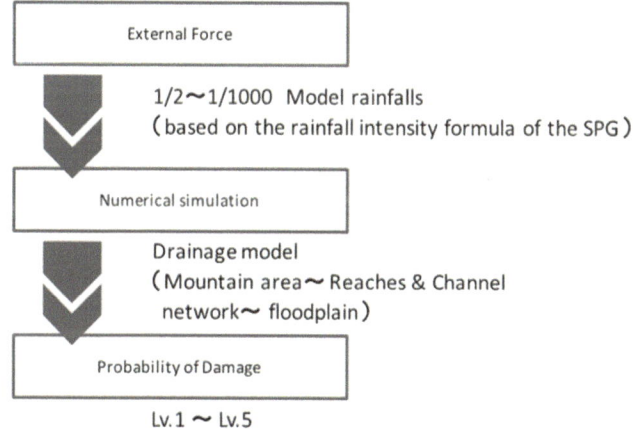

Fig 4.7 Flow of calculations (Taki et al. 2010)

Some of the calculation results are shown in Figs. 4.8 and 4.9. For official prefectural site safety level data, see the Shiga Prefecture disaster preparedness information map (https://shiga-bousai.jp/dmap). GIS data (shapefiles) are available along with maps of areas of possible inundation for each river.

The flood hazard maps published by local governments across the country usually show the depth of inundation for each river in the event of a design flood as an assumed external force. Recent amendments to the Flood Fighting Act in 2015 mandated the designation of areas with a high probability of inundation in the

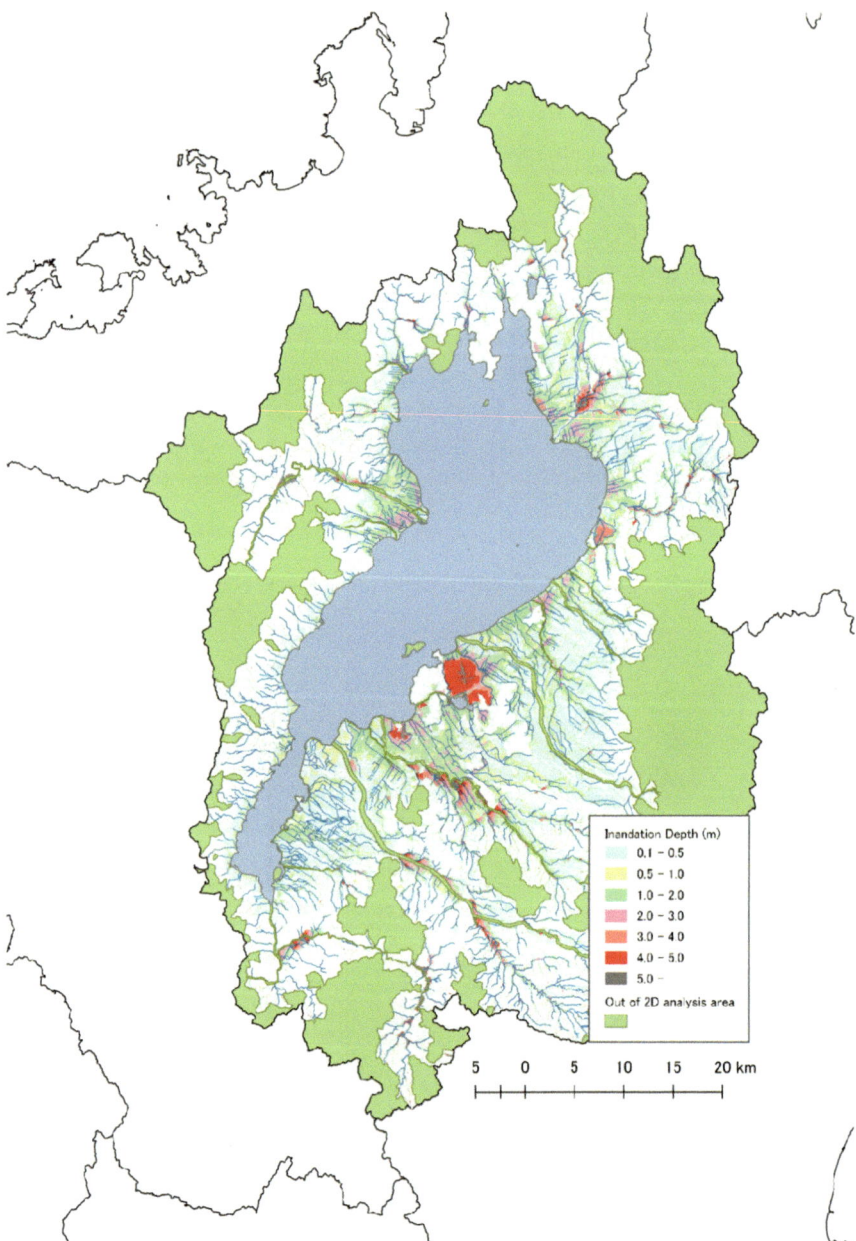

Fig 4.8 Maximum inundation depth (1000-year flood) (Source: Taki et al. 2019)

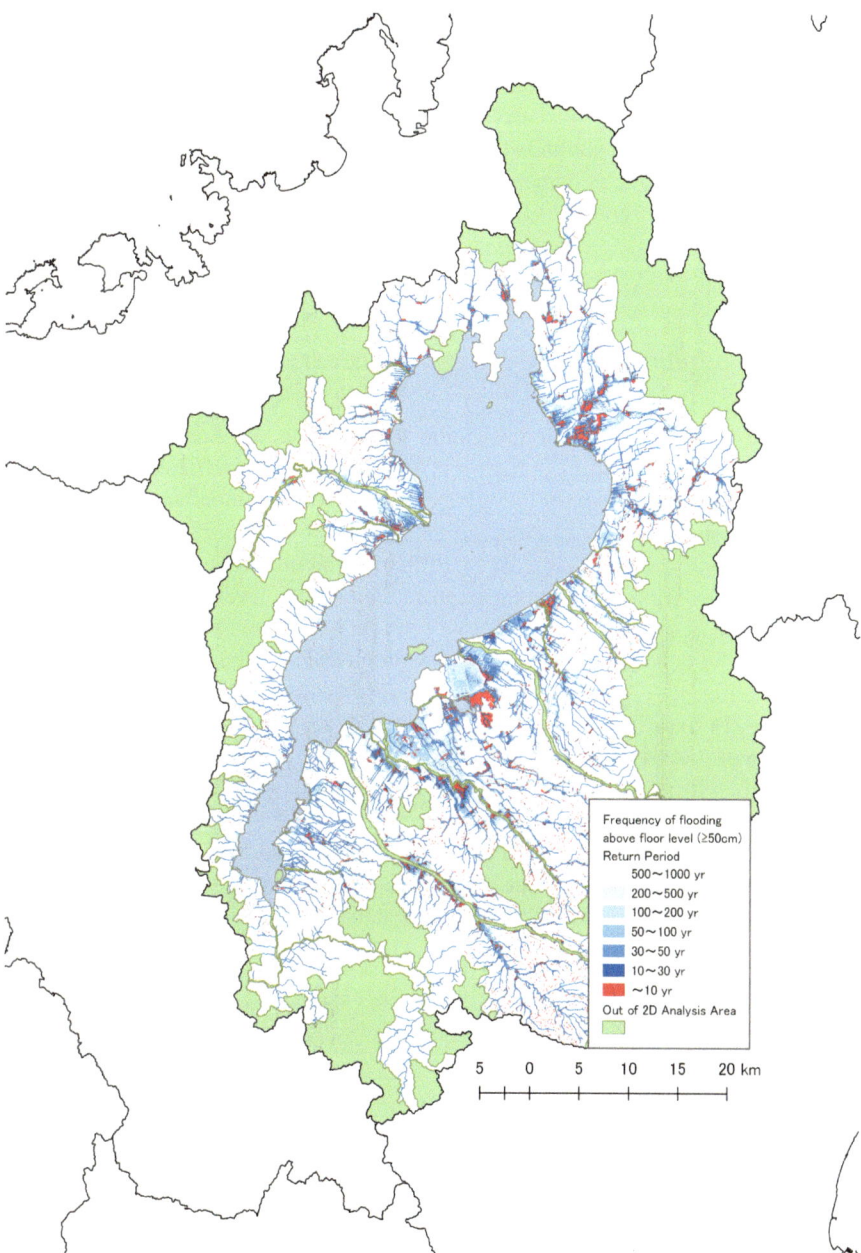

Fig 4.9 Frequency of inundation above floor level (≥50 cm) (Source: Taki et al. 2019)

maximum expected scale of flooding and led to an urgent need to review inundation assumptions and hazard maps across the country. The assumed maximum size is often calculated as the equivalent of an event that has a probability of occurring once in 1000 years. Despite this increased awareness of excessive flooding in inundation assumptions, it is not yet common to express flood risk in a matrix of "frequency of inundation" and "degree of damage" as in the case of "site safety level."

Based on this "site safety level," SPG has developed disaster mitigation measures for floodplains.

4.2.4 Floodplain Disaster Mitigation Measures

One of the floodplain disaster mitigation measures in the Basic Policy was to regulate land use and building in high-risk locations. As shown in Fig. 4.10, Area A is "in principle not included in the urbanization promotion area" in order to avoid "severe property damage." In Area A, inundation on the floor occurs frequently, causing economic damage that makes it difficult to rebuild lives. Area B requires "a structure with a floor higher than the expected inundation level to which evacuation is possible" or "a strong structure that will not be washed out by the expected fluid force" for building permits to avoid house washout and submersion that directly lead to human suffering (regulations based on fluid force are not yet in operation (as of July 2021)). In Area B, even if the frequency is rather low, human damage such as submersion or loss of houses will occur.

However, with the enactment of the Omnibus Decentralization Act in 2000, both of the above notifications were redesignated as legally nonbinding technical advice. Therefore, SPG enacted a Flood Management Ordinance (March 31, 2014, SPG Ordinance 55) to clarify the legal basis for regulation. In addition, Flood-Resistant Building Guidelines (2015) were established to provide building permit requirements under the Flood Management Ordinance.

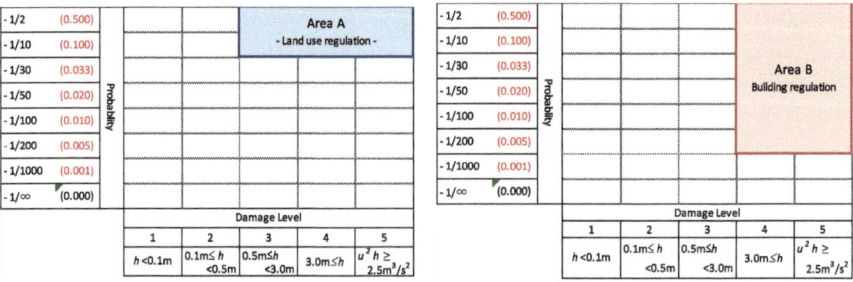

Fig 4.10 Scope of risk covered by land use (Area **a**) and building regulations (Area **b**) (Source: Taki et al. 2010; Shiga Prefectural Government 2012)

Land use and building regulations are nothing more than restrictions on the private property rights guaranteed by the Constitution. In other words, the strength of the regulations must be kept to a minimum in light of the socially accepted norms at the time and after due consideration of the public interest to be secured, such as the protection of life and property. Policy and legal considerations are critical when applying a regulatory approach. Simply discussing the need without fully identifying institutional common ground does not lead to social implementation.

Furthermore, floodplain disaster reduction measures impose a duty of care on developers to ensure that the installation of continuous embankment structures, such as road projects, does not cause significant changes in flood zones, for example, by making bridges over floodplain. The "Guidelines for installation of embankment structures pursuant to Article 25 of the Flood Management Ordinance (2015)" have been released as technical standards. The preservation of open levees, which serve as flood flow control facilities, and the development of armor levees are included in the statutory river improvement plans for each area in the prefecture.

4.3   Floodplain Management and Green Infrastructure

4.3.1   Establishing a "Floodplain Manager": Social Implementation and Points of Practice

Among the four measures for flood risk management (Table 4.2), river improvements (river works) and basin storage measures (retention measures) primarily aim to reduce the frequency of flood inundation through structural measures (levees and flood control facilities) in the basin and river channel area. For river improvements in particular, the River Law and related laws and regulations, as well as major court cases, such as the Daito flood lawsuit and the Tama River flood lawsuit, have clarified the scope of obligatory liability for facility management. Most basin storage measures are installed by developers in accordance with prefectural ordinances and guidelines, but some are positioned in river plains within the framework of comprehensive flood control projects and installed by river administrators such as public works offices of the Ministry of Land, Infrastructure, Transport and Tourism or prefectural governments.

Measures to improve local disaster preparedness (preparation measures) have been developed mainly by local governments with the support of river administrators under various systems related to crisis management, such as the Flood Fighting Law and the Disaster Countermeasure Basic Law, as a complementary measure to river maintenance and basin storage measures.

In contrast, floodplain disaster mitigation measures (containment measures) such as open levees, two-line levees, forest protection, and land use and building

Table 4.2 Classification of flood management measures (Shiga Prefectural Government 2012)

River works Measures to ensure safe flow of design floods in river channels	Measures to improve rivers and waterways in order to prevent floods from river overflowing. These include artificial levees, reservoirs, and flood control dams
Watershed works Basin retention measures	Measures to mitigate rapid flood runoff into rivers and waterways. These include rainwater retention in ponds, grounds, forest soils, paddy fields, and reservoirs
Floodplain works Flood disaster reduction measures	Urban planning measures to minimize damage in the event of flooding that exceeds the capacity of river and waterway facilities. These include ring levees, secondary levees, flood prevention forests, land use regulations, and building flood proofing
Community works Disaster resilience improvement	Measures to support evacuation and flood prevention activities. These include disaster drills and the dissemination of disaster preparedness information

regulations have long been adopted as risk reduction measures but have not been fully taken into account in modern flood control systems. The open levee, which is said to have been developed by Shingen Takeda (1521–1657, a famous territorial lord), was seen in many places in the past, but many of them have now disappeared without finding a place in the current institutionalization. With regard to land use, the risk of flooding in the area has not been treated as a major determinant during subsequent annexation of an existing residential zone into an urbanized zone or during the conversion of agricultural land.

Under the current legal system, the primary responsibility of river management is to handle design floods within river channels. Therefore, when floodplain disaster mitigation measures are implemented by river administrators, they have to be implemented as an extension of river management for the purpose of handling design floods. This means that part of the floodplain is considered a river channel (river management facility) and used to handle design floods. For example, ring levees and secondary levees that are constructed as part of a land use integrated water disaster prevention project fall under this category. That is, the current position is different from the idea that disaster mitigation is accomplished by responding throughout the floodplain (floodplain management) rather than by using river management facilities.

However, the current flood control system has been deeply adapted to the socioeconomic needs of Japan since the enactment of the River Law in 1898, with modifications added as necessary. Dramatically altering the system would cause social friction and confusion. A practical and appropriate approach to floodplain disaster mitigation practices, therefore, is to respect the various plans implemented by river administrators as a given condition and add a separate system of administration (floodplain managers) to complement them.

To enable the establishment of a complementary relationship, floodplain managers should be given the objective of minimizing damage from external forces beyond the capacity of facilities, regardless of the stage of river maintenance. A separate system of administrative procedures (legal, organizational, and budgetary measures) of river management should also be added.

Based on this idea, SPG established the goals and system of flood risk management as mentioned in Sect. 4.2. Specifically, an independent department in charge of river basin-wide flood management (Floodplain Management Office) was established, which is separate from the department in charge of rivers (River and Port Division). This was successful, and floodplain disaster mitigation measures can now be promoted in parallel with river maintenance, rather than as a binary choice.

4.3.2 Artificial Structures and Green Infrastructure

The Science Council of Japan (2014) describes the benefits of artificial structures as "achieving a single function that contributes to a specific and distinct purpose with a high degree of precision (accurately providing the performance required by the society)," and the benefits of green infrastructure as "maintaining and creating diverse spaces that can provide a variety of ecosystem services and contribute to the conservation of biodiversity." Green infrastructure should be understood as complementing rather than replacing the function of artificial structures.

In terms of the sole function of flood protection, green infrastructure is nowhere near the standard of artificial structures. Nonetheless, although green infrastructure may be less effective in a specific function, it exhibits diverse functions in normal times. For example, there are still some flood prevention forests along the Azumi River, which runs through the western part of Shiga Prefecture. Many are bamboo groves, which not only mitigate flooding but also support rich riverine biota, provide food, and have been used as materials to make the frames of traditional Japanese fans and the like. In this way, green infrastructure, which is closely related to lifestyle and culture, is itself a local resource and forms the character of the region and its original landscape. This is a role that could not easily be played by functionally oriented concrete embankments.

In advancing flood control policy, it is essential to find the best combination of artificial structures and green infrastructure in accordance with the legal system and regional characteristics, considering the characteristics of each.

4.3.3 Green Infrastructure: The Trump Card in Floodplain Management

To ensure that administrative responsibilities are fulfilled under the river management system based on River Law, artificial structures such as levees and dams that reliably perform the required functions are essential. In contrast, attempts to build artificial structures to prevent excess floods are met with irrefutable policy arguments from the fiscal authority that the responsibility of dealing with design floods should first be executed before engaging in excess flood control measures. Therefore, in this author's view, floodplain management must focus on a response with green infrastructure that can perform a variety of functions during normal times. From the above, the scope of defense by artificial structures and green infrastructure can be summarized as follows.

River management: continues to play a role in ensuring the safe flow of design floods within river channels, based on reliably functioning artificial structures to reduce the frequency of flood inundation. Promote nature-oriented river development to the extent that it does not impair flood control functions with respect to design floods.

Floodplain management: addresses floods that exceed the capacity of artificial structure complexes through Eco-DRR-oriented green infrastructure, including land use, with a focus on "avoiding exposure" and "reducing vulnerability".

As we have seen, Eco-DRR-oriented green infrastructure has a strong affinity with floodplain disaster mitigation measures, but we cannot expect to see a large amount of public investment on the so-called "extra" measures when river management is still in its infancy. In addition, many of the current laws, such as the Rivers Act, the Cropland Act, and the Forest Act, are organized according to purpose. Hence, it is inherently difficult to develop something as multifaceted as green infrastructure within a single legal framework. It is an issue of vertical compartmentalization.

However, in the face of a declining population and climate change, we must establish methods to solve problems by going beyond the vertical division of the legal system and the administrative structure. For example, the disaster mitigation effects and multifaceted functions of Eco-DRR-oriented green infrastructure should be evaluated, and the ways to raise funds directly from financial markets should be considered. I expect research for social implementation to rapidly advance from now on.

References

ADRC (2015), Total disaster risk management - good practices

Hori T, Furukawa S, Fujita A, Inazu K, Ikebuchi S (2008) An optimal design framework of a flood control system including in-floodplain countermeasures based on distributed risk assessment - basic concepts and methodology. Jpn Soc Civil Eng 64(1):1–12. [in Japanese with English abstract]

Natural Environment Bureau, Ministry of the Environment (2016) Approaches to disaster reduction and mitigation using ecosystems [in Japanese]

Science Council of Japan (2014) Proposal: recommendations for the use of ecosystem infrastructure in reconstruction and national resilience [in Japanese]

Shiga Prefectural Government (2012) Flood management basic policy in Shiga prefecture [in Japanese]

Shiga Prefectural Government (2014) Flood management ordinance in Shiga prefecture. Shiga Prefecture Ordinance No 55, March 31, 2014 [in Japanese]

Taki K, Matsuda T, Ukai E, Ogasawara Y, Nishijima T, Nakatani K (2010) Design for flood disaster-reduction system in a floodplain of small-medium sized rivers. Adv River Eng 64:477–482. [in Japanese with English abstract]

Taki K, Matsuda T, Ukai E, Nishijima T, Egashira S (2013) Method for evaluating flood disaster reduction measures in alluvial plains. J Flood Risk Manag 6:210–218

Taki K, Yamashita K, Hirayama N, Takanishi S (2019) A study of regional disaster preparedness improvement strategy based on hydraulic analysis of a small and medium-sized river system. Adv River Eng 25:79–84. [in Japanese]

Chapter 5
Toward Social Infrastructure: Typological Idea for Evaluating Implementation Potential of Green Infrastructure

Takeshi Osawa and Takaaki Nishida

Abstract Green infrastructure (GI) comprises widely distributed objects in human residential communities. However, because of the variety of certain objects, it is sometimes difficult to improve public awareness and enhance social implementation of GI. To expand the idea of GI and apply it widely in our society, we should understand clearly what exactly GI is and where and how it can be applied. In this article, we classify the types of GI and present a basic approach to evaluate their implementation potential as the first step for expanding the application of GI in human society. First, based on the definition of GI, we classified it as the infrastructure involving the natural ecosystem, seminatural ecosystem, and artificial ecosystem in each. The essential differences among these types arise from their installability depending on human activities. Then, we considered the principle of evaluation of the implementation potential of GI for the three types GI based on three dimensions, natural condition, top-down regulation, and bottom-up regulation, in human society. Additionally, appropriate ideas for the evaluation of each dimension were considered. Relative importance of the natural condition, top-down regulation, and bottom-up regulation differs among the types of GI. We believe that the findings of this work will be useful for decision makers dealing with the application of GI in their administrative areas.

Keywords Artificial ecosystem · Decision support · Driving force pressure state impact response (DPSIR) framework · Installability · Seminatural ecosystem

T. Osawa (✉)
Graduate School of Urban Environmental Sciences, Tokyo Metropolitan University, Tokyo, Japan

T. Nishida
Kyoto Sangyo University, Kyoto, Japan

Mitsubishi UFJ Research and Consulting Co. Ltd., Osaka, Japan

F. Nakamura (ed.), *Green Infrastructure and Climate Change Adaptation*,
Ecological Research Monographs, https://doi.org/10.1007/978-981-16-6791-6_5

5.1   Introduction

Green infrastructure (GI) is defined as the natural, seminatural, and artificial networks of multifunctional ecological systems within, around, and between human residential communities on all spatial scales (Tzoulas et al. 2007; Pearlmutter et al. 2017). This concept can cover a wide range of elements such as intact forests (Svensson et al. 2019), wetland (Nakamura et al. 2019), agricultural land (Osawa et al. 2020a), and urbanized green spaces (Tzoulas et al. 2007; Matsunaga and Osawa 2021). Thus, GI can exist widely in human residential communities. However, such a broad definition sometimes makes it difficult for decision makers, such as government personnel who are the stakeholders for the development of infrastructures, to perceive GI. For example, a policy maker who is responsible for the mitigation of climate change tends to focus on the GI that is useful for carbon fixation (e.g., Chen 2015). Health-care professionals, such as public health department personnel and hospital operators, tend to focus on the GI that can benefit the human health, both physical and mental (e.g., Tzoulas et al. 2007). Although many categories of people have a potential interest in GI, integrating their interests to link various GI implementation practices is a challenge. In addition, local residents often cannot perceive the existing GI itself, i.e., the ecosystem with ecosystem services existing around them (Zhang et al. 2010; Cortinovis and Geneletti 2018). These challenges hamper the application of GI as social infrastructure for the human society. To expand the idea and application of GI and to contribute the human society, a clear understanding of what GI comprises and where and how it can be applied is necessary.

Proposing a typology for conceptual term is one of the effective ways to perceive future challenges for research and management (Eggermont et al. 2015). Thus, we can consider that clarifying and classifying the concept of GI are important steps to discuss the feasibility of the application of GI as social infrastructure. In this work, we expanded the basic concept of GI to evaluate its implementation potential for the human society and to improve its applicability as social infrastructure, especially for land managers and decision makers. First, based on the basic definition of GI and the concept of human society–environment interactions, we classified the GI for three types of ecosystems, i.e., the natural, seminatural, and artificial ecosystems. Second, we proposed the underlying concept of the components of GI implementation potential for GI and considered basic methods to evaluate it. Finally, we discussed the next challenges and perspectives for the application of GI.

5.2   Classification Basis for GIs

One of the challenges in expanding the idea and implementing GI in the human society is to link with concept of that already widely accepted in other fields (Romano et al. 2015; Spanò et al. 2017). Spanò et al. (2017) indicated the usefulness of the driving force–pressure–state–impact–response (DPSIR) framework to integrate knowledge between diverse disciplines and GI (Svarstad et al. 2008). This framework considers the driving forces (D) (e.g., human activity) that exert pressure (P) (e.g., land-use change), leading to changes in the state (S) (e.g., ecological processes) and thereby cause impacts (I) and the human health and society that may elicit a societal response (R) (Spanò et al. 2017). At the heart of this framework is a simple incorporation of the interaction between human society and environment. The GI should support the human society, and thus, GI should interact with human activities. That is, for an ecosystem to be classified as GI, humans should take appropriate efforts (driving force and pressure) to regulate that ecosystem as required by humans. Interaction between the human society and environment is an essential factor in determining the GI, and hence, this factor should be considered when classifying GI.

5.3 Classification of GI

Based on the conceptual definition of GI considering the interactions between the human society and environment, we can specify at least three types of elements, namely, the natural ecosystem, seminatural ecosystem, and artificial ecosystem as GI (Fig. 5.1). Examples of natural ecosystems as GI are coral reefs, which can reduce the impact of ocean waves (Martin and Watson 2016), and natural wetland, which can reduce flooding occurrences (Nakamura et al. 2019). Examples of seminatural ecosystems as GI are paddy fields, which have a variety of functions such as food production, wetland habitat support (Natuhara 2013; Osawa et al. 2020b), and disaster prevention (Osawa et al. 2020a, b). Artificial ecosystems as GI include rain gardens, which are urban green spaces that store rain water and prevent urban flooding (Ishimatsu et al. 2017). These three types have clear differences in terms of their installability depending on human activities, i.e., the driving force and pressure in the DPSIR framework (Fig. 5.1). To be specific, we cannot create a natural ecosystem, and it is difficult to create a seminatural ecosystem by ourselves; nevertheless, we can support these ecosystems and create an artificial ecosystem in any area in theory. Thus, the implementation potentials for natural and seminatural ecosystem-based GI are strongly regulated by the location of the target area, whereas there are few such limitations on GI based on artificial ecosystems—this is an important aspect related to the implementation potential of GI. Of course, it may be possible to combine these types of GI. However, for simplicity, we have classified them clearly in this work.

This classification is similar to the classification of nature-based solution (NbS), proposed by Eggermont et al. (2015). They proposed that NbS could be applied along with two evaluation axes. Those are (1) the extent of involvement of biodiversity and ecosystems' engineering in NbS and (2) the number of ecosystem services and stakeholder groups that are targeted by a given NbS. Moreover, they discussed that low level of axis 1 combined with high level of axis 2 constitutes

Natural ecosystem   Seminatural ecosystem   Artificial ecosystem

Installability by humans (Driving force and Pressure)

Fig. 5.1 Essential differences among the three types of green infrastructure (GI)

ecosystem with no or minimal intervention; both mid-levels are connected with human-mediated ecosystem, namely, seminatural; and high level of axis 1 combined with low level of axis 2 constitutes artificial ecosystem. Although they categorized GI as the artificial ecosystem, which comprises high level of axis 1 combined with low level of axis 2 (Eggermont et al. 2015), the basic idea established by interaction between the human society and environment is common.

5.4 Three Dimensions for the Implementation Potential of GI

We propose three dimensions for evaluating implementation potential of GI (Fig. 5.2). The first is natural condition. Implementation potential of natural and seminatural GI is basically regulated by their geographical setting. We cannot introduce a coral reef as GI in a mountainous or hilly area. We cannot introduce a forest as GI for the sea. Thus, implementation potential of a GI type should be defined based on the geographical setting, i.e., the natural condition in the target area. This is one of essential dimensions for evaluating the implementation potential of GI.

Second is top-down regulation (Fig. 5.2). Artificial ecosystem such as urban green spaces can be created by humans. In other words, the implementation potential of this type of GI is not as strongly regulated by the geographical setting as in the case of the other two types. That is, in theory, artificial GI can be introduced anywhere. However, such GIs are generally introduced at a large scale by public work departments, not by individuals. For example, roadside trees are often established and maintained by local governments. Thus, to introduce such GI, the public administration should take interest. Therefore, the implementation potential of this type of GI can be evaluated based on the situation of human society, mainly the top-down regulation such as that by the government. This administrative regulation is the second essential dimension for evaluating the implementation potential of GI.

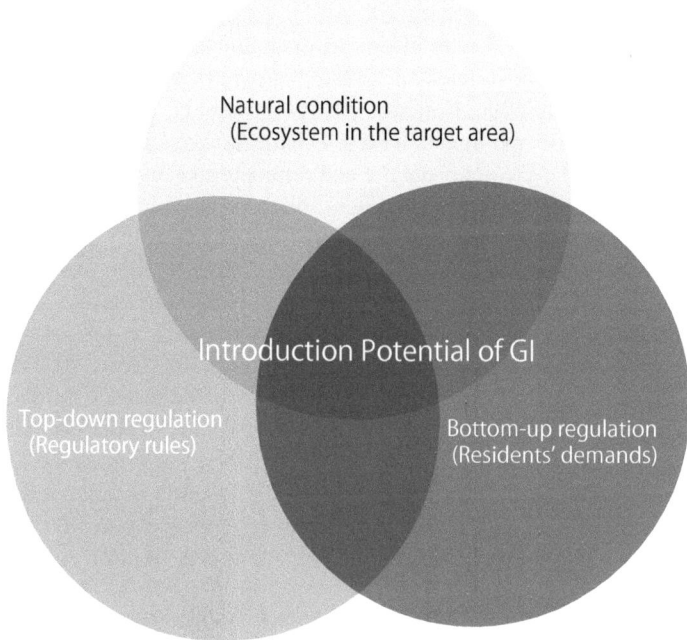

Fig. 5.2 Three essential dimensions to estimate the implementation potential for GI

Third is bottom-up regulation (Fig. 5.2). To decide the specific public work, public administrations should respect the will of the residents. Thus, to introduce GI in residential areas, the residents too should take interest in the GI. This demand/requirement, i.e., bottom-up regulation, is the third essential dimension for estimating the implementation potential of GI. Thus, the implementation potential of GI is based on three essential dimensions that are often related to each other (Fig. 5.2). We can clearly define and evaluate these aspects independently. Next, we discuss these three dimensions more in detail.

5.4.1   Natural Condition

The first dimension, i.e., the natural conditions, is the easiest dimension to perceive and evaluate because it is essentially the ecosystem existing in the target area. Thus, the expected benefits of the GI are almost the same as the existing ecosystem services. The aim of introduction of the GI should be to employ and/or enhance their existing ecosystem services in the target area. For example, the Ministry of Agriculture, Forestry and Fisheries, Japan, has adopted a GI-related policy to enhance the ecosystem services produced from agricultural areas using direct payment systems (MAFF web site https://www.maff.go.jp/j/nousin/kanri/tamen_ siharai.html, accessed on February 10, 2021). The Forestry Agency, Japan, has made efforts to promote GI with efforts to maintain forests (Maeda 2017). Both agricultural and forest ecosystems are already known to provide several ecosystem services (Matsuno et al. 2006; Natuhara 2013; Osawa et al. 2020a, b); hence, these GI strategies aim to use and maintain these existing ecosystem services. Therefore, if a decision maker plans to apply natural and seminatural ecosystem GI in a specific area, they must consider the local ecosystem and the type of ecosystem services that can be obtained in this target area as the basis of GI. This is a very simple condition. Additionally, the practitioner should focus on the benefits they expect from that GI. For example, a paddy field is an extremely useful GI because of its multiple functions besides food production: ecosystem services such as regulating service, cultural service, and supporting service (Natuhara 2013). However, the relative values of the services strongly depend on their local conditions. A previous study reported a trade-off relationship between the provisioning service and supporting service in a paddy field (Osawa et al. 2016). Further, the regulating service, in specific, flood prevention, strongly depends on the location (Osawa et al. 2020a). Therefore, to introduce a natural and seminatural ecosystem-based GI, practitioners should consider the types of the existing ecosystems and the services that can be extracted from them.

5.4.2 Top-Down Regulation

The second dimension, i.e., top-down regulation, refers to the social regulatory rules, such as laws, administrative programs, and related individual rights. We cannot consider any ecosystem as GI without considering such top-down regulation even when the focusing ecosystem has high potential. For example, a protected area such as a national park that is strongly regulated for any development work can be a potential GI (Benedict and McMahon 2006), but its uses are restricted according to the regulatory roles. In other words, within the protected area, the expected ecosystem services, i.e., benefits to humans, may be regulated. For example, it is difficult to receive any provisioning services from a closed zone in a protected area. The decision maker should not consider these unavailable ecosystem services in the focusing area as the benefit of GI. However, the regulatory roles could promote the application of the area as GI. For example, according to the Urban Park Act, Japan, urban green spaces are expected to provide the temperature reduction effects in summer (https://www.kantei.go.jp/jp/singi/tiiki/kokusentoc_ wg/hearing_s/150123siryou03_1.pdf, in Japanese, accessed on February 10, 2021). This type of ecosystem services is easy to use as the function of GI for urban green space (Matsunaga and Osawa 2021). Both the introduction advisability and expected ecosystem services are basically decided by the role of the human society. Therefore, one effective way to evaluate the implementation potential of GI is to review administrative documents. Administrative documents include laws, local codes, and administrative programs such as city development, environment, and water usage. One specific example of administrative document in the environment is the establishment of the biodiversity strategy by the national government and some local governments based on the recommendation by the Convention on Biological Diversity. Any mentions of both the protected area and ecosystem services expected from the current ecosystems in the document can be a key factor for introducing GI in the area. These documents provide us hints regarding the implementation potential of GI.

The top-down regulation is not only administrative but can also include individual rights such as property rights. To simplify, if a practitioner wants to apply GI for a given forest land, the permission of the landowners is essential. Of course, this type of regulation can exist for intact ecosystem. Such top-down regulations are essential for applying GI because almost of all land, sea, and rivers, i.e., ecosystems, are owned and/or managed by someone.

5.4.3 Bottom-Up Regulation: Residential Demands and Requirements

The third dimension is also a part of social regulation, but this arises from bottom-up regulation, i.e., the demand/requirements of local residents. If a government wants to introduce GI in an administrative area, they should first estimate and visualize the merits of the GI for residents because public works must be undertaken only if it has public benefits. Unfortunately, estimating the benefits of GI in detail is difficult because of the uncertainty of the multiple functions of GI, at least partly owing to the limited knowledge regarding them at present. Thus, residential acceptability, i.e., the appreciation and understanding of ecosystem functions and/or services, is one of the key factors determining the implementation potential of GI.

Residential appreciation and understanding of ecosystem function and/or services are difficult to estimate because they are qualitative factors. Recently, these qualitative factors were estimated quantitatively via interview and questionnaire surveys (Cheng et al. 2019). People who have interest and/or affinity toward ecosystem or natural environment may have a relatively high acceptability for the introduction of GI as a part of the ecological system.

5.5 Evaluation of the Potential for Introduction of GI

To summarize, the three essential dimensions of evaluating the implementation potential of GI are the natural condition, top-down, and bottom-up social regulations (Fig. 5.2). Notably, relative importance of three dimensions among GI types could differ (Fig. 5.3). For example, the natural ecosystem-based GI is strongly

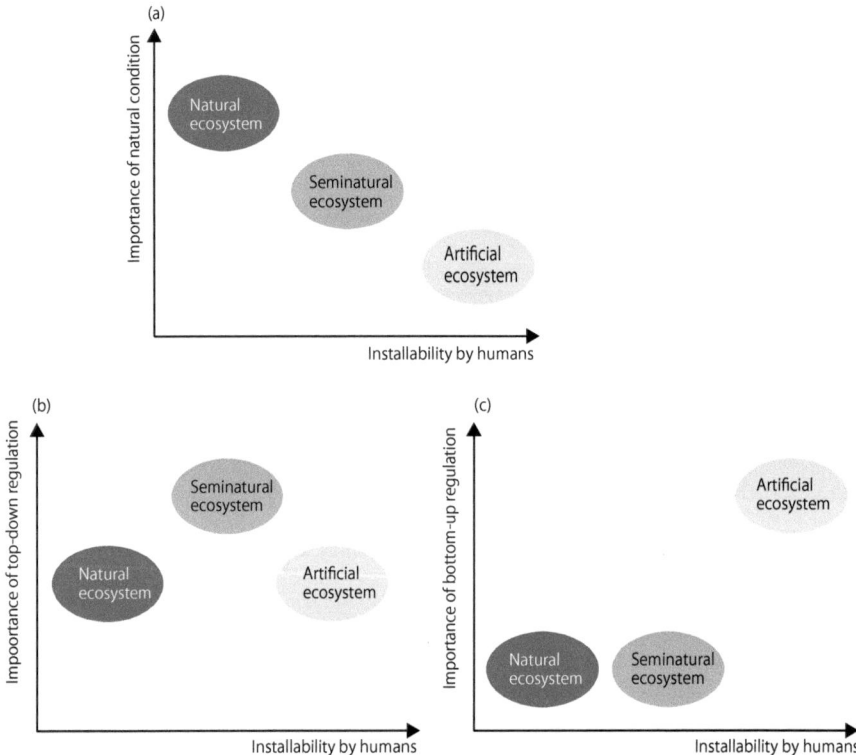

Fig. 5.3 Relative importance of three dimensions: (**a**) natural condition, (**b**) top-down regulation, and (**c**) bottom-up regulation for each GI type

Table 5.1 Relative importance of each dimension in each type of GI

GI type	Natural condition	Top-down regulation	Bottom-up regulation
Natural ecosystem	High	Moderate	Low
Seminatural ecosystem	Moderate	High	Low
Artificial ecosystem	Low	Moderate	High

regulated by the natural condition, i.e., the first dimension (Fig. 5.3a). A seminatural ecosystem-based GI is strongly regulated by top-down regulation, i.e., the second dimension (Fig. 5.3b). An artificial ecosystem is strongly regulated by bottom-up social regulation, i.e., the third dimension, but rarely regulated by the first dimension (Fig. 5.3a, c). A decision maker should consider all the three dimensions when introducing GI in the planning stage. Table 5.1 summarizes the relative importance of each dimension for each type of GI, which can be used by the practitioners as a checklist in the local planning of GI; at least the items marked "high" should be carefully considered to introduce GI suitably for the target area.

5.6 Conclusion and Perspective

We classified the GI into three types on the basis of ecosystems and considered the basic concept for evaluating the implementation potential of each type of GI in three dimensions. Essentially the ubiquitous objects, i.e., ecosystems in our society, can function as GI, but there exists a large variety in such elements. Some types of GI can cover a wide area and provide several benefits, while others may exist locally and provide fewer benefits. GI is a comprehensive concept of the use of an ecosystem for the human society. Thus, the basic idea of GI itself is not new. However, the word "infrastructure" refers to the basic systems and services for human society that are managed by a country or an organization. Thus, the idea of GI may include the concept that the human society should use natural environment more systematically and effectively. Expanding the idea of GI is almost the same as establishing a society that is in harmony with nature. Estimating the implementation potential of GI is an important step to expand the GI for our society.

Acknowledgments This work was supported by the Environment Research and Technology Development Fund (4-1805) of the Ministry of the Environment, Japan, and partially supported by the JSPS KAKENHI Grant Number 20 K06096. Profs. F. Nakamura and N. Furuta provide us many useful suggestions.

References

Benedict MA, McMahon ET (2006) Green infrastructure. Island, Washington, DC
Chen WY (2015) The role of urban green infrastructure in offsetting carbon emissions in 35 major Chinese cities: a nationwide estimate. Cities 44:112–120
Cheng X, Van Damme S, Li L, Uyttenhove P (2019) Evaluation of cultural ecosystem services: a review of methods. Ecosyst Serv 37:100925
Cortinovis C, Geneletti D (2018) Ecosystem services in urban plans: what is there, and what is still needed for better decisions. Land Use Policy 70:298–312
Eggermont H, Balian E, Azevedo JMN et al (2015) Nature-based solutions: new influence for environmental management and research in Europe. GAIA Ecol Perspect Sci Soc 24:243–248
Ishimatsu K, Ito K, Mitani Y et al (2017) Use of rain gardens for stormwater management in urban design and planning. Landsc Ecol Eng 13:205–212. https://doi.org/10.1007/s11355-016-0309-3
Maeda S (2017) Green infrastructure: forestry and trends in securing new Funcing sources. Q J Public Policy Manag 2017:47–57. (in Japanese)
Martin TG, Watson JEM (2016) Intact ecosystems provide best defence against climate change. Nat Clim Chang 6:122–124. https://doi.org/10.1038/nclimate2918
Matsunaga K, Osawa T (2021) A method for evaluating temperature reduction effect in urban green space: case study in the Central Tokyo metropolitan. Jpn J Conserv Ecol. (in Japanese)
Matsuno Y, Nakamura K, Masumoto T et al (2006) Prospects for multifunctionality of paddy rice cultivation in Japan and other countries in monsoon Asia. Paddy Water Environ 4:189–197. https://doi.org/10.1007/s10333-006-0048-4
Nakamura F, Akasaka T, Ishiyama N et al (2019) Adaptation to climate change and conservation of biodiversity using green infrastructure. River Res Appl:1–13. https://doi.org/10.1002/rra.3576

Natuhara Y (2013) Ecosystem services by paddy fields as substitutes of natural wetlands in Japan. Ecol Eng 56:97–106

Osawa T, Kohyama K, Mitsuhashi H (2016) Trade-off relationship between modern agriculture and biodiversity: heavy consolidation work has a long-term negative impact on plant species diversity. Land Use Policy 54. https://doi.org/10.1016/j.landusepol.2016.02.001

Osawa T, Nishida T, Oka T (2020a) High tolerance land use against flood disasters: how paddy fields as previously natural wetland inhibit the occurrence of floods. Ecol Indic 114:106306

Osawa T, Nishida T, Oka T (2020b) Paddy fields located in water storage zones could take over the wetland plant community. Sci Rep 10:1–8

Pearlmutter D, Calfapietra C, Samson R et al (2017) The urban forest: cultivating green infrastructure for people and the environment. Springer, Cham. https://doi.org/10.1007/978-3-319-50280-9

Romano G, Dal Sasso P, Trisorio Liuzzi G, Gentile F (2015) Multi-criteria decision analysis for land suitability mapping in a rural area of Southern Italy. Land Use Policy 48:131–143. https://doi.org/10.1016/j.landusepol.2015.05.013

Spanò M, Gentile F, Davies C, Lafortezza R (2017) The DPSIR framework in support of green infrastructure planning: a case study in Southern Italy. Land Use Policy 61:242–250. https://doi.org/10.1016/j.landusepol.2016.10.051

Svarstad H, Petersen LK, Rothman D et al (2008) Discursive biases of the environmental research framework DPSIR. Land Use Policy 25:116–125. https://doi.org/10.1016/j.landusepol.2007.03.005

Svensson J, Andersson J, Sandström P et al (2019) Landscape trajectory of natural boreal forest loss as an impediment to green infrastructure. Conserv Biol 33:152–163

Tzoulas K, Korpela K, Venn S et al (2007) Promoting ecosystem and human health in urban areas using green infrastructure: a literature review. Landsc Urban Plan 81:167–178

Zhang B, Li W, Xie G (2010) Ecosystem services research in China: progress and perspective. Ecol Econ 69:1389–1395

Part II
Forest Ecosystem

Chapter 6
Riparian Forests and Climate Change: Interactive Zone of Green and Blue Infrastructure

Futoshi Nakamura

Abstract Riparian forests are recognized as green/blue infrastructure that provides various ecosystem services, including water transport, shade from sunlight energy, supply of leaf litter, input of terrestrial insect, delivery of large wood, maintenance of water quality, and corridors for wildlife. However, these forests have already experienced negative impacts from various anthropogenic stressors, such as forest cutting, agricultural development, river regulation, and dam construction. Moreover, the climate projections for the twenty-first century in Japan indicate that mean precipitation may increase by more than 10%, and other projections predict an increase in the frequency of high-magnitude floods and a reduction in the discharge of snowmelt floods. In this chapter, we describe probable changes in the structure and function of riparian forests that might result and describe adaption strategies to reduce the potential impacts of climate change on stream and riparian ecosystems exacerbated by anthropogenic stressors.

Keywords Riparian zone · Climate change adaptation · Green/blue infrastructure · Ecological function · Ecosystem services · Restoration

6.1 Introduction

A "riparian forest" refers to a forest that is transversely distributed from a riverbank to its floodplain at the foot of adjacent hillslopes (Fig. 6.1). Riparian forests include not only forests associated with stream/river courses but also those around wetlands and lakeshores. Riparian forest communities can survive intensive and frequent disturbance regimes through physiological and structural adaptations while utilizing the materials (sediment, nutrients, and organic matter) trapped by these disturbance events in complex river geomorphic surfaces (i.e., secondary channels, point bars,

F. Nakamura (✉)
Laboratory of Ecosystem Management, Graduate School of Agriculture, Hokkaido University, Sapporo, Japan
e-mail: nakaf@for.agr.hokudai.ac.jp

© The Author(s) 2022   73
F. Nakamura (ed.), *Green Infrastructure and Climate Change Adaptation*,
Ecological Research Monographs, https://doi.org/10.1007/978-981-16-6791-6_6

Fig. 6.1 Riparian forest of Satsunai River, Japan

and backswamps) as habitats for their growth and maintenance (Nakamura and Inahara 2007).

A riparian forest longitudinally extends from headwater streams to lowland rivers at the catchment scale (Niiyama 1987). The disturbance regime also changes from headwaters to low-gradient alluvial rivers (Nakamura et al. 2000; Nakamura and Swanson 2003) (Fig. 6.2). A dominant disturbance in headwater streams is mass movement, such as landslides and debris flows, characterized by their intensity but rare occurrence. In downstream, braided reaches, seasonal flood disturbances generate frequent high flows that form extensively developed geomorphic surfaces on floodplains. Low-gradient alluvial rivers in developing peat marshes rarely experience disturbances that result in landform alterations, although prolonged inundation in a marsh can cause substantial physiological stress to trees. Thus, a variety of disturbance types, frequencies, and intensities can be found at a catchment scale and may have diverse ecological consequences for riparian communities. In general, the structure and composition of riparian forests are determined by numerous environmental gradients, such as longitudinal and cross-sectional elevation, substrate (or soil) conditions, and disturbance frequency (Nakamura et al. 1997; Shin and Nakamura 2005; Sakio 1997).

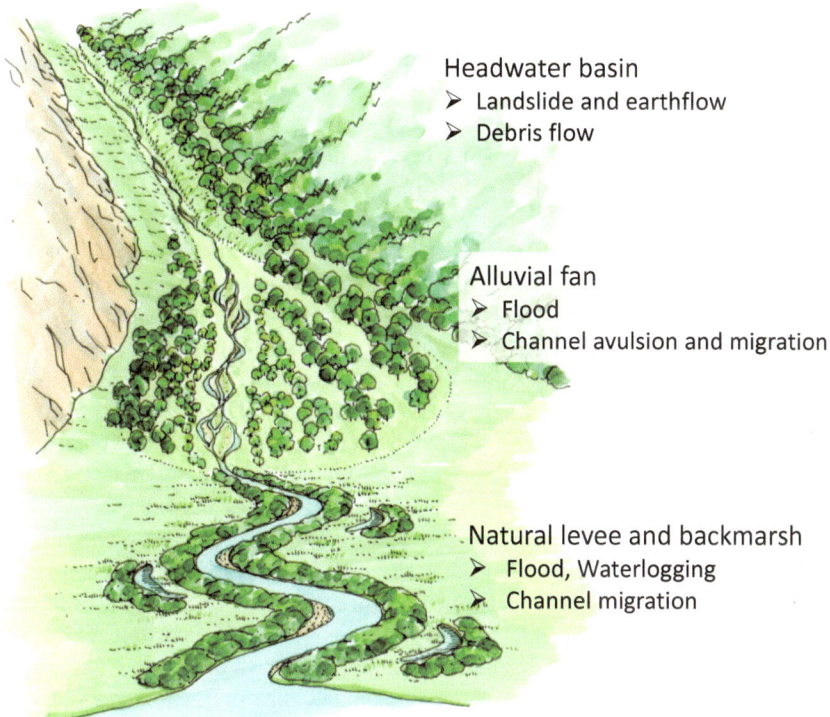

Headwater basin
➢ Landslide and earthflow
➢ Debris flow

Alluvial fan
➢ Flood
➢ Channel avulsion and migration

Natural levee and backmarsh
➢ Flood, Waterlogging
➢ Channel migration

Fig. 6.2 Riparian forests and dominant disturbance regime from headwater basins to lowland alluvial rivers

Moreover, riparian forests and ponds in floodplains provide essential habitats for various terrestrial and aquatic organisms in all or specific stages of their life cycles. Thus, biodiversity in a riparian forest is generally higher than that in upland ecosystems, and riparian habitat specialists increase overall regional biodiversity (gamma diversity) at the catchment scale. Species richness, especially that of plants, is commonly higher in riparian forests than in upland forests, mainly because complex and diverse mosaics of habitat patches are created through land-water interactions (Amoros and Bornette 2002).

The species richness and diversity of macroinvertebrates and fishes in floodplain waterbodies vary along the gradients of hydrological connectivity (Gallardo et al. 2014; Schomaker and Wolter 2011; Ishiyama et al. 2014). Some waterbodies close to the main channel in a floodplain experience frequent floods, while other waterbodies more distant from the main channel may be disturbed only during high flood events or extremely large floods. Some waterbodies may be connected by drainage channels and/or small tributaries with other ponds or mainstems. These differences in connectivity alter the potential frequency of immigration and emigration of individuals and the physical and chemical characteristics of waterbodies. Consequently, taxon richness and diversity as well as species traits vary with connectivity gradients (Ishiyama et al. 2014). Riparian forests are recognized as a key habitat for the conservation of bird communities. Bird assemblages in riparian zones are distinctively different from those in neighboring areas, owing to differences in the plant communities (Yabuhara et al. 2019). Additionally, the species richness and abundance of birds are generally higher in riparian zones than in adjacent upland habitats due to the structural complexity of the riparian habitat.

Riparian forests provide essential ecological functions through a variety of mechanisms, including reducing sunlight and influencing in-stream primary production and water temperature, supplying food resources for aquatic invertebrates (e.g., fine organic matter, leaves, wood), supplying large wood to streams, filtering fine sediment and nutrients before their entry into streams, adding large roughness elements that reduce water velocity and protect stream banks, and providing migratory corridors for animals (Gregory et al. 1991; Naiman and Decamps 1997) (Fig. 6.3). A riparian forest is a unique landscape element that consists of diverse plant species, provides essential habitats for various terrestrial and aquatic organisms, and plays a pivotal role in the functioning of ecological processes between terrestrial and aquatic ecosystems. Thus, we can refer to a riparian forest as an "interactive zone of green and blue infrastructure" that should be preserved and restored for the twenty-first century. Here, green/blue infrastructure is defined as "a strategically planned network of natural and seminatural areas with other environmental features designed and managed to deliver a wide range of ecosystem services such as water purification, air quality, space for recreation and climate mitigation and adaptation" (European Commission 2016). Green infrastructure refers to terrestrial elements like trees, parks, hedgerows, and riparian vegetation, while blue infrastructure refers to aquatic elements such as rivers, wetlands, ponds, estuaries, and coasts.

However, in lowland areas of river basins, riparian forests occur on wet, fertile soil with flat topography. As a result, they have been exploited historically to harvest timber, construct forestry roads, cultivate farmlands, and develop urban

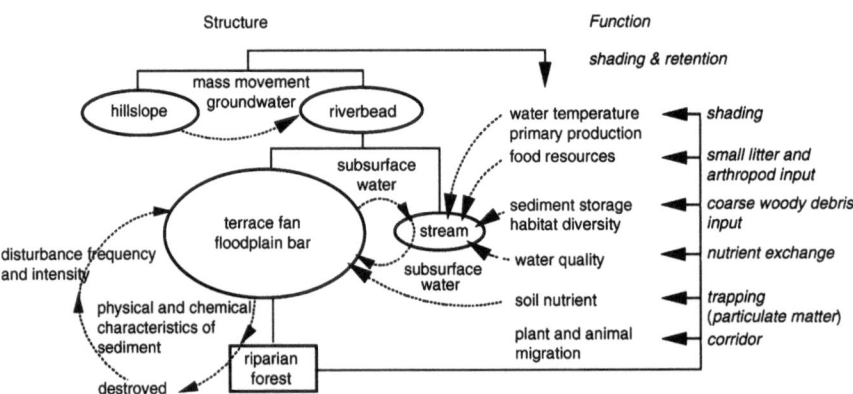

Fig. 6.3 Structures and functions of the riparian zone (from Nakamura and Yamada 2005)

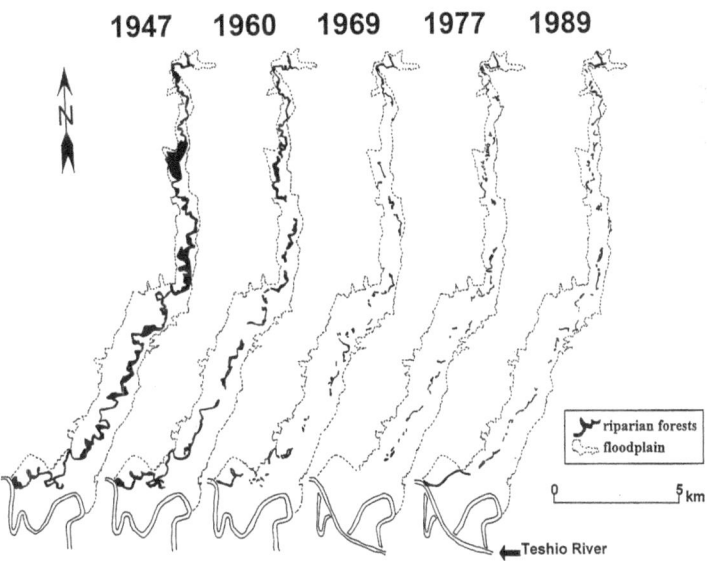

Fig. 6.4 Fragmentation of the riparian forests along the Toikanbetsu River, Japan (from Nagasaka and Nakamura 1999)

land uses (Sedell and Froggatt 1984; Nagasaka and Nakamura 1999) (Fig. 6.4). Dams greatly alter flow, sediment, and large wood regimes and thereby riparian vegetation (Nakamura 2017). The loss and degradation of riparian forests cause serious environmental problems; therefore, scientists and land managers have developed management approaches to conserve and restore riparian zones (e.g., riparian buffers, reforestation, channel remeandering, floodplain reconnection).

Another key driver that may greatly alter riparian forest ecosystems and their functions is climate change (Capon et al. 2013). In this chapter, we explore new roles of riparian forests under climate change scenarios and discuss a process for incorporating green/blue infrastructure for adaptation strategies to conserve biodiversity and address effects of climate change. Although this chapter focuses on riparian forests in Japanese landscapes, the derived adaptation strategies and principles are applicable to other parts of the world.

6.2 Riparian Forest as an Interactive Zone of Green and Blue Infrastructure

River and floodplain ecosystems are among the most sensitive of all ecosystems to climate change because they are tightly linked to atmospheric thermal regimes and global hydrological cycles, and they are affected by interactions between climate change and non-climatic stressors (e.g., farmland development, river regulation, impoundments, water pollution, urbanization, hydropower development) (Ormerod 2009; Thomas et al. 2016). According to the World Wide Fund for Nature (WWF) Living Planet Report 2014 Living Planet Index (LPI: a measure of the state of the world's biological diversity based on population trends in vertebrate species), the freshwater index has shown the greatest decline of any of the biome-based indices (WWF 2014).

Maintenance and enhancement of ecosystem services provided by riparian forests (Fig. 6.3) are important issues in biodiversity conservation and adaptation to climate change (Palmer et al. 2009). Stream temperature is the primary water quality variable that affects animal species distribution, phenology, and ecosystem processes and is expected to increase with rising air temperature under climate change (Thomas et al. 2016). Riparian forest canopies significantly reduce the amount of solar radiation reaching a stream surface. The removal of a riparian forest greatly increases insolation and thereby increases stream temperatures. A field study in northern Hokkaido streams suggested that the abundance of masu salmon is primarily limited by water temperature (Inoue et al. 1997). Masu salmon density in lowland grassland reaches, where the summer maximum water temperature is generally higher than 20 °C, was lower than density in nearby forest reaches. Effects of riparian canopies on stream water temperature have been examined in streams flowing through deciduous broad-leaved forests (Nakamura and Dokai 1989; Sugimoto et al. 1997). Nakamura and Dokai (1989) developed a heat budget to analyze the effect of riparian canopies and concluded that 86% of the daily total input of solar radiation is intercepted by the riparian canopy and its removal would result in a 4 °C increase in water temperature. Thus, the use of riparian forest shade to mitigate potential increases in stream temperature with climate change is a fundamental action in adaptation strategies for protecting cold-water species (Seavy et al. 2009; Wilby et al. 2010; Suzuki et al. 2021).

A riparian forest supplies various types of litterfall to streams (Benfield 1997). Litterfall is defined as allochthonous organic matter that enters streams from riparian vegetation. It includes leaves and leaf fragments, floral parts, bark, wood (stems, branches, twigs), cones and nuts, fruits, and other plant parts. Among these, the food quality of leaf litter is substantially greater than the others. Litter may reach a stream by directly falling into the stream or by laterally moving into the stream. Plant litter supplied from riparian forests is consumed by stream invertebrates. Terrestrial invertebrates fall into streams from riparian forests, providing important food resources for stream-dwelling fishes (Kawaguchi and Nakano 2001). In addition to the supply of leaf litter and terrestrial invertebrates into streams, large wood and twig accumulation, gravel bars, pools, and complex stream edges retain transported food resources for aquatic invertebrates and fishes. However, we expect more frequent, high flood peaks with climate change. In Japan, the fluctuation in yearly precipitation has increased since the 1970s, and the frequency of hourly heavy rains of 50 mm or more has increased (Ministry of the Environment et al. 2018). Rivers in developed countries have been regulated for over a century, and naturally meandering rivers have been altered to form straighten channelized rivers (Nagayama and Nakamura 2018). Thus, organic matter supplied from riparian forests into streams may be flushed downstream without retention on riverbeds, making it less available to stream biota.

Large wood is supplied from riparian forests into streams and rivers by various processes, such as landslides, debris flow, floods, windthrow, and bank erosion (Harmon et al. 2004). Large wood alters the structural and functional characteristics of stream channels, thereby influencing stream biota. In particular, the formation of pools by large wood has been investigated with reference to fish habitats (Fausch and Northcote 1992). Another important function of in-stream wood is modifying the movement and transformation of energy, nutrients, and food for stream-dwelling organisms. Step-pool sequences created by large wood provide storage sites for sediment, organic matter, and nutrients (Nakamura and Swanson 1993). It is widely recognized that trees growing along streambanks protect against bank erosion through their root system and create complex habitat along river margins (Gregory et al. 2019). Trees, however, will eventually fall into streams when flood power exceeds their tolerances. In this situation, riparian forests may be eroded and uprooted and transported downstream by floods. Wood-laden floods may destroy artificial levees, bridges, and houses along the river course, and the intensity and possibility of such disasters will increase with climate change. Thus, river managers have to balance the ecological benefits and geomorphic risks, both of which are important issues in a changing environment. The risks generated by floated wood should be avoided in urban areas, whereas the ecological functions of large wood should be sustained in rural and seminatural areas.

The effects of a riparian zone on water quality are well-known functions, particularly the reduction of sediment and nutrients before their entry into streams (Nakamura and Yamada 2005). Fine sediment eroded in uplands is efficiently removed by riparian vegetation that is 10–30 m in width, while nutrients such as nitrogen and phosphorus are efficiently removed by riparian vegetation that is 10–50 m in width. However, their effects vary with a number of local conditions, such as sedimentation rates, surface and subsurface drainage characteristics, soil characteristics (e.g., particle size composition and groundwater-level regulating redox potential), organic matter content, temperature, vegetation type (e.g., grass or forest, forest age, and density), nutrient loading from uplands, and slope steepness. Soil erosion and nutrient leaching from uplands are predicted to increase with increasing rainfall intensity under climate change (Jeppesen et al. 2009). Thus, the filtering function of riparian forests is a critical process in maintaining the health of river and wetland ecosystems.

Riparian forests are corridors in fragmented landscapes that connect healthy intact habitats and allow animals to move between them. Corridor use by bird and mammal species has been well investigated relative to that of other types of wildlife (Machtans et al. 1996; Hilty and Merenlender 2004). Terrestrial wildlife moves preferentially along or inhabits riparian forests. Wide, well-preserved corridors maintain a nearly complete species assemblage of the region. The home range of animals may change with climate change, driving animals to migrate more appropriately to higher-elevation areas. Thus, the current connectivity of riparian and stream corridors should be protected to facilitate climate-induced movement and restored when land use and dam construction impede their connectivity.

6.3 Adaptation Strategies to Climate Change Using Riparian Green/Blue Infrastructure

Riparian forests have been acknowledged as green/blue infrastructure that provides the various ecosystem services explained above. However, these forests have already experienced negative impacts from various anthropogenic stressors, such as forest cutting, agricultural development, river regulation, and dam construction. These stressors generally have detrimental effects on stream and floodplain biota, and climate change may magnify negative impacts (Palmer et al. 2009). Thus, strategies to conserve biodiversity at the landscape level must reduce or eliminate negative impacts of existing anthropogenic stressors on riparian ecosystems and anticipate the possible interactions of past impacts with climate change are key management needs for the future.

6.3.1 Maintain River Dynamics Using Artificial Floods

The structure and composition of riparian vegetation have been maintained by frequent migration of river channels and the active movement of sediment (Nakamura and Inahara 2007). Plant and animal species in streams and riparian zones are adapted to the dynamic features of rivers and floodplains (Poff et al. 1997). Additionally, the large wood produced by forests plays a critical role in providing a wide array of habitats for plants, invertebrates, and fishes (Nakamura et al. 2012; Nakano et al. 2018; Fausch and Northcote 1992). Thus, the dynamics of water, sediment, and large wood are three key components that maintain the regeneration of riparian plant species and thereby riparian ecosystems (Nakamura et al. 2017). However, water, sediment, and large wood regimes are greatly altered by dams and land use in Japan and other developed countries which results in drastic changes in the dynamic and interactive features of rivers and floodplains. These regime shifts of physical conditions convert diverse river habitats to stable, single-thread channels and increased forest establishment on gravel bars and floodplains (Nadler and Schumm 1981; Johnson 1994; Bejarano et al. 2011). Riparian forests play a pivotal role in maintaining stream and floodplain ecosystem, but they should be sustained by dynamic feature of the above three components. In natural rivers, pioneer tree species dominate gravel-bed rivers, and unvegetated gravel bars are essential sites for these species to germinate. In contrast, mid- and late-successional tree species start to colonize under pioneer trees and on gravel bars in dam-regulated rivers. The spread of late-successional

Fig. 6.5 Forest expansion over the gravel bars and floodplains of the Satsunai River

trees over gravel bars potentially threatens the survival of native species that are dependent on unvegetated habitat (Tiedemann and Rood 2015; Nakamura and Shin 2001).

One of the main causes of forest expansion in gravel rivers is dam construction and reservoir management (Takahashi and Nakamura 2011). Dams regulate water discharge by storing floodwater in their reservoirs and releasing water resources for power generation and agricultural use, which generally results in leveling the fluctuation in water discharge. The stabilization of water discharge and substantial reduction in floodwater discharge promote colonization of tree species on gravel bars and forest expansion over valley floors (Fig. 6.5). Dams and other human activities, such as channelization and revetment or spur-dike construction, promote channel incision and forest colonization over gravel bars and floodplains. Forest expansion causes two major problems for river management. One issue is an increase in channel roughness, which reduces its floodwater transport capacity, and the other issue is a reduction in gravel-bed habitat on which native, rare plant, and animal are dependent species (Nakamura et al. 2020; Yabuhara et al. 2015).

The climate projections for the twenty-first century in Japan indicate that mean precipitation may increase by more than 10% (Kimoto et al. 2005), and other projections predict an increase in the frequency of high-magnitude floods and a reduced discharge from snowmelt floods. With forest expansion and climate change, we expect more frequent inundation with floodwaters due to the increase in roughness elements and an increase in log-laden floods as riparian forest trees

Fig. 6.6 Wood-laden flood caused by heavy rain in northern Kyushu in 2017

are removed by extreme floods (Fig. 6.6). Additionally, reduced discharge from snowmelt floods may cause recruitment failure of pioneer tree species such as *Populus* and *Salix* spp., which time their seed dispersal periods to the descending period of snowmelt floods (Nakamura and Inahara 2007).

Environmental flow regimes, including artificial floods, can be used to mitigate magnified effects of dams and climate change. In the Satsunai River, the Japanese government launched a restoration project in 2012 to partially restore its riparian ecosystems, releasing a maximum water volume of 120 m^3/s, with a 2-year return-period flood before dam construction (Nakamura et al. 2020). This was a large-scale experiment developed jointly with an interdisciplinary science team and river managers, who conducted monitoring and evaluations under an adaptive management scheme. Artificial floods have been initiated once a year at the end of June since 2012, and they have been synchronized with the seed dispersal period of *Salix arbutifolia*, which is endangered species and a high conservation priority for the project (Fig. 6.7). The project is successful so far in restoring a shifting mosaic of floodplain habitat patches and gravel bars and thereby in regenerating *S. arbutifolia* (Nakamura et al. 2020). With this project, an increase in dense forest cover has been minimized within the flood-disturbed area and regulated by the current flow regime, including artificial floods.

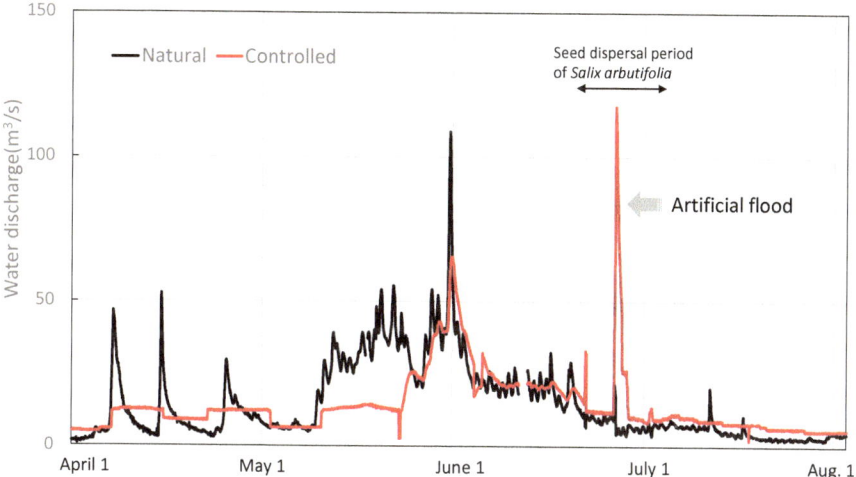

Hydrographs (water discharge) of natural flows (flows into the dam from the upper basin) and controlled flows (flows out of the dam)

Fig. 6.7 Hydrographs (water discharge) of natural flows (flows into the dam from the upper basin) and controlled flows with artificial floods (flows out of the dam) in the Satsunai River (from Nakamura et al. 2020)

6.3.2   *Maintain Riparian Forests Continuously from Headwater Streams to Lowland Rivers*

In Japan, considerable areas of hills and plains that were originally covered by forests have been converted to farmland, with streams and rivers being channelized to develop and conserve farmlands (Nagasaka and Nakamura 1999). One of the critical roles of riparian forests is the maintenance of cold water through their canopies; therefore, cold-water species, such as salmonids, are able to inhabit streams in Hokkaido, northern Japan. A laboratory experiment and field study indicated that feeding and growth of masu salmon decline when water temperatures reached 24 °C (Sato et al. 2001). Summer maximum water temperature is significantly related to the longitudinal extent of canopy removal (Sugimoto et al. 1997). This study concluded that removal of 1 km of canopy would be associated with a 2–4 °C increase in water temperature, and a 6-km canopy removal would result in a 5–8 °C increase. Based on these data, they estimated a historical change in the summer maximum water temperature in a small stream and determined that stream temperature increased by 6 °C from 1947 (22 °C) to 1990 (28 °C), with a sharp increase during the 1950s as a result of rapid agricultural development. The results suggest that suitable habitat for masu salmon decreased historically as a result of these land use changes.

Water temperature is expected to increase with increasing air temperature with climate change. However, studies that predict an increase in water temperature are scarce and limited by uncertainties associated with catchment geohydrology. In particular, porous volcanic rock is extensive across the Japanese archipelago, enhancing the contribution of groundwater discharge. This scenario insulates areas from exposure to sunlight energy and exchanges of latent and sensible heats and creates spring-fed streams with water temperatures that are cool in summer and warm in winter compared with those of runoff-fed streams. Spring-fed streams may provide refuges for cold-water fishes in hot summers under climate change conditions. Riparian forests will reduce the sunlight energy reaching the stream surface; therefore, maintaining continuous belts of riparian forests from headwater streams to lowland rivers is an adaptation strategy for climate change. In particular, the shading effect of a riparian forest canopy dominates in headwater streams where the stream width is relatively smaller than the size of the riparian forest canopy. In the case of Japanese streams, shading by riparian canopies is effective in streams less than 30 m in width (Nakamura and Dokai 1989). Thus, riparian cover should be protected along headwater streams as well as spring-fed streams to mitigate increases in air temperature. Such continuous belts of riparian forests will provide corridors for terrestrial birds and mammals to migrate to cooler environments.

6.3.3 Remove or Improve Check Dams for Material Transport and Fish Migration

Management agencies have constructed many check dams in headwater streams in Japan to prevent human lives and infrastructure from being affected by sediment-related disasters (e.g., landslides, debris flows, and hyperconcentrated flows) and to control erosion. Some check dams are not equipped with fish ladders, and prevent or reduce upstream migration of socially and economically important fish. Even if check dams have fish ladders, all ladders reduce passage to some degree, the continuity and dynamics of sediment transport are interrupted, and some organic matter produced from riparian forests is trapped by dams and buried by sediment. Check dams substantially alter the natural sediment regime, which sustains riparian forests in headwater basins (Nakamura et al. 2017).

Check dams greatly alter channel morphology and degrade salmonid habitats (Kishi and Maekawa 2009). In general, the channel above check dams becomes wider and shallower due to the wide dam face, and riparian trees become established farther away from the wide stream channel. Thus, the riparian canopy cover over the stream surface is reduced, and greater amounts of sunlight reach the stream surface and increase water temperature. On the Shiretoko Peninsula, currently designated as a World Natural Heritage Site, forest canopy cover over the stream surface is less than 10% in dam-installed reaches in contrast to 80–90% in natural reaches, and the maximum water temperature is 15–21 °C in dam-installed reaches versus 12–17 °C in natural reaches (Kishi and Maekawa 2009).

Check dams may exacerbate the thermal effects of global warming, resulting in greater heat energy exchange between air and waterbodies and even greater increase in stream temperature. To reduce the potential impacts of climate change on stream ecosystems exacerbated by anthropogenic stressors, check dams should be improved and modified to allow fish migration, restore the natural regime of sediment and organic matter, and ensure stream temperatures remain cool. Partial removal of check dams is a modification alternative that recovers quasi-natural systems and maintains disaster prevention functions. The center of the dam is cut to create a rectangular opening or slit (Nakamura and Komiyama 2010; Nagayama et al. 2020). The width of the opening is set to equal the low-flow channel width, thereby maintaining the natural transport of water, sediment, and organic matter under ordinary conditions and allowing fish to migrate upstream and downstream (Fig. 6.8). When debris flows arrive at the dam site, all sediment and turbid water cannot pass through the slit simultaneously, depositing sediment behind the dam due to the backwater effect. Subsequently, the trapped sediment behind the dam is gradually released from the slit during the descending limb of hydrographs and/or subsequent floodwaters. Riparian forests establish along the low-flow channel and their canopies cover the stream surface to a greater extent.

| 2014.10 | 2018.11 | 2019.6 |

| 2019.8 | 2020.8 | 2020.10 |

Fig. 6.8 Dam modification and recovery of up- and downstream linkages at the Shiretoko World National Heritage Site

6.3.4 Broaden Riparian Forests and Remeander Rivers Where Possible

Based on the climate change projection for 2040 under the RCP2.6 scenario in Japan, the average air temperature will increase by 2 °C from the current level, rainfall with a 100- to 200-year recurrence interval will increase by 10%, and floodwater discharge will increase by 20%. Thus, we must prepare for megafloods and associated environmental problems. We can expect an increase in the discharge of sediment, organic matter, and nutrients from upper basins. A study conducted in the Kushiro River basin, northern Japan, under the harshest climate change scenario (RCP 8.5), predicted that suspended sediment, total nitrogen, and total phosphorus will increase by 4.3–8.3-, 2.3–3.3-, and 2.6–4.2-fold, respectively, by the end of the twenty-first century (Ministry of the Environment 2020).

Adaptation actions to expand riparian vegetation cover on floodplains is a potential strategy to increase floodwater retention capacity and reduce floodwater velocity, which will reduce peak discharge of megafloods and increase the retention of transported materials (e.g., sediment, organic matter, and nutrients) in floodplains (Dybala et al. 2019). Remeandering artificially straightened rivers associated with the extension of floodplain areas will enhance the efficacy of this management strategy by increasing inundation and retention of carbon and nutrients in floodplains (Dahm et al. 1995; Nakamura et al. 2014) (Fig. 6.9). Movement of large wood into streams will increase with bank erosion associated with gradual movement of meandering river courses. Furthermore, large wood pieces that are transported by floodwater from upstream to downstream and that may destroy downstream infrastructure and houses during megafloods can be efficiently trapped on the outside of meandering rivers due to the inertial force of floodwater.

The adaptation strategies of extending floodplains and remeandering rivers also create various habitats for aquatic and terrestrial plants and animals. This approach is similar to the "Room for Rivers" concept in the EU (Buijse et al. 2002). The historical development of agricultural and urban land use converts naturally meandering rivers to straight river channels with revetments, disconnecting stream from their floodplain ecosystems. Adaptation measures will restore connectivity between rivers, ponds, and floodplain wetlands with increasing floodwater inundation. Ishiyama et al. (2014) investigated the effects of wetland connectivity on the species richness of aquatic invertebrates and fishes in agricultural landscapes. They found that the species richness of highly mobile groups of insects and fishes increased with increasing spatial connectivity. Thus, the frequent inundation of floodwater over riparian forests increases temporal and spatial connectivity among various waterbodies in floodplains and rivers.

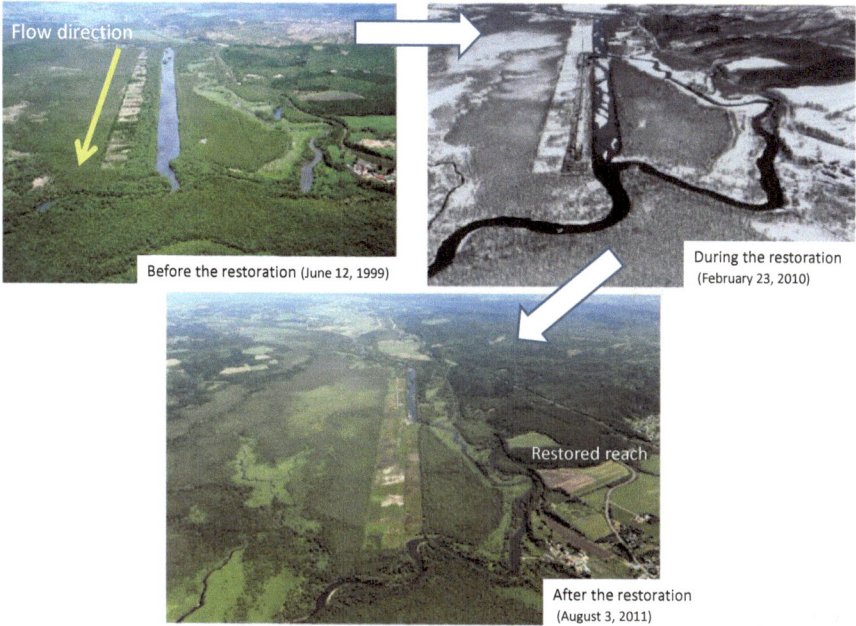

Fig. 6.9 Meander restoration in the Kushiro wetlands. The top left-hand photograph shows a 1.6-km stretch of the main channel before it was restored by reconnecting the remaining former cutoff channel and backfilling the straightened reach (from Nakamura et al. 2014)

6.4 Conclusion

A riparian forest is regarded as an "interactive zone of green and blue infrastructure," and should be preserved and restored for the twenty-first century because it provides essential habitats for plant and animals and precious ecological functions in a changing environment. However, forests have already experienced negative impacts from various anthropogenic stressors, such as forest cutting, agricultural development, and river regulation due to its high productivity and geomorphic setting close to rivers. However, predictions of future climate change and associated impacts on biodiversity and ecosystem services project that the interaction of anthropogenic stressors and climate change will magnify those impacts. If those stressors are not reduced in the future, biodiversity and riparian ecosystems would not be sustained and important ecosystem services would be lost. Thus, active management and conservation strategies should be implemented to lessen non-climate stressors. Ecological restoration projects conducted throughout the world to protect and restore biodiversity and ecosystem services will be essential components of adaptation strategies for climate change in the future.

Acknowledgments The author deeply appreciates all thoughtful comments and suggestions made by Dr. Stanley Gregory that significantly improve the quality of the original manuscript. This research was funded by the Environment Research and Technology Development Funds (4-1504 and 4-1805) of the Ministry of the Environment of Japan.

References

Amoros C, Bornette G (2002) Connectivity and biocomplexity in waterbodies of riverine floodplains. Freshw Biol 47:761–776

Bejarano DM, Nilsson C, del Tanago GM, Marchamalo M (2011) Responses of riparian trees and shrubs to flow regulation along a boreal stream in northern Sweden. Freshw Biol 56:853–866

Benfield EF (1997) Comparison of litterfall input to streams. J N Am Benthol Soc 16:104–108

Buijse AD, Coops H, Staras M, Jans LH, Van Geest GJ, Grift RE et al (2002) Restoration strategies for river floodplains along large lowland rivers in Europe. Freshw Biol 47(4):889–907

Capon SJ, Chambers LE, Mac Nally R, Naiman RJ, Davies P, Marshall N et al (2013) Riparian ecosystems in the 21st century: hotspots for climate change adaptation? Ecosystems 16(3):359–381

Dahm CN, Cummins KW, Valett HM, Coleman RL (1995) An ecosystem view of the restoration of the Kissimmee River. Restor Ecol 3:225–238

Dybala KE, Matzek V, Gardali T, Seavy NE (2019) Carbon sequestration in riparian forests: a global synthesis and meta-analysis. Glob Chang Biol 25(1):57–67

European Commission (2016) Supporting the implementation of green infrastructure, Rotterdam

Fausch KD, Northcote TG (1992) Large woody debris and salmonid habitat in a small coastal British Columbia stream. Can J Fishery Aquat Sci 49:682–693

Gallardo B, Doledec S, Pallex A et al (2014) Response of benthic macroinvertebrates to gradients in hydrological connectivity: a comparison of temperate, subtropical, Mediterranean and semiarid river floodplains. Freshw Biol 59:630–648

Gregory SV, Swanson FJ, McKee WA, Cummins KW (1991) An ecosystem perspective of riparian zones. Bioscience 41:540–551

Gregory SV, Wildman R, Hulse D, Ashkenas L, Boyer K (2019) Historical changes in hydrology, geomorphology, and floodplain vegetation of the Willamette River, Oregon. River Res Appl 35(8):1279–1290

Harmon ME, Franklin JF, Swanson FJ, Sollins P, Gregory SV, Lattin JD et al (2004) Ecology of coarse woody debris in temperate ecosystems. Adv Ecol Res 34:59–234

Hilty JA, Merenlender AM (2004) Use of riparian corridors and vineyards by mammalian predators in northern California. Conserv Biol 18:126–135

Inoue M, Nakano S, Nakamura F (1997) Juvenile masu salmon (Oncorhynchus masou) abundance and stream habitat relationships in northern Japan. Can J Fish Aquat Sci 54:1331–1341

Ishiyama N, Akasaka T, Nakamura F (2014) Mobility-dependent response of aquatic animal species richness to a wetland network in an agricultural landscape. Aquat Sci 76:437–449

Jeppesen E, Kronvang B, Meerhoff M, Søndergaard M, Hansen KM, Andersen HE et al (2009) Climate change effects on runoff, catchment phosphorus loading and lake ecological state, and potential adaptations. J Environ Qual 38(5):1930–1941

Johnson WC (1994) Woodland expansion in the Platte River, Nebraska: patterns and causes. Ecol Monogr 64:45–84

Kawaguchi Y, Nakano S (2001) Contribution of terrestrial invertebrates to the annual resource budget for salmonids in forest and grassland reaches of a headwater stream. Freshw Biol 46:303–316

Kimoto M, Yasutomi N, Yokoyama C, Emori S (2005) Projected changes in precipitation characteristics around Japan under the global warming. SOLA 1:085–088. https://doi.org/10.2151/sola.2005-023

Kishi D, Maekawa K (2009) Stream-dwelling Dolly Varden (Salvelinus malma) density and habitat characteristics in stream sections installed with low-head dams in the Shiretoko Peninsula, Hokkaido, Japan. Ecol Res 24(4):873–880

Machtans CS, Villard M, Hannon SJ (1996) Use of riparian buffer strips as movement corridors by forest birds. Conserv Biol 10:1366–1379

Ministry of the Environment, Ministry of Education, Culture, Sports, Science and Technology, Ministry of Agriculture, Forestry and Fisheries, Ministry of Land, Infrastructure, Transport and Tourism and Meteorological Agency of Japan (2018) Kikou henndou no kannsoku yosoku eikyou hyouka nikannsuru tougou report 2018 nihonn no kikouhenndou to sono eikyou [Consolidated report on observations, projections and impact assessments of climate change: climate change and its impacts in Japan]. http://www.env.go.jp/earth/tekiou/report2018_full.pdf

Ministry of the Environment of Japan (2020) Final reports of regional adaptation consortium project. 1-6

Nadler CT, Schumm SA (1981) Metamorphosis of South Platte and Arkansas Rivers, Eastern Colorado. Phys Geogr 2:95–115

Nagasaka A, Nakamura F (1999) The influences of land-use changes on hydrology and riparian environment in a northern Japanese landscape. Landsc Ecol 14:543–556

Nagayama S, Nakamura F (2018) The significance of meandering channel to habitat diversity and fish assemblage: a case study in the Shibetsu River, northern Japan. Limnology 19:7–20

Nagayama S, Ishiyama N, Seno T, Kawai H, Kawaguchi Y, Nakano D, Nakamura F (2020) Time series changes in fish assemblages and habitat structures caused by partial check dam removal. Water 12(12):3357

Naiman RJ, Decamps H (1997) The ecology of interfaces: riparian zones. Annu Rev Ecol Syst 28:621–658

Nakamura F, Dokai T (1989) Estimation of the effect of riparian forest on stream temperature based on heat budget. J Jpn For Soc 71:387–394. (in Japanese with English summary)

Nakamura F, Swanson FJ (1993) Effects of coarse woody debris on morphology and sediment storage of a mountain stream system in western Oregon. Earth Surf Process Landf 18:43–61

Nakamura F, Yajima T, Kikuchi S (1997) Structure and composition of riparian forests with special reference to geomorphic site conditions along the Tokachi River, northern Japan. Plant Ecol 133:209–219

Nakamura F, Swanson FJ, Wondzell SM (2000) Disturbance regimes of stream and riparian systems - a disturbance-cascade perspective. Hydrol Process 14:2849–2860

Nakamura F, Shin N (2001) The downstream effects of dams on the regeneration of riparian tree species in northern Japan. Geomorphic processes and riverine habitat (Dorava J M, Montgomery D R, Palcsak B and Fitzpatrick F A eds.). AGU Water Sci Appl 4:173–181. ISBN: 0-87590-353-3

Nakamura F, Swanson FJ (2003) Dynamics of wood in rivers in the context of ecological disturbance. The ecology and management of wood in world rivers (Gregory S V, Boyer K L and Gurnell AM eds.). Symp 37:279–297. ISBN: 1-888569-56-5

Nakamura F, Yamada H (2005) The effects of pasture development on the ecological functions of riparian forests in Hokkaido in northern Japan. Ecol Eng 24:539–550

Nakamura F, Inahara S (2007) Fluvial geomorphic disturbances and life history traits of riparian tree species. In: Johnson EA, Miyanishi K (eds) Plant disturbance ecology: the process and the response. Academic Press, pp 283–310. ISBN: 978-0-12-088778-1

Nakamura F, Komiyama E (2010) A challenge to dam improvement for the protection of both salmon and human livelihood in Shiretoko, Japan's third natural heritage site. Landsc Ecol Eng 6:143–152

Nakamura F, Fuke N, Kubo M (2012) Contributions of large wood to the initial establishment and diversity of riparian vegetation in a bar-braided temperate river. Plant Ecol 213:735–747

Nakamura F, Ishiyama N, Sueyoshi M, Negishi J, Akasaka T (2014) The significance of meander restoration for the hydrogeomorphology and recovery of wetland organisms in the Kushiro River, a lowland river in Japan. Restor Ecol 22:544–554

Nakamura F, Seo JIL, Akasaka T, Swanson FJ (2017) Large wood, sediment, and flow regimes: their interactions and temporal changes caused by human impacts in Japan. Geomorphology 279:176–187

Nakamura F, Watanabe Y, Negishi J, Akasaka T, Yabuhara Y, Terui A, Yamanaka S, Konno M (2020) Restoration of the shifting mosaic of floodplain forests under a flow regime altered by a dam. Ecol Eng 157:105974

Nakano D, Nagayama S, Kawaguchi Y, Nakamura F (2018) Significance of the stable foundations provided and created by large wood for benthic fauna in the Shibetsu River, Japan. Ecol Eng 120:249–259

Niiyama K (1987) Distribution of Salicaceae species and soil texture of habitats along the Ishikari River. Jpn J Ecol 37:163–174. (in Japanese with English summary)

Ormerod SJ (2009) Climate change, river conservation and the adaptation challenge. Aquat Conserv 19:609–613

Palmer MA, Lettenmaier DP, Poff NL, Postel SL, Richter B, Warner R (2009) Climate change and river ecosystems: protection and adaptation options. Environ Manag 44(6):1053–1068

Poff LR, Allan JD, Bain MB, Karr JR, Prestegaard KL, Richter BD, Sparks RE, Stromberg JC (1997) The natural flow regime. Bioscience 47:769–784

Sakio H (1997) Effects of natural disturbance on the regeneration of riparian forests in a Chichibu Mountains, Central Japan. Plant Ecol 132:181–195

Sato H, Nagata M, Takami T, Yanai S (2001) Shade effect of riparian forest in controlling summer stream temperature: impact on growth of masu salmon juveniles (Oncorhynchus masou Brevoort). J Jpn For Soc 83:22–29. (in Japanese with English Summary)

Schomaker C, Wolter C (2011) The contribution of long-term isolated water bodies to floodplain fish diversity. Freshw Biol 56:1469–1480

Seavy NE, Gardali T, Golet GH, Griggs FT, Howell CA, Kelsey R et al (2009) Why climate change makes riparian restoration more important than ever: recommendations for practice and research. Ecol Restor 27(3):330–338

Sedell JR, Froggatt JL (1984) Importance of streamside forests to large rivers: the isolation of the Willamette River, Oregon, U. S. A., from its floodplain by snagging and streamside forest removal. Verh Int Ver Limnol 22:1828–1834

Shin N, Nakamura F (2005) Effects of fluvial geomorphology on riparian tree species in Rekifune River, northern Japan. Plant Ecol 178:15–28

Sugimoto S, Nakamura F, Ito A (1997) Heat budget and statistical analysis of the relationship between stream temperature and riparian forest in the Toikanbetsu River basin, northern Japan. J For Res 2:103–107

Suzuki K, Ishiyama N, Koizumi I, Nakamura F (2021) Combined effects of summer water temperature and current velocity on the distribution of a cold-water-adapted sculpin (*Cottus nozawae*). Water 13:975

Takahashi M, Nakamura F (2011) Impacts of dam-regulated flows on channel morphology and riparian vegetation: a longitudinal analysis of Satsunai River. Jpn Landsc Ecol Eng 7:65–77

Thomas SM, Griffiths SW, Ormerod SJ (2016) Beyond cool: adapting upland streams for climate change using riparian woodlands. Glob Chang Biol 22(1):310–324

Tiedemann RB, Rood SB (2015) Flood flow attenuation diminishes cotton-wood colonization sites: an experimental test along the Boise River, USA Ecohydrology. https://doi.org/10.1002/eco.1619

Wilby RL, Orr H, Watts G, Battarbee RW, Berry PM, Chadd R et al (2010) Evidence needed to manage freshwater ecosystems in a changing climate: turning adaptation principles into practice. Sci Total Environ 408(19):4150–4164

WWF (2014) Living planet report 2014. https://www.wwf.or.jp/activities/data/WWF_LPR_2014.pdf

Yabuhara Y, Yamaura Y, Akasaka T, Nakamura F (2015) Predicting long-term changes in riparian bird communities in floodplain landscapes. River Res Appl 31:109–119

Yabuhara Y, Yamaura Y, Akasaka T, Yamanaka S, Nakamura F (2019) Seasonal variation in patch and landscape effects on forest bird communities in a lowland fragmented landscape. For Ecol Manag 454:117–140

Chapter 7
Improvement of the Flood-Reduction Function of Forests Based on Their Interception Evaporation and Surface Storage Capacities

Takao Tamura

Abstract Forests have a flood-reduction function that reduces flood peak flow and delays the flood peak time. In the mountains of Japan, artificial forests planted between the 1950s and 1970s are widespread; however, many of these forests are not well managed. The effective use of the flood-reduction function of forests as a remarkable approach for river basin management has been discussed for several years. In this study, two aspects of the water cycle in forests were explored: the interception evaporation process in the forest canopy and the groundwater storage process on the forest slope. A runoff model was applied to the hydrological data obtained in several forest basins with different characteristics to evaluate the effects of the processes. In the case of the Japanese cedar plantations studied, it was suggested that the improvement of interception evaporation capacity and surface storage capacity by conversion to mixed forests and selective logging would significantly reduce the peak flood discharge on a timescale of approximately 20–30 years.

Keywords Forest · Flood-reduction function · Interception evaporation · Surface runoff · Selective cutting

7.1 Introduction

The influence of climate change is conspicuous in Japan. Severe flood disasters occur almost annually in various regions of Japan. Forests in Japan account for approximately 70% of the land, while 40% of the forests are artificial forests comprising coniferous trees. These forests are distributed at mountainous areas,

T. Tamura (✉)
Department of Civil and Environmental Engineering, Graduate School of Technology, Industrial and Social Science, Tokushima University, Tokushima, Japan
e-mail: tamura@ce.tokushima-u.ac.jp

© The Author(s) 2022
F. Nakamura (ed.), *Green Infrastructure and Climate Change Adaptation*,
Ecological Research Monographs, https://doi.org/10.1007/978-981-16-6791-6_7

which are the source areas of streams. Therefore, heavy rain in mountainous forests has become a major cause of flood disasters. Forests have a flood control function for storing a part of the rainfall and delaying runoff discharge from their basins. This contributes toward decreasing the quantity of flood flow. As forests are part of the green infrastructure, an effective use of the flood-reduction function for disaster prevention has been previously discussed.

Generally, it is recognized that forest soil plays a principal role in the flood-reduction function. Since forest soil has a large void structure, forests can store a significant amount of rainwater as compared with other land cover types. Therefore, methods used to strengthen the flood-reduction function, such as thinning of forests and conversion of vegetation structure, have been discussed. However, it has been pointed out that the flood-reduction function of forests is not sufficiently effective for large floods (Laurance 2007). It is also extremely difficult to increase the thickness of the soil layer, which determines the water storage capacity. The soil formation speed has been reported in the range of 0.05–0.2 mm/year (Amundson et al. 2015); it is estimated that a significant amount of soil is washed away by heavy rainfall in the steep Japanese mountains. Therefore, the increase in the thickness of the surface soil layer will not be able to adapt to the large-scale heavy rainfall that is predicted to accompany the rapid progression of climate change. In addition, it is difficult to incorporate this method in flood control plans, since the plans in Japan are developed for the subsequent 20–30 years. Two methods that can improve the flood-reduction function of forests in a relatively short period of time include enhancement of interception evaporation capacity and enhancement of ground surface storage capacity. These two processes are examined and their effects are discussed using a rainfall-runoff model.

7.2   Rainwater Runoff Mechanisms in Mountainous Forests and Measures to Improve the Flood-Reduction Functions of Forests

Figure 7.1 shows the rainwater runoff processes in a mountainous forest. First, the interception evaporation on the forest crown or canopy affects the rainfall. Some raindrops are intercepted by the branches and leaves, wherein they become microscopic particles and drift through the air. The interception evaporation process was strongly affected by the multilayered structure around the crown. The rainwater lost through interception evaporation reduces the flood amount and its peak flow. This forest property can be effectively improved by forest management through planting, thinning, and felling of trees.

Thereafter, the rainwater that reaches the ground surface will then infiltrate the soil layer if its intensity is below the infiltration and storage capacity of the soil. Rainwater that infiltrates the soil becomes groundwater and contributes to subsurface runoff components, flowing out relatively slowly as compared to the

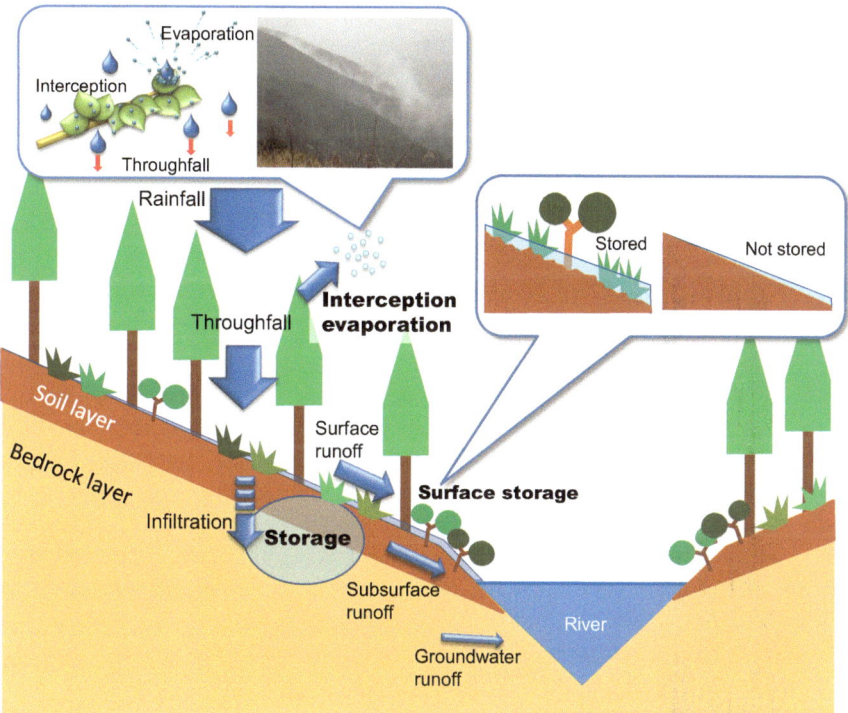

Fig. 7.1 Rainwater runoff processes in mountainous forest

surface runoff components. This process is generally recognized as the key role of the flood-reduction function of forests. However, rainwater becomes a surface runoff component when the rainfall intensity increases beyond the infiltration and storage capacity of the soil layer. The surface runoff component immediately flows down a slope, resulting in a flood. The flood-reduction function of the forest then reaches its limit. The infiltration and storage capacity are determined by the thickness of the forest soil layer. Therefore, it would be extremely difficult to increase the storage capacity of the soil because a period greater than 100 years is necessary for recovery of soil layer thickness (Ogawa et al. 2011).

Surface flows account for most of the peak flood discharge during heavy rainfall events that can cause flood disasters. It flows down forest slopes under the influence of the amount of understory vegetation and the roughness of the ground surface. Therefore, a forest with significant ground surface roughness can decrease the speed of surface runoff flow. Consequently, the surface runoff component is stored on the ground surface for a short time. The amount of water storage on the ground contributes to the reduction and delay of the flood peak flow. For instance, the management of understory vegetation is important for increasing the ground surface roughness. This can be actualized by daily forest management, and the effects are expected to manifest rapidly.

In this study, two aspects were explored: the interception evaporation process in the forest canopy and the water storage process on the ground surface. Based on hydrological observations and analysis using a rainfall-runoff model for two forest areas with different forest characteristics, the possibility of improving the flood-reduction function of forests by enhancing the operation of these two hydrological processes is discussed.

7.3   Runoff Model for Evaluating Flood-Reduction Function

The tank model is used as the runoff model in this study, as shown in Fig. 7.2. The author refers to it as the two-stage tank model with separated surface flow (Tamura et al. 2006). The model is broadly divided into a surface tank (soil layer) and a groundwater tank (bedrock layer) to represent the infiltration, storage, and runoff processes of rainwater in forest soils, as well as the various runoff components: surface runoff, subsurface runoff, and groundwater runoff. Rainfall input to the model is either the amount of throughfall that reaches the ground surface or the amount of rainfall multiplied by the percentage of throughfall (the reciprocal of the interception evaporation rate). A characteristic feature of this model is that the surface runoff coefficient is calculated from the average slope, average slope

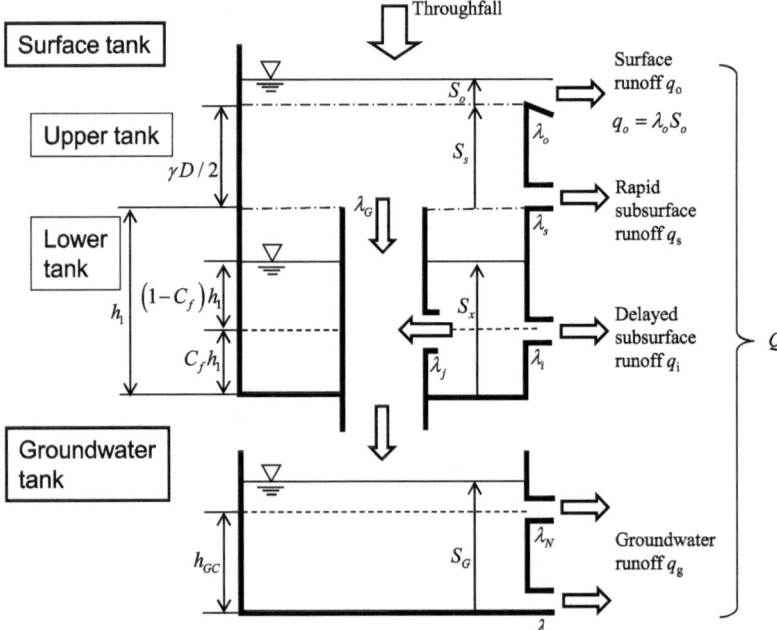

Fig. 7.2 Two-stage tank model with separated surface flow

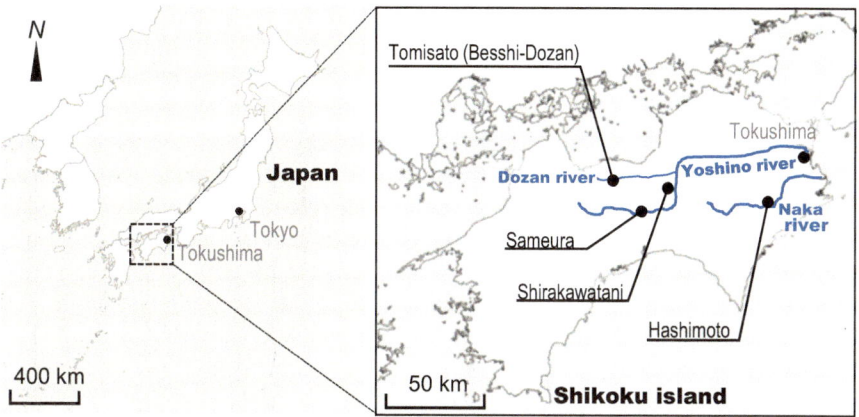

Fig. 7.3 Location of the study sites

length, and roughness of the forest slope (Eq. 7.1). The coefficients of the tank model, including the surface roughness, are identified as a set of parameters that can uniformly reproduce multiple flood hydrographs obtained from field observations; these are used in flood simulations to compare the flood-reduction functions of forest basins.

$$\lambda_o = 2.52 \times 10^{-3} \cdot I_s / \left\{ r_{\max}^{0.8} (N \cdot L_s)^{1.8} \right\} \tag{7.1}$$

where λ_o is the surface runoff coefficient (/h), L_s is the slope length, I_s is the average slope (m), N is the surface roughness ($m^{-1/3}$ s), and $r_{\max}$ is the observed maximum rainfall intensity considered as the maximum surface runoff intensity (mm/h).

7.4 Enhancing Interception Evaporation Capacity Through Afforestation

In the upper basin of the Dozan River, which is one of the prevailing tributaries of the Yoshino River on Shikoku Island (Fig. 7.3), a large copper mine (Besshi-dozan) had been operated from 1691 to 1973. A significant number of trees had been felled for copper refinement business, and a substantial quantity of soil had been washed away from the mountains. A severe flood was caused by a heavy rain-related typhoon in August 1899, and several people lost their lives. Thereafter, a large-scale tree plantation was created, and the vegetation was restored (Sumitomo group Public Affairs Committee 2021).

The flood-reduction function of the Tomisato dam basin, including the Besshi Copper Mine area, was evaluated at the beginning of the twenty-first century when the vegetation had recovered for approximately 100 years after the beginning of large-scale planting. A runoff model was used in the evaluation to describe the water cycle processes in a forested basin (Tamura et al. 2008). In this study, the water storage capacity of the soil layers and the interception evaporation effect were estimated; Figs. 7.4 and 7.5 show the calculation results. The characteristics of the flood-reduction function of the basin were examined by comparing its runoff results with those of the Sameura dam basin, because it is located next to the Tomisato dam basin and has similar geology and vegetation. The role of interception evaporation is very important during the forest recovery stage (Fig. 7.4). The simulation results demonstrate that the flood peak flow would increase by 50%, compared to the current condition, if clear-cutting was performed in the Tomisato dam basin. The rate of increase of the flood peak flow is higher than that of the Sameura dam.

This is because the runoff model estimated that the interception evaporation rate of the current Tomisato dam basin is larger than that of the Sameura dam basin, and the thickness of the surface soil layer of the current Tomisato dam basin is much smaller than that of the Sameura dam. In other words, if clear-cutting is implemented in the Tomisato dam basin, the large amount of rainfall lost due to interception evaporation will reach the ground surface, but the thin soil layer will quickly become saturated, resulting in an increase in peak flood discharge. Figure 7.5 shows the maximum groundwater storage volume for the two basins under the clear-cut conditions (the situation is shown in Fig. 7.4b). It shows the incremental storage from the start of rainfall to the time of peak flood flow in the surface and groundwater tanks shown in Fig. 7.2. The possibility that the thickness of the soil layer at the Tomisato dam basin has not recovered sufficiently since the start of the planting program is shown in Fig. 7.5. Nevertheless, the fact that the peak flood discharge in the Tomisato dam basin is lower than that in the Sameura dam basin (Fig. 7.4a) suggests that the interception evaporation in the flood-reduction function of the current Tomisato dam basin is quite effective. This can be considered as the greatest effect of the large-scale afforestation project in terms of flood reduction.

As a result, it can be concluded that the enhancement of interception evaporation capacity is more effective than the enhancement of soil layer thickness as a means of improving the flood-reduction function of forests in a short period of time. In addition, it is important to avoid excessive thinning and clear-cutting in forest management when applying this method to basins with poor forest soil layer thickness, such as the Tomisato dam basin.

7.5 Enhancing the Interception Evaporation Capacity and Surface Storage Capacity by Vegetation Conversion

The upper basin of the Naka River is notable for the production of Kito-sugi Japanese cedar. This basin is known to be a high precipitation area, with an annual precipitation of approximately 3000 mm. The Hashimoto forest is located in the upper district of the Naka River (Fig. 7.3), wherein selection cutting forestry was

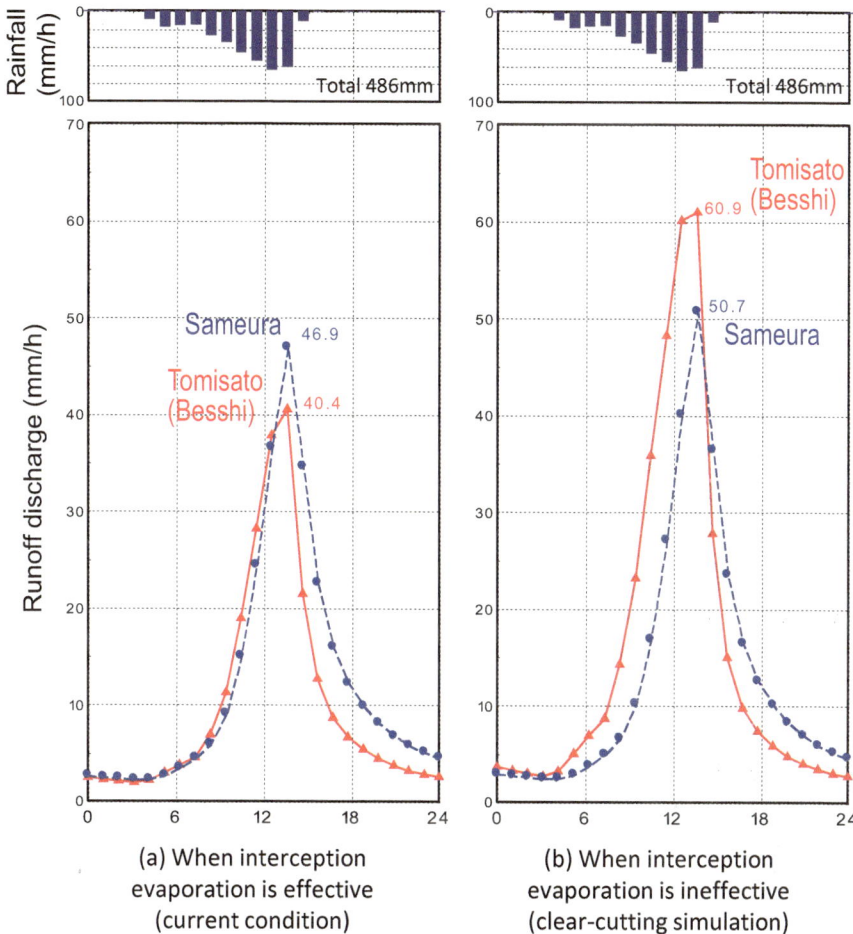

Fig. 7.4 Comparison of flood hydrographs under current and clear-cut conditions to evaluate the impact of interception evaporation considering Typhoon No. 23 in 2004 (summarized from Tamura et al. 2008)

performed from the 1980s and broadleaf trees were preserved. Its forest type is a mixed forest consisting of conifers and broadleaf trees, with a multilayered crown structure. The standard thinning rate is approximately 30% in common cedar artificial forests in Japan, but the rate of the Hashimoto forest ranges from 15% to 20%. However, due to the mixture of various aged trees and species, the canopy is not dense and the understory vegetation is rich (Fig. 7.6a).

The flood-reduction function of the Hashimoto forest was evaluated using a runoff model (Tamura et al. 2020). The effects of the interception evaporation capacity and the ground surface storage capacity in the Hashimoto forest, a mixed conifer and broadleaf forest, were discussed and compared with those in the

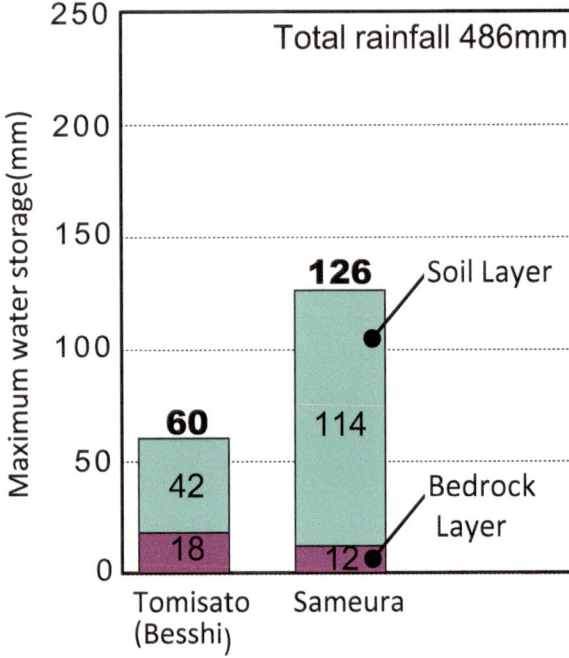

Fig. 7.5 Estimation of maximum groundwater storage in forested basins using the two-stage tank model with separated surface flow considering Typhoon No. 23 in 2004 (summarized from Tamura et al. 2008)

(a) Hashimoto forest   (b) Shirakawatani experimental forested basin

Fig. 7.6 Trees and understory vegetation at the observation sites

Shirakawatani experimental forested basin. Shirakawatani is a Japanese cedar forest where general Japanese forestry (large-scale planting and felling) is performed (Fig. 7.6b), and its geological features and topography are similar to those of the Hashimoto forest. The model parameters of the interception evaporation rate and surface roughness obtained from the runoff analysis of the Hashimoto forest

Table 7.1 Main parameters of the in-line two-stage tank model for simulating the conversion of Japanese cedar artificial forest to mixed needle and broadleaf forests

Parameter	Hashimoto	Shirakawatani	
	Mixed needle and broadleaf forests	Artificial cedar forest (current condition)	Mixed needle and broadleaf forests (simulation)
Interception evaporation rate I (%)	25	14	25
Ground surface roughness N (m$^{-1/3}$ s)	1.8	1.2	1.8
Apparent surface thickness γD (mm) Groundwater supply coefficient	137	100	100
λ_g (/h)	0.05	0.01	0.01
Final infiltration rate f_* (/h)	6.45	1.45	1.45
Soil saturation capacity h_1 (mm)	168	40	40

were applied to the Shirakawatani experimental forested basin model to study the improvement of flood-reduction function of a general cedar artificial forest by converting it into a mixed needle-hardwood with only selective logging. The model parameters for the two basins, as well as the parameters when the Shirakawa valley is changed to a mixed needle-hardwood forest, are listed in Table 7.1.

The flood-reduction function of the Hashimoto forest is greater than that of the Shirakawatani. Runoff simulation using the same rainfall pattern as shown in Fig. 7.7 for both runoff models showed that the peak flow rate in Hashimoto forest area was 22% lower than that in Shirakawatani (Fig. 7.8). When the interception evaporation capacity rate of the Hashimoto forest model was applied to the Shirakawatani model, the peak flood discharge in the Shirakawatani decreased from 66.8 mm/h to 56.3 mm/h. In addition, when the surface roughness parameter of the Hashimoto forest was applied to the Shirakawatani model, it was further reduced to 50.9 mm/h. As a result, the peak flood discharge was estimated to be 24% lower than that for the current Shirakawatani model.

The flood-reduction function of the Hashimoto forest is higher than that of the Shirakawatani owing to several factors. In the Hashimoto forest, cedar trees of different ages and broad-leaved trees of various species are mixed; the canopy is multilayered. The density of the canopy of the multistoried forest seems to be sparse in the vertical space. The thinning rate in the Hashimoto forest is low, but sunlight reaches the forest floor easily. This is a favorable condition for understory vegetation. Therefore, the interception evaporation rate and the surface roughness are high, and it is assumed that a high flood-reduction function is demonstrated in the Hashimoto forest.

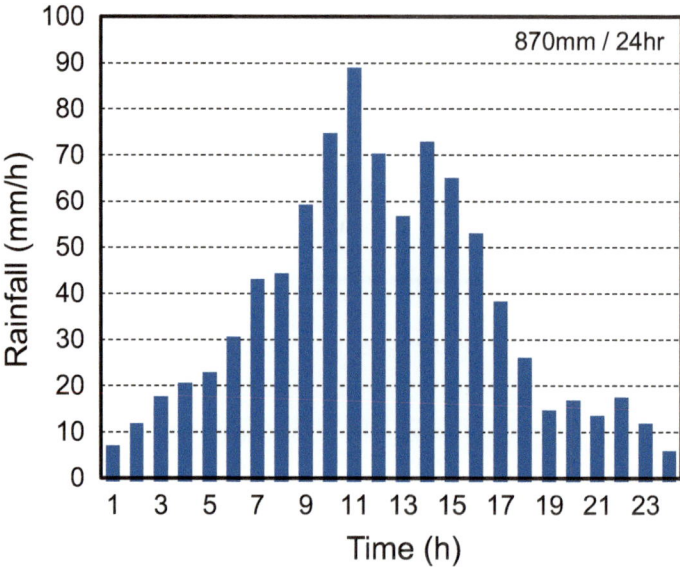

Fig. 7.7 Rainfall patterns used to evaluate the Hashimoto forest and Shirakawatani experimental forested basin (Tamura et al. 2020)

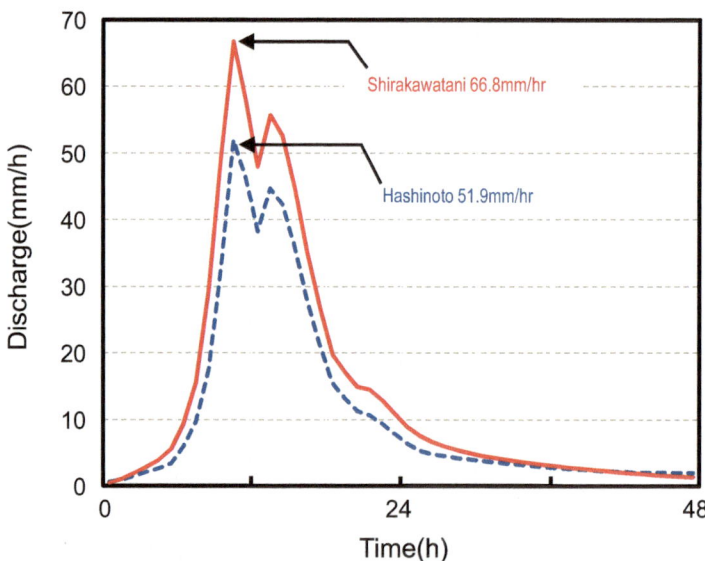

Fig. 7.8 Comparison of flood hydrographs between Hashimoto forest and Shirakawatani experimental forested basin under current conditions (summarized from Tamura et al. 2020)

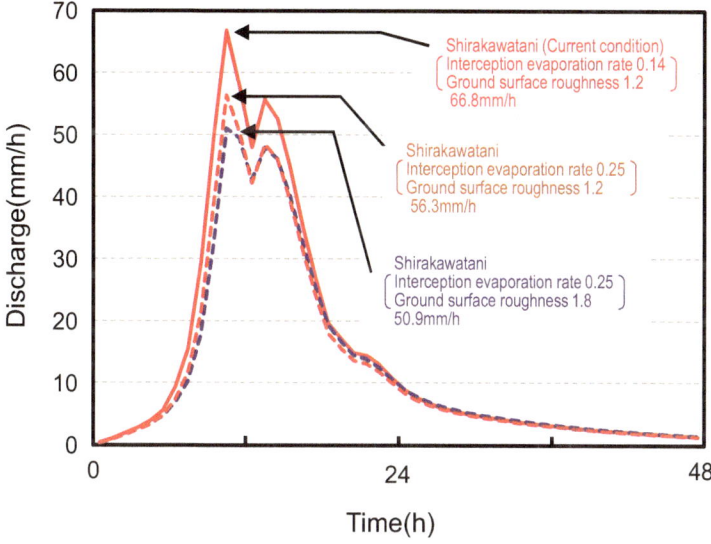

Fig. 7.9 Flood hydrographs of the Shirakawatani experimental forested basin under the condition that the interception evaporation rate and surface roughness are changed by the conversion to mixed needle and broadleaf forest (summarized from Tamura et al. 2020)

The results suggest that the flood-reduction function of cedar forests with poor understory vegetation, such as those in the Shirakawatani experimental forested basin, can be improved on a timescale of approximately 20–30 years by conversion to mixed forests with only selective logging (Fig. 7.9).

7.6 Methods and Limitations of Early Enhancement of Flood Mitigation Functions of Forests

In this study, the reinforcement measure for the flood-reduction function of forests and the expected effects were discussed in terms of interception evaporation in the forest canopy and water storage on the forest ground surface. These methods have two advantages compared to the restoration of soil layer thickness. First, they are achieved easily via daily forest operations. Second, the effects can be expected to be realized in 20–30 years. However, the expected effect is still restrictive (10–20% reduction of the flood peak runoff) for heavy rainfall events that are targeted by the flood control plan. In addition, because the interception evaporation rate observed in forests has changed owing to a rainfall event (Tanaka et al. 2005), it is believed that the flood-reduction function of forests fluctuates due to forest site and structure. For utilizing the flood-reduction function of a forest is in river basin management, a thorough understanding the characteristics of this function is required.

References

Amundson R, Berhe AA, Hopmans JW, Olson C, Sztein AE, Sparks DL (2015) Soil science. Soil and human security in the 21st century. Science 348:1261071. https://doi.org/10.1126/science.1261071

Laurance WF (2007) Environmental science: forests and floods. Nature 449:409–410

Ogawa K, Tamura T, Takigawa N, Kuwahara M, Takanishi S (2011) Flood runoff characteristic change over 50 years by tree species conversion in upper Dozangawa River Basin. Ann J Hydraul Eng JSCE 55:571–576

Sumitomo group Public Affairs Committee (2021). https://www.sumitomo.gr.jp/history/besshidouzan/index04.html

Tamura T, Hashino M, Tachibana D (2006) Parameter identification of runoff model using rainfall and water-level data. Ann J Hydraul Eng JSCE 50:355–360

Tamura T, Hashino M, Anamizu H, Araki T (2008) Evaluation of flood reduction function of the forested basins located upstream in the Yoshino river. Ann J Hydraul Eng JSCE 52:379–384

Tamura T, Ueda S, Muto Y, Kamada M (2020) Improvement of flood reduction function of forested basin by reinforcement of evaporation rate and ground roughness. J Jpn Soc Civ Eng Ser B1(Hydraul Eng) 76(2):127–132

Tanaka N, Kuraji K, Shiraki K, Masanori S, Masakazu S, Ohta T, Makoto S (2005) Throughfall, stemflow and rainfall interception at mature Cryptomeria japonica and Chamaecyparis obtusa stands in Fukuroyamasawa watershed. Bull Tokyo Univ For 113:197–240

Chapter 8
Forest Green Infrastructure to Protect Water Quality: A Step-by-Step Guide for Payment Schemes

Thomas R. Nisbet, Maria-Beatrice Andreucci, Rik De Vreese, Lars Högbom, Sonja Kay, Mary Kelly-Quinn, Alessandro Leonardi, Mariyana I. Lyubenova, Paola Ovando Pol, Paula Quinteiro, Ignacio Pérez Silos, and Gregory Valatin

---

T. R. Nisbet (✉) · G. Valatin
Forest Research, Alice Holt Lodge, Farnham, UK
e-mail: tom.nisbet@forestresearch.gov.uk

M.-B. Andreucci
Department of Planning, Design, Technology of Architecture, Faculty of Architecture, Sapienza University of Rome, Rome, Italy

R. De Vreese
Independent Researcher and Consultant in Urban Forestry and Ecosystem Services, Ghent, Belgium

L. Högbom
Forestry Research Institute of Sweden (Skogforsk), Uppsala, Sweden

S. Kay
Agroscope, Research Division Agroecology and Environment, Research Group Agricultural Landscapes and Biodiversity, Zurich, Switzerland

M. Kelly-Quinn
School of Biology and Environmental Science, University College Dublin, Dublin 4, Ireland

A. Leonardi
ETIFOR | Valuing Nature, Padova University Spin-off, Padova, Italy

M. I. Lyubenova
Department of Ecology and Environmental Protection, University of Sofia "St. Kliment Ohridski", Sofia, Bulgaria

P. O. Pol
The James Hutton Institute, Aberdeen, UK

P. Quinteiro
Centre for Environmental and Marine Studies (CESAM), Department of Environment and Planning, University of Aveiro, Aveiro, Portugal

I. P. Silos
Environmental Hydraulics Institute, Universidad de Cantabria, Santander, Spain

Abstract This chapter describes how to design appropriate and cost-effective forest green infrastructure for water payment schemes to protect and improve water quality. It is structured by the main steps involved in establishing a payment scheme, starting with identifying the water issues and how tree planting and forest management can help, managing potential disbenefits and exploring multiple benefits, followed by scheme design, monitoring and communication. The approach is relevant to all actors involved in sustainable water management, farming and forestry, from policy makers, catchment planners and land managers to private investors, practitioners and local communities. We provide a common language and framework to help ensure schemes are successful in delivering water and other ecosystem services while minimising possible trade-offs (such as the potential for tree planting to reduce water resources).

Keywords Forest green infrastructure · Forests for water · Diffuse pollution · Payments for ecosystem services · Sustainable water management

8.1 Introduction

The main aim of European Union (EU) water policy is to ensure that a sufficient quantity of good-quality water is available for both people's needs and the environment. Despite ongoing efforts by Member States to improve water status, only 40% of surface waters (rivers and lakes) are in good ecological status or potential (EEA 2018). Diffuse pollution is a major pressure, dominated by agricultural sources in the form of excessive emissions of nutrients (nitrates and phosphates), pesticides, sediment and faecal indicator organisms (FIO) (Fig. 8.1). Agriculture is estimated to contribute to 25% of surface water bodies failing good ecological status, and is the main cause of groundwater bodies failing to achieve good chemical status (EEA 2018).

Member States continue to develop and invest in best practice farming measures to reduce diffuse pollution from agriculture, such as farm-level nutrient planning, reduced tillage and the use of catch crops. These have improved water quality but in most cases by an insufficient margin to meet environmental quality standards. Notably, there has been limited improvement in the proportion of water bodies achieving good ecological status since the first River Basin Management Plans were published in 2009 (EEA 2018). While there is scope for further improvement in the effectiveness and uptake of farm measures, there is a growing recognition that wider delivery of good ecological status will only be achieved by a significant degree of land-use change (Stutter et al. 2012). This chapter focuses on the water benefits of forest green infrastructure (FGI), although it is recognised that the use of tree-based measures such as tree planting and management should be part of a wider framework of integrated catchment management (e.g. including opportunities for peatland and wetland restoration) and associated strategies and plans.

NUTRIENTS

Wash-off and leaching of nitrate and phosphate from fertiliser and manure applications.

SEDIMENT

Soil disturbance and sediment runoff due to land management practices and stock grazing.

PESTICIDES

Aerial drift, wash-off or leaching of pesticides following applications.

FAECAL INDICATOR ORGANISMS

Runoff of faecal bacteria from animal manure and slurry applications.

Fig. 8.1 Catchment sources and pathways of agricultural diffuse pollution impacting on the water environment and water users

The multiple benefits provided by FGI in the form of trees, woodlands and forests (the terms woodlands and forests are used throughout the document and describe land predominantly covered by trees; woodland means a relatively small area of trees, while forest refers to a large tract of trees) are increasingly recognised and valued for society. Benefits for the water environment include the ability to protect aquatic habitats and species from disturbance, preserve the quality of drinking water, alleviate flooding and guard against erosion, landslides and the loss of soil (Nisbet et al. 2011). Tree planting provides a very effective and relatively secure measure for tackling agricultural diffuse pollution, in addition to helping with carbon storage and providing other environmental benefits. Small-scale, targeted use of FGI such as planting of woodlands on or around pollutant sources, or along pollutant pathways in the form of 'woodland buffers', offers a smart way of attenuating or eliminating pollutant delivery to surface waters and groundwaters while minimising land take and impacts on food security.

Although the benefits of FGI for water are well known (Creed and Noordvijk 2018), progress is highly constrained by the significant cost to landowners and managers in terms of reduction in land value and agricultural income resulting from land-use change. This is especially the case for tree planting on the more productive and intensively managed agricultural land that represents the greatest source of diffuse pollutants. Achieving enough tree planting to make a difference for water body status will require better incentives in the form of payments for the water and other ecosystem services provided. Maintaining and protecting the water benefits provided by existing FGI may also require funding support, especially if changes to forest design and management are needed to address threats posed by climate change.

There are many different types of payment for ecosystem services (PES) schemes, and the approach and definitions continue to evolve (Forest Europe 2019). The main purpose of PES is to protect and enhance the provision of ecosystem services for environmental gain and better management of natural resources by incentives (Gatto et al. 2009). Ideally, five conditions should be met in a PES scheme, which are: (1) the identification of a well-defined ecosystem service to be exchanged, in this case principally targeted tree planting and the appropriate management of new or existing woodlands and forests to improve water quality; (2) the presence of at least one service buyer and (3) at least one seller; (4) the voluntary nature of the marketing of the ecosystem service; and finally, (5) the conditionality of the payment, requiring the seller/provider to ensure that the expected benefit is delivered and sustained over time. Often these conditions are not met, especially condition 4, such as where schemes are implemented within a compulsory regulatory framework. Such cases are often referred to as 'PES-like' schemes.

We adopt a broad definition of a FGI for water payment scheme based upon three criteria: (1) a transfer of resources between at least two stakeholders; (2) a transaction explicitly targeted at obtaining water-related services; and (3) a payment for actions related to trees, either primarily for water services or for bundled (including water) ecosystem services.

8.2 Identifying the Problem

As a consequence of the introduction in 2000 of the EU Water Framework Directive (WFD), there is now far greater knowledge about the water environment across Member States. Regular monitoring and assessment of surface- and groundwaters at more than 130,000 sites has generated a detailed understanding of the condition of Europe's water bodies, as well as of the pressures that are preventing the majority achieving the targets of good ecological status or potential (Fig. 8.2).

National water regulators compile and regularly update datasets and maps showing which water bodies are at less than good status, the causal activities and progress made with introducing programmes of measures to achieve target status.

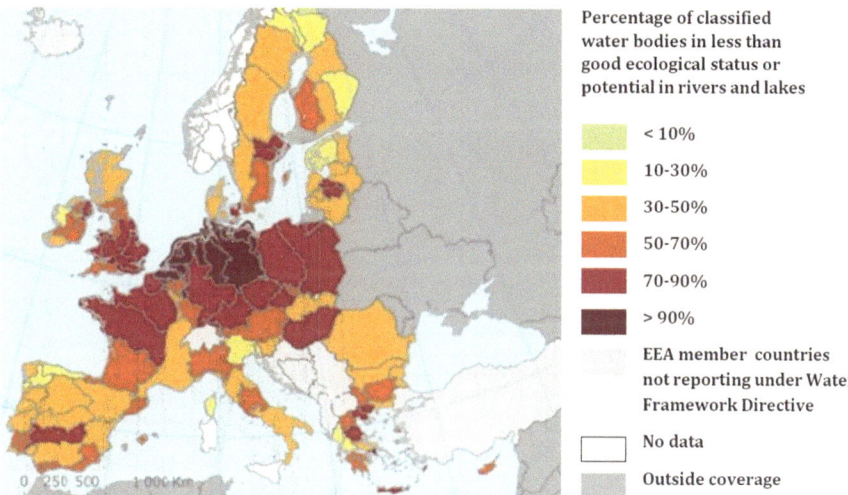

Fig. 8.2 Proportion of surface water bodies by Member State at less than good ecological status. Reproduced from EEA (2016) © European Environment Agency, 2016

This includes knowledge of which water bodies are failing good status due to diffuse pollutants such as nutrients, sediment and pesticides derived from agriculture, as well as impacted by other anthropic pressures on water (Fig. 8.3). Water regulators also have information on the location and condition of particularly sensitive waters, such as drinking water protected areas and high-status waters supporting priority habitats and species. These data are submitted to the European Environment Agency (EEA) at regular intervals to allow an assessment of the status and pressures acting on European waters, with the last assessment published in 2018 (EEA 2018).

Addressing water pressures and achieving water targets requires co-ordinated and long-term actions at the level of the catchment or sub-catchment of the water body. This is particularly the case with managing diffuse pollution, which often has a variety of sources spread across the landscape and land ownerships. In some regions and countries, pollutant models have already identified pollutant sources and pathways to aid targeting of measures (Collins et al. 2018; Mockler and Bruen 2018). Catchment partnerships have been formed in many Member States to adopt an integrated, catchment-based approach to tackling polluting activities and delivering improvements. Partnerships are often led by trusted intermediaries who are better able to achieve change on the ground supported by co-ordinated funding bids (ribblelifetogether.org/improve/woodlands, wrt.org.uk/project/3rivers-project, www.woodlandsofireland.com, www.etifor.com/en/studies-and-research).

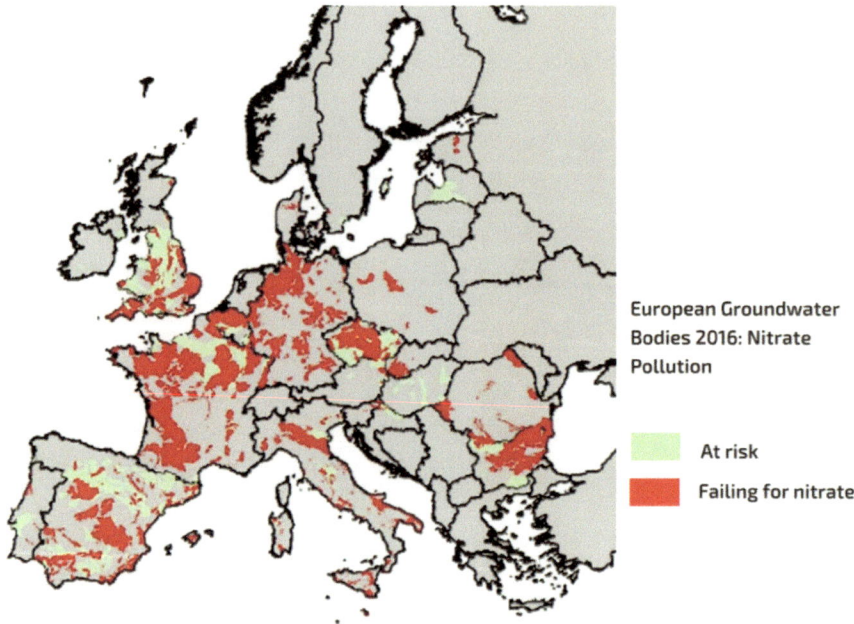

European Groundwater
Bodies 2016: Nitrate
Pollution

At risk

Failing for nitrate

Fig. 8.3 Distribution of groundwater bodies across Member States failing for nitrate

8.3 Role of Opportunity Mapping

Opportunity mapping can help identify and prioritise water bodies and component areas of land for targeting FGI to reduce water pressures (Broadmeadow and Nisbet 2012). Evidence-based planning supports integrated catchment management, and the maps guide and underpin the development of FGI for water PES (PESFOR-W) schemes. This approach is based on using geographic information systems (GIS) and integrates a wide range of spatial datasets to determine the most effective locations for changing land use and management, to meet WFD targets and generate multiple benefits for society (Box 8.1).

The key steps involved in evidence-based planning are summarised below:

1. Use WFD datasets to identify boundaries of surface and groundwater bodies failing good ecological or chemical status due to diffuse pollution from agriculture; determine which and how many diffuse pollutants are causing failure, either through WFD measurements in relation to water chemical or biological standards or an assessment of risk.
2. Draw on any available statistics (e.g. from agricultural inventories, or surveys of fertiliser, or pesticide use), site surveys and measured or modelled pollutant data to identify and rank the spatial sources and pathways of each diffuse pollutant draining to the contributing catchments.

Box 8.1 Opportunity Mapping to Reduce Diffuse Pollution and Flood Risk in England and Wales

Opportunity mapping was applied to England and Wales in 2014 to better target grant aid and private investment for FGI to help deliver positive outcomes for water quality and flood risk management. The mapping used national datasets of modelled pollutant loads and pressures at a 1 km^2 scale for each of phosphate, sediment, nitrate, total pesticides and faecal indicator organisms. These were overlaid with datasets on the risk of flooding from rivers, including on the propensity of soils to generate rapid runoff. Target areas for woodland creation were identified based on the scope to reduce one or more diffuse pollutants and contribute to flood risk management (the colours on the map and the associated values in the key refer to the number of diffuse pollutants that tree planting could benefit in a given location).The maps were subsequently used to score water benefits to inform planting applications and Rural Development Programme grant support.

For further information, see www.forestresearch.gov.uk/research/forest-hydrology/opportunity-mapping.

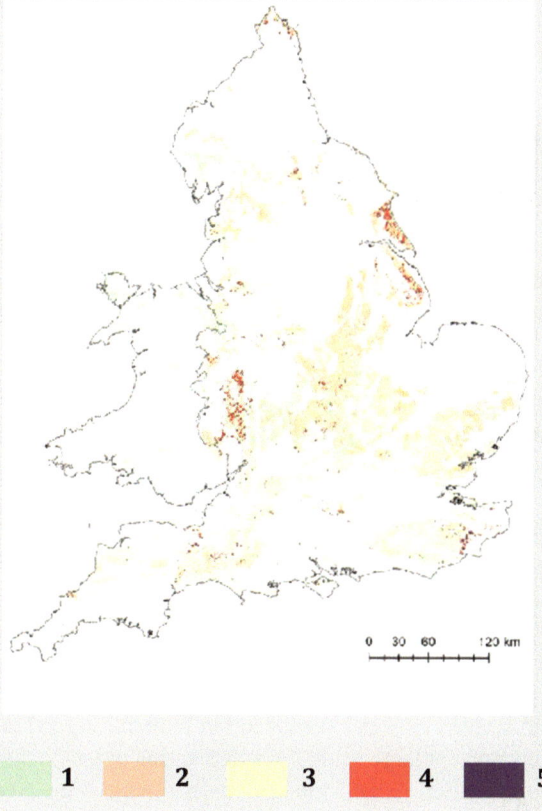

0   30   60   120 km

1   2   3   4   5

3. Map any spatial constraints (e.g. designated open habitats or archaeological features) and sensitivities (e.g. landscape views) to FGI within the catchments; overlay spatial datasets to identify pollutant hotspots free of constraints to FGI where there are opportunities for woodland creation to reduce one or more diffuse pollutants in failing water bodies.
4. Consider and map any other water issues that could benefit from FGI (e.g. local downstream communities or assets at flood risk) and overlay these to determine scope to deliver multiple benefits through tree planting where there is greatest need.
5. Map any potential water trade-offs associated with FGI (e.g. water bodies failing due to poor quantitative status or inadequate flows) and use mapped sensitivities to guide tree planting design and management to minimise disbenefits (e.g. by changing tree type or species to reduce tree water use).
6. For existing FGI, map data on tree type, species and age to determine opportunities for forest redesign and management, to reduce future risks to forest water protection functions (e.g. from climate change and related effects on the incidence of fires, storms and drought, as well as pest and disease outbreaks).
7. Use findings to amend and integrate FGI, water, flood and related strategies and plans to deliver a more effective, catchment-based approach to tackling diffuse pollution and achieving WFD objectives.

8.4   How Can FGI Help?

Forests are widely recognised as the preferred land cover for protecting water supplies. This reflects a range of attributes, including the ability of forest canopies to moderate rainfall inputs due to wet canopy evaporation; the well-structured nature of forest soils resulting from sustained organic matter inputs, tree rooting and lack of soil disturbance, reducing erodibility and promoting slope stability; active uptake and tight canopy recycling of nutrients; and the generally very low level of chemical inputs to forests such as fertilisers or pesticides (Nisbet et al. 2011; Creed and Noordvijk 2018). Consequently, waters draining forests are typically of high quality and good ecological condition, requiring little or no treatment for public water supply.

Historic clearance of forests for agriculture has resulted in the widespread loss of these water benefits and a shift to a more intensive land use often associated with frequent soil disturbance, soil damage, increased erosion and high inputs of nutrients and chemicals. Despite recent improvements to farming practice, many agricultural activities typically generate significant losses of sediment, nitrate, phosphate, pesticides and/or FIO to the water environment. These result in diffuse pollution and cause a large number of water bodies to fail to achieve good ecological status. Food scarcity may prevent large-scale forest replanting to tackle the issue, but there is significant scope for FGI in the form of targeted, small-scale, woodland planting on agricultural land to make a difference. This includes the use of agroforestry, shelterbelts and tree-lined hedges to help capture and remove diffuse pollutants from adjoining arable crops or livestock pasture.

Targeted planting works because the sources of pollutants, the pathways by which they move to watercourses and the vulnerability of downstream water users are spatially variable (Fig. 8.4). For example, soils vary in their vulnerability to damage, ability to retain nutrients and chemicals, propensity to generate rapid surface runoff and degree of connectivity to watercourses. Once pollutants are mobilised in water or the air, they tend to move along preferred pathways such as surface channels, drains/ditches and the prevailing wind direction. Water receptors such as groundwater boreholes draw water from distinct areas and depths of ground. Tree planting on, around, across or along these key pollutant sources, pathways

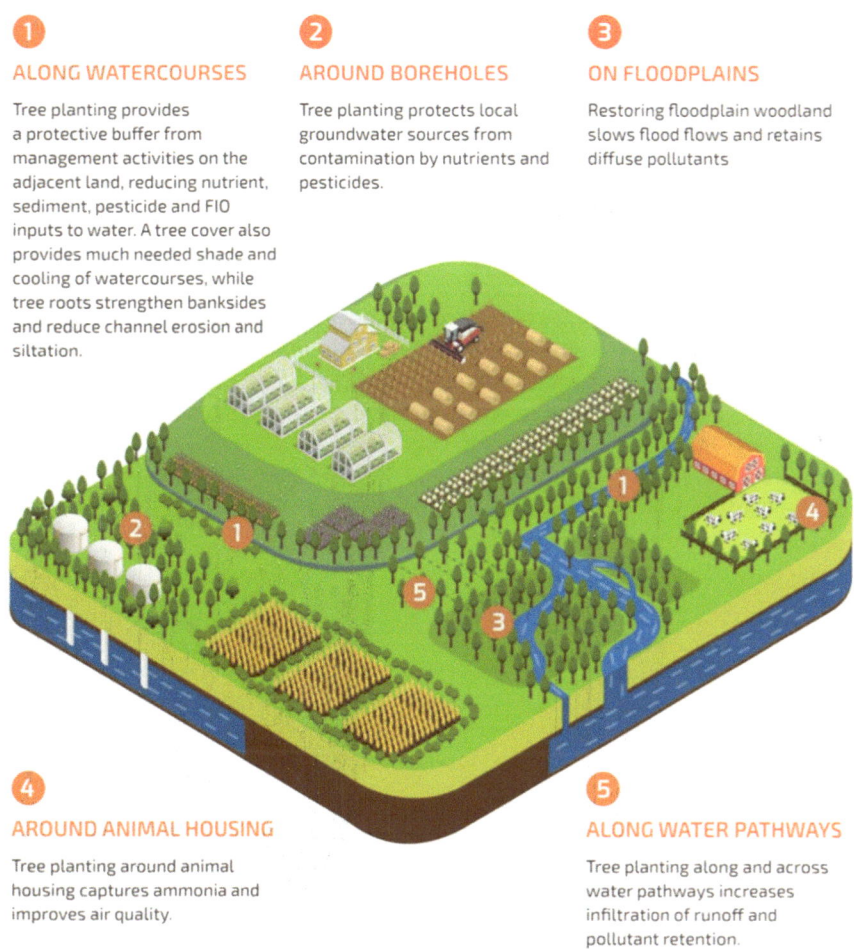

1 ALONG WATERCOURSES

Tree planting provides a protective buffer from management activities on the adjacent land, reducing nutrient, sediment, pesticide and FIO inputs to water. A tree cover also provides much needed shade and cooling of watercourses, while tree roots strengthen banksides and reduce channel erosion and siltation.

2 AROUND BOREHOLES

Tree planting protects local groundwater sources from contamination by nutrients and pesticides.

3 ON FLOODPLAINS

Restoring floodplain woodland slows flood flows and retains diffuse pollutants

4 AROUND ANIMAL HOUSING

Tree planting around animal housing captures ammonia and improves air quality.

5 ALONG WATER PATHWAYS

Tree planting along and across water pathways increases infiltration of runoff and pollutant retention.

Fig. 8.4 Preferred locations in a farmed landscape for tree planting to reduce diffuse pollution

and receptors can potentially be very effective at reducing pollutant delivery to watercourses and water supplies, thereby markedly improving water quality for a limited land take.

Planting across or along pollutant pathways in the form of buffer areas or strips offers a dual water quality benefit (Fig. 8.4). Firstly, the pollutant input associated with the previous agricultural activity on this sensitive area of ground will be removed. Secondly, there is a significant opportunity for the planted trees to act as a barrier to the movement of pollutants from upslope or upwind (Ucar and Hall 2001). Pollutants can be retained or removed by runoff being encouraged to infiltrate into the better structured soil of the buffer; by filtration and surface deposition as surface runoff passes through the leaf litter layer or is held in surface depressions created by tree roots; by root uptake and incorporation into growing trees; or by interception and capture as the polluted airflow passes through the tree canopy. Riparian woodland buffers have the added benefit of removing pollutant inputs and reducing damage to this very vulnerable and connected area of land, as well as providing scope for planted trees to remove pollutants carried downstream within the main watercourse during out-of-bank flows.

WFD monitoring data show a marked difference in water quality between forest and agricultural land uses, with the magnitude of difference depending on the intensity and quality of land management. The quantity of pollutant inputs in the form of fertiliser, organic amendments and chemicals and typical exports in surface runoff or leaching to groundwater are well known for each land use and can be used by pollutant models to estimate the impact and effectiveness of a given area of land-use change (Table 8.1). In contrast, it is more difficult to predict the barrier effect of buffer areas since this is influenced by many design and management factors, as well as by the nature and type of pollutant and the scale of intervention. However, studies have shown that with good design and appropriate management, tree buffer areas can be highly effective at reducing pollutant delivery in surface runoff from upslope land, with efficiencies of up to 100% possible for certain pollutants (Perez-Silos 2017).

A review of 65 studies found buffer width to be a dominant factor, with pollutant removal generally decreasing with declining buffer width (Perez-Silos 2017). There are a number of important factors that act to reduce the efficiency by which tree buffers can remove diffuse pollutants from upslope land. These include increasing volume of runoff, increasing pollutant load (especially if the quantity of pollutant draining from upslope land exceeds the capacity of the trees and soil to remove or process it), the presence of newly established/very young or old trees, poor tree condition or weak tree growth, wider tree spacing and the presence of any bypass channels such as drains. Great care is therefore required in the design and management of buffer areas to cope with local pollutant loads and to efficiently achieve and maintain high levels of pollutant removal. This may require productive woodland management and regular harvesting to sustain nutrient uptake (see Sect. 8.5).

Landowner pressure to minimise land take from agriculture acts as a major constraint on buffer width that can result in suboptimal performance. As a rough

Table 8.1 Nutrient loads and modelled export coefficients to water for different crops *vs* woodland in Great Britain. Nutrient loads taken from the British Survey of Fertiliser Practice for 2000–2011 (BSFP 2013) and export coefficients based on the same data modelled for the UK National Ecosystem Assessment Follow-on Report (Bateman et al. 2014)

	Permanent grassland	Rough pasture	Wheat	Barley	Maize	Oil seed rape	Woodland
Nitrogen input (kg/ha/year)	94–135	10	131–167	120–132	46–62	155–189	20
Nitrate-N export (kg/ha/year)	0.86–10.58	0.02–0.05	1.54–19.72	1.54–19.72	1.52–19.72	3.29–17.4	0.02–0.1
Phosphate input (kg/ha/year)	6–16	0	13–35	18–41	27–43	15–37	0
Phosphate export (kg/ha/year)	0.012–0.169	0.008	0.038–0.458	0.038–0.458	0.038–0.458	0.15–1.834	0.008

Table 8.2 Per cent reduction in diffuse pollutant concentration in surface runoff from upslope land to watercourses achievable from a well-designed and managed woodland buffer of variable width. Interpolated from relationships derived from review by Perez-Silos (2017)

Buffer width	5 m	10 m	20 m	50 m	100 m
Nitrate-N	20%	30%	40%	80%	90+%
Phosphate-P	10%	20%	30%	60%	90+%
Suspended sediment	80%	90+%	90+%	90+%	90+%

guide, Table 8.2 shows what could be expected in terms of reduction in different diffuse pollutants in surface runoff from upslope land by an increasing width of a well-designed and managed woodland buffer, based on a recent review (Perez-Silos 2017). Data are lacking for FIO and many pesticides, with removal efficiency for the latter greatly depending on the type of pesticide and its pathway of movement.

While the above tables can be used to explore the environmental effectiveness of woodland planting compared to other measures and help underpin rough cost-effectiveness estimates, the design of a scheme and especially one that seeks to make a difference at the catchment level is best informed by hydrological modelling. Understanding soil and hydrological processes is crucial to identifying pollutant sources and pathways for correct targeting of measures, as well as for quantifying and upscaling their environmental effectiveness. It is important to use a spatially distributed model that can be incorporated within GIS to determine the most effective placement and integration of measures, accounting for dynamics in pollutant movement and the evolution of measures. The physically based Soil and Water Assessment Tool (SWAT) model is often preferred for such applications, although it is highly data demanding, is unsuited to very small catchments (<150 ha) and requires great care to ensure correct model parameterisation, especially in relation to forest processes (Baksic 2018). A groundwater model would be needed to predict the movement of pollutants to deep boreholes. Where resources and data are lacking to allow process modelling, more simple tools can be applied such as the Integrated Valuation of Ecosystem Services and Tradeoffs (InVEST) suite of models (Kareiva et al. 2011).

8.5 Importance of Design and Management of FGI

The benefits of tree planting for water can evolve over a number of years due to the time taken for trees to grow, a forest canopy to form and the associated effects on water use and soil improvements to become fully established. There can also be a significant lag before soil stocks of nutrients or contaminants such as pesticides from the previous land-use breakdown, or are washed out of the soil-rock system. The delay and lag time will vary with type of pollutant, the depth of water pathways (being much slower for deeper groundwater) and the nature of tree planting (e.g. tree type, species and stocking density). For example, reductions in suspended sediment, adsorbed/total phosphate and FIO can be rapidly delivered (within 1–3 years) by the cessation of soil disturbance, removal of livestock and improvement in soil infiltration with tree rooting following planting. In contrast, while there will be a rapid step-change in nitrate and pesticide inputs following land-use change, soil and groundwater stocks of these chemicals can take decades to be removed from groundwater supplies.

A particular advantage of tree planting is the semi-permanent nature of the land-use change such that water benefits can be secured in the long term. However, this relies on planted trees and established woodlands and forests being sustainably managed and replanted if felled or lost to fire, storms or pest and disease. Productive management can provide economic benefits to landowners and managers in the form of timber and wood fuel but poses risks of pollution due to forestry operations such as timber harvesting and extraction, as well as the temporary loss of water benefits until trees regrow after felling. These risks can be minimised through good forestry practice but require great care in forest planning, design and management, especially involving sensitive locations such as riparian buffers (Forestry Commission 2019).

The targeted planting of woodland buffers for intercepting diffuse pollutants from adjacent agricultural land requires more detailed design and active management to sustain and promote pollutant removal. The technical specification will vary between sites and needs to be determined on a case-by-case basis, with woodland design tailored to site type, the nature of the pollutant and its pathway of movement. For example, sediment trapping is enhanced by creating a vegetated and rough ground surface that benefits from less canopy shade and wider tree spacing. In contrast, nitrate removal either requires wet woodland with waterlogged soils to promote nitrate loss by denitrification (note that this increases nitrous oxide emissions, a potent greenhouse gas (GHG)) or planting close-spaced, faster growing tree species to maximise nitrate uptake. Where nitrate inputs in surface runoff from upslope land are very high, regular harvesting of timber or wood fuel may be necessary to avoid nitrate saturation and overloading buffers. In such cases, wider buffers would facilitate phased harvesting of strips to maintain some degree of nitrate removal, although particular care is required to avoid ground damage during operations.

Other pollutants such as ammonia and pesticides that are dispersed via aerial pathways require special attention to the design of the structure of the woodland canopy to maximise aerial deposition and pollutant trapping. A typical example is the design of woodland buffers around animal housing to reduce ammonia emissions (Bealey et al. 2016). Another issue can be the presence of drains or soil fissures that allow pollutants in drainage waters to bypass the soil and the potential for soil retention and root uptake. This may require physical interventions to disrupt these pathways such as by drain blocking, although blockage can also occur naturally over time by tree rooting or soil shrinkage and swelling (Stutter et al. 2020).

Some view leaving FGI unmanaged to be a more attractive option for securing water benefits, but this is increasingly challenged by climate change and associated risks of storm, fire and disease outbreaks. Managing these risks is driving the need for greater intervention to increase tree species and age diversity or install fire breaks to enhance forest resilience, especially in forests where historic management has left relatively even-aged or single species stands. Examples include extensive forest conversion from conifer to broadleaves in areas of Germany as spruce and pine stands are increasingly impacted by pests and disease (Schuller et al. 2011), and the installation of forest-fire prevention measures in parts of France. However, the absence of productive management in some forests makes such interventions uneconomic for forest owners necessitating economic support.

8.6   Managing Potential Disbenefits

While FGI is generally very good for protecting water quality, there is one common potential disbenefit. This relates to the ability of trees to use more water than shorter types of vegetation, resulting in less water runoff or recharge (e.g. due to interception/wet canopy evaporation and/or potentially higher transpiration rates sustained by deeper rooting) (Nisbet 2005). The subject is complex, is widely researched and still attracts debate. Much depends on a wide range of site factors, especially geographical scale, climate, altitude, geology, soil type, forest type, tree species, tree age and the counterfactual land cover. In general: conifers reduce water yield more than broadleaves; differences between individual species tend to be small (although with a few exceptions); reductions are much less for very young and old trees; and the impact on catchment water yield is relatively limited (difficult to measure) when less than 20% of a catchment is planted or cleared of forest (Creed and Noordvijk 2018).

In some locations, forests can have the opposite effect and increase water yield. Notable examples include high altitude forests that are effective in trapping cloud water, the planting of broadleaved forest on grassland overlying chalk geology and where forests replace irrigated agriculture or crops with a high water use (Creed and Noordvijk 2018; Roberts and Rosier 2005). The water use of an existing forest can be reduced by changing forest type from conifers to broadleaves, diversifying forest age and introducing more open space, although these represent a significant cost to forest owners.

The water-use issue becomes further complicated concerning the impact of forests on dry weather flows, when water supplies are most limiting. The generally greater water use by trees can be expected to reduce low flows but much depends on the nature of local soils and geology. Permeable geologies are the most vulnerable, while reductions can be expected to be small or even reversed on impermeable geologies with poorly structured soils. Here, tree planting can improve soil infiltration leading to a greater proportion of net rainfall draining to depth and supplementing low flows. Another exception concerns the presence of riparian and floodplain woodland, which can enhance the storage of flood waters and their subsequent release, helping maintain dry season flows. A further complicating factor is that the higher water use and potential water yield reduction by forests can be beneficial for reducing flood flows in flood risk catchments and tackling soil salinisation issues in dryland environments.

Models (e.g. SWAT, WaSSI-C, Hydro-JULES) are available to estimate the effect of FGI on water yield, but these vary in their ability to represent and handle key forest hydrological processes (e.g. wet canopy evaporation), as well as forest design and management factors. Great care is therefore required in selecting appropriate models and parameter values, as well as making a serious effort to test and validate model predictions. Modelling impacts on extreme flows is the most difficult.

Other potential water disbenefits tend to be more localised in extent and include the risk of increased water acidification and mobilisation of metals such as aluminium due to forest canopy scavenging of acid deposition. This issue is largely limited to acid-sensitive, upland geologies and rapidly declining due to EU success in emission control (Nisbet and Evans 2014). Another issue is excessive canopy shade and poor channel morphology resulting from conifer plantations located too close to watercourses. Care is also required to avoid the build-up of captured pollutants in tree biomass or in the soil, which could be released back to the water environment. Lastly, while tree planting and forest cover generally act to reduce flood risk, there are some localised exceptions, such as the backing-up of floodwaters upstream of floodplain woodland and the blocking of culverts and bridges by the washout of woody material (Nisbet et al. 2011).

All above potential disbenefits can be effectively managed by good design and management of FGI, especially by planting the most appropriate type and species of tree in the right place based on site factors and ecological requirements.

8.7 Identifying and Assessing Multiple Benefits

While the emphasis of this chapter is on how FGI can protect and improve water quality, it is important to recognise that this strategy will also increase natural capital and deliver multiple benefits (from the so-called ecosystem services) for other policy agendas. There is particular scope for FGI to reduce downstream flood risk as well as ameliorate rising water temperatures through shade provision, which are topics of growing concern in the context of climate change (Burgess-Gamble et al. 2017). FGI will also directly contribute to climate change mitigation through carbon sequestration and building soil carbon, thereby helping to offset agricultural GHG emissions (Morison et al. 2012). Other notable benefits include improving biodiversity and tackling the decline in woodland birds by increasing woodland habitat and linking-up fragmented woodlands within agricultural landscapes; the provision of timber and wood fuel to diversify agricultural businesses; and the potential to improve open landscapes and provide increased access for recreation (Bateman et al. 2014).

8.8 How to Design a PES Scheme

There are eight operational steps involved in designing a successful PES scheme, which are described below in the context of using FGI to improve or protect water quality. Schemes can be simple or complex and large or small, depending on the location of interest and nature of the water issue. Three case studies summarising successful schemes are referenced in accompanying text boxes (Boxes 8.2, 8.3, and 8.4), and more information on these can be found via the PESFOR-W web site (www.forestresearch.gov.uk/research/pesforw/case-studies).

1. Defining the water quality issue: This can take a number of forms. For an agricultural area, the water issue could typically be an excessive level of diffuse pollutants generated by farming activities resulting in a failure to achieve water quality standards and good water status. For an existing forest, it could be that its inherent water protection function is threatened by an environmental pressure such as an increased risk of storm damage or fire due to climate change, or spread of a pest or disease. Alternatively, the threat could be posed by agricultural intensification or urban expansion. Whatever the issue, the starting point is to clearly define its nature, including its spatial extent and temporal dimension. In the case of diffuse pollution, there is a need to determine which pollutants (e.g. nitrate, phosphate, sediment, pesticides or FIO) are involved, their sources (e.g. which fields, areas or soils) and pathways of movement (e.g. surface runoff or groundwater). The water regulatory authority will be a key partner in defining the water quality issue.

2. Identifying local actors: This involves identifying all stakeholders linked to and affected by the water issue. In theory, PES schemes can be limited to single buyers and sellers but are more likely to involve a broad range of actors, especially for more extensive water issues. Local actors can be categorised into five main groups: regulatory bodies, suppliers and sellers, beneficiaries and buyers, intermediaries and designers (Fig. 8.5). Regulatory bodies or beneficiaries and buyers are more likely to take the lead in developing a PES scheme, while work will be required to raise awareness of the water issue amongst some actors and to persuade them to fully engage.

3. Assessing the feasibility of a PES scheme: Bring together stakeholders to explore the water quality issue and consider potential solutions and opportunities, drawing on wider experience and examples of different PES schemes. Assess the existing water quality baseline and margin of improvement needed to meet a water quality target or standard, or in the case of the loss of an existing forest protection function, the degree of damage likely to be caused. Check that the issue cannot be addressed by good management practices, or regulatory

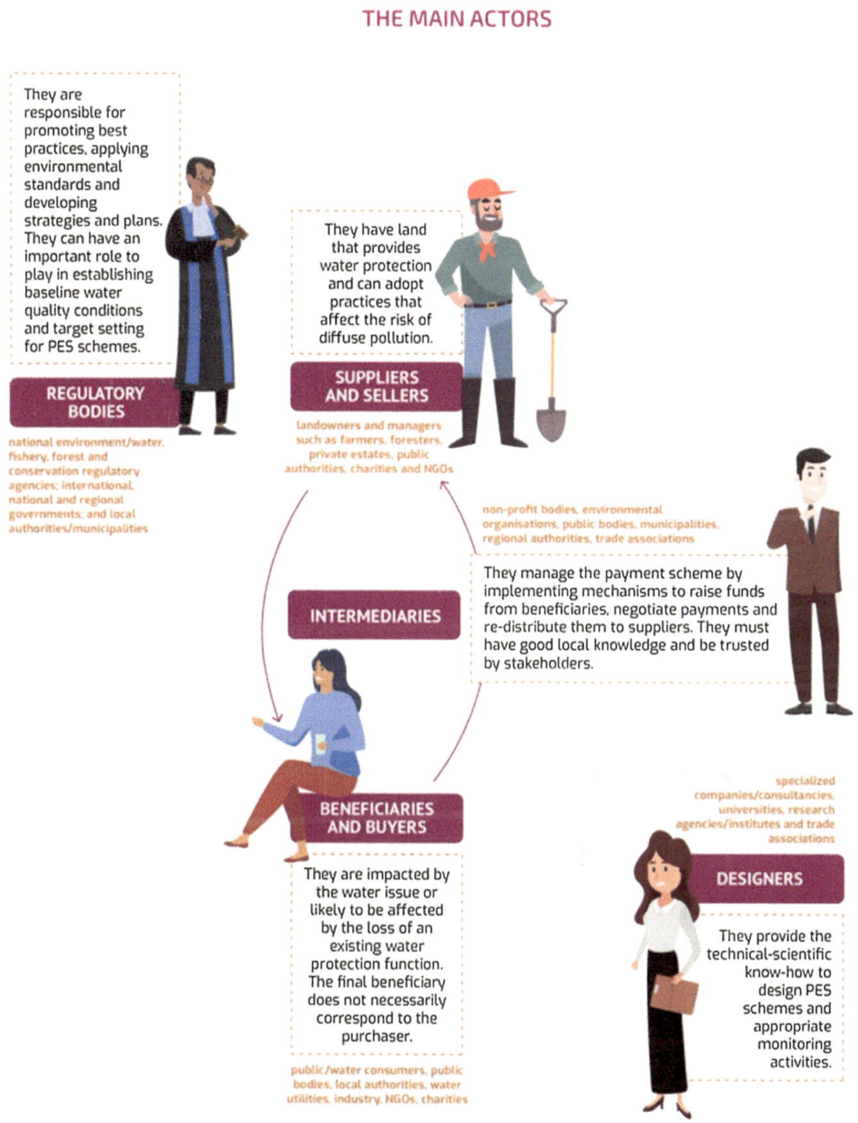

Fig. 8.5 Main actors who may be involved in designing a PES scheme

mechanisms. Examine the alternative measures that could be adopted, exploring how costs, benefits and avoided risks differ between them and according to the extent to which they are used, and identify the least cost and most acceptable option. Establish whether there are willing buyers and sellers to implement and finance the preferred measures, a desire to collaborate amongst actors, and trusted intermediaries to assist with organising and managing a PES scheme. If so, assess transaction, management and monitoring costs.

4. Exploring potential win-win solutions: Consider whether the identified option(s) will deliver additional benefits from ecosystem services (e.g. for carbon sequestration, flood risk management, recreation and biodiversity [see CICES (Haines-Young and Potschin 2017)] and if so, whether there is a market for these. Where there is a willing buyer, try to quantify the potential benefits and assess the scope to develop an integrated scheme that also considers these other ecosystem services and impacts on natural capital. To help design the PES scheme, underpin investment decisions and increase public support, a comparative cost-effectiveness analysis of the per unit environmental improvement can be undertaken by intermediaries or designers where data are available (Box 8.5).

5. Defining roles and responsibilities: Providing there is local support for developing a PES scheme, define roles and responsibilities of key actors. This should include setting clear spatial boundaries for the scheme and agreeing measures, associated costs, payments and timelines. Seek and draft agreements.

6. Resolving or minimising potential legal issues: Consider legal, fiscal and regulatory issues for key actors, such as implications for taxes, property rights and pollution control, especially for those making or receiving payments. Where necessary, legal advice should be sought to aid decision making.

7. Drawing-up technical specifications: Technical specifications should be developed and agreed for the design and management of the selected measure(s) to address the water quality issue. Regulatory bodies and designers are usually best placed to advise on the required specification. For tree planting, this will vary from site to site and include consideration of location (e.g. local climate, geology, soils and topography), extent/area or width of planting (see Table 8.2), tree type, species mix, planting density, timing and required management practices to ensure effective tree establishment and growth (e.g. ground preparation, weed control and fencing). For an existing forest, it could involve elements of forest redesign to reduce risks or improve resilience, such as changing forest type or species mix, altering age structure or introducing fire breaks. The specification should build-in some leeway (e.g. extra tree planting) to allow for uncertainty in the effectiveness of a given measure.

8. Formalising scheme contract: A formal contract should be drawn up between buyers and sellers, covering the technical specification for measures to be implemented, timelines for delivery, baseline water quality conditions, success criteria, monitoring needs, staged payments and scheduled reviews. However, it is best to incorporate a degree of flexibility in the Terms and Conditions to allow for future adjustments informed by monitoring and evaluation. Care is required to avoid a high level of bureaucracy and transaction costs, ensuring that scheme management and monitoring are fit for purpose.

Box 8.2 Case Study: Tree Planting to Secure Water Benefits

Globally, many water utilities are increasingly recognising the growing threats to water supplies and rising cost of water treatment. Consequently, attention is shifting away from grey to green infrastructure to better secure future water quality and quantity in source catchments. PES schemes are emerging as a more sustainable approach to water management whereby landowners such as farmers and forest owners are incentivised to change land use or management to better protect water supplies. Targeted measures such as tree planting within safeguard zones potentially offer a more cost-effective way of addressing diffuse water pollution issues compared to water treatment. At the European Level, Article 9 of the WFD stipulates that 'Member States shall take account of the principle of recovery of the costs of water services, including Environmental and Resource Costs'. This has led to some water utilities in the UK (South West Water and United Utilities), Germany (Saxony and Hanover) and Italy (ETRA and Romagna Acque) to charge consumers for the costs incurred in source area protection, with payments passed onto landowners and managers. These schemes are sometimes characterised as 'PES-like' as consumers are not voluntarily paying for the benefit/bill; however, they have proven to be the most effective systems at EU level for improving water quality at the catchment level (UNECE 2018).

Box 8.3 Case Study: Tree Planting to Protect Groundwater Quality

There are three notable woodlands for water PES schemes in Denmark. Two of these are located near Odense on Funen and the third near Aalborg on Jutland. All are designed to tackle the growing issue of groundwater pollution by agricultural practices, especially the contamination of drinking water by rising nitrate and/or pesticide levels. The scheme at Aalborg is one of the oldest and established in 1991 with funding from EU LIFE and the Aalborg Municipality to purchase land from farmers within vulnerable groundwater recharge zones. 900 ha of intensive farmland were converted into 500 ha of broadleaved woodland and 400 ha of low-input pasture, primarily to reduce nitrate levels. The drinking water benefit was estimated at a minimum of €489/ha/year and the net social benefit (excluding drinking water) at €189/ha/year, which included the provision of local recreation and carbon gain. The two schemes near Odense are Elmelund Skov and Brylle Water, both of which involve woodland creation to reduce pesticide pollution of local groundwater supplies. This is achieved by a voluntary process of land consolidation whereby agricultural land is purchased in low vulnerable areas and used to encourage land swaps with farmers for land within vulnerable groundwater recharge zones. The land is transferred to public or private partners at a reduced price for woodland planting and management, with a permanent change from farmland to forest legally guaranteed. At Elmelund Skov, 380 ha of farmland have been converted to woodland since 2001 under a partnership agreement between the local water utility, the Odense Municipality and the state forestry agency. The Brylle Water scheme is the most recent and commenced in 2014. 156 ha of farmland were purchased and planted with woodland by a private foundation, who met 40% of the cost, with the other 60% funded by the local water utility. The land consolidation process involved a significant transaction cost in negotiating agreements with farmers and building trust. Public access for recreation was a strong component of the schemes and underpinned municipality funding and support.

Box 8.4 Case Study: Tree Planting to Improve Groundwater Infiltration
This PES scheme is located on the edge of the town of Carmignano di
Brenta, near Padua in northern Italy. It was established in 2012 as a 'Forested
Infiltration Area' (FIA) to help replenish and improve groundwater resources
in the area. Overexploitation of the aquifer had led to the disappearance
of local springs and streams, while agricultural activities had degraded
groundwater quality. A 2.5 ha broadleaved woodland was planted on arable
(maize) land and a system of trenches dug to channel surface water (at a
rate of ~1 million cubic metres per hectare per year) onto the site during
periods of excess flow in the nearby River Brenta. The establishment of the
woodland helped to facilitate water infiltration into the aquifer and enhanced
phyto-purification, removing nutrients and other contaminants. The woodland
also provided a carbon gain to the landowner and woodland products such as
firewood, biomass and timber in the longer term, as well as benefiting the
local community as a valuable habitat and opportunities for recreation and
education. A group of local and regional stakeholders, including municipalities and local companies, formed a partnership to bid for supporting funds that
were used to design and set up the PES scheme on private farmland. Around
80% of implementation costs were financed by LIFE+ and RDP funds. The
loss of income to the landowner from the change in use from maize cropping
to woodland was exceeded by payments from the Brenta Land Reclamation
Board for the infiltration water service (€1200/ha/year), the municipality for
community access and related recreation and education events (€1500/year),
plus the value of generated wood products and carbon gain.

Box 8.5 Cost-Effectiveness Analysis (CEA)

Cost-effectiveness analysis (CEA) is a technique that can be used to compare the cost of FGI with that of alternative measures for improving water quality. It can be important both in making a case to adopt woodland planting and in selecting which woodland (and/or other) measures to implement. The main challenge is to estimate how much planting and/or the extent of an alternative measure is required to achieve a given water quality target. It is easiest to focus on a specific diffuse pollutant or set of pollutants such as reducing the concentration of nitrate in a watercourse by X% or to below an environmental standard, rather than adopting a broader water quality index like 'water status' or using a biological metric. The assessment is likely to require the use of a spatially distributed model such as SWAT to estimate the effect of different levels and locations of planting on water quality, as described in Sect. 8.5. An alternative, simpler approach would be to limit consideration to the effect of alternative measures on pollutant loads, with values for the latter relatively well known for different land covers and crop types (Table 8.1). In cases where measures are proposed in order to reduce the risk of infringing water quality targets that are currently met, the improvement would need to be considered in terms of the reduced risk of exceeding this target.

Cost-effectiveness analysis (CEA) involves dividing the cost of the woodland planting or other measure by the improvement/gain made in water quality. Depending on whether a reduction in concentration or load is the focus, the cost-effectiveness ratio (RCE) computed may be expressed in units such as €/mg/l or €/kg/ha of a given pollutant. The aggregate cost should be computed by discounting costs in future years over the lifespan of the FGI. The costs need to include the revenue forgone from the change in land use and any transaction costs. Where a private sector perspective is adopted, the costs should also include any changes in financial incentives involved. However, where CEA is undertaken from a societal perspective, changes in subsidy payments (e.g. associated with EU agri-environmental schemes) should be excluded because these are considered transfer payments. Although the focus here is on water quality, the value of other benefits generated by the woodland can also be included in the calculation and used to offset the cost, so reducing the RCE. This can be a challenging task for some benefits such as habitat gain but easier for others like carbon sequestration.

Key steps involved in CEA are:

- Identify the water quality issue and level of required improvement, such as in the concentration or load of a specific diffuse pollutant to meet or contribute to achieving a given water quality standard.
- Estimate the amount of FGI and other alternative measures needed to deliver the desired reduction in pollutant concentration or load using mod-

(continued)

elling (e.g. SWAT) or pollutant budgeting. Calculate the aggregate cost of implementing the measures, including revenue forgone, transaction costs and (for a private sector perspective) incentives, discounting management and other costs expected to arise in the future over the lifespan of the measures.

- In cases where the expected changes in measure effectiveness over time differ between measures, one of two approaches should be adopted: either divide the aggregate costs by an environmental benefit index that weights future environmental improvements according to when they are expected to arise (e.g. by applying a discount rate to improvements in future years) or compare measures using a benchmark cost-effectiveness cost-comparator that takes account of when improvements arise (an approach sometimes used in comparing options for climate change mitigation—e.g. DBEIS 2019).
- Consider other benefits and any disbenefits provided by the measures or by the original land use and where feasible, estimate the value or costs of these (similarly by applying a discount rate to those arising in future years), to compute the net cost of each measure.
- Taking into account the above net costs and environmental effectiveness of potential measures, including an allowance for risk and uncertainty, calculate the average and range of the RCE for each measure and compare performance. Marginal Abatement Cost Curves can be used to explore the relationship between the cost-effectiveness of different measures and the total amount of diffuse pollution abated at specific points in the future.
- Use results to design an appropriate payment scheme to deliver the least-cost measure or mix of measures to meet and secure water quality target(s) within a given timeframe, taking account of local suitability and acceptability of measure(s).
- Where attracting a range of investors for FGI is important, consider computing cost-effectiveness from other perspectives too (e.g. climate change mitigation and/or flood risk attenuation).
- It is important to keep all relevant documentation detailing calculations to inform future review and learning.

8.9 Monitoring, Evaluation and Review

Monitoring can take many forms that vary greatly in cost. Firstly, implemented measures will require a certain level of monitoring to ensure that they are designed as planned and managed appropriately to deliver and maintain their effectiveness in pollutant reduction. For FGI, this includes checking that the trees become fully established (e.g. replacing any losses and managing grazing and potential weed and pest issues) and that the established woodland is sustainably managed and, where appropriate, the woodland replanted to secure long-term effectiveness. In the case of more targeted planting such as in the form of riparian woodland buffer areas, there is a need to check that these are designed and managed appropriately to deal with the diffuse pollution issue (and to deliver any co-benefits) over the expected lifespan of the buffer. This includes checking that tree growth rates are sufficient and sustained to deal with nutrient runoff from adjacent land, the buffer does not get overloaded/saturated by the pollutant(s) and any management interventions do not damage the site.

Secondly, in many cases, there will be a need to monitor the water quality response to check that the FGI is having the desired effect. It may be possible to rely on existing monitoring networks (e.g. for WFD assessments) operated by water regulators for this purpose, but these will usually be undertaken at the water body scale and so may need to be supplemented by local measurements. Monitoring needs to be tailored to the nature of the implemented measure, how it will affect diffuse pollutant sources and pathways and the specific pollutant(s) involved. For example, small-scale woodland planting is less likely to justify river water quality monitoring and better suited to conducting plot-based measurements such as changes to soil conditions. There should be agreement on the location, type, frequency and cost of monitoring; data handling, storage and ownership; and on data analysis, reporting and publication of results.

Tackling water body scale diffuse pollution issues is likely to require extended areas of targeted FGI involving multiple landowners and taking multiple years to deliver. This will necessitate strategic planning and integrated catchment management, with regular review of progress and actions. Progress made in achieving water quality targets can be directly linked to contracted payments, although the high temporal variability of water quality parameters (e.g. due to variability in weather conditions and river levels) can make this very challenging, especially over short-medium timescales. Consideration should be given to undertaking a periodic evaluation of the cost-effectiveness of the scheme to inform the need for any changes, share lessons and ultimately ensure long-term success.

8.10   Spreading the Word

Successful communication, dissemination and marketing of PES actions are dependent on messages being appropriately targeted to the relevant actors. Levels of awareness and understanding of the benefits of FGI for water will vary greatly between actors, and therefore different approaches and stories may be required. Be aware of different learning styles and tailor information accordingly. Academic approaches that apply in the classroom may not work out in the field. Be sensitive to local conditions and relationships.

Sources and pathways of diffuse pollution are likely to be disputed, especially amongst landowners and managers, necessitating an open discussion of the evidence and drawing on expert opinion. A number of alternative options for tackling a given water issue will often be available, and those involving land-use change and particularly woodland creation will face inherent resistance by some. Focus on careful targeting of FGI to enhance effectiveness and minimise impact on existing land use.

Bringing different groups together, discussing contrasting viewpoints and resolving differences are critical. Consideration should be given to using a facilitator such as a trusted intermediary, especially where there are deep-seated differences between actors. The use of maps and model outputs can be very helpful but must be handled sensitively and not used to apportion blame. Many will not trust these products and question their ability to reflect reality. Consequently, it is very important to ground truth such information, which is best done by key actors meeting and discussing the issues and proposed countermeasures in the field.

Use should be made of regional and wider case studies to develop options and build confidence and consensus, including by visiting existing demonstration sites where possible (check relevant case studies at www.forestresearch.gov.uk/research/pesforw/case-studies). These studies can be a great way of showcasing the water and wider benefits of FGI, as well as the protection function of existing forests and threats faced.

Acknowledgements This chapter is based upon work from COST Action CA15206, supported by COST (European Cooperation in Science and Technology). COST is a funding agency for research and innovation networks. Our Actions help connect research initiatives across Europe and enable scientists to grow their ideas by sharing them with their peers. This boosts their research, career and innovation. See www.cost.eu

EUROPEAN COOPERATION
IN SCIENCE & TECHNOLOGY

COST is supported
by the Horizon 2020
Framework Programme
of the European Union

References

Baksic N (2018) Reviewing available pollutant models and decision support tools for informing the design and management of woodland creation measures for reducing agricultural diffuse pollution. STSM Scientific Report. COST Action 15206. Forest Research, Surrey. 12 pp

Bateman I, Day B, Agarwala M, Bacon P, Bad'ura T, Binner A, De-Gol A, Ditchburn B, Dugdale S, Emmett B, Ferrini S, Carlo Fezzi C, Harwood A, Hillier J, Hiscock K, Hulme M, Jackson B, Lovett A, Mackie E, Matthews R, Sen A, Siriwardena G, Smith P, Snowdon P, Sünnenberg G, Vetter S, Vinjili S (2014) UK National Ecosystem Assessment Follow-on. Work package report 3: economic value of ecosystem services. UNEP-WCMC, LWEC, UK, 246 pp

Bealey WJ, Dore AJ, Dragosits U, Reis S, Reay DS, Sutton MA (2016) The potential for tree planting strategies to reduce local and regional ecosystem impacts of agricultural ammonia emissions. J Environ Manag 165:106–116

Broadmeadow S, Nisbet T (2012) National map of woodland creation opportunities: targeting eWGS to help meet the objectives of the WFD and reduce flood risk in England. Final Report to Forestry Commission England. Forest Research, Alice Holt Lodge, Farnham, Surrey. 29 pp

BSFP (2013) British survey of fertiliser practice dataset. From Department for Environment, Food & Rural Affairs, London

Burgess-Gamble L, Ngai R, Wilkinson M, Nisbet T, Pontee N, Harvey R, Kipling K, Addy S, Rose S, Maslen S, Jay H, Nicholson A, Page T, Jonczyk J, Quinn P (2017) Working with natural processes—evidence directory. Environment Agency, Bristol. 298 pp

Collins AL, Newell Price JP, Zhang Y, Godday R, Naden PS, Skirvin D (2018) Assessing the potential impacts of a revised set of on-farm nutrient and sediment 'basic' control measures for reducing agricultural diffuse pollution across England. Sci Total Environ 621:1499–1511

Creed IF, Noordvijk Mv (eds) (2018) Forest and water on a changing planet: vulnerability, adaptation and governance opportunities. A global assessment report, IUFRO World Series, vol 38. IUFRO, Vienna. 192 pp

DBEIS (2019) Valuation of energy use and greenhouse gas. Supplementary guidance to the HM Treasury Green Book on Appraisal and Evaluation in Central Government. Department of Business, Energy and Industrial Strategy, London. https://assets.publishing.service.gov.uk/government/uploads/system/uploads/attachment_data/file/794737/valuation-of-energy-use-and-greenhouse-gas-emissions-for-appraisal-2018.pdf

EEA (2016) Annual indicator report series (AIRES): surface waters. EEA Environmental Indicator Report No 30/2016. EEA, Copenhagen

EEA (2018) European waters—assessment of status and pressures 2018. European Environment Agency EEA Report 7/2018. EEA, Copenhagen. https://www.eea.europa.eu/publications/state-of-water

Forest Europe (2019) Valuation and payments for forest ecosystem services in the pan-European region. Final report of the forest Europe Expert Group on valuation and payments for forest ecosystem services. Forest Europe, Bratislava. https://foresteurope.org/publications/

Forestry Commission (2019) Managing forest operations to protect the water environment. Forestry Commission Practice Guide, Forestry Commission, Edinburgh, 48 pp

Gatto P, Pettenella D, Secco L (2009) Payments for forest environmental services: organisational models and related experiences in Italy. iForest-Biogeosci Forest 2(4):133–139

Haines-Young R, Potschin MB (2017) Common International Classification of Ecosystem Services (CICES) V5.1 and guidance on the application of the revised structure. www.cices.eu

Kareiva P, Tallis H, Ricketts TH, Daily GC, Polasky S (2011) Natural capital: theory and practice of mapping ecosystem services. Oxford University Press, Oxford

Mockler AEM, Bruen M (2018) Support tools for characterisation and evaluation of programmes of measures. Report no. 249. Environment Protection Agency, Dublin

Morison J, Matthews R, Miller G, Perks M, Randle T, Vanguelova E, White M, Yamulki S (2012) Understanding the carbon and greenhouse gas balance of forests in Great Britain. Forestry Commission research report. Forestry Commission, Edinburgh. 149 pp

Nisbet TR (2005) Water use by trees. Forestry Commission Information Note 65. Forestry Commission, Edinburgh

Nisbet TR, Evans CD (2014) Forestry and surface water acidification. Forestry Commission research note 16. Forestry Commission, Edinburgh

Nisbet T, Silgram M, Morrow K, Broadmeadow S (2011) Woodland for water: Woodland measures for meeting Water Framework Directive objectives. Forest Research Monograph 4. Forest Research, Surrey. 156 pp

Perez-Silos I (2017) Assessing the effectiveness of woodland creation for reducing agricultural diffuse pollution—developing value ranges to create look-up tables. STSM scientific report. COST action 15206. Forest Research, Surrey. 11 pp

Roberts J, Rosier P (2005) The impact of broadleaved woodland on water resources in lowland UK: III. The results from Black Wood and Bridgets Farm compared with those from other woodland and grassland sites. Hydrol Earth Syst Sci 9:614–620

Schuller G, Pfister L, Vohland M, Seeling S, Hill J (2011) Large scale approaches to forest and water interactions. In: Bredemeier M, Cohen S, Godbold D, Lode E, Pichler V, Schleppi P (eds) Forest management and the water cycle: an ecosystem-based approach, Ecological Studies, vol 212. Springer, New York, pp 435–452

Stutter MI, Chardon WJ, Kronvang B (2012) Riparian buffers as a multifunctional management tool in agricultural landscapes: introduction. J Environ Qual 41:297–303

Stutter M, Wilkinson M, Nisbet TR (2020) Improving the benefits from watercourse field margins using 3-D buffers. Environment Agency report. Environment Agency, Bristol

Ucar T, Hall FR (2001) Windbreaks as a pesticide drift mitigation strategy: a review. Pest Manag Sci 57:663–675

UNECE (2018) Forests and water—valuation and payments for forest ecosystem services. UNECE, Geneva. 108 pp

Part III
River and Floodplain Ecosystem (Including Paddy Field and Other Farmlands)

Chapter 9
Wetland Paddy Fields as Green Infrastructure Against Flood

Yasunori Muto and Ryo Yokokawa

Abstract Inundation simulation was carried out for an inland depression mainly covered by paddy fields to estimate its flood retention volume. In addition to surface water flooding, river flooding combined with it was also explored in a wide range of precipitation including a 1000-year return period. The results showed that the paddy fields in the studied area can retain approximately 60% of the total inundation volume. Judging from the relationship between the flood retention volume in the paddy fields and the number of inundated buildings above floor level, we could estimate two reference values regarding the potential volume of floodwater retention in paddy fields: one at which damage to the buildings starts to appear and the other where damage seriously expands in the main residential area. By using these reference values, we demonstrated the importance of land-use strategies in flood alleviation. In other words, by transferring buildings inundated between these levels to safer places, the potential volume of floodwater retention in paddy fields can be increased to a higher level, which well agrees with the concept of green infrastructure. The idea of flood alleviation deduced from this study can be applied to similar near-shore depressions cut off by sand dunes, which are common in Japan.

Keywords Wetland paddy fields · Flood retention · Surface water flooding · River flooding · Combined flooding

Y. Muto (✉)
Research Center for Management of Disaster and Environment, Tokushima University, Tokushima, Japan

Department of Civil and Environmental Engineering, Graduate School of Technology, Industrial and Social Sciences, Tokushima University, Tokushima, Japan
e-mail: muto_yas@ce.tokushima-u.ac.jp

R. Yokokawa
Water and Land Sector, CTI Engineering Co., Ltd., Osaka, Japan
e-mail: ry-yokokawa@ctie.co.jp

© The Author(s) 2022
F. Nakamura (ed.), *Green Infrastructure and Climate Change Adaptation*,
Ecological Research Monographs, https://doi.org/10.1007/978-981-16-6791-6_9

9.1 Introduction

Paddy fields are well known to have many functions as green infrastructure, including flood risk reduction, water resources conservation, food production for both humans and animals, landscape improvement, community maintenance through agricultural activities, etc. (Ministry of the Environment, Japan 2016). As for flood risk reduction, although this has been pointed out qualitatively in early days (e.g. Masumoto 1998), there are few studies that quantitatively evaluate the flood control capacity of paddy fields. Recently in Japan, the so-called *tambo*-dam, using an orifice plate in a drain to improve the flood control volume of paddy fields, has been widely explored and has been adopted in some places (e.g. Yoshikawa et al. 2010). However, in contrast to such a micro device, exploring the macro function of paddy fields to floods, in connection with land-use planning, is still not popular.

From another aspect, owing to recent advances in numerical simulation techniques, inundation simulation models have been widely applied to urban flooding and have been proven to reproduce well flooding processes of surface water flooding (e.g. Thang et al. 2004). Such models can also be applied to wetlands to evaluate their flood control capacity through inundated volume and its time variation (e.g. Miyazu et al. 2012). However, most of the simulations considered surface water flooding and river flooding separately. Since a drainage network system consists of channels of various sizes, from small streams to large rivers, irrespective of either urban or rural areas, their design flood levels are different according to the channel size. In other words, particularly in a low and depressed area, flooding is likely to start from a small stream due to surface water flooding, and then inundation by river flooding from a larger river is combined.

Based on these exposed points, we conducted inundation simulations for an inland depression. The studied area is a small plain near the sea, cut off by sand dunes and presumably used to be a lagoon. Lower areas are mainly used as paddy fields and residential areas are located in higher places. Such topography and land use are typically seen along the shore in Japan (Kusaka 2020); thus, it is valuable to highlight the inundation characteristics of the area. The outputs will be anticipated to show a standard for flood alleviation measures in such areas.

9.2 Studied Area

The studied area is located in the southern part of Tokushima Prefecture, Japan (see Fig. 9.1). It is surrounded by the Osato beach (East), Kaifu River (South) and a mountainous area (North and West), and forms a depression of 2.86 km^2 in an inland area. The elevation of the studied area is shown in Fig. 9.2 in terms of the altitude distribution drawn from the 5-m mesh DEM data. As shown in the figure, the eastern part along the beach is relatively higher because of geological sand dunes; thus, the inland part, even away from the beach, is lower than the dunes. The lowest parts are

Fig. 9.1 The studied area: the Zenzo, Osato and Nishinosawa rivers basin, Tokushima, Japan

located along the three small streams, the Zenzo, Osato and Nishinosawa rivers. The land-use classification for the same mesh size is shown in Fig. 9.3. Nearly 35% of the area is utilised as paddy fields, mainly in the lower western and southern parts. The three streams together with irrigation channels that spread over the wetland are eventually connected to the Kaifu River, and build a network drainage system for the area. In this study, the downstream boundary of the simulation area was set at the confluence point of the Kaifu River. Although floodgates are set at the confluence, they are open in normal time; thus, the water level in the streams is affected by the tide level through the Kaifu River and the confluence, particularly in their downstream reaches. During floods in the Kaifu River, floodgates are closed and a pump of 5 m^3/s is operated for drainage when the water level in the Zenzo River exceeds a certain value.

9.3 Inundation Simulation

9.3.1 Simulation Model

A commercial inundation simulation software, AFREL (© NITA Consultant Co., Ltd. n.d.), was employed in this study. As shown in Fig. 9.4, AFREL can treat not only the surface flow but channel and pipe networks for sewerage overlapped on the surface. Moreover, flow exchanges among these three forms (surface, channel and sewerage pipe flows) can also be simulated; therefore, both inundation into the drainage and flooding from a river into the ground surface can be considered. In the

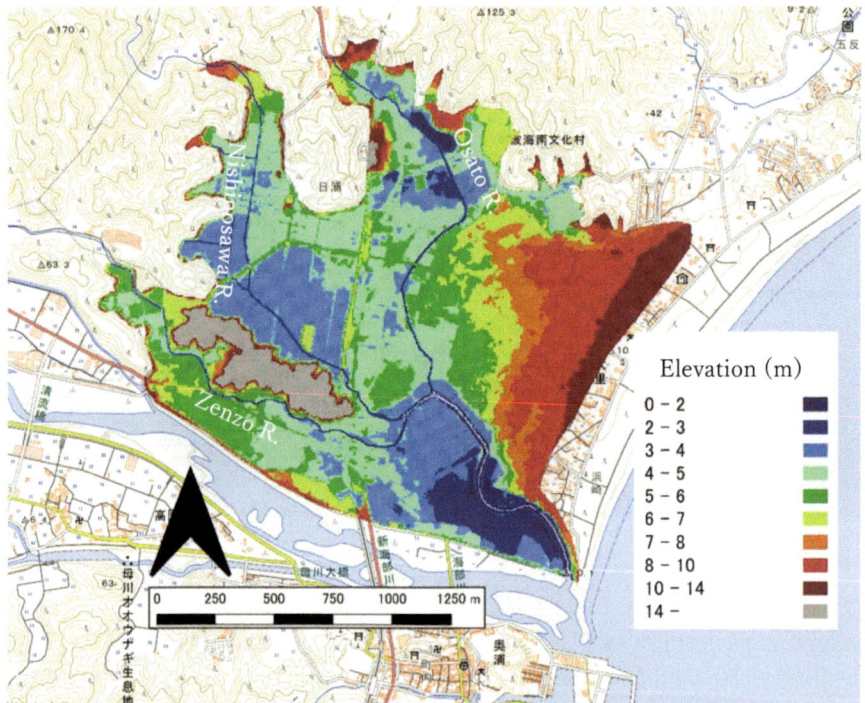

Fig. 9.2 The elevation of the studied area

studied area, a sewerage pipe network is not installed; thus, it was omitted in the simulation.

The governing equations for the 2-D surface flow based on Iwasa and Inoue (1982) are as follows:

$$\frac{\partial h}{\partial t} + \frac{\partial M}{\partial x} + \frac{\partial N}{\partial y} = r(t) + q_{CH} + q_{SW} \tag{9.1}$$

$$\frac{\partial M}{\partial t} + \frac{\partial}{\partial x} uM + \frac{\partial}{\partial y} vM = -gh\frac{\partial H}{\partial x} - \frac{\tau_{bx}}{\rho} \tag{9.2}$$

$$\frac{\partial N}{\partial t} + \frac{\partial}{\partial x} uN + \frac{\partial}{\partial y} vN = -gh\frac{\partial H}{\partial y} - \frac{\tau_{by}}{\rho} \tag{9.3}$$

$$\frac{\tau_{bx}}{\rho} = \frac{gn^2 M\sqrt{u^2 + v^2}}{h^{4/3}}, \quad \frac{\tau_{by}}{\rho} = \frac{gn^2 N\sqrt{u^2 + v^2}}{h^{4/3}} \tag{9.4}$$

The governing equations for the channel flow are basically the same as those for the surface flow but omitting the advection terms in the momentum equations, as

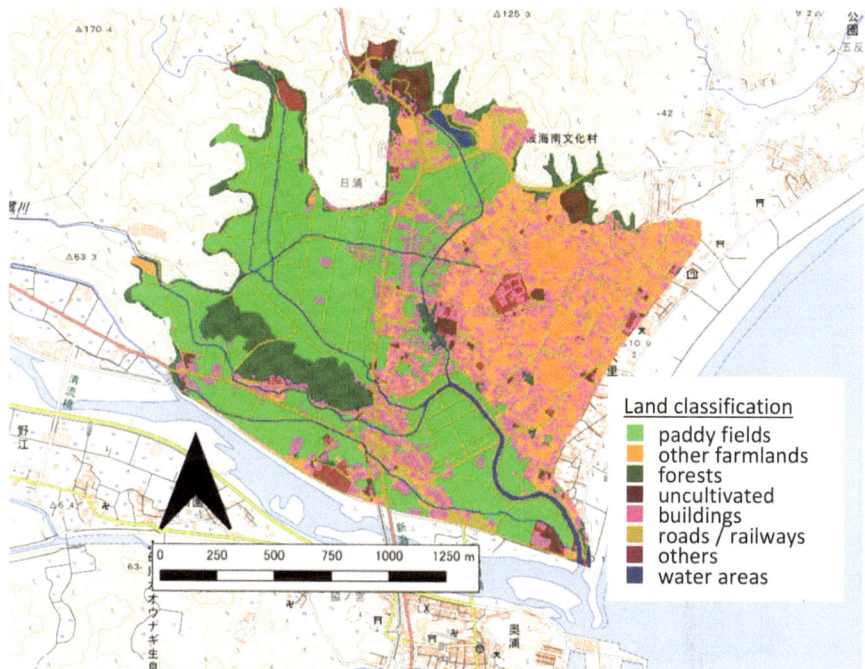

Fig. 9.3 The land-use classification of the studied area

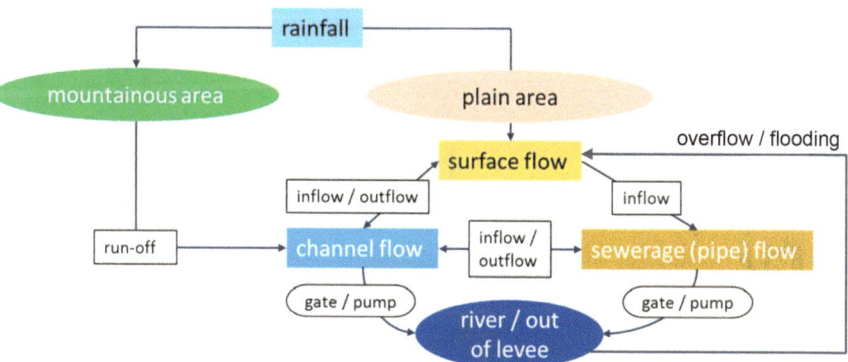

Fig. 9.4 Structure of the inundation simulation model, AFREL (©NITA Consultant Co., Ltd. n.d.)

follows:

$$\frac{\partial h}{\partial t} + \frac{\partial q_x}{\partial x} + \frac{\partial q_y}{\partial y} = q_{GR} + q_{SW} \tag{9.5}$$

$$\frac{\partial q_x}{\partial t} = -gh\frac{\partial H}{\partial x} - \frac{\tau_x}{\rho} \tag{9.6}$$

Table 9.1 Roughness coefficients and infiltration coefficients adopted in this study

Land-use type	Roughness coefficient	Infiltration coefficient
Paddy fields	0.060	0.70
Other farmlands	0.050	0.60
Forests	0.060	0.60
Uncultivated	0.050	0.20
Buildings	0.050	0.80
Roads/railways	0.047	0.85
Others	0.050	0.80
Water areas	0.030	1.00

$$\frac{\partial q_y}{\partial t} = -gh\frac{\partial H}{\partial y} - \frac{\tau_{xy}}{\rho} \tag{9.7}$$

$$\frac{\tau_x}{\rho} = \frac{gn^2 R_x^{1/3} |q_x| q_x}{h^2}, \quad \frac{\tau_{xy}}{\rho} = \frac{gn^2 R_y^{1/3} |q_y| q_y}{h^2} \tag{9.8}$$

where q_{CH}, q_{SW} and q_{GR} are the interacting discharges with the channel, sewerage pipe and ground surface, respectively. q_x and q_y are defined at the same grid point as those for M, N and h.

These equations are numerically solved explicitly on the staggered grid in space, and by the leapfrog method in time. Runoff discharge from the surrounding mountainous areas of 2.78 km^2 is estimated by the rational method and given at the upstream end of a channel connected to the relevant mountain stream. Table 9.1 shows the roughness coefficients and infiltration coefficients adopted here in accordance with land-use classification.

9.3.2 Precipitation and River Discharge Setting

The model hyetograph was generated based on the records at the nearest observation point to the studied area during Typhoon Nakri 1412, August 2014, as shown in Fig. 9.5. At this event, the total rainfall was approximately 640 mm within 72 h; however, for the studied area, 48 h is more suitable as a design rainfall duration, because the main part of consecutive rainfall in this event was observed from 6:00 on August 2 to 5:00 on August 4. Consequently, the annual maximum rainfall in 48 h of 590.5 mm recorded during that time was selected. In addition, probability analysis was conducted using 45 annual maximum 48-h rainfall data recorded at the same observation point. The results of the probability analysis (LogP3) are shown in Table 9.2. As an input for the simulation, the shape of the hyetograph was identical

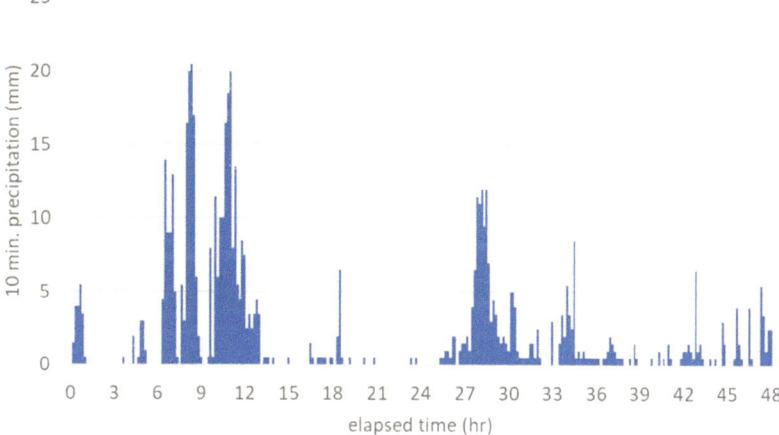

Fig. 9.5 The original hyetograph (recorded in Typhoon 1412, August 2014)

Table 9.2 Probability analysis of 48 h precipitation (LogP3)

Return period (year)	Total precipitation (mm)	Magnification ratio
2	332.0	0.56
5	443.8	0.75
10	508.5	0.86
20	564.5	0.96
25	581.1	0.98
27 (original)	590.5	1.00
30	594.4	1.01
50	629.7	1.07
80	660.3	1.12
100	674.3	1.14
150	698.8	1.18
200	715.5	1.21
400	754.0	1.28
500	765.9	1.30
1000	801.3	1.36

(see Fig. 9.5), but the unit precipitation (for 10 min) was uniformly enlarged in accordance with the calculated ratio shown in the table.

Before estimating the flood discharge from the Kaifu River into the studied area, the model hydrograph of river discharge was calculated as follows: first, the peak discharge at a certain probability in the river was estimated using the rational method, with the corresponding probability precipitation as an input. Next, using the stage-discharge curve and the recorded stage hydrograph during Typhoon Nakri, the discharge hydrograph was worked out at the upstream boundary of the 2-D river flow simulation model, 12.4 km upstream from the river mouth. Then, a 2-D

Table 9.3 River discharge conditions at the upstream boundary

Return period (year)	River discharge (m^3/s)	Magnification ratio
27 (original)	2230	1.00
50	2490	1.12
80	2600	1.17
100	2680	1.20
1000	4480	2.01

unsteady flow simulation was performed to obtain a stage hydrograph at an arbitrary levee breaching point. The peak discharge at the upstream boundary was changed in accordance with the required probability precipitation; however, the shape of the hydrograph was fixed, just enlarging the unit discharge. Table 9.3 summarises the peak discharges at the upstream boundary. Since the studied area is located close to the sea, the water stages in both the Kaifu and Zenzo rivers are affected by tidal levels. Thus, the water stage recorded at the river mouth of the Kaifu River during Typhoon Nakri was given as the downstream boundary condition.

Levee breaching points were set at 0.4 km and 1.8 km from the river mouth, as those were indicated as weak points against floods in the simulation in terms of relation between the levee height and the peak stage during floods. The levee breaching level was set at the record maximum water level (RMWL) as the primary condition. We assumed that river water starts to flood into the adjacent plains when the water stage in the river reaches the given breach level. In addition, two more breaching levels, the design high water level (HWL) and HWL + 0.6 m, were also considered as cases of levee reinforcement. The relation between water stage hydrograph and the three levee breaching levels considered here is schematised in Fig. 9.6. Since RMWL was lower than HWL, breaching at RMWL provides more flooding volume into the studied area than at HWL. The river flooding discharge as an input into the studied area in the inundation simulation was calculated using a modified Homma's formula for overflow (JSCE 1999).

Two inundation scenarios were explored: one considering only surface water flooding and the other river flooding in addition to surface water flooding. In the latter, the probability of precipitation for river discharge estimated by the aforementioned method and that for surface water flooding was set identically.

9.4 Results and Discussions

9.4.1 Model Verification

In order to check whether the model could adequately reproduce an actual inundation event, model verification was performed with the inundation records in Typhoon Nakri. The given precipitation hyetograph is shown in Fig. 9.5. Table 9.4

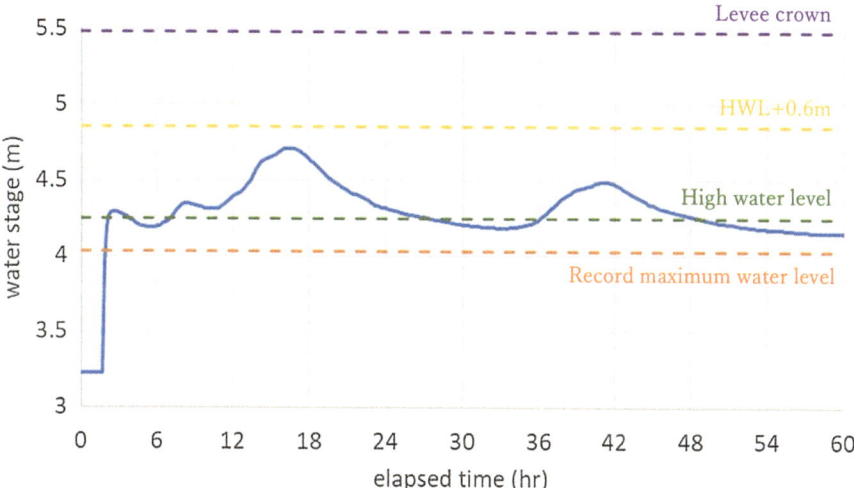

Fig. 9.6 A schematic image of water stage hydrograph and levee breaching levels

Point no.	Maximum inundation depth (m)		Errors (m)
	Observation	Simulation	
1	0.2	0.21	+0.01
2	0.7	0.69	−0.01
3	0.3	0.25	−0.05
4	0.3	0.21	−0.09
5	0.2	0.20	±0
6	0.8	0.72	−0.08
7	0.8	0.88	+0.08
8	0.4	0.32	−0.08
9	0.3	0.37	+0.07
10	0.4	0.36	−0.04
11	0.4	0.42	+0.02
12	0.6	0.62	+0.02
13	0.2	0.22	+0.02

Table 9.4 Model verification: error analyses for the point depths

shows a comparison between the water depths read from the flood marks at 13 points in the field (see Fig. 9.7) and the simulation results. The table shows that the errors were less than 10 cm at all points, and their average was 1 cm, which indicates that the model shows quite high reproducibility.

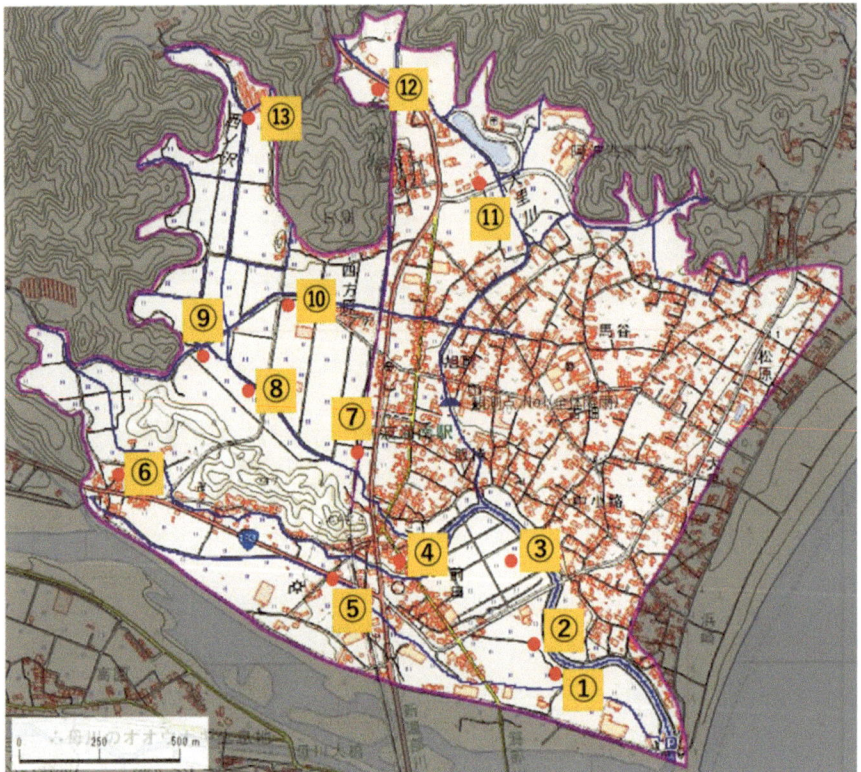

Fig. 9.7 The locations of inundation depth verification

9.4.2 Surface Water Flooding

Figure 9.8 shows the relationship between the return period of total precipitation and the peak inundation volume for surface water flooding. The dotted line shows their regression. The relation considering the logarithm of the return period shows almost linear.

Figure 9.9 shows inundation depth distributions at the peak volume for various return periods. As shown in the figure, the depth gradually increases as the return period increases; however, the inundated area does not considerably spread. Referring to the land-use classification in Fig. 9.3, the inundated area consists mostly of paddy fields. In particular, the southern part along the downstream of Zenzo River and the western part around the Nishinosawa River start to be inundated even in a high frequent flood like a 2-year return period. Most of the flooding water remains within paddy fields in these parts even in low frequent floods.

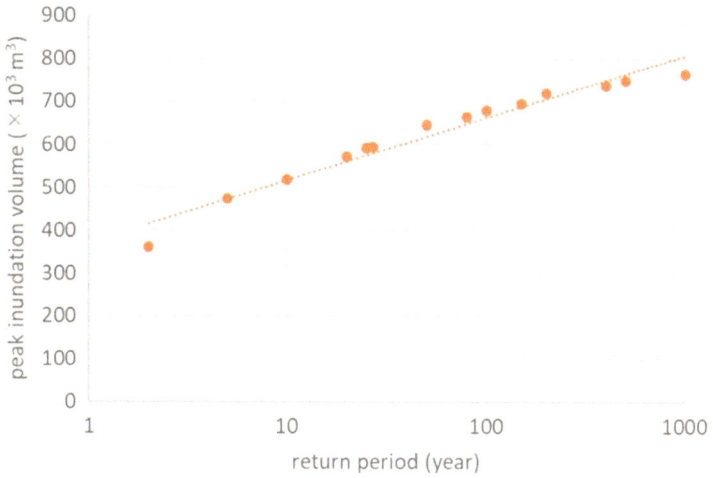

Fig. 9.8 The relationship between the return period of total precipitation and the peak inundation volume for surface water flooding

Figure 9.10 shows the relation between the return period of total precipitation and the area of buildings inundated above floor level. Here inundation above floor level is defined as the water depth on a building mesh over 45 cm. Similar to the relation between the return period and the inundation volume shown in Fig. 9.8, a linear function can be drawn for the inundated buildings. Nevertheless, its ratio to the total building area, 48.2 ha, is utmost 3.1% among the studied cases. This can also be recognised in Fig. 9.11, where locations of the inundated buildings are shown. The figure shows that most of the inundated buildings suffer from 2-year return period flooding, and the number of buildings dose not considerably increase as the return period increases. In other words, only a quite limited amount of buildings is damaged by surface water flooding in the studied area.

Figures 9.12 and 9.13 show the distributions and ratio of damaged paddy fields in response to the precipitation return period, respectively. Here, the threshold of damaged/non-damaged is defined as the water depth 30 cm. Although the number of damaged paddy fields becomes larger as the return period increases, from 25% to 50%, they are located within a limited area, mainly around the Nishinosawa River and the upstream of Osato River. Referring to the land elevation in Fig. 9.2, those areas are identical to the lower area. In such areas, lots of endangered hygrophytes can be found (Muto et al. 2020). It should be stressed that 25% of the paddy fields are damaged even in a high frequent flood like 2-year return period.

Figure 9.14 shows the estimated inundation volumes stored only in the paddy fields, plotted as a function of the total inundation volume at its peak. The graph shows that the paddy fields retain 64% of the inundation volume, and this value is nearly constant over the tested return period.

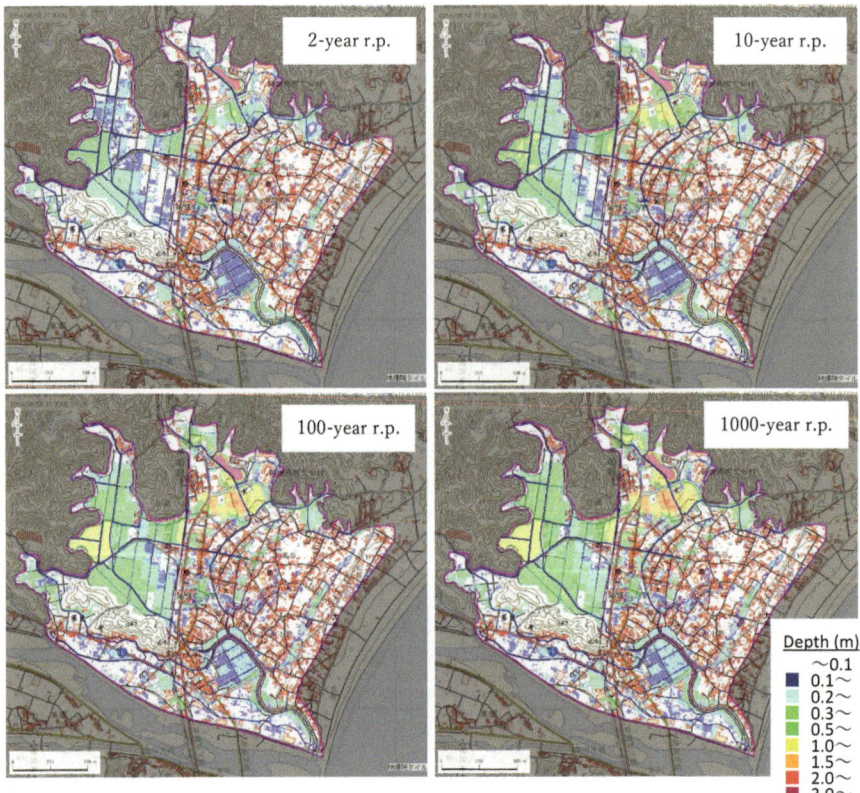

Fig. 9.9 The depth distributions at the peak inundation volume for surface water flooding

9.4.3 River Flooding Combined with Surface Water Flooding

Figure 9.15 shows the relation between the return period of total precipitation and the peak inundation volume for the river flooding combined with surface water flooding. Here the breaching level was set at RMWL, and the results for the surface water flooding are also plotted. The dotted lines in the figure indicate tendencies. By adding the river flooding, the inundation volume drastically increases compared with the surface water flooding only. In addition, the inundation volume is different as the breaching point changes. In this study, flooding from 1.8 km supplies more inundation volume than from 0.4 km. Thus, in sequence, the results shown are mainly for breaching at 1.8 km.

Figure 9.16 shows the inundation depth distributions at the peak volume for various return periods. In the figure, the breaching point is denoted by red X. Considering the depth distribution only for surface water flooding shown in Fig. 9.9, the combined flooding in a 50-year return period matches as the surface water flooding in a 1000-year return period regarding inundated area and depths. This

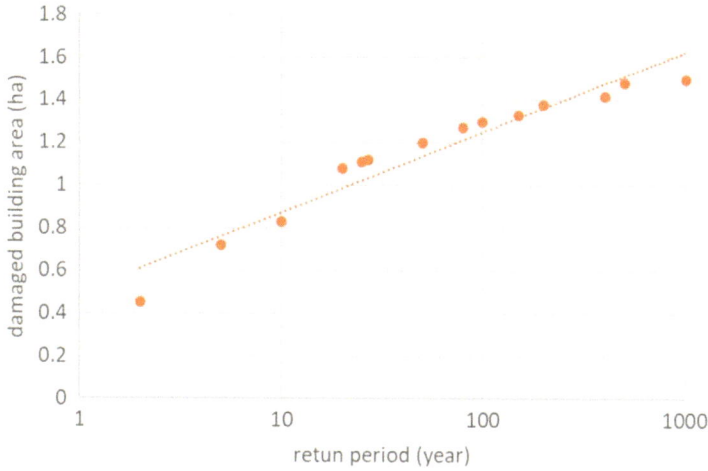

Fig. 9.10 The relation between the return period of total precipitation and the area of buildings inundated above floor level for surface water flooding

can also be seen in Fig. 9.15, where the peak inundation volumes are more or less similar, 700×10^3 m^3. As the precipitation return period increases, the inundation depth initially increases gradually, and then both depth and inundated area largely increase from 100-year to 1000-year return period. Such expansion of inundation is influenced by the breaching point location. When comparing the same return period, breaching at 1.8 km causes more severe flooding damage than at 0.4 km. This is probably due to not only larger flooding volume (see Fig. 9.6) but wider spreading flooding water because breaching occurs upstream.

In order to estimate the depth increment due to river flooding only, the inundation depth due to surface water flooding, shown in Fig. 9.9, was subtracted from the depth by combined flooding, as shown in Fig. 9.16, in each mesh, that is:

$$H_{rf} = H_{cf} - H_{sf} \tag{9.9}$$

where H_{rf} is the depth increment attributed solely to river flooding and H_{cf} and H_{sf} are the maximum water depths in each mesh in combined and surface water flooding, respectively. The distributions of H_{rf} for various return periods are shown in Fig. 9.17. Not surprisingly, the area near the breaching point shows large increments. By using H_{rf}, the expanding process of the inundated area in accordance with the return period becomes clearer. For example, the downstream of the Zenzo River is also extensively affected, although apart from the breaching point to some distance. This indicates that the flooding water from 1.8 km point runs easterly between the Kaifu and Zenzo rivers. Another interesting feature is that paddy fields in the western and northern parts take charge of floodwater through the three streams. At least up to 100-year return period, floodwater remains within the paddy fields, except for the area near the breaching point. However, in the 1000-year return

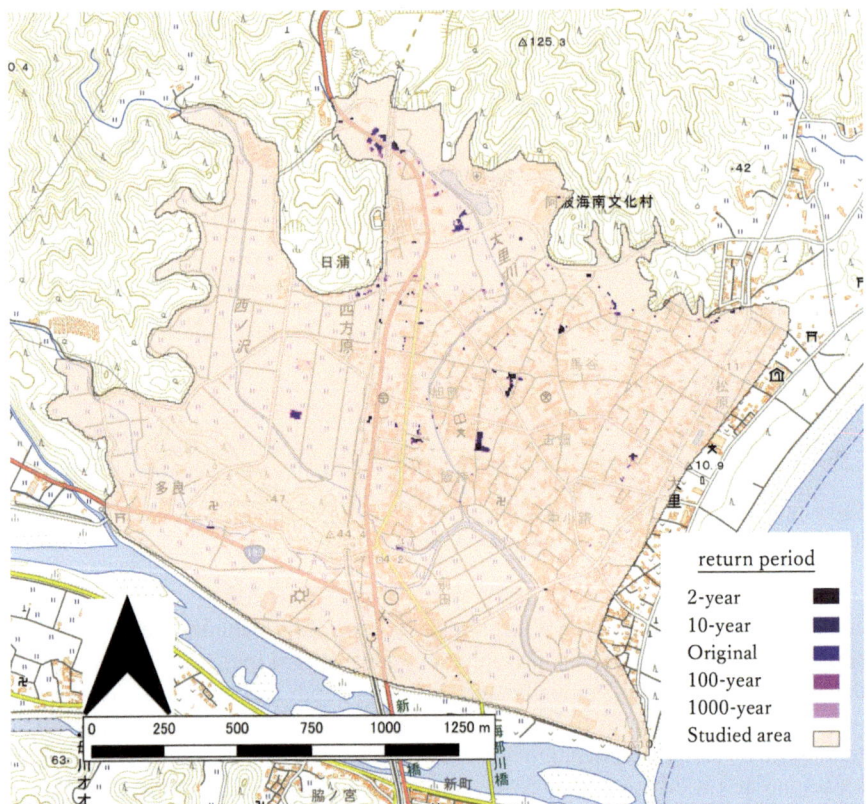

Fig. 9.11 The locations of buildings inundated above floor level in response to the precipitation return period for surface water flooding

period, the inundated area spreads devastatingly, and the maximum depth increment reaches up to 3–4 m near the downstream end of the studied area.

Figure 9.18 shows the relationship between the return period of total precipitation and the area of buildings inundated above the floor level. Similar to the peak inundation volume in Fig. 9.15, the number of inundated buildings increases drastically in the combined flooding. Figure 9.19 shows the locations of the inundated buildings. As shown in the figure, the number of inundated buildings increases gradually as the return period increases. In addition, there are inundated buildings even in the central part including the town office and railway station. In the worst case, the ratio of inundated buildings reaches 46.3%. Such an expanding process of the damaged area is well coincident with the expansion of the inundated area mainly attributed to river flooding, as shown in Fig. 9.17.

Figures 9.20 and 9.21 show the distributions and ratio of damaged paddy fields in response to the precipitation return period, respectively. As shown in these figures, the ratio of damaged paddy fields in combined flooding is fairly higher than that in

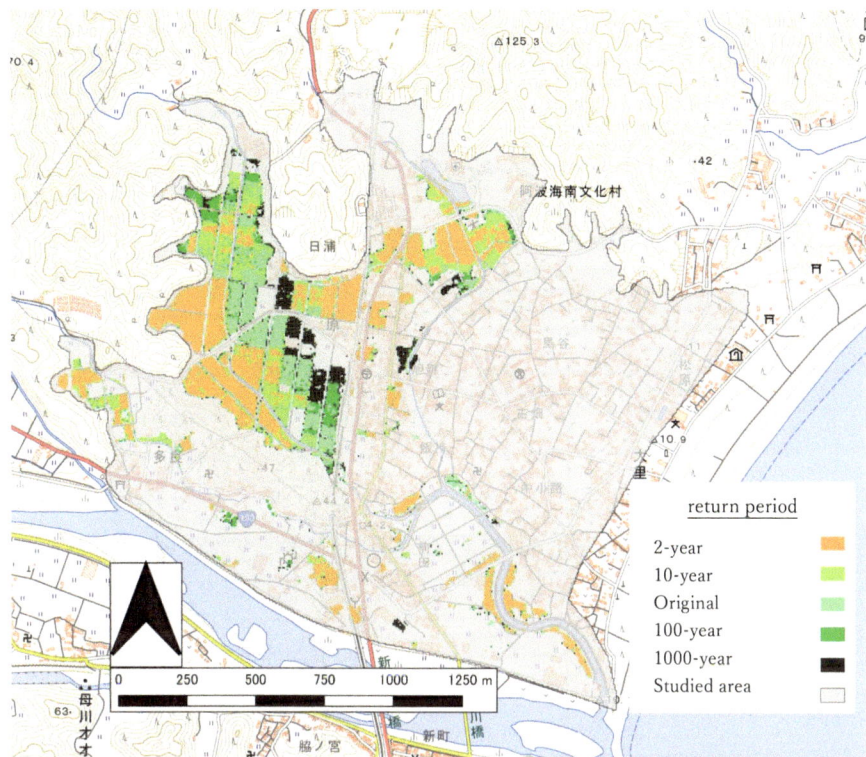

Fig. 9.12 The distributions of damaged paddy fields in response to the precipitation return period for surface water flooding

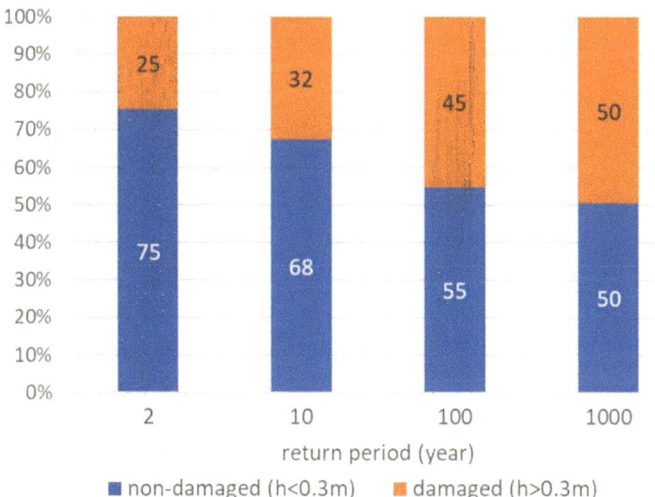

Fig. 9.13 The ratio of damaged/non-damaged paddy fields for surface water flooding

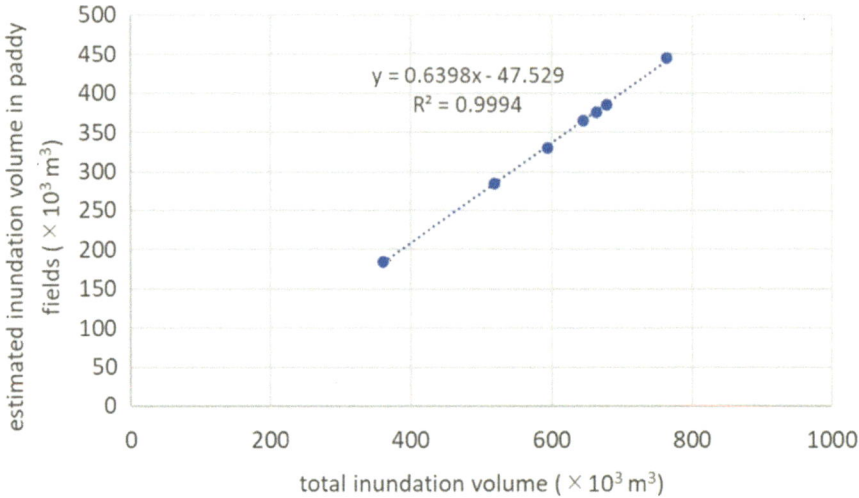

Fig. 9.14 The estimated inundation volume stored in the paddy fields for surface water flooding

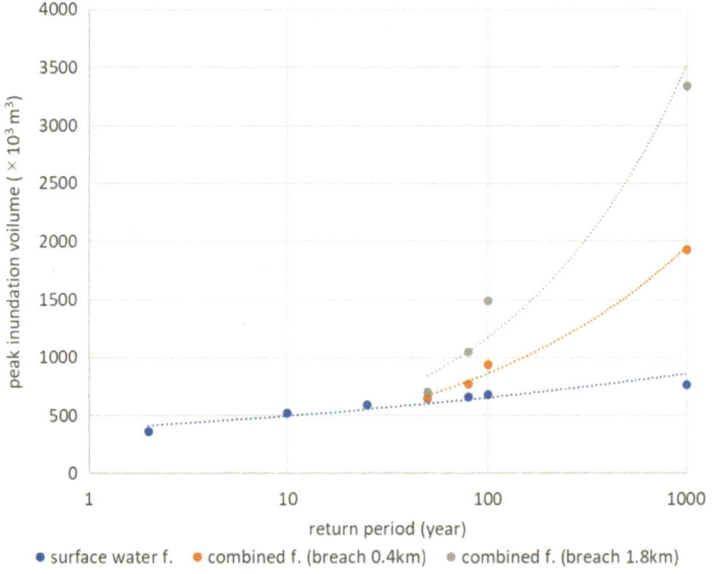

Fig. 9.15 The relation between the return period of total precipitation and the peak inundation volume for combined flooding

surface water flooding. The ratio in the 1000-year return period becomes as high as 94%. Comparing Figs. 9.20 and 9.12, in the western and northern parts, the locations of damaged paddy fields are almost the same under a 100-year return period. Consequently, this increment of damaged paddy fields is mainly brought

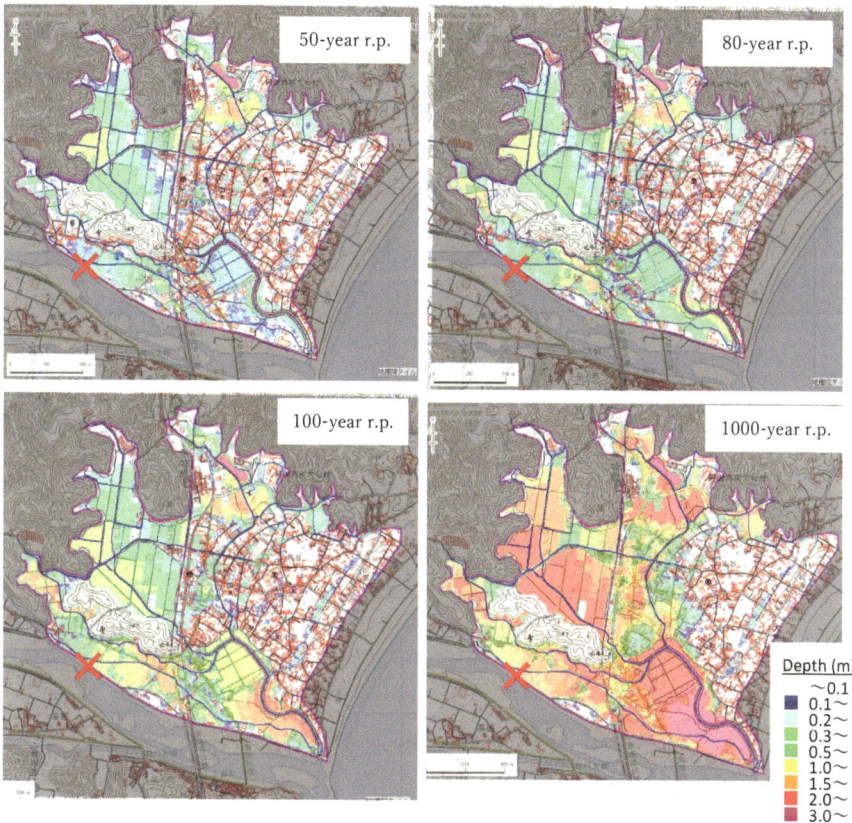

Fig. 9.16 The depth distributions at the peak inundation volume for combined flooding (breaching at 1.8 km denoted by red X, breaching level at RMWL)

by inundation close to the breach point by river flooding. In the 1000-year return period, almost all the western and northern paddy fields are damaged.

Figure 9.22 shows the estimated inundation volumes only in the paddy fields, plotted in the same manner as in Fig. 9.14. The graph shows that the paddy fields retain 60% of the inundation volume for combined flooding, which slightly smaller than that for surface water flooding shown in Fig. 9.14. Again, this value is nearly constant regardless of the return period and breaching point.

9.4.4   Discussions

As described in Sect. 9.4.2, for surface water flooding, damaged (inundated above floor level) buildings are quite rare, representing only 3.1% of the total buildings even in the 1000-year return period. The total inundation volume in the 1000-year

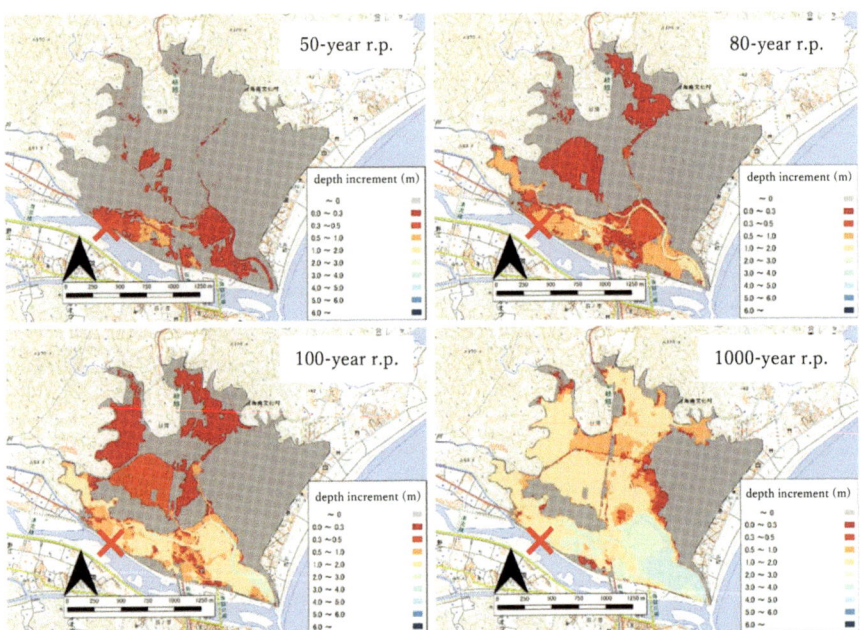

Fig. 9.17 The inundation depth increment due to river flooding onto surface water flooding (breaching at 1.8 km denoted by red X, breaching level at RMWL)

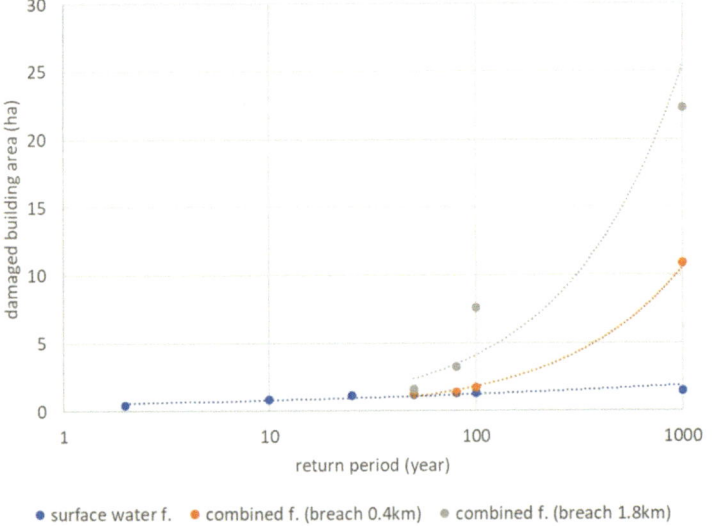

● surface water f.   ● combined f. (breach 0.4km)   ● combined f. (breach 1.8km)

Fig. 9.18 The relationship between the return period of total precipitation and the area of buildings inundated above floor level for combined flooding

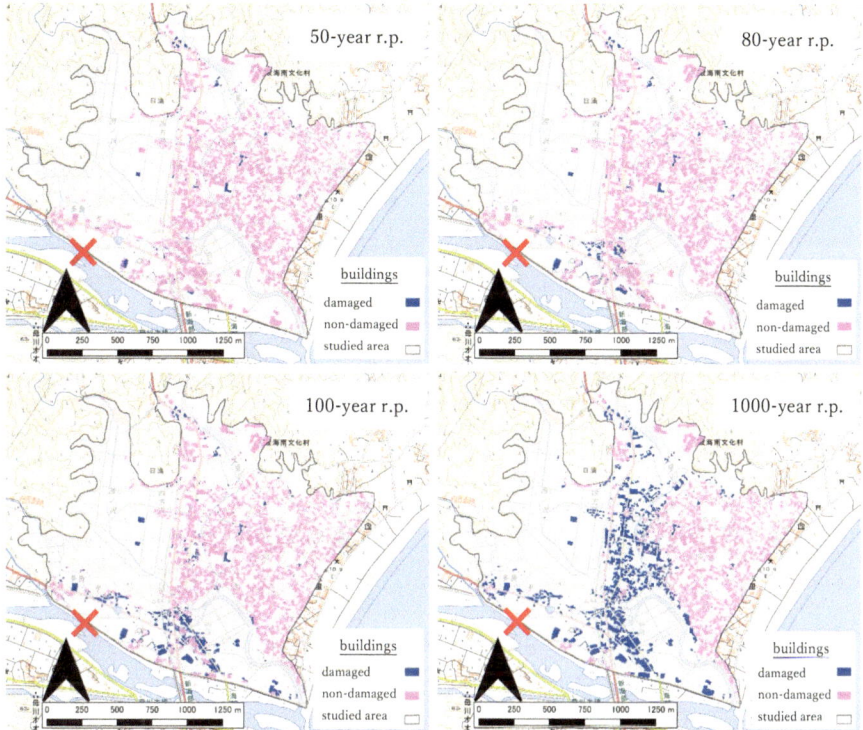

Fig. 9.19 The locations of buildings inundated above floor level for combined flooding (breaching at 1.8 km denoted by red X, breaching level at RMWL)

return period is approximately 764×10^3 m^3, and 64% of the volume is allocated to the paddy fields. Although some damaged paddy fields become apparent, disaster risk directly affecting mankind is still low. In other words, paddy fields secure human lives and properties by retaining most of the flooding volume. In the studied area, as a result, the current land-use classification can make sense in terms of disaster risk reduction against surface water flooding.

For river flooding combined with surface water flooding, as shown in Sect. 9.4.3, flooding from the river increases the inundated area and depth. However, the floodwater is primarily deposited in the vicinity of the breaching point, and in the western and northern paddy fields through the streams, surprisingly a bit far from the breach point. The studied area again can retain the floodwater volume to remain human risk relatively low, up to 80- to 100-year return period for combined flooding. In addition, the area between the Kaifu and Zenzo rivers is mainly used as paddy fields, directly receiving the floodwater from the river; thus, it can store the floodwater without increasing the threat to human lives to some extent. This should also be appreciated as a wise example of land use against floods.

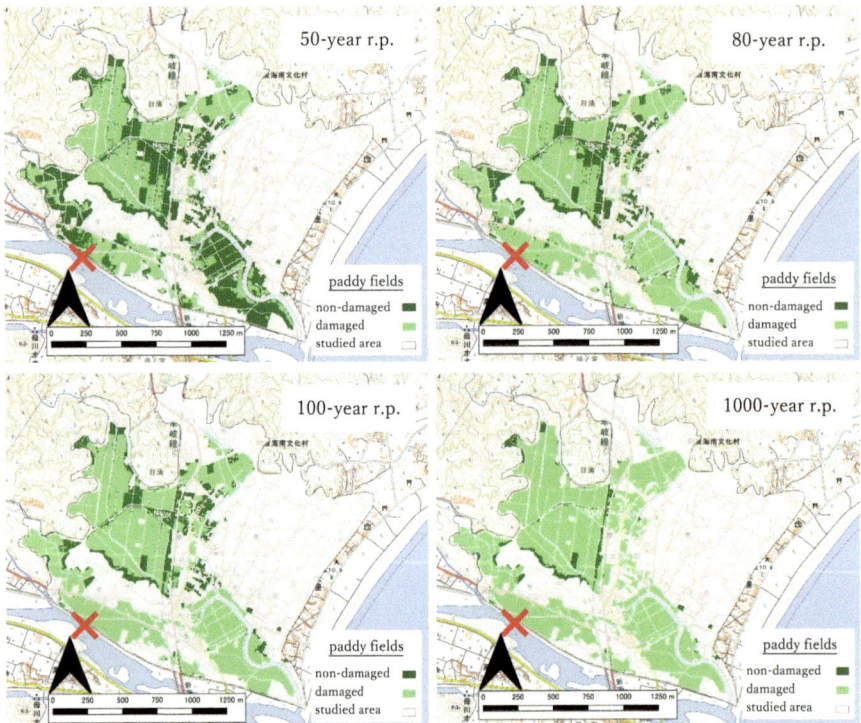

Fig. 9.20 The distributions of damaged paddy fields for combined flooding (breaching at 1.8 km denoted by red X, breaching level at RMWL)

Under a quite low-frequency event such as a 1000-year return period, damage to housing areas in the eastern part, mainly on high sand dunes along the beach, is inevitable. As mentioned above, inundation in the studied area starts from the southern (near the river), western and northern (paddy fields) parts, then it spreads from these three directions toward the central part as the precipitation becomes larger, and finally the relatively high eastern part gets inundated. Here, we can consider two reference values regarding damages: one is when a building in the central part near the town hall gets inundated and the other does other buildings on the higher sand dune. The former is recognised as damage to buildings starts to expand, and the latter is considered as seriously enlarge damage by inundating the residential area on high and relatively safer parts. Figure 9.23 shows the calculated inundation volumes only in the paddy fields plotted as a function of the return period, together with the locations of these two reference values. Judging from the simulation results, the former lies within 80- (1.8 km breaching) and 100- (0.4 km breaching) year return periods, indicated with a blue band in the figure, whereas the latter lies over a 100-year return period, shown as a green band. By doing this, we can estimate the potential volume of floodwater that can be retained by the paddy

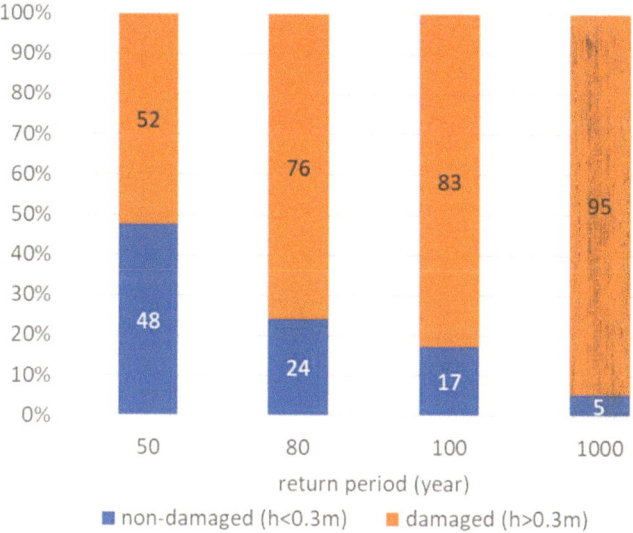

Fig. 9.21 The ratio of damaged/non-damaged paddy fields for combined flooding (breaching at 1.8 km, breaching level at RMWL)

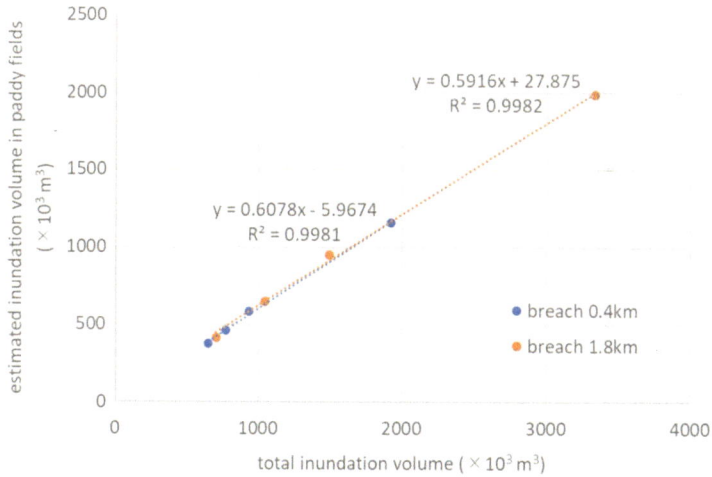

Fig. 9.22 The estimated inundation volume stored in the paddy fields for combined flooding

fields until damage to buildings and houses starts to occur. From the figure, the paddy fields in the studied area can retain 590 to 640 $\times$ 10^3 m^3 of floodwater before damage starts to expand in the central part, and of 950 to 1150 $\times$ 10^3 m^3 of floodwater before huge damage occurs in the eastern part. The figure can also be used to explain why damage to buildings is small for surface water flooding.

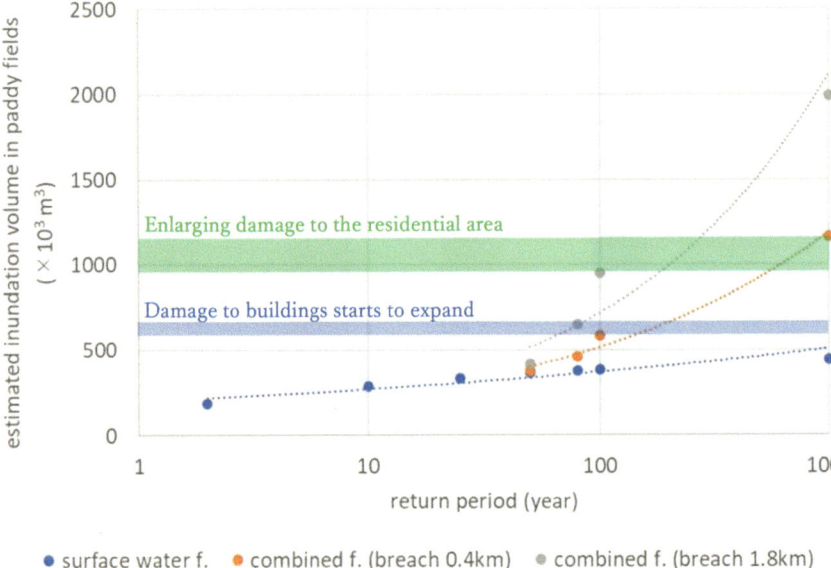

● surface water f.   ● combined f. (breach 0.4km)   ● combined f. (breach 1.8km)

Fig. 9.23 Two reference values regarding damage to buildings with reference to the estimated inundation volumes in the paddy fields

Considering the expanding process of floodwater together with the two reference values described above, it is important to secure the buildings in the central part, which start to be inundated between two reference levels. If these buildings are protected in some ways, a critical inundation volume in which damage to buildings starts can be enlarged. One of traditional methods is levee reinforcement. Figure 9.24 shows how much levee reinforcement reduces damaged buildings. Here three additional breaching scenarios, changing the breaching level at (1) HWL, (2) HWL +0.6 m and (3) at the same breaching level (RMWL) but overflow starting from the hydrograph peak, were explored for a 1000-year return period. The third condition is supposed to reinforce the levee in some ways, to avoid breaching immediately after reaching the setting level and to endure against flow until the hydrograph peak. The figure indicates that these levee reinforcements are all effective since they can reduce the total flooding volume. From the viewpoint of hybrid infrastructure, that is, the best matching of grey and green infrastructures, such considerations will be useful to seek how much risk the grey should take. On the other hand, because the levee reinforcement does not change alluvial plains and their drainage system at all, once flooding exceeds a certain level and pours into the plains, the same expansion process of flood flow as without levee reinforcement will be reproduced. Rather, in order to make full use of the flood control function of paddy fields, it would be better to consider land-use alteration, including building transference and wetland restoration. By doing this, the potential

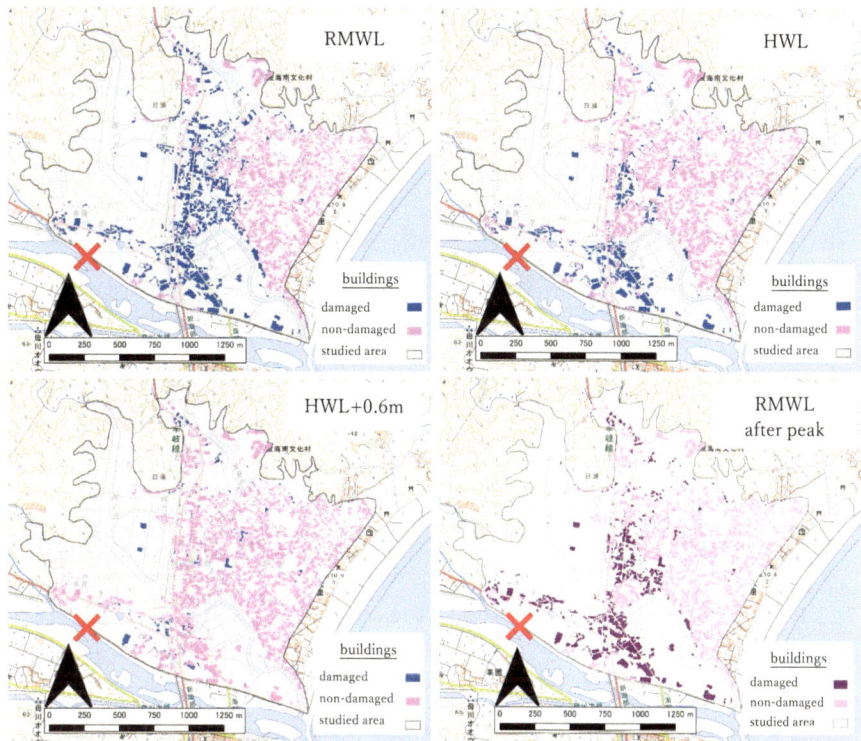

Fig. 9.24 The effect of the breaching level on locations of damaged buildings for combined flooding (1000-year return period, breaching at 1.8 km denoted by red X)

volume of floodwater retention in paddy fields can be extended at least up to the green band level, which well meets the concept of green infrastructure.

9.5   Concluding Remarks

By performing the inundation simulation, not only the potential storage volume but the expanding process of inundation in the studied area can be clarified. This also provides insights into the advantage of current land-use towards flooding, and leads to present a land-use strategy to be in the future against low-frequency floods.

Our recommendation for future land use is to transfer the buildings from the vulnerable central part to floods and to turn it back to the wetland. This may sound to require considerable efforts; however, with reference to a past aerial photo shown in Fig. 9.25, this part used to be mainly utilised as paddy fields. The potential volume of floodwater retention in paddy fields in the studied area should presumably be larger in the past. This can be a good reference for considering land-use strategies.

Fig. 9.25 An aerial photo of the studied area in the mid-1960s

Recently, Watanabe and Muto (2021) also confirmed the advantage of building transfer from the central part, from the viewpoint of town planning.

We do not take a position criticising the current land use, since it has been realised as a result of various factors, such as population growth, high economic growth, technological innovation against floods and so on. These factors sometimes allowed, and sometimes pushed, to move residential areas onto low and vulnerable parts. Nevertheless, the tendencies have now changed, particularly regarding the population. We can now do manage to live outside the vulnerable area. It must be a good time to consider withdrawing from the risk exposure.

In this study, we do not quantify damage to inundated paddy fields, but we intend to maximally utilise their flood retention capacity to reduce damage in residential areas as the most important aspect. However, this concept will make controversy among the concerned persons. To realise our proposal, not only explaining its economic advantages but consensus building is crucially important.

References

Iwasa Y, Inoue K (1982) Mathematical simulation of channel and overland flood flows in view of flood disaster engineering. J Nat Dis Sci 4(1):1–30

Japan Society of Civil Engineers (edited and published) (1999) Formulary of Hydraulics. 713 pp. (in Japanese)

Kusaka M (2020) Plains and Japanese history. Kadokawa. 209 pp. (in Japanese)

Masumoto T (1998) Paradigm shift in the evaluation of water storage function of paddies and in watershed management. J Japan Soc Hydrol Water Res 11(7):711–722. (in Japanese)

Ministry of the Environment, Japan (2016) Ecosystem-based disaster risk reduction in Japan. In: A handbook for practitioners. 24 pp

Miyazu S, Yoshikawa N, Abe S, Misawa S, Yasuda H (2012) Development of application of evaluation model for interior flood damage mitigation effect of a paddy field dam. IDRE J 282:15–24. (in Japanese)

Muto Y, Imai Y, Kamada M (2020) Ecosystem services of paddy fields and basinwide comprehensive flood alleviation. In: Green infrastructure—its practices. Nikkei BP, pp 308–317. (in Japanese)

NITA Consultant Co., Ltd (n.d.) Introduction to AFREL. http://www.nita.co.jp/index.php/software/afrel. (in Japanese)

Thang NT, Inoue K, Toda K, Kawaike K (2004) A model for flood inundation analysis in urban area: verification and application. Ann Disast Prev Res Inst Kyoto Univ 47B:13 pp

Watanabe K, Muto Y (2021) A study on future land use considering flood disaster in Osato district in Kaiyo Town, Tokushima. In: Proceedings of Awagakkai, no 63, pp 145–153 (in Japanese)

Yoshikawa N, Miyazu S, Koide H, Misawa S, Yasuda H (2010) Evaluation of flood mitigation effect of a paddy field dam in an unimproved paddy field area. Adv River Eng 16:507–512. (in Japanese)

Chapter 10
Change in Floodwater Retention Function of a Paddy Field Due to Cultivation Abandonment in a Depopulating Rural Region in Japan

Yota Imai, Yasunori Muto, and Mahito Kamada

Abstract Flood control concept at watershed scale is used as an adaptation strategy for climate change; rainwater and floodwater should be stored in forests and agricultural lands in the watershed, rather than relying on only river systems; and paddy fields are important ecosystems for floodwater retention. The abandonment of paddy cultivation, however, is increasing year after year in Japan due to depopulation in rural regions. Through two-dimensional numerical inundation analysis, we evaluated the negative effects of topographical changes after abandonment on the floodwater retention function. The bipolarization of the inundation depth is apparent in the current topography. The areas with deepwater inundation are wider, particularly in the most downstream areas, whereas in the past, areas with shallow water were widely distributed. The peak water level in the current land use is 0.4 m higher than in the past. After the peak level is achieved, the water level decreases in a shorter time in the present land use than in the past. Owing to hydraulic changes after the abandonment of rice cultivation, the risk of flooding disasters for people living downstream has increased. To maintain floodwater retention function, one of the regulating services, by artificial wetlands, such as paddy fields, continuous management is essential to prevent vegetation succession. Thus, it is important to provide incentives for local people to perform the daily maintenance activities.

Keywords Abandoned paddy field · Flood management · Rural area · Depopulation · Green infrastructure

Y. Imai (✉)
Department of Civil and Environmental Engineering, Graduate School of Advanced Technology and Science, Tokushima University, Tokushima, Japan

Department of Civil Engineering, Kobe City College of Technology, Kobe, Hyogo, Japan

Y. Muto · M. Kamada
Research Center for Management of Disaster and Environment, Tokushima University, Tokushima, Japan

Department of Civil and Environmental Engineering, Graduate School of Technology, Industrial and Social Sciences, Tokushima University, Tokushima, Japan

161

F. Nakamura (ed.), *Green Infrastructure and Climate Change Adaptation*,
Ecological Research Monographs, https://doi.org/10.1007/978-981-16-6791-6_10

10.1 Introduction

Owing to climate change, rainfall is increasing every year along with the magnitude and frequency of flooding (http://www.env.go.jp/earth/tekiou/report2018_full.pdf). Although floods have been controlled by strengthening river embankments and dam construction, solving this issue solely by adopting structural measures is becoming difficult from a financial viewpoint due to changes in the social condition of Japan: decreasing tax income associated with a rapid decrease in population and increasing maintenance costs for existing infrastructures (https://www.mlit.go.jp/sogoseisaku/maintenance/_pdf/research01_02_pdf02.pdf). Therefore, the Ministry of Land, Infrastructure, Transport and Tourism steered toward "Ryuiki-chisui" which refers to a new flood control concept at the watershed scale as an adaptation strategy for climate change (https://www.mlit.go.jp/river/kasen/ryuiki_pro/index.html): the rainwater and floodwater should be stored in forests and agricultural lands in the watershed, rather than relying on only river systems.

The use of paddy fields to reduce flood damage has been proposed as one of the flood control methods at the watershed scale (https://www.mlit.go.jp/river/kasen/suisin/pdf/renkei_siryou02/siryou04.pdf). Paddy fields are a land use for rice production in Japan and provide several ecosystem functions (Natuhara 2013). For example, paddy fields provide a water storage function as well as a habitat for several wetland species (Katayama et al. 2015; Osawa et al. 2020), and the function has been utilized for flood disaster risk reductions (Teramura and Shimatani 2021).

The integrated management of paddies, ditches, and ridges by farmers has contributed to maintaining the water storage function of paddy fields (Yoshikawa 2014). Previous studies have been using various approaches to evaluate the water storage function of paddy fields in different regions (Matsumoto et al. 2013; Yoshikawa 2014; Muto et al. 2018; Teramura and Shimatani 2021).

In Japan, the abandonment of paddy cultivation is increasing every year due to depopulation (Osawa et al. 2013, 2016); this may have a negative effect on water storage functions due to the collapse of the ditches-ridges structure (Yoshida et al. 2013). Although several studies have evaluated changes in runoff volume from paddy fields after abandonment of cultivation, few have described the hydraulic characteristics of abandoned paddy fields. Hydraulic characteristics can be linked to sediment deposition, followed by the development of woody communities through vegetation succession. If the hydraulic characteristics of abandoned paddy fields were described, a technique which control floodwater may be developed for improvement of water storage functions. Such a technique can only be realized with a lower cost of introduction and maintenance, particularly in a depopulated rural area.

On the other hand, abandoned paddy fields have often restored to wetland/biotope and are maintained as a wildlife habitat. In wetland restoration, construction of ridges and removing shrub are conducted for keeping topographical condition flat, for retention of floodwater and rainwater. Thus, the water storage function is likely to be improved at restored abandoned paddy fields.

In this paper, first, we clarify the negative effects of the abandonment of rice cultivation on water storage function in Shitaru District, Tsushima Island in Nagasaki Prefecture, Japan. The hydraulic characteristics of paddy fields before and after abandonment are described, compared, and discussed in terms of the process of functional decline. Second, a method of restoring the function is discussed based on the case of Tai District, Toyooka City in Hyogo Prefecture, where abandoned paddy fields have been used as wildlife habitat wetlands (Imai et al. 2020). Finally, we referred to the changes in water storage function with structural change and wetland restoration at paddy fields and to the management of abandoned paddy fields based on the comparison of two districts.

10.2   Material and Methods

10.2.1   Study Area

The study was conducted in Shitaru District in Kamiagata, Tsushima Island, in Nagasaki Prefecture (34°34′N, 129°18′E; Fig. 10.1). The watershed comprises forests, paddy fields, and villages (Fig. 10.2b). The highest monthly average temperature is 26 °C in August, while the lowest is 6 °C in January. The monthly average precipitation is the highest in July (350 mm), followed by June (320 mm; (http://www.city.tsushima.nagasaki.jp/deta/post.html). The population of Shitaru District in 2020 was 73, and is estimated to decrease to 21 by 2050.

The area used to be under paddy cultivation until the 1960s (Shigehara and Shibata 2018); however, almost all paddies have been abandoned and transformed to wetland vegetation (Fig. 10.2c). The rainwater in the watershed spreads across the floodplain and eventually accumulates downstream (Fig. 10.2d).

Figure 10.3a shows the past distribution of paddy fields in the watershed. Sediments that flowed into the channels and paddy fields were removed by local farmers to maintain cultivation after flooding (Imai et al. 2019). Figure 10.3b shows the current vegetation map. Almost all paddy fields in the watershed have transformed to natural vegetation. The vegetation was classified into four types of plant communities. According to an inundation analysis, the spatial patterns of plant communities corresponded to the velocity of water and inundated depth during flooding (Imai et al. 2019).

10.2.2   Two-Dimensional Inundation Analysis

To obtain the current topographical conditions, aerial and field surveys were conducted from December 5 to 8, 2017. For aerial surveys, unmanned aerial vehicles (UAVs; DJI Phantom3 Pro and Phantom4 Pro) were used. To minimize errors in

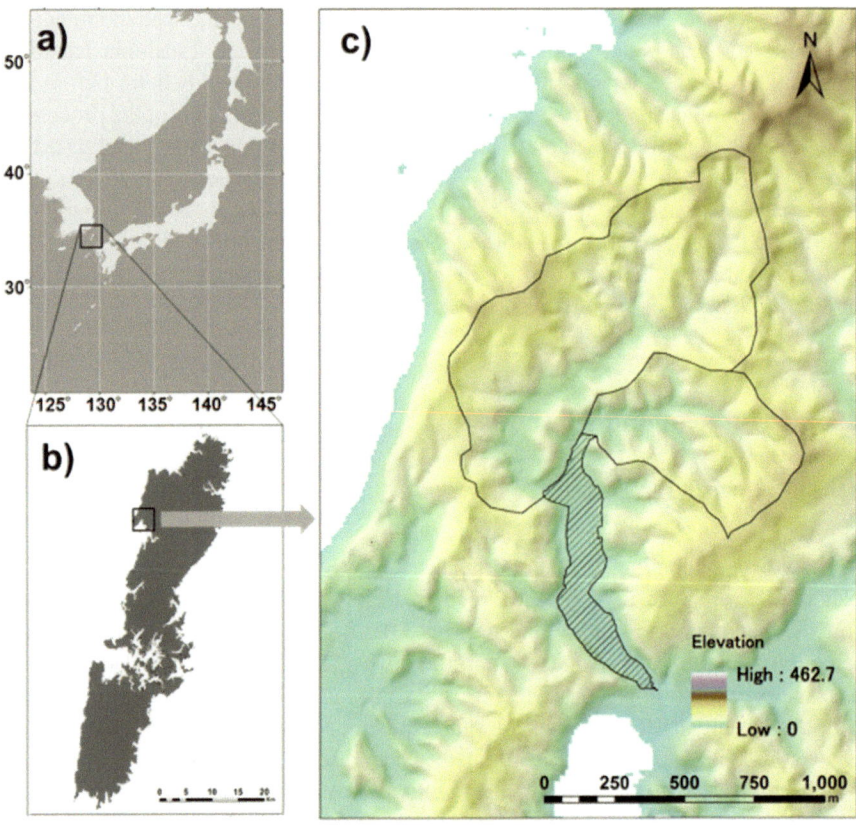

Fig. 10.1 (a) Location of Tsushima Island, Japan. (b) Map of Tsushima Island. (c) The study floodplain (hatched area in c), in Shitaru District, Tsushima Island, Japan. The black borders shown in (c) are sub-watersheds (Imai et al. 2019)

the digital terrain model, flight routes were planned with 80% overlap and 60% sidelap. A digital surface model (DSM) with a 0.8 m grid resolution was generated from 3447 images using the Agisoft Photoscan Pro v1.0.0 and ArcGIS v10.2.2. The parameters of Align Photos and Build Dense Cloud, structure from motion-multi view stereo (SfM-MVS) photogrammetry using Agisoft Photoscan Pro v1. 0. 0 were high, respectively. Errors in position coordinates were corrected using the coordinates from 19 ground control points (GCPs: Fig. 10.4), which were set in the watershed and measured using RTK-GNSS (TOPCON GR-2100) and Total Station. Because almost all areas of the watershed were covered by vegetation, the representative height was measured by using a handy measure for each vegetation type by field surveys on December 9, 2017; then, the digital terrain model (DTM) was obtained by subtracting the vegetation heights from the DSM.

Fig. 10.2 Shitaru District. (**a**) The study area, (**b**) distribution of village and floodplain, (**c**) woody area, and (**d**) downstream area

When rice was cultivated, the farmers also maintained the ridges, and therefore, rainwater and floodwater were stored in every paddy field before overflowing into the lower fields in a step-by-step manner. Thus, the past topography was reconstructed as follows: the mean current elevation in each polygon of the paddy field was calculated from the DTM using zonal statistics (ArcGIS 10.2.2, Spatial Analyst). Then, 0.3 m, representative height of ridges remained at study floodplain was added to the elevation of cells surrounding each paddy field to include the height of the ridge.

Differences in hydraulic characteristics before and after abandonment of cultivation were clarified by two-dimensional inundation analyses in conditions with annual maximum rainfall for past and current topography using iRIC Nays2DFlood Solver (https://i-ric.org/en/solvers/nays2dflood/). Nays2DFlood Solver enables the user to easily set the inflow conditions and topographical conditions for flood flow analysis and obtains hydraulic characteristics such as water velocity and depth at river and floodplain.

(a)   **(b)**

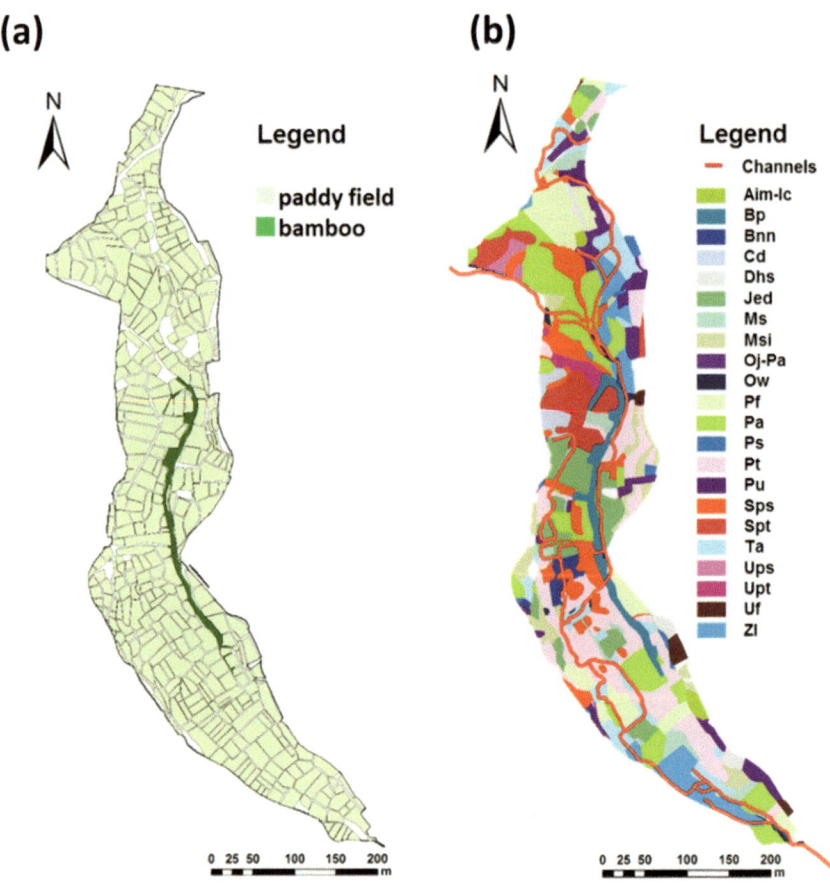

Fig. 10.3 (**a**) Past distribution of paddy fields and (**b**) current distribution of plant communities Abbreviations: Aim-Ic, *Artemisia indica* var. *maximowiczii – Imperata cylindrica* community; Bp, bamboo plantation; Bnn, *Boehmeria nivea* var. *nipononivea* community; Cd, *Carex dispalata* community; Dhs, dried herbaceous community (short); Jed, *Juncus effusus* var. *decipiens* community; Ms., *Miscanthus sacchariflorus* community; Msi, *Miscanthus sinensis* community; Oj-Pa, *Oenanthe javanica – Phalaris arundinacea* community; Ow, open water; Pf, paddy field; Pa, *Phragmites australis* community; Ps, pioneer shrub community; Pt, *Polygonum thunbergii* community; Pu, *Pueraria montana* community; Sps, *Salix pierotii* community (short); Spt, *Salix pierotii* community (tall); Ta, *Typha angustifolia* community; Ups, *Ulmus parvifolia* community (short); Upt, *Ulmus parvifolia* community (tall); Uf, upland field; Zl, *Zizania latifolia* community (Imai et al. 2019)

For inundation analyses, the model hyetograph was generated based on records from the nearest observation points (34°33′N, 129°18′E; http://www.kasen-sabo. pref.nagasaki.jp/nagasaki/main/index.php) on Tsushima Island: rainfall of 14 h, from 12:00 on June 30 to 2:00 on July 1, 2018. The total rainfall was 73 mm, and

Fig. 10.4 Flight routes of
UAV (yellow circles) were
planned for 80% overlap and
60% sidelap and 19 GCPs
were set (white circles). Black
line shows study floodplain

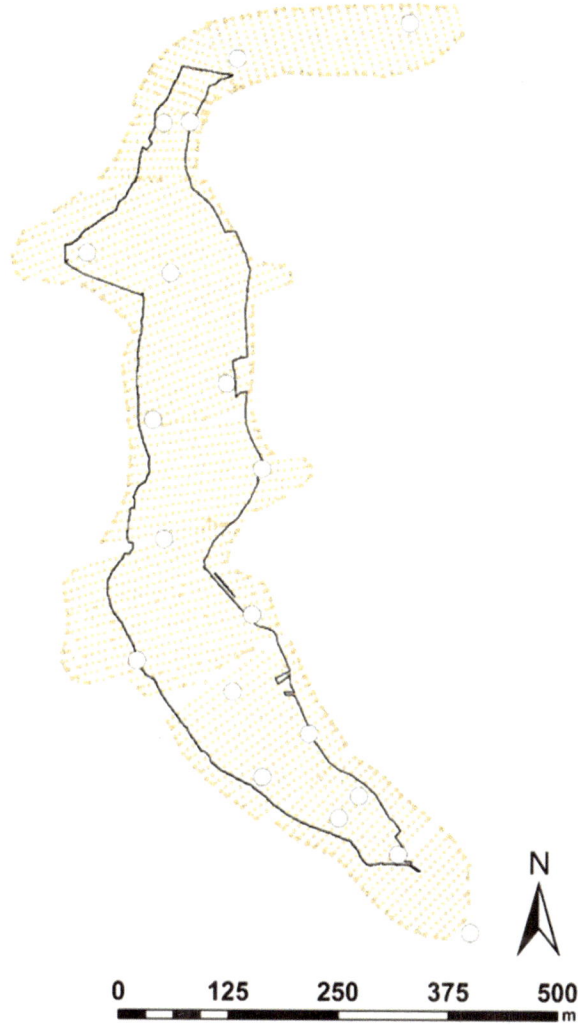

it was estimated to be 1/1-year probability based on 20-year records. The model
hyetograph was produced by uniformly enlarging the annual maximum rainfall
of 342 mm (Fig. 10.5). For the model hydrograph, discharge flowing into the
target watershed from the northern and northwestern sub-watersheds was considered
(Fig. 10.5).

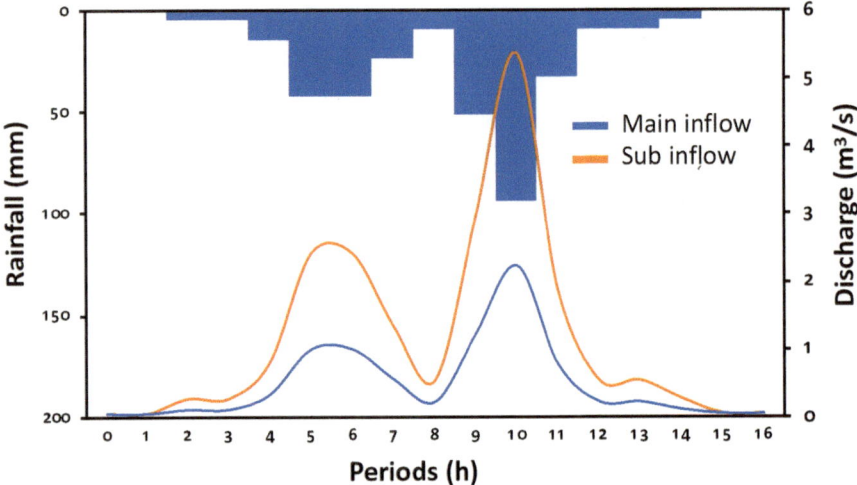

Fig. 10.5 Model hyetograph and hydrograph. Main inflow and sub-inflow show inflows from the northern and north-western sub-watersheds (Fig. 10.1(c)), respectively

10.3 Results

Figure 10.6 shows the hydraulic characteristics in different topographies—in the present (Fig. 10.6a) and before the abandonment of cultivation (Fig. 10.6b). The bipolarization of inundation depth is apparent in the current topography; areas without inundation and with deepwater depths (red) are wider, particularly in the most downstream area, whereas in the past, areas with shallow water (blue) were widely distributed. These results indicate that floodwater now accumulates in the most downstream area connecting to the residential area.

Figure 10.7 shows the water level at the most downstream point (see Fig. 10.2d). The peak water level in the current land use is 0.4 m higher than in the past. After the peak level is achieved, the water level decreases in a shorter time in the present land use than in the past. These results indicate that the water storage function has been decreased by the abandonment of cultivation.

10.4 Discussion

The peak water level at the downstream in the current land use is 0.4 m higher than in the past topographical condition (Fig. 10.7). This result shows that the hydraulic characteristics have changed, and the risk of flooding has been increased through abandonment of paddy cultivation.

The abandonment of cultivation leads to halting of farmer activities to maintain paddy field structures, and ridges collapse due to erosion (Yoshikawa et al. 2009).

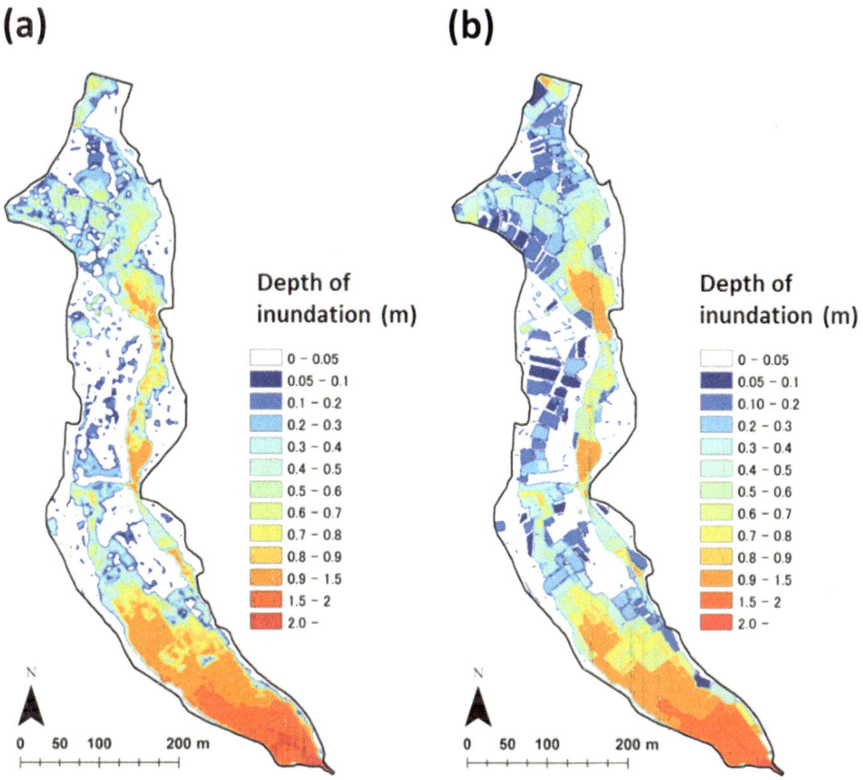

(a)   **(b)**

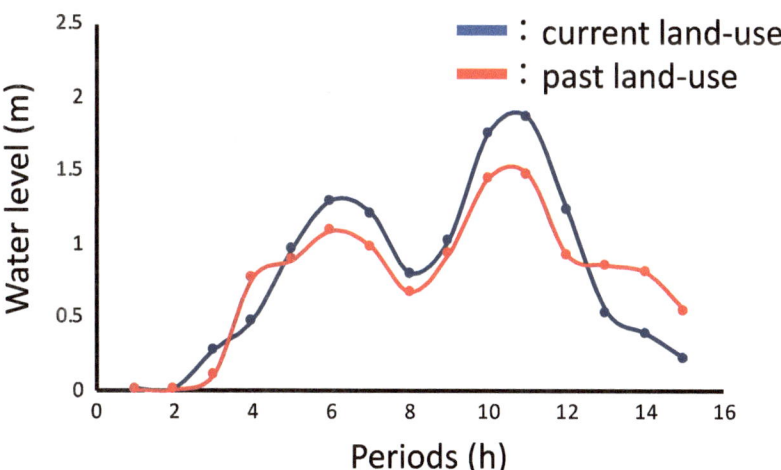

Fig. 10.6 Depth of inundation immediately after peak discharge for maximum rainfall: (**a**) present and (**b**) past land use

Fig. 10.7 Water level change at downstream area

Once the ridges collapse, a water channel is formed as a natural process (Yoshida et al. 2012, 2013), and rainwater and floodwater flow into the channel as runoff. Thus, the water storage function is reduced, and the moisture condition of the area changes (Ohkuro et al. 1996).

Because of hydraulic changes after the abandonment of rice cultivation, natural vegetation has established (Imai et al. 2019), but the risk of flooding disasters for people living in downstream areas has increased. The same situation has occurred in many rural regions in Japan undergoing depopulation. It is necessary to adopt appropriate measures against flood disasters in these regions, where little administrative support can be obtained.

10.5 Mitigating Flood Risk Using Abandoned Paddy Field Through Wetland Restoration

At Tai District in Toyooka City, Hyogo Prefecture, Japan (35°38′37.9″N 134°50′41.5″E; Figs. 10.7 and 10.8), where the population is estimated to decrease from 136 in 2020 to 73 by 2050, abandoned paddy fields have restored to

Fig. 10.8 Tai District. (**a**) Study site, (**b**) distribution of village and floodplain, and (**c, d**) weirs and flashboards maintained by local people and volunteers for wetland improvement

wetland/biotope and are maintained as feeding sites for the Oriental white stork (*Ciconia boyciana*). This is an endangered bird protected by the government and local people, through collaborative activities by local people and volunteers from outside the district.

The wetlands have been used as a floodwater retention area. These structures reduce the flow velocity during flooding. In addition to voluntary work, overflow levees and small water reservoirs were constructed in the uppermost stream of the wetland with the support of Toyooka City.

The results of two-dimensional flood flow analysis for 1/30-year rainfall showed that 22% of peak discharge was stored in abandoned paddy fields (Fig. 10.9; Imai et al. 2020) and 25% of the total discharge was stored in the wetland (Fig. 10.10; Imai et al. 2020). Thus, managed abandoned paddy fields can be used as floodwater retention areas, and water storage function can be improved by wetland restoration for flood disaster reduction.

Fig. 10.9 Temporal changes in discharge (Imai et al. 2020). Outflow shows the discharge at the downstream boundary and inflow shows the discharge at the upstream boundary

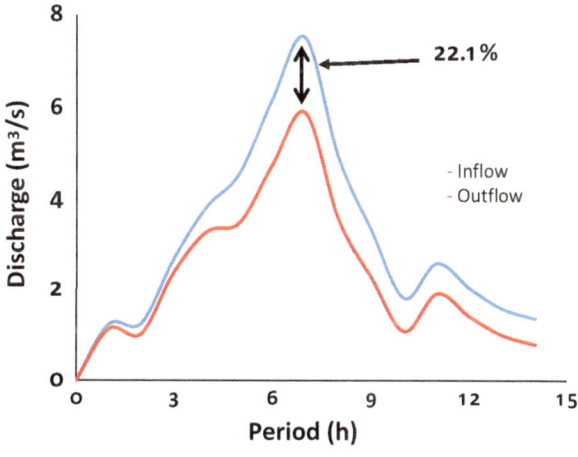

Fig. 10.10 Comparison of floodwater volume between the calculated area (total) and abandoned paddy fields on its area during flooding (Imai et al. 2020)

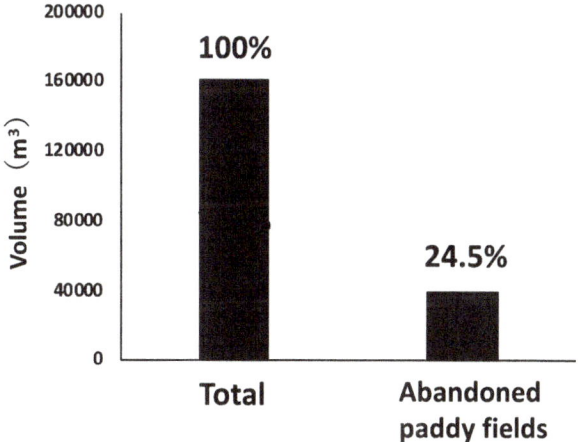

10.6 Changes in Water Storage Function with Structural Change and Their Management at Abandoned Paddy Fields

Paddy fields provide provisioning services for rice cultivation as well as regulating services via water storage functions through maintenance of ditches and ridges structure. However, the water storage function of paddy fields is decreased by structural changes after abandonment of cultivation; the quality of regulating services worsens after relinquishing provisioning services, as in the case of Shitaru. In the case of Tai, the water storage function was provided at the restored abandoned paddy fields with weirs and flashboards introduced by the local people due to habitat improvement for supporting Oriental white storks. Thus, the water storage function decreases from cultivated paddy fields to abandoned paddy fields, but its function of abandoned paddy fields could be improved through wetland restoration.

At the Tai District, supporting Oriental white storks was an incentive for local people and volunteers from outside the district, rather than flood control. To maintain regulating services of artificial wetlands, such as abandoned paddy fields, continuous management to prevent vegetation succession is essential. Hence, it is important to provide monetary compensation and to emerge incentives such as recovery of biodiversity for local people to contribute to this management (Satake 2014). A technique that emerges both water storage function and biodiversity conservation should be developed with adaptive management approach.

Acknowledgments We are grateful to the local people of the Shitaru and Tai districts, who allowed us to carry out our survey and provided useful information in this regard. This study was supported by the Environment Research and Technology Development Fund (4-1504, 4-1805) of the Ministry of the Environment, Japan.

References

Imai Y, Shigehara N, Shibata S, Muto Y, Kamada M (2019) Spatial pattern of plant communities in relation to hydraulic conditions at wetland emerged at abandoned paddy fields. In: E-proceedings of the 38th IAHR World Congress, 1–6 September, 2019, Panama City, Panama, pp 3311–3320

Imai Y, Mitsuhashi H, Kamada M, Muto Y (2020) Fundamental study on water storage function of abandoned paddy field on small river floodplain. Annu J Hydraulic Eng JSCE 65:793–798; [In Japanese with English Abstract]

Katayama N, Baba YG, Kusumoto Y, Tanaka K (2015) A review of post-war changes in rice farming and biodiversity in Japan. Agric Syst 132:73–84

Matsumoto T, Fukuoka S, Sumi T (2013) Estimation of flood storage volume in the three retarding basins along the tone river. Annu J Hydraulic Eng JSCE 69(4):793–798; [In Japanese with English Abstract]

Muto Y, Kotani S, Miyoshi M, Kamada M, Tamura T (2018) Retarding capacity change of wetland paddy fields due to house land development -utilizing paddy fields as green infrastructure against flood. In: Proceedings of the 21th IAHR-APD congress 2018, 8pp

Natuhara Y (2013) Ecosystem services by paddy fields as substitutes of natural wetlands in Japan. Ecol Eng 56:97–106

Ohkuro T, Matsuo K, Nemoto M (1996) Vegetation dynamics of abandoned paddy fields and their levee slopes in mountainous regions of Central Japan. Jpn J Ecol 46:245–256; [In Japanese with English Abstract]

Osawa T, Kohyama K, Mitsuhashi H (2013) Areas of increasing agricultural abandonment overlap the distribution of previously common, currently threatened plant species. PLoS One 8(11):e79978

Osawa T, Kohyama K, Mitsuhashi H (2016) Multiple factors drive regional agricultural abandonment. Sci Total Environ 542:478–483

Osawa T, Nishida T, Oka T (2020) Paddy fields located in water storage zones could take over the wetland plant community. Sci Rep 10(1):1–8

Satake S (2014) Habitat creation and restoration in collaboration with local residents. Reintroduction 3:25–27; [In Japanese with English Abstract]

Shigehara N, Shibata S (2018) The transition of biological resources use and land use in the 1950s and later in Shitaru, Tsushima city, Nagasaki prefecture. Jpn Inst Landscape Architect 81(5):699–702; [In Japanese with English Abstract]

Teramura J, Shimatani Y (2021) Advantages of the open levee (Kasumi-Tei), a traditional Japanese river technology on the Matsuura river, from an ecosystem-based disaster risk reduction perspective. Water 13(4):480

Yoshida T, Masumoto T, Horikawa N (2012) Changes in rainfall-runoff characteristics of small watersheds due to cultivation conditions of hilly paddies. Irrigation Drainage Rural Eng J 80(2):117–124; [In Japanese with English Abstract]

Yoshida T, Masumoto T, Horikawa N, Minakawa H (2013) Modelling of short-term runoff processes in watersheds dominated by hilly paddics. Irrigation Drainage Rural Eng J 81(3):235–244; [In Japanese with English Abstract]

Yoshikawa N (2014) Can paddy fields mitigate flood disaster? Possible use and technical aspects of the paddy field dam. In: Social-ecological restoration in Paddy-dominated landscapes. Springer, Tokyo, pp 197–207

Yoshikawa N, Nagao N, Misawa S (2009) Watershed scale evaluation of flood mitigation function of paddy fields installed with runoff control devices. Irrigation Drainage Rural Eng J 261(3):44–48; [In Japanese with English Abstract]

Chapter 11
Paddy Fields as Green Infrastructure: Their Ecosystem Services and Threatening Drivers

Takeshi Osawa, Takaaki Nishida, and Takashi Oka

Abstract The paddy field is a typical ecosystem in monsoon Asia. The main purpose of a paddy field is food production (i.e., a provisioning ecosystem service), but it also has several other ecosystem services. Therefore, we can consider paddy fields as components of green infrastructure (GI) that are already widely distributed. To maintain and improve the value of paddy fields as GI, we should strive to maintain and improve other ecosystem services besides provisioning services. However, issues that reduce the value of paddy fields as GI remain, and these are related to the consolidation and abandonment of paddy fields in recent years. In this paper, we focus on two ecosystem services of paddy fields, other than the provisioning service: (1) the habitat service and (2) the regulating service. Both these services are relatively well-evaluated ecosystem services of paddy fields in Japan. We discuss the effects of both consolidation and abandonment on these ecosystem services as well as their corresponding countermeasures. Based on this, we also discuss the challenges in applying and expanding the idea of paddy fields as GI.

Keywords Agricultural abandonment · Consolidation · Ecosystem-based disaster risk reduction · Ecosystem service · Habitat service · Wetlands

T. Osawa (✉)
Graduate school of Urban environmental Sciences, Tokyo Metropolitan University, Tokyo, Japan

T. Nishida
Mitsubishi UFJ Research and Consulting Co. Ltd., Osaka, Japan

Kyoto Sangyo University, Kyoto, Japan

T. Oka
Kyoto Sangyo University, Kyoto, Japan

© The Author(s) 2022
F. Nakamura (ed.), *Green Infrastructure and Climate Change Adaptation*,
Ecological Research Monographs, https://doi.org/10.1007/978-981-16-6791-6_11

11.1 Introduction

Paddy fields are a typical agricultural land type in monsoon Asia, mainly for rice crops. In Japan, paddy fields occupy much of the country's coastal plain (Natuhara 2013), with an area of approximately 24,000 km^2, which represents 6% of the total land area (Ministry of Agriculture, Forestry and Fishery [MAFF], Japan; https://www.maff.go.jp/j/tokei/sihyo/data/10.html; accessed on March 10, 2021). Paddy fields are the second largest land cover type in Japan after forests and are distributed throughout the country (Osawa et al. 2013, 2016a, b). Thus, paddy fields represent one of the most common seminatural ecosystems in Japan.

The essential purpose of maintaining paddy fields is for agricultural food production, i.e., for a provisioning ecosystem service. It is well known that paddy fields can provide several ecosystem services other than their provisioning service. For instance, paddy fields provide a supporting (habitat) service for several wetland species (Yoon 2009; Natuhara 2013; Katayama et al. 2015) in addition to providing a regulating service, such as flood control (Matsuno et al. 2006; Natuhara 2013; Osawa et al. 2020a), greenhouse gas emission control (Ahmad et al. 2009; Yang et al. 2012), landslide prevention (Natuhara 2013), and decreasing pests (insects) (Jincai et al. 1994). Japan's MAFF has claimed that paddy fields provide not only food production, i.e., a provisioning service, but also diverse multiple services, referred to as "multiple functions of agricultural lands." Thus, the MAFF has supported some of these functions through a grant system (https://www.maff.go.jp/j/nousin/kanri/tamen_siharai.html; accessed on March 10, 2021).

Green infrastructure (GI) has been defined as natural, seminatural, and artificial networks of multiple functional ecological systems within, around, and between human residential communities at all spatial scales (Tzoulas et al. 2007). At the same time, GI should have multiple ecosystem services based on its ecosystem functions (Sandstrom 2002; Tzoulas et al. 2007). Thus, based on these definitions, we can also consider paddy fields as GI (Fig. 11.1). However, in recent years, the situations of many agricultural lands, including paddy fields, have changed. Changes in agricultural land mainly consist of either modernization or abandonment (Osawa et al. 2013; Katayama et al. 2015). Both these changes in land use can exert negative effects on the ecosystem functions of the land (Benayas et al. 2007; Yoshida et al. 2012; Osawa et al. 2013). Consequently, such changes in paddy fields might cause their inherent ecosystem functions and services to collapse (Osawa et al. 2020a, b).

The concept of GI, particularly in monsoon Asia (including Japan), should be expanded. Paddy fields represent one of the most important ecosystems with potential as GI because they are very widespread and could thus offer multiple ecosystem services. Therefore, it is important to understand the threats and current status of multiple ecosystem services of paddy fields for sustainable use. In this paper, we focus on two ecosystem services—habitat and regulating services—of paddy fields that have been relatively well-evaluated in Japan recently and discuss their current status and threats as well as the countermeasures against these threats.

Green Infrastructure

Ecosystem services A
Ecosystem services B
Ecosystem services C

Green Infrastructure should have multiple ecosystem services

Paddy field

Provisioning services
(Food production)

Regulating services
(Flood control)

Habitat services
(Wetland habitat)

Paddy fields have several ecosystem services other than food production

Fig. 11.1 The relationship between the definition of green infrastructure (GI) and paddy fields

Then, we discuss the challenges in applying and expanding the idea of paddy fields as GI.

11.2 Habitat Service: Biodiversity Conservation in Paddy Fields

The idea that paddy fields can provide habitats for several wetland species is already well-established (Toral et al. 2012; Katayama et al. 2015; Osawa 2017; Osawa et al. 2020b). The review of the biota of Japanese paddy fields by Katayama et al. (2015) revealed that paddy fields can harbor several small aquatic organisms (Yamazaki et al. 2001), at least 184 weed (i.e., non-crop) species, 27 aquatic invertebrate species (not including rice pests), 15 species and 2 subspecies of amphibians, 11 fish species, and 49 bird species, even considering underestimation due to data availability (Katayama et al. 2015). The value of paddy fields as habitats for several wetland species and in contributing to regional biodiversity is clear. However, in recent years, a drastic decline in species has been observed in Japanese paddy fields (Katayama et al. 2015; Osawa 2017). This trend indicates that paddy fields have been recently losing their ecosystem services, namely, their habitat service. Losing an ecosystem service is linked to losing GI value. Thus, maintaining the habitat service is an important issue when considering paddy fields as GI.

Agricultural modernization is one of the most severe drivers of habitat degradation in paddy fields (Katayama et al. 2015; Osawa 2017). After World War II, Japanese agriculture increased both chemical usage and land consolidation, both related to mechanization (Katayama et al. 2015; Osawa 2017). These practices have been clearly successful in increasing the effectiveness of agricultural activities such as time efficiency and yield per area (Katayama et al. 2015). In other words, this strategy specifically improved the provisioning service of paddy fields. However, there is often a trade-off between such agricultural modernization and biodiversity in paddy fields (Katayama et al. 2015; Osawa et al. 2016b; Osawa 2017; Osawa and Mitsuhashi 2017). In particular, the effect of land consolidation, which changes the ecosystem structure from a wetland habitat to a dryland habitat, is severe because paddy fields act as alternative wetland habitats (Osawa et al. 2016b, 2020b). Wetlands are one of the habitats that are on the decline worldwide (Denny 1994; Lougheed et al. 2008). This is one of the reasons that could affect the value of paddy fields as GI. However, paddy fields that have experienced land consolidation provide little or no habitat service for wetland species (Osawa et al. 2016b, 2020b). Thus, because such paddy fields lack multiple ecosystem services, their GI value might be relatively low.

Agricultural modernization can be accompanied by agricultural abandonment, mainly in areas that are unsuitable for modernization, such as hilly and mountainous areas (Katayama et al. 2015). Agricultural abandonment signifies losing the provisioning service in that area, thus reducing the value of paddy fields as GI. Moreover, previous studies have suggested that agricultural abandonment could degrade the habitat service of paddy fields (Osawa et al. 2013; Uchida and Ushimaru 2014). Abandoned paddy fields have relatively little value as GI because they do not have any ecosystem services, i.e., at least provisioning and/or habitat services. Therefore, changes in agricultural land use, either through consolidation or abandonment, might reduce the land's value as GI. In particular, these could reduce the variety of the ecosystem service, because abandonment is a severe negative driver that causes the loss of not only habitat services but also provisioning services, which are essential ecosystem services in paddy fields.

11.3 Regulating Service: Disaster Risk Reduction (DRR) by Paddy Fields

In recent years, interest from both researchers and practitioners has grown in ecosystem services that reduce the frequency and severity of natural disasters (Dudley et al. 2015; Renaud et al. 2016; Ministry of Environment, Japan; https://www.env.go.jp/nature/biodic/eco-drr.html; accessed on April 28, 2021). These types of ecosystem services are often known as Ecosystem-based DRR (Eco-DRR) (Martin and Watson 2016; Furuta and Shimatani 2017; Scarano 2017; Nakamura et al. 2019; Osawa et al. 2020a). The use of GI for Eco-DRR has gained much research interest because of the low costs of introduction and maintenance and its provision of other ecosystem services (Sudmeier-Rieux et al. 2013; Furuta and Shimatani 2017; Onuma and Tsuge 2018; Osawa et al. 2020a). Paddy fields have been expected to function mainly for flood control, i.e., by increasing the water storage capacity of river basins, reducing the peak flow of rivers, and increasing groundwater recharge (Matsuno et al. 2006), all of which can help mitigate the risk of flood disaster. As an ecosystem service, Eco-DRR is a regulating service (Bennett et al. 2009), and thus we should also consider Eco-DRR to be relevant when considering paddy fields as GI.

Recently, studies on Eco-DRR for flooding in paddy fields have progressed. For instance, we have shown that paddy fields located in water storage zones, i.e., paddy fields that are located in previous wetland habitats, are very effective in preventing flooding (Osawa et al. 2020a). A simulation study revealed that paddy fields with runoff control devices installed in their drainage boxes effectively draw down flood by the storing rainwater (Yoshikawa et al. 2009, 2010). Thus, there is some evidence that paddy fields could have Eco-DRR potential against floods.

However, a previous study indicated that abandoned paddy fields could reduce the Eco-DRR function against flooding, both due to decreased penetration capability and water runoff from degraded ridges (Yoshida et al. 2012). Thus, abandonment could result in the loss of not only the provisioning service but also the regulating service of a paddy field. Moreover, as we have already shown in the previous section, abandonment could also result in loss of the habitat service (Osawa et al. 2013; Uchida and Ushimaru 2014). Conversely, to the best of our knowledge, there are no studies that indicate a negative effect of agricultural modernization, i.e., consolidation, on Eco-DRR. Consolidation might actually have a positive effect on Eco-DRR because concrete ridges could prevent water runoff from paddy fields, which could increase the capacity of water storage during heavy rain (Yoshikawa et al. 2010). Although both abandonment and consolidation are major changes in agricultural land use, their effects on paddy fields (in relation to GI) differ.

11.4   A Strategy to Use and Maintain Paddy Fields as GI

GI is expected to provide multiple ecosystem services (Sandstrom 2002; Tzoulas et al. 2007). Therefore, to use an ecosystem as GI, land managers and decision makers should focus on certain expected ecosystem services and then make the appropriate efforts to maintain and enhance these services. However, managing the target ecosystem services can be difficult, as ecosystem services can be related to each other (Bennett et al. 2009). Moreover, Bennett et al. (2009) reported that some ecosystem services could be positively or negatively influenced by one driver: human activity (Fig. 11.2). Thus, if land managers and decision makers put some effort into maintaining and enhancing only one ecosystem service, it might result in a reduction in other ecosystem services. Agricultural modernization of paddy fields is a typical example of this because it improves the provisioning service but reduces the habitat service (Osawa et al. 2016b, 2020b). Thus, if land managers and decision makers expect the habitat service from paddy fields when they use them as GI, it will be difficult to achieve the high yield and work efficiency provided by modernization. Conversely, previous studies indicate that organic farming, which is an environmentally friendly farming system, could contribute to the biodiversity of several taxa (Fuller et al. 2005; Hole et al. 2005; Katayama et al. 2019). Thus, organic farming could even be compatible with both provisioning and habitat services that have relatively low yields or low work efficiencies compared to fully modernized agricultural areas. This suggests that certain related ecosystem services that have a trade-off relationship could be compatible at moderate levels.

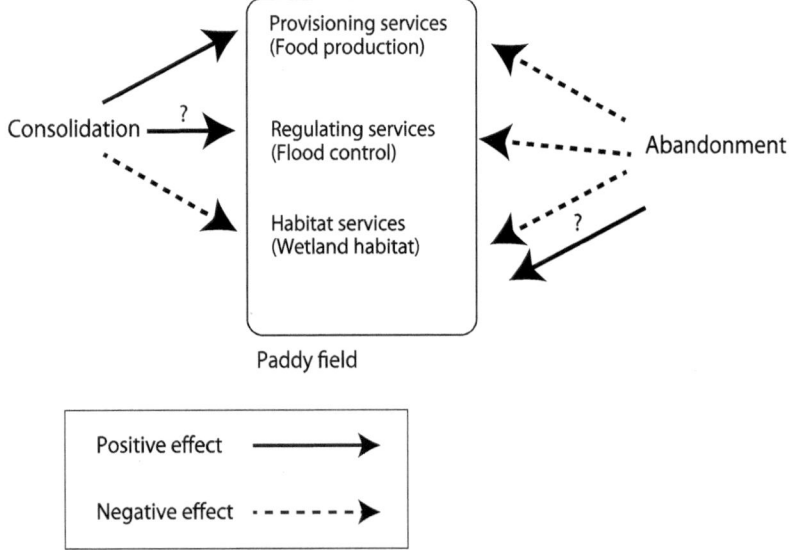

Fig. 11.2 Positive and negative drivers for ecosystem services in paddy fields

To achieve multiple ecosystem services from a target ecosystem to be considered as GI, land managers and decision makers should carefully manage these ecosystem services by considering the relationship between focused ecosystem services and the management method, i.e., the driver for ecosystem services.

However, abandoned paddy fields have already lost their essential role, namely, their provisioning service. Thus, abandoned paddy fields cannot act as agricultural GI land that should provide multiple ecosystem services including the provisioning service. In addition, we believe that abandonment not only results in the loss of the provisioning service but also reduces the regulating and habitat services (Yoshida et al. 2012; Osawa et al. 2013). Nevertheless, recent studies suggested that some abandoned farmlands, including paddy fields, can harbor the biodiversity of several taxa, i.e., provide habitat services (Morimoto et al. 2017; Yamanaka et al. 2017; Hanioka et al. 2018a, b), depending on the location, condition, and historical land use of that area (Osawa and Mitsuhashi 2017). As habitat services (i.e., supporting services) are prerequisites for all ecosystem services (Sarukhán et al. 2005), we could expect that areas with high habitat services could provide several ecosystem services. Therefore, for the use of abandoned paddy fields as GI, land managers and decision makers should change their strategy to focus on ecosystem services other than provisioning services, such as using areas like non-farmland. If an abandoned paddy has a high level of biodiversity, converting it into a conservation area could represent a practical option.

11.5   Conclusion and Perspective

In this paper, we have focused on two ecosystem services and have discussed the potential of paddy fields as GI as well as changing land uses as a driver threatening this potential. The most important point in this paper is that the significance of a paddy field as GI might be different from that of a paddy field as a specific agricultural area, i.e., an intense food production zone. As paddy fields that represent GI should have multiple ecosystem services, concentrating on the provisioning service alone through modernization is not an effective way to promote GI. If land managers and decision makers want to apply the idea of GI to paddy fields, they should adopt a different strategy that focuses on combining paddy fields with provisioning services and other ecosystem services. Traditional agricultural promotion basically concentrates on food production; however, this does not suit a strategy that considers a paddy field as GI. Moreover, many countries, including Japan, are currently experiencing a phase of population decline (Osawa 2017; https://population.un.org/wpp/; accessed on April 29, 2021). Given this situation, a strategy that concentrates on food production might not be an effective strategy for maintaining paddy fields because of decreasing food demands. Other than the provisioning service (i.e., food production), we assessed two ecosystem services, providing wetland habitat and Eco-DRR, that are appropriate for paddy fields as GI; however, it remains unclear which service could coexist with the provisioning service. If a trade-off exists between food production and other services, not focusing on food production might be one option for land managers and decision makers to promote paddy fields as GI, considering the situation of population decline. Fortunately, previous studies have suggested that flood control (i.e., Eco-DRR) and food production could coexist (Kramer et al. 1997; Bennett et al. 2009), whereas there might be a trade-off between the habitat service and agricultural modernization, which contributes to food production (Katayama et al. 2015; Osawa et al. 2016b; Osawa and Mitsuhashi 2017) (Fig. 11.2). To balance food production and the habitat service, these ecosystem services should be maintained at a moderate level.

Nevertheless, one driver could enhance multiple ecosystem services (Bennett et al. 2009). Thus, focusing on compatible ecosystem services could be another effective option to promote GI. Flood control and food production is one potential combination. However, in our opinion, enhancing human activities such as consolidation is not a smart way to expand on the concept of GI because ecosystem services should depend on inherent ecosystem functions. Ideas of ecological intensification already exist and are defined as realizing the intensive and smart use of ecosystem functions for multiple ecological services (Bommarco et al. 2013; Tittonell 2014). This should represent a nature-based solution, which is defined as a transition toward ecosystem services with decreased input of nonrenewable natural capital and increased investment in renewable natural processes (Maes and Jacobs 2017). Ecosystem- or nature-based approaches should be one of the most important ideas to be applied to the idea of GI.

In addition, we should emphasize that abandonment is a driver of the loss of essential ecosystem services, namely, the provisioning service in paddy fields (Fig. 11.2). Therefore, land managers and decision makers should strive to prevent agricultural abandonment when they want to promote paddy fields as GI. However, when abandonment does occur, land managers and decision makers have different options, including recovering the paddy field (i.e., provisioning service) or recovering its natural ecosystem. Regardless of the option selected for restoration, efforts should be made to support the recovery of any ecosystem service (Osawa et al. 2016b). In abandoned paddy fields, food production is probably not the most important ecosystem service. Thus, returning the paddy fields to an intact ecosystem is one of the options. Given the importance of GI, human society should invest in maintaining such infrastructure. Moreover, promoting GI is a cost-effective strategy because of its multiple benefits to people. Overall, paddy fields have good potential as GI because of their multiple ecosystem services and widespread distribution.

Acknowledgments This work was supported by the Environment Research and Technology Development Fund (4-1805) of the Ministry of the Environment, Japan, and partially supported by the JSPS KAKENHI Grant Number 20K06096. Drs. F. Nakamura and N. Katayama provide us many useful suggestions.

References

Ahmad S, Li C, Dai G et al (2009) Greenhouse gas emission from direct seeding paddy field under different rice tillage systems in Central China. Soil Tillage Res 106:54–61

Benayas JMR, Martins A, Nicolau JM, Schulz JJ (2007) Abandonment of agricultural land: an overview of drivers and consequences. CAB Rev Perspect Agric Vet Sci Nutr Nat Resour 2:1–14

Bennett EM, Peterson GD, Gordon LJ (2009) Understanding relationships among multiple ecosystem services. Ecol Lett 12:1394–1404

Bommarco R, Kleijn D, Potts SG (2013) Ecological intensification: harnessing ecosystem services for food security. Trends Ecol Evol 28:230–238. https://doi.org/10.1016/j.tree.2012.10.012

Denny P (1994) Biodiversity and wetlands. Wetl Ecol Manag 3:55–611. https://doi.org/10.1007/BF00177296

Dudley N, Buyck C, Furuta N et al (2015) Protected areas as tools for disaster risk reduction. A handbook for practitioners. IUCN, Gland

Fuller RJ, Norton LR, Feber RE et al (2005) Benefits of organic farming to biodiversity vary among taxa. Biol Lett 1:431–434

Furuta N, Shimatani Y (2017) Integrating ecological perspectives into engineering practices - perspectives and lessons from Japan. Int J Disaster Risk Reduct 32:87–94. https://doi.org/10.1016/j.ijdrr.2017.12.003

Hanioka M, Yamaura Y, Senzaki M et al (2018a) Assessing the landscape-dependent restoration potential of abandoned farmland using a hierarchical model of bird communities. Agric Ecosyst Environ 265:217–225. https://doi.org/10.1016/j.agee.2018.06.014

Hanioka M, Yamaura Y, Yamanaka S et al (2018b) How much abandoned farmland is required to harbor comparable species richness and abundance of bird communities in wetland? Hierarchical community model suggests the importance of habitat structure and landscape context. Biodivers Conserv 27:1831–1848. https://doi.org/10.1007/s10531-018-1510-5

Hole DG, Perkins AJ, Wilson JD et al (2005) Does organic farming benefit biodiversity? Biol Conserv 122:113–130

Jincai W, Guowen H, Jian T et al (1994) Studies on the regulation effect of natural insect on the community food web in paddy field. Acta Ecol Sin 14:381–386

Katayama N, Baba YG, Kusumoto Y, Tanaka K (2015) A review of post-war changes in rice farming and biodiversity in Japan. Agric Syst 132:73–84

Katayama N, Osada Y, Mashiko M et al (2019) Organic farming and associated management practices benefit multiple wildlife taxa: a large-scale field study in rice paddy landscapes. J Appl Ecol 56:1970–1981

Kramer RA, Richter DD, Pattanayak S, Sharma NP (1997) Ecological and economic analysis of watershed protection in Eastern Madagascar. J Environ Manag 49:277–295

Lougheed VL, Mcintosh MD, Parker CA, Stevenson RJ (2008) Wetland degradation leads to homogenization of the biota at local and landscape scales. Freshw Biol 53:2402–2413. https://doi.org/10.1111/j.1365-2427.2008.02064.x

Maes J, Jacobs S (2017) Nature-based solutions for Europe's sustainable development. Conserv Lett 10:121–124

Martin TG, Watson JEM (2016) Intact ecosystems provide best defence against climate change. Nat Clim Chang 6:122–124. https://doi.org/10.1038/nclimate2918

Matsuno Y, Nakamura K, Masumoto T et al (2006) Prospects for multifunctionality of paddy rice cultivation in Japan and other countries in monsoon Asia. Paddy Water Environ 4:189–197. https://doi.org/10.1007/s10333-006-0048-4

Morimoto J, Shibata M, Shida Y, Nakamura F (2017) Wetland restoration by natural succession in abandoned pastures with a degraded soil seed bank. Restor Ecol 25:1005–1014. https://doi.org/10.1111/rec.12516

Nakamura F, Akasaka T, Ishiyama N et al (2019) Adaptation to climate change and conservation of biodiversity using green infrastructure. River Res Appl 1–13. https://doi.org/10.1002/rra.3576

Natuhara Y (2013) Ecosystem services by paddy fields as substitutes of natural wetlands in Japan. Ecol Eng 56:97–106

Onuma A, Tsuge T (2018) Comparing green infrastructure as ecosystem-based disaster risk reduction with gray infrastructure in terms of costs and benefits under uncertainty: a theoretical approach. Int J Disaster Risk Reduct 32:22–28. https://doi.org/10.1016/j.ijdrr.2018.01.025

Osawa T (2017) Agricultural land use policy in Japan in an era of declining population. Wildlife Hum Soc 5:17–27; (in Japanese)

Osawa T, Mitsuhashi H (2017) Trial for area zoning in Japanese agricultural area based on ecological function. Ecol Civ Eng 19:211–220

Osawa T, Kohyama K, Mitsuhashi H (2013) Areas of increasing agricultural abandonment overlap the distribution of previously common, currently threatened plant species. PLoS One 8:e79978. https://doi.org/10.1371/journal.pone.0079978

Osawa T, Kohyama K, Mitsuhashi H (2016a) Multiple factors drive regional agricultural abandonment. Sci Total Environ 542:478–483. https://doi.org/10.1016/j.scitotenv.2015.10.067

Osawa T, Kohyama K, Mitsuhashi H (2016b) Trade-off relationship between modern agriculture and biodiversity: heavy consolidation work has a long-term negative impact on plant species diversity. Land Use Policy 54:78–84. https://doi.org/10.1016/j.landusepol.2016.02.001

Osawa T, Nishida T, Oka T (2020a) High tolerance land use against flood disasters: how paddy fields as previously natural wetland inhibit the occurrence of floods. Ecol Indic 114:106306

Osawa T, Nishida T, Oka T (2020b) Paddy fields located in water storage zones could take over the wetland plant community. Sci Rep 10:1–8

Renaud FG, Sudmeier-Rieux K, Estrella M, Nehren U (2016) Ecosystem-based disaster risk reduction and adaptation in practice. Springer, Cham. https://doi.org/10.1007/978-3-319-43633-3

Sandstrom U (2002) Green infrastructure planning in urban Sweden. Plan Pract Res 17:373–385

Sarukhán J, Whyte A, Hassan R et al (2005) Millennium ecosystem assessment: ecosystems and human well-being

Scarano FR (2017) Ecosystem-based adaptation to climate change: concept, scalability and a role for conservation science. Perspect Ecol Conserv 15:65–73. https://doi.org/10.1016/j.pecon.2017.05.003

Sudmeier-Rieux K, Ash N, Murti R (2013) Environmental guidance note for disaster risk reduction. IUCN, Gland, Switzerland. https://www.iucn.org/sites/dev/files/content/documents/2013_iucn_bookv2.pdf

Tittonell P (2014) Ecological intensification of agriculture-sustainable by nature. Curr Opin Environ Sustain 8:53–61. https://doi.org/10.1016/j.cosust.2014.08.006

Toral GM, Stillman RA, Santoro S, Figuerola J (2012) The importance of rice fields for glossy ibis (Plegadis falcinellus): management recommendations derived from an individual-based model. Biol Conserv 148:19–27

Tzoulas K, Korpela K, Venn S et al (2007) Promoting ecosystem and human health in urban areas using green infrastructure: a literature review. Landsc Urban Plan 81:167–178

Uchida K, Ushimaru A (2014) Biodiversity declines due to abandonment and intensification of agricultural lands: patterns and mechanisms. Ecol Monogr 84:637–658. https://doi.org/10.1890/13-2170.1

Yamanaka S, Akasaka T, Yabuhara Y, Nakamura F (2017) Influence of farmland abandonment on the species composition of wetland ground beetles in Kushiro, Japan. Agric Ecosyst Environ 249:31–37. https://doi.org/10.1016/j.agee.2017.07.027

Yamazaki M, Hamada Y, Ibuka T et al (2001) Seasonal variations in the community structure of aquatic organisms in a paddy field under a long-term fertilizer trial. Soil Sci Plant Nutr 47:587–599

Yang S, Peng S, Xu J et al (2012) Methane and nitrous oxide emissions from paddy field as affected by water-saving irrigation. Phys Chem Earth Parts A/B/C 53:30–37

Yoon CG (2009) Wise use of paddy rice fields to partially compensate for the loss of natural wetlands. Paddy Water Environ 7:357–366. https://doi.org/10.1007/s10333-009-0178-6

Yoshida T, Masumoto T, HOrikawa N (2012) Changes in rainfall-runoff characteristics of small watersheds due to cultivation conditions of hilly paddies. Trans Jpn Soc Irrig Drain Rural Eng 80:117–124; (in Japanese)

Yoshikawa N, Nagao N, Misawa S (2009) Watershed scale evaluation of flood mitigation function of Paddy fields installed with runoff control devices. Trans Jpn Soc Irrig Drain Rural Eng 77:273–280. (in Japanese)

Yoshikawa N, Nagao N, Misawa S (2010) Evaluation of the flood mitigation effect of a paddy field dam project. Agric Water Manag 97:259–270

Part IV
Wetland Ecosystem (Including Flood-Control Pond)

Chapter 12
Flood-Control Basins as Green Infrastructures: Flood-Risk Reduction, Biodiversity Conservation, and Sustainable Management in Japan

Nobuo Ishiyama, Satoshi Yamanaka, Keita Ooue, Masayuki Senzaki, Munehiro Kitazawa, Junko Morimoto, and Futoshi Nakamura

Abstract Green infrastructure (GI) is a strategic planning infrastructure that uses the functions of ecosystems. Under an increased river flood risk, flood-risk management utilizing GI is gaining attention from managers and ecologists in Japan. Flood-control basins are facilities that temporarily store river water in adjacent reservoirs to mitigate flood peaks and gradually drain the water back to the main channels after a flood. GI is expected to provide multiple functions, such as flood-risk reduction and habitat provisions. However, there are limited studies on the ecological functions of flood-control basins. In this article, we first introduce the characteristics of flood-control basins constructed in Japan. Next, we show the ecological importance of flood-control basins in terms of wetland organism biodiversity conservation. Finally, to aid the integration of GI into conventional flood-control measures, we highlight ecological and social issues about introducing and managing flood-control basins.

Keywords Climate change · Biodiversity conservation · Ecosystem-based adaptation · Ecosystem services · Flooding · Habitat connectivity · Sustainable development

N. Ishiyama (✉)
Forest Research Institute, Hokkaido Research Organization, Bibai, Hokkaido, Japan

S. Yamanaka
Hokkaido Research Center, Forestry and Forest Products Research Institute, Sapporo, Japan

K. Ooue
Environmental Policy Division, Ministry of Land, Infrastructure, Transport and Tourism, Tokyo, Japan

M. Senzaki
Faculty of Environmental Earth Science, Hokkaido University, Sapporo, Hokkaido, Japan

M. Kitazawa · J. Morimoto · F. Nakamura
Graduate School of Agriculture, Hokkaido University, Sapporo, Hokkaido, Japan

© The Author(s) 2022
F. Nakamura (ed.), *Green Infrastructure and Climate Change Adaptation*,
Ecological Research Monographs, https://doi.org/10.1007/978-981-16-6791-6_12

12.1 Introduction

Among disasters triggered by natural hazards, flood disasters have been most frequently reported worldwide. Under a changing climate, an increased flood risk is predicted to affect human and economic losses globally (Dottori et al. 2018). Historical records from 1962 to 2011 in the central United States demonstrated an increase in flooding frequency (Mallakpour and Villarini 2015). In Europe, peaks of 1/100 river floods are projected to double in frequency within the next three decades (Alfieri et al. 2015). Jongman et al. (2012) reported that the amount of the global population exposed to a 1/100 river flood reached 800 million by 2010, of which 73% was living in Asia. Moreover, Dottori et al. (2018) estimated that the population exposed to flooding will increase as a result of anthropogenic warming; an average increase of more than 120% is expected in a 3 °C-warming scenario.

Considering the increased disaster risk, adaptation efforts for flood-risk management are urgently needed. In Japan, heavy rainfall events (e.g., those above 50 mm/h) have increased in the last half a century (Fig. 12.1), and a flood disaster with the most severe economic damage occurred in 2019 (Fig. 12.2). Typhoon Hagibis in 2019 bore down on central Japan and caused 19 billion USD in economic damage. The heavy rainfall event caused 142 levee collapses and overflowed along rivers managed by the Japanese government, resulting in ca. 25,000 ha of inundated land. Flood-control measures in Japan have focused on the construction of dams and artificial levees for the last century. These conventional gray infrastructures usually

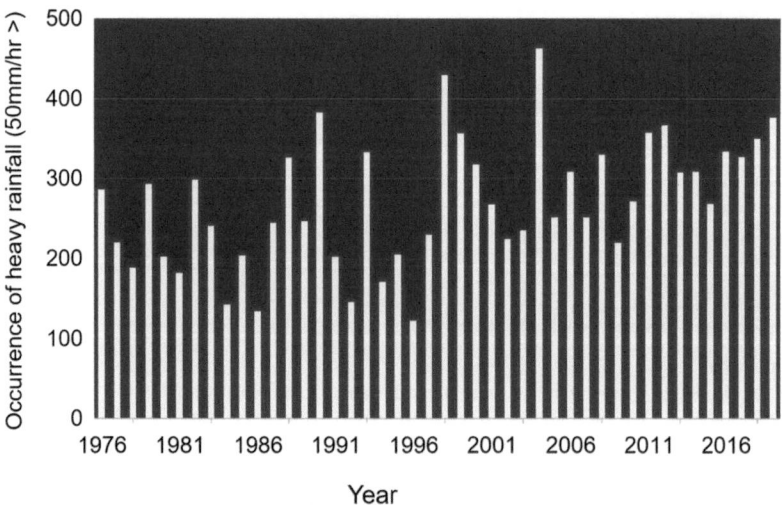

Fig. 12.1 Historical occurrence of heavy rainfall events in Japan. The number of monitoring sites is 1300. In the first decade (1976–1985) and last decade (2010–2019), the mean annual occurrences were 226 and 327, respectively. (Data are provided by the Japan Meteorological Agency (https://www.data.jma.go.jp/cpdinfo/extreme/extreme_p.html))

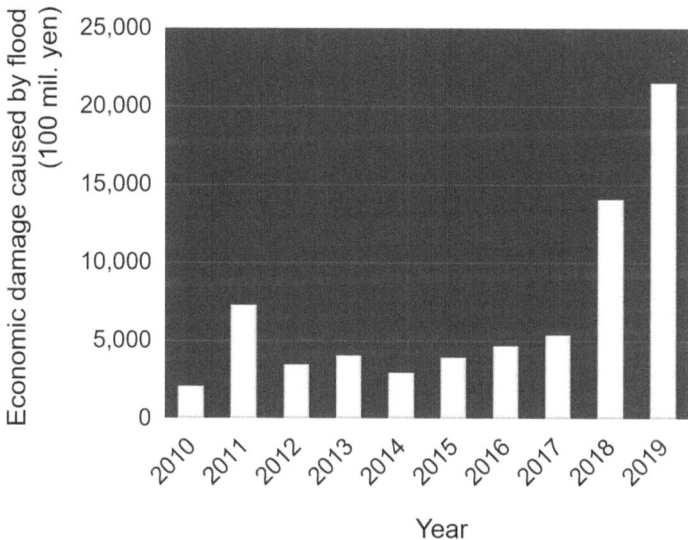

Fig. 12.2 Historical economic damage caused by flood disasters in Japan. (Data are provided by the Japanese Ministry of Land, Infrastructure, Transport and Tourism (https://www.mlit.go.jp/report/press/content/001359046.pdf))

assure 100% disaster protection until the magnitude of the disaster reaches an upper limit determined by the prevention plan, but the function will completely fail once the magnitude exceeds the upper limits (Nakamura et al. 2020). Against the back-drop of global warming, it has been widely recognized that conventional measures that depend highly on gray infrastructures are no longer adequate for flood-risk management. The Ministry of Land, Infrastructure, Transport and Tourism (MLIT) has started to shift conventional measures to basin-wide flood-risk management that focuses on both river and floodplain management using green infrastructure (GI).

GIs are conceptually classified into fundamental GI (GI-1) and multilevel GI (GI-2) (Nakamura et al. 2020; see Chap. 2 for details). In flood-risk management, GI-1 are natural ecosystems, and GI-2 are seminatural basins that reduce or delay river flooding. In 2016, Hokkaido Island was hit by three typhoons in one summer, which was the first time this occurred since records began in 1952. Nakamura et al. (2020) analyzed Kushiro wetland's water retention function (GI-1), the largest remnant wetland in Japan. Their study reported that a hydrograph of the simulation case of a partial loss of wetlands (ca. 55% loss) showed 1.5 times higher peak discharge and a 2-day-faster peak arrival. However, natural wetlands have been globally reduced by human activities (Davidson 2014), and the importance of GI-2 would increase in regions where the function of natural ecosystems is degraded. Considering the effectiveness of GI, MLIT tries to integrate both gray and green infrastructures (GI-1 and GI-2) into basin-wide flood-risk management to adapt to the increasing hazard risk. This combination of different GIs is referred to as "hybrid infrastructures" in Nakamura et al. (2020) (see Chap. 2).

In Japan, flood-control basins are rapidly gaining attention as a practical multilevel GI (GI-2). Multifunctionality is a fundamental property of sustainable GI (Lovell and Taylor 2013; Wang and Banzhaf 2018). Biodiversity conservation in a changing climate and landscape is another key challenge, and GI can be a tool for mitigating or restoring declined biodiversity (Nakamura et al. 2020). In this article, we first introduce the characteristics of flood-control basins constructed in Japan. Next, we show the ecological importance of flood-control basins in terms of wetland organism biodiversity conservation. Finally, to aid the integration of GI into conventional flood-control measures in Japan, we would like to raise ecological and social issues about introducing and managing flood-control basins. We believe that this chapter provides insight into the future management of flood-control basins in other Asian countries experiencing rapid urbanization and increased flooding risk.

12.2   Flood-Control Basins in Japan

There are two types of flood-risk management using retention/detention ponds. The first is to store rainwater in floodplain areas to reduce runoff into the main channels and mitigate the flood peak in farmland and urban areas. Urban retaining ponds are examples, and rice paddies also may serve a similar function. Another is to temporarily store river water in adjacent reservoirs to mitigate the flood peak and gradually drain the water back into the main channels after the flood. The latter type is called "*Yusuichi*" in Japanese. In this chapter, we focus on *Yusuichi* and represent it as a flood-control basin.

Flood-control basins generally consist of reservoirs surrounded by artificial levees and an overflow embankment or sluice gate adjacent to rivers (Fig. 12.3). Reservoirs of flood-control basins are used for various purposes, such as sports

a)   b)

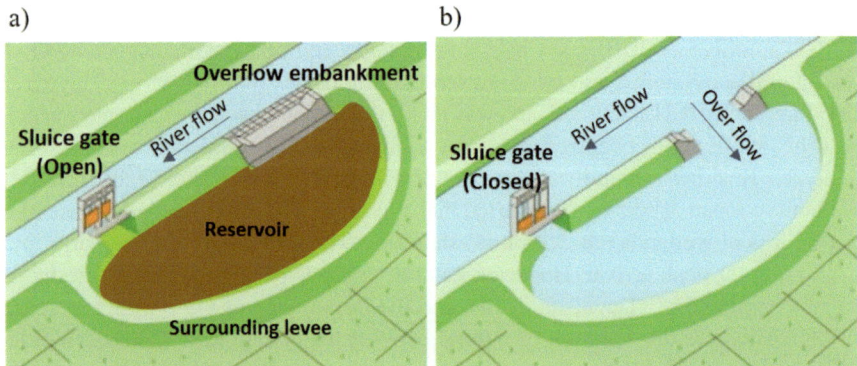

Fig. 12.3 Illustration of a flood-control basin. (Adapted from materials provided by the Hokkaido Regional Development Bureau (https://www.hkd.mlit.go.jp/sp/kasen_keikaku/kluhh40000001qfy. html)). (**a**) Normal flow. (**b**) High flow

grounds, farmland, urban parks, and wildlife habitat, and the presence or absence of permanent water varies depending on the region and the type of usage. In Japan, flood-control basins have been constructed nationwide (Suwa and Nishihiro 2020), and their construction is also planned in several regions to reduce flood risks (e.g., Hitachi River and National Highway Office 2020; Kumamoto Prefecture 2020). These flood-control basins are reported to have mitigated the disaster risk level. For example, the four flood-control basins located in the Tone and Watarase rivers in central Japan are estimated to have stored 250 million cubic meters of river water during Typhoon Hagibis in 2019, resulting in mitigating flood damage downstream (Tone River Upstream Office 2020).

Flood-control basins are also assumed to provide alternative habitats for wetland species. Suwa and Nishihiro (2020) demonstrated that most flood-control basins in Japan are located in floodplains and that 88% of them have natural observation areas, paddy fields, reed marshes, or water surfaces, suggesting that they can potentially provide wetland environments. Therefore, there is a possibility that flood-control basins can be used not only for disaster risk reduction but also for regional biodiversity conservation.

12.3 Case Study: Biodiversity Conservation in Flood-Control Basins

Flood-control basins are assumed to contribute to maintaining regional biodiversity and thus can work as GI. For example, the Watarase flood-control basin in central Japan was designated as a Ramsar site having the largest reed bed and provides wetland habitats for many species, including endangered species (Ministry of Environment of Japan 2020b). However, there are limited studies assessing the importance of flood-control basins for regional biodiversity conservation. Therefore, we examined wetland species in flood-control basins recently implemented in the Chitose River basin in Hokkaido, northern Japan (Yamanaka et al. 2020).

In this region, river flooding often occurred because of the gentle riverbed slope (1/7000 on average). In particular, a heavy rain event in August 1981 led to flooding and caused severe damage to urban areas and farmlands in this region (Hokkaido Regional Development Bureau 2018; Segawa et al. 2008). To mitigate flood risk, MLIT drove the construction of six flood-control basins near the main stream and tributary of the Chitose River (Fig. 12.4). Construction started in 2008 and finished in 2020. The reservoirs of the basins comprise a total of 1150 ha, and they have wetland environments where river water accumulates, except in areas used for pasture and other purposes (Fig. 12.5). In the Chitose River basin, agricultural land use, such as rice paddies and cropland, dominates, and there are many watercourses for irrigation. The expansion of agricultural land in this area began approximately 100 years ago, and most of the natural wetlands have been converted to farmland (GSI 2000), resulting in a massive decrease in habitats

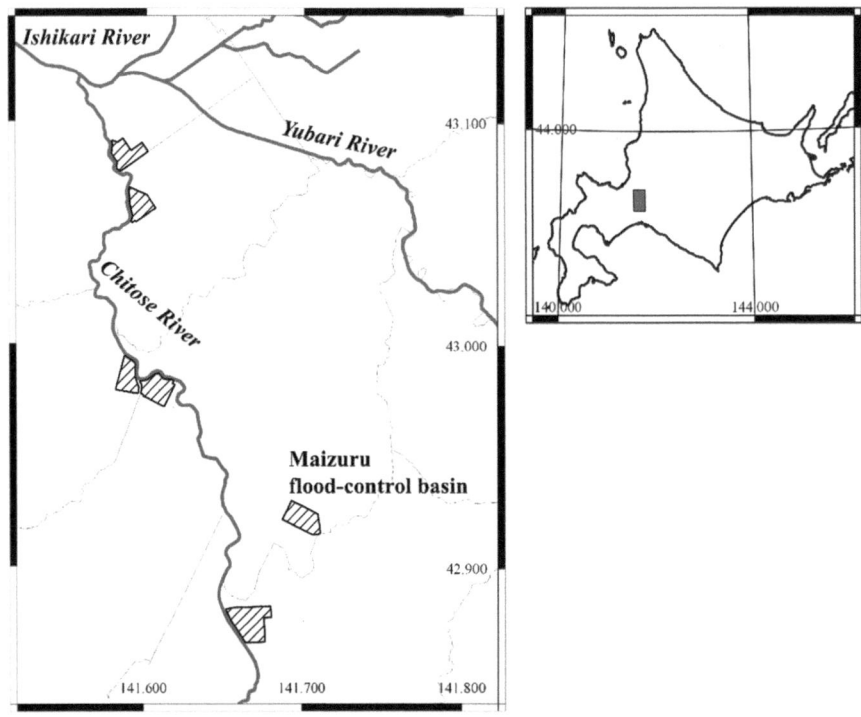

Fig. 12.4 Location of the Chitose flood-control basins

for wetland species. Therefore, flood-control basins consisting of new wetland environments are expected to provide alternative habitats for wetland biota.

We examined the species composition of four wetland taxa, fish, aquatic insects, birds, and plants, in five flood-control basins in summer 2016 (Yamanaka et al. 2020). In 2016, the construction of the basins was not completed except for the Maizuru basin (the construction of all basins was completed and operations started in 2020). Thus, we surveyed the above taxa in a part of the reservoir of each flood-control basin. We also examined the species composition of three other waterbodies in this area (channelized watercourses, drainage pumping stations, and remnant floodplain ponds) to compare the species compositions with those of flood-control basins.

We found that flood-control basins have a comparable or higher species richness and abundance of wetland species than other waterbodies (e.g., remnant floodplain ponds and drainage pumping stations) (Fig. 12.6). We also found that flood-control basins were characterized by some pioneer species that preferred shallow water or adapted to fluctuations in water levels (e.g., herbivorous insects, shorebirds, and hygrophyte plants). However, channelized watercourses, which are widely distributed in the study region, have lower species richness and abundance (Fig. 12.6). This result could be because of their simplified habitats led by channelization.

Fig. 12.5 Chitose flood-control basins

We also found some red list species of each taxon in flood-control basins although there was lower abundance and richness than those in drainage pump stations and remnant floodplain ponds. These results suggest that flood-control basins provide alternative habitats for wetland species, including endangered species. Nevertheless, for fishes, we observed a high abundance of nonnative species, such as *Pseudorasbora parva* and *Rhodeus ocellatus ocellatus*, in some flood-control basins (Fig. 12.6b).

In the Maizuru flood-control basin, whose construction finished in 2016, the breeding of the red-crowned crane, *Grus japonensis*, was observed in 2020 (Ministry of Environment of Japan 2020a). The red-crowned crane is an endangered species whose population in Japan experienced a significant decrease approximately 100 years ago, and is one of the flagship wetland species in Japan. The Ministry of the Environment has implemented conservation measures for this species, such as promoting breeding, and the distribution area is expanding from the east to the west of Hokkaido Island. Therefore, flood-control basins are also expected to contribute to the dispersal and recolonization of this species into uncolonized areas.

Our findings suggest that newly created environments in the Chitose flood-control basins provide suitable habitats for wetland species. However, there is room for future research to evaluate the ecological function of flood-control basins.

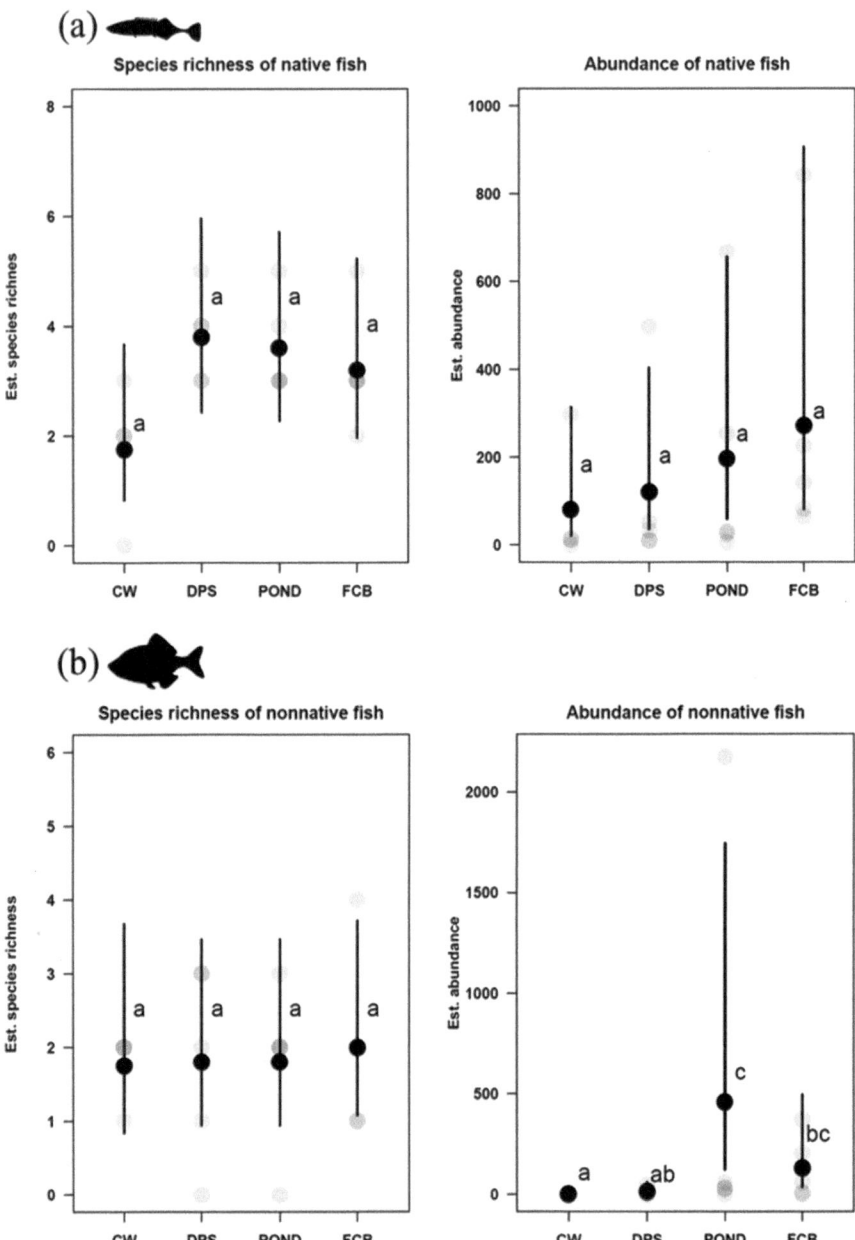

Fig. 12.6 (continued)

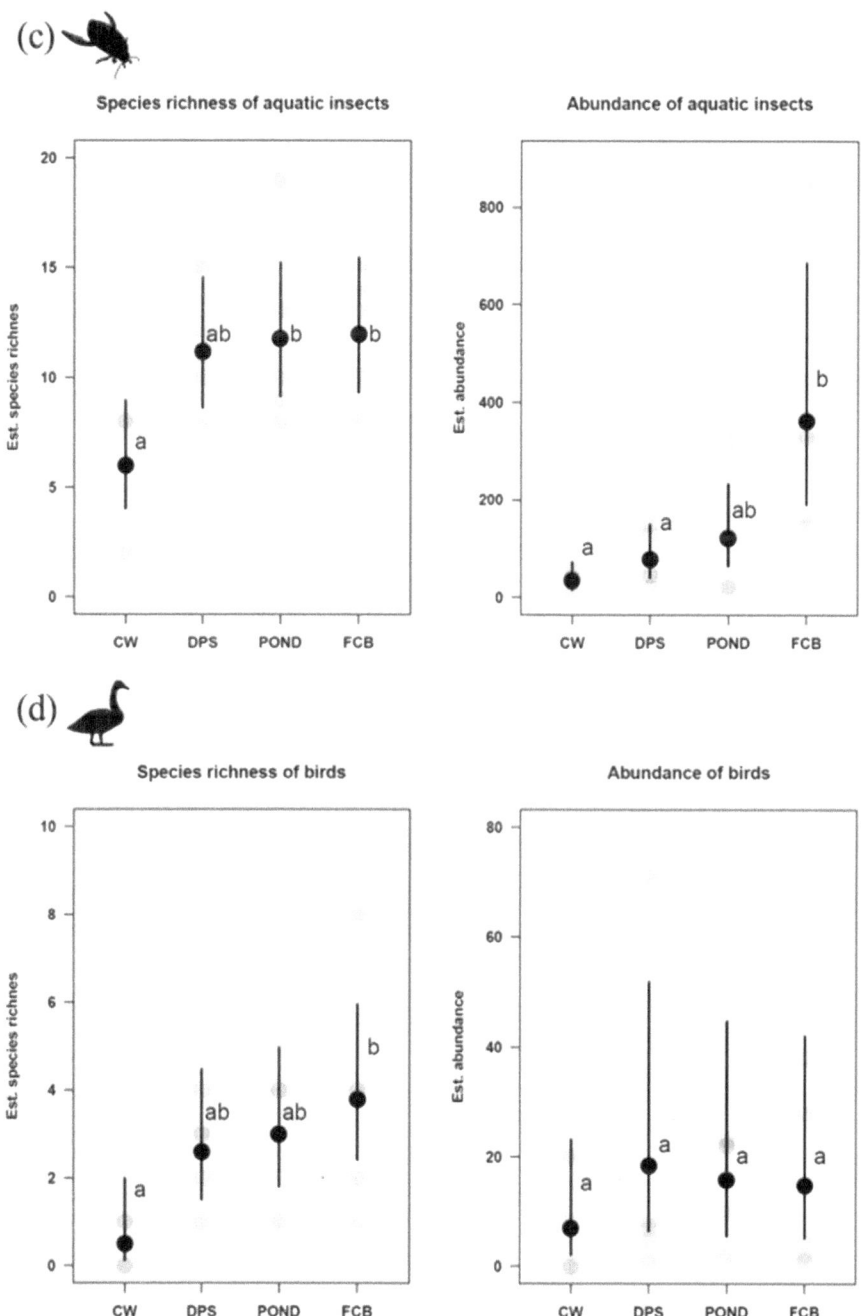

Fig. 12.6 (continued)

(e)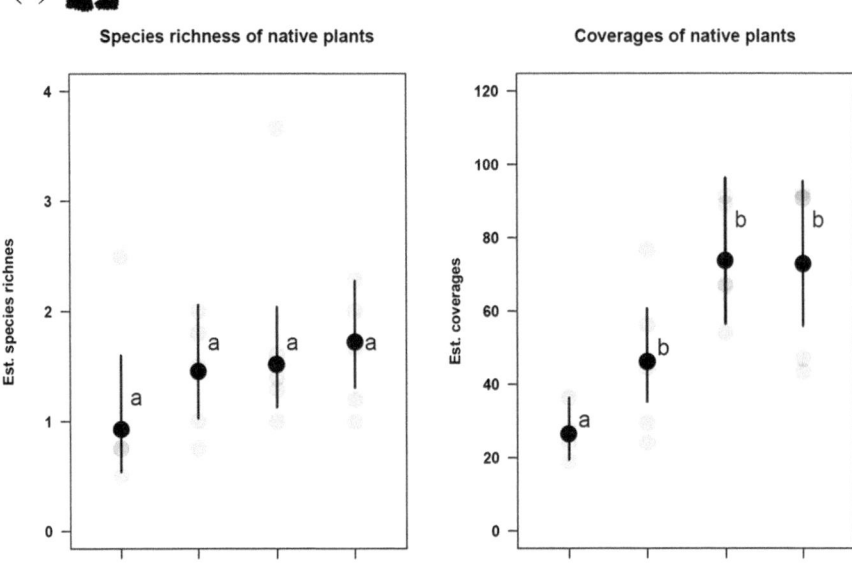

Fig. 12.6 Estimated species richness and abundance of four taxa; (a) native fishes, (b) nonnative fishes, (c) aquatic insects, (d) birds and (e) plants. *CW* channelized watercourse, *DPS* drainage pumping station, *POND* remnant pond, *FCB* flood-control basin. Black circles denote values estimated by generalized linear models (GLMs). The whiskers indicate the 95% confidence interval (CI). Gray circles denote each observed value. Different letters indicate significant differences in the multiple comparison analysis ($p < 0.05$). The values for species richness and coverage of vegetation indicate values per quadrat (2×2 m). (See Yamanaka et al. (2020) for details)

First, Yamanaka et al. (2020) limited the study season to summer. The importance of flood-control basins for biodiversity conservation can vary with the studied season because wetland species use different environments in different seasons. For example, the importance of flood-control basins as spawning sites for some fishes in spring and as stopover sites for immigrant birds in spring and autumn was not examined in a previous study. Second, Yamanaka et al. (2020) did not consider wetland-plant succession. Accumulated sediment delivered with flooding in reservoirs will change the water level, which could change the vegetation types from hydrophytic to terrestrial plants. Such changes in vegetation will affect habitat qualities for other wetland organisms. For effective biological conservation using flood-control basins, further studies are needed to assess the species composition of flood-control basins with different seasons, and the succession should be monitored over a long-term period.

12.4 Future Issues for the Construction and Management of Flood-Control Basins

12.4.1 Social Issues for Construction

To adapt to increased disaster risks caused by climate change, the planning and construction of flood-control basins are urgently needed in Japan. Nevertheless, the construction of flood-control basins in floodplain areas tends to be costly in terms of both time and money. The construction of flood-control basins requires a large space around rivers; however, most areas of floodplains in Japan are used for residential areas and farmland. Managers therefore spend much time negotiating with landowners to acquire land for construction. For the Chitose flood-control basins, lands for each basin were owned by approximately 50 people, and it took 3 years for the land acquisition process. Managers can leave private lands, such as farmland, within reservoirs, but the approach could be costly because they need to establish easements for private lands and provide momentary compensation to landowners. Therefore, a pre-investigation of information on landowners is essential for construction to start quickly. It is not always possible to construct a large flood-control basin due to many landowners and stakeholders (e.g., nature conservation groups), while a large storage capacity is needed for flood-risk management. In such cases, managers need to consider more feasible plans, such as selecting sites for multiple small basins to achieve the total desired storage capacity.

Another solution for rapid construction is utilizing unused land, such as abandoned farmland. Farmland was abandoned after the period of economic growth in the 1950s and 1960s in developed countries, including Japan (e.g., Kobayashi et al. 2020). The use of such degraded lands with depopulation is increasingly recognized as a tool for biodiversity conservation (Ishiyama et al. 2020a; Nakamura et al. 2020). For example, Nakamura et al. (2020) proposed a prioritization technique for construction sites by estimating the distribution of abandoned farmland and the biodiversity of wetland organisms (e.g., birds and plants). In their study, created maps overlapped with a flood hazard map for selection and revealed a financially and ecologically high-priority area.

12.4.2 Ecological Issues for Constructions

GI is defined as "a strategically planned network of natural and seminatural areas with other environmental features designed and managed to deliver a wide range of ecosystem services" (European Commission 2013). In human-modified floodplains, the contemporary migration of wetland organisms has been spatially restricted (e.g., Ishiyama et al. 2015b). However, artificial watercourses and remnant wetlands create a seminatural wetland network in current landscapes, and landscape connectivity supports the high biodiversity of wetland organisms (Ishiyama et al.

2014, 2015a). Flood-control basins provide large open water spaces compared to other lentic waterbodies in human-modified floodplains (Yamanaka et al. 2020), suggesting that the construction of flood-control basins potentially contributes to enhancing a habitat network of wetland organisms by interacting with the existing lentic waterbodies (Ishiyama et al. 2017). Spatial network analyses can be a solution for creating "a strategically planned network" (Hermoso et al. 2020). One of the simple methods is site-scale measures that consider direct connections between the focal habitat patch and surrounding patches. The nearest neighbor connectivity measure (e.g., the distance between the focal habitat patch and the nearest patch) and the buffer measure (e.g., the summed area of the habitat patch within a circle around the focal habitat patch) are typical examples (Moilanen and Nieminen 2002). The site-scale measures require little data because the method is simple, as mentioned above, which is a positive aspect of the method for practitioners in the spatial planning of flood-control basins. However, the measures potentially evaluate the connectivity of candidate sites with low precision, especially for highly mobile animals such as insects and waterfowls. This is because the dispersal ranges of such animals are not restricted to the area around the focal habitat patch; they seasonally use multiple patches in the landscape by using stepping-stone patches. That is to say, both direct and indirect connections among habitat patches should be assessed for such animals. Alternatively, regional-scale measures can prioritize candidate sites for flood-control basin construction by considering the importance of stepping-stone habitats (e.g., Saura and Rubio 2010). For instance, a graph-theoretical approach can calculate overall connectivity (i.e., regional-scale habitat availability) by considering the spatial position and size of habitat patches (Saura and Torné 2009). By using the regional-scale measure, practitioners can assess the importance of individual patches in maintaining the entire habitat network (e.g., Ishiyama et al. 2014, 2015a). However, the regional-scale measures may be more strenuous for practitioners compared to buffer measures due to more complicated operations for creating a landscape graph and matrix calculations. More effective sites for improving existing wetland networks can be selected by using the network measures according to the species traits of conservation targets and available conservation resources.

However, improving a habitat network can have adverse effects on indigenous species. A key reason for the species decline is expanding introduced species (Rudnick et al. 2012). We therefore should carefully select the construction sites of flood-control basins and the connections with other waterbodies in consideration of the potential negative impacts. In northern Japan, many floodplain waterbodies, such as oxbow lakes and backswamps, are invaded by a small invaded cyprinid, *Pseudorasbora parva*. Cyprid abundance is higher in waterbodies with high hydrologic connectivity (Ishiyama et al. 2020b). Ishiyama et al. (2020b) showed that populations of the endangered minnow *Rhynchocypris percnurus sachalinensis*, which has an ecological niche similar to that of *P. parva*, decline with the invasion of *P. parva*, suggesting that small isolated waterbodies function as refuges for the endangered minnow (Fig. 12.7). Nevertheless, small isolated habitats are easily diminished and degraded by human activities. Considering the importance of small-

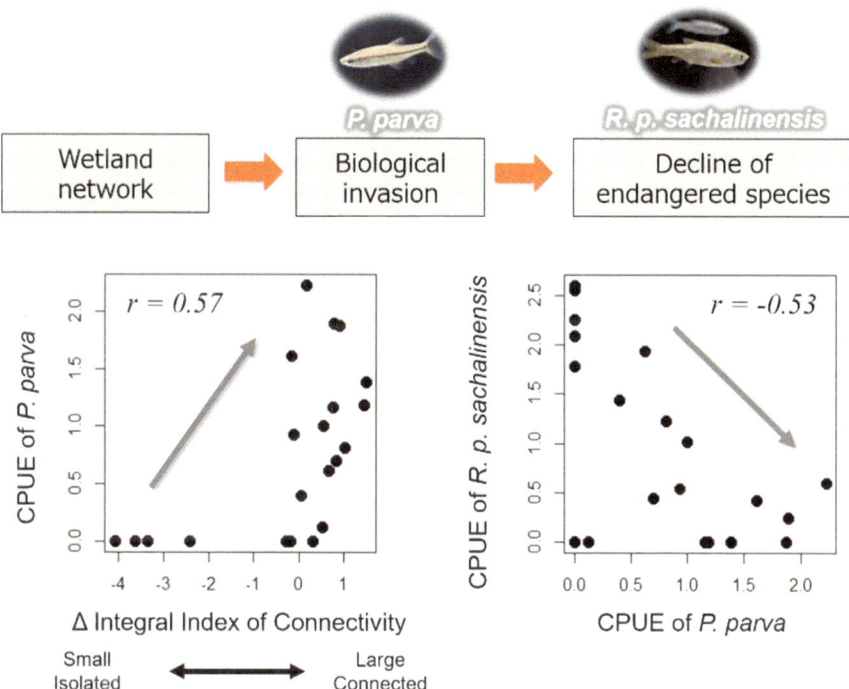

Fig. 12.7 Biological invasion in a wetland network and decline in endangered species. Example of *P. parva* and *R. p. sachaliensis* in the Tokachi floodplain, northern Japan. (See Ishiyama et al. (2020b) for details)

isolated waterbodies, such remnants should be preserved when managers plan the construction of flood-control basins (Fig. 12.8). As previously mentioned in the third section, we should also remember that the creation of flood-control basins can support the establishment of nonnative species and create new sources for the secondary spread of these species to connected waterbodies in wetland networks.

12.4.3 Sustainable Management of Flood-Control Basins

Various management practices are needed to fulfill both functions of flood-control basins (i.e., disaster risk reduction and habitat provision). In the Maizuru and Watarase flood-control basins, managers have taken measures to conserve endangered species, such as making a breeding site and eliminating invasive species (Chino and Mizuno 2019; MLIT Kantou Regional Development Bureau 2020a). In addition, the Maizuru flood-control basin is legally protected as a wildlife reserve area by the Hokkaido government. Long-term maintenance of wetland environments is also a vital issue for sustaining functions. The accumulation of dead reed plants

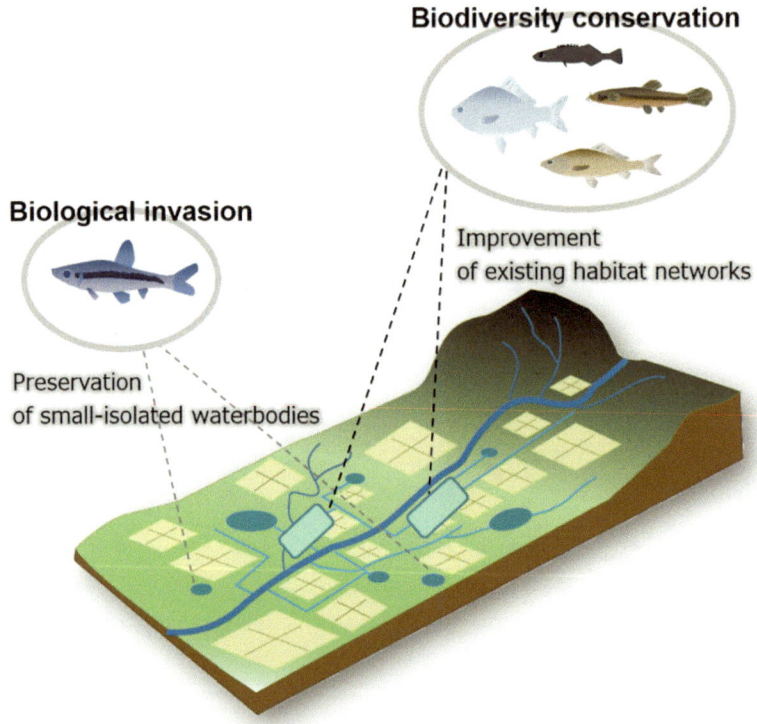

Fig. 12.8 Network thinking in biodiversity conservation using flood-control basins. Improvement of existing wetland networks and preservation of small isolated wetlands should be strategically conducted

and the establishment of pioneer tree species can change the setting in flood-control basins from fen to swamp dominated by *Salix* spp., which will decrease the flood-control capacity and amount of habitat for wetland organisms. For example, wetland environments in the Asahata flood-control basins, Shizuoka prefecture, have dried, and the habitat used by endangered wetland species is shrinking (Shizuoka prefecture 2017). Control burning is one of the historical measures for preventing vegetation succession in Japan. In some Japanese flood-control basins, from winter to early spring, control burning is conducted by local communities, including the government, nonprofit organizations (NPOs), and residents, to maintain the function (e.g., MLIT Kantou Regional Development Bureau 2020c). Creating open water and an extended hydroperiod by dredging accumulated sediment and organic material is another option for wetland management (e.g., Stevens et al. 2003). In the Watarase flood-control basin, dredging works were conducted to restore a heterogenous wetland landscape (MLIT Kantou Regional Development Bureau 2020d). Wetlands with different flood frequencies and multiple open waters were created by this work, which contributed to restoring floodplain vegetation (Ishii et al. 2011) and foraging sites for a top avian predator, the eastern marsh harrier *Circus spilonotus* (Hirano 2015).

As mentioned above, various types of management are necessary for maintaining the functions of flood-control basins. However, the running cost for deteriorated gray infrastructure is increasing (Council for Social Infrastructure 2020), and it is difficult for the government alone to sustain GI functions. Cooperative management of flood-control basins among the government, researchers, the private sector, NPOs, and residents will be essential to maintain the multiple functions of flood-control basins. For the Watarase flood-control basins, the government has established a council for management that aims to conserve and sustainably use wetland environments. The council is composed of multiple organizations, such as NPOs and local governments, and conducts biological monitoring and environmental education to help local residents to understand the importance of wetland environments and participate in maintenance (MLIT Kantou Regional Development Bureau 2020b). The attraction of private investment to flood-control basins has received growing attention as a useful tool for sustainable management. The council of the Watarase flood-control basin established the fund *"Watarase-Mirai-Kikin"* in 2001 and raises funds from private companies that are interested in environmental, social, and governance (ESG) investment and corporate social responsibility (CSR). This fund is used for sustainable management, such as restoration of wetland environments and human resource development. In March 2020, MLIT launched the GI Public-Private Partnership Platform to manage GI, including flood-control basins, more sustainably. The platform is organized by multiple stakeholders, such as national and local governments, private companies, research institutes, NPOs, and citizens. Working toward sustainable and attractive city development using flood-control basins, the platform is working on promoting GI, developing construction and management technologies, and financing techniques.

The gap between conservation science and real-world action is a genuine phenomenon (Knight et al. 2008; Osawa and Ueno 2017). The research-implementation gap will impede the success of the sustainable management of flood-control basins regardless of how many academic papers are published. Knight et al. (2008) observed that "conservation planners must facilitate a solution to a specific practitioner's need; it is generally not effective to conduct a conservation assessment and then attempt to promote it post hoc to a practitioner." For the Watarase flood-control basins, MLIT established a committee for conserving wetland environments, composed of both practitioners and researchers, that held multiple meetings to develop the restoration plan. One of the key benefits of establishing the committee is facilitating interactions between researchers and practitioners. In such a committee, researchers can formulate research questions collaboratively with stakeholders and understand specific practitioner's needs and implementation constraints. Indeed, the information on potential impacts of topsoil dredging on the distribution of threatened plant species assessed by one of the committee members (Obata et al. 2012) delivered a management plan for the flood-control basin; practitioners selected dredging sites by referring to the academic evidence.

12.4.4 Importance of Multifunctionality

In this chapter, we introduced that flood-control basins can have environments similar to those of natural wetlands, providing important functions for humans and living organisms (i.e., flood-risk reduction and habitat provision). However, natural wetlands (GI-1) are multiple-value systems (Mitsch and Gosselink 2000) and provide more functions for humans, such as water quality improvement and recreational activities. In future flood-risk management, we should also focus on such functions and wisely utilize flood-control basins (GI-2) to increase residents' quality of life.

In Japan, some flood-control basins located around cities are used for public parks with green spaces (e.g., the Tsurumi River flood-control basins). A growing body of literature has emphasized the importance of people's green space use for mental health during the COVID-19 pandemic (e.g., Soga et al. 2020). Considering lifestyle changes after the global pandemic, flood-control basins around cities can serve as good health and physical resources for citizens. As previously mentioned in the third and fourth sections, spatial planning of flood-control basins would benefit wetland organisms in terms of improving habitat networks. Humans will also profit from the strategic restoration of habitat networks through recreational activities such as fisheries and birdwatching.

Flood-control basins with wetland environments can be a mitigation tool for improving water quality in river networks. Recent work in the USA has shown that a spatially targeted increase in wetland area by 10% (i.e., wetlands are preferentially placed in areas with the highest nitrate surplus) would double wetland nitrate removal (Cheng et al. 2020). For example, farmland expansions in northern Japan cause a decline in juvenile *Oncorhynchus masou*, which is one of the key fishery resources, through nutrient enrichment (Ishiyama et al. 2020a). Strategic spatial planning of flood-control basins can mitigate such degradation in river ecosystems and their delivered services.

As mentioned above, flood-control basins can bring various benefits inside and outside of the GI and possibly contribute to improving residents' quality of life. However, ecosystem services can have trade-off relationships with each other (Bennett et al. 2009). To maximize the multiple functions of flood-control basins and manage them sustainably, scientists and managers should further understand the relationships among the provided functions and the mechanisms behind the relationships.

References

Alfieri L, Burek P, Feyen L, Forzieri G (2015) Global warming increases the frequency of river floods in Europe. Hydrol Earth Syst Sci 19(5):2247–2260
Bennett EM, Peterson GD, Gordon LJ (2009) Understanding relationships among multiple ecosystem services. Ecol Lett 12(12):1394–1404

Cheng F, Van Meter K, Byrnes D, Basu N (2020) Maximizing US nitrate removal through wetland protection and restoration. Nature:1–6

Chino SK, Mizuno H (2019) Environmental monitoring in the Maiduru flood-control basin (in Japanese). https://www.hkd.mlit.go.jp/ky/jg/gijyutu/splaat000001t2er-att/splaat000001t2op.pdf. Accessed 15 Jan 2021

Council for Social Infrastructure (2020) 44th Planning committee: socioeconomic conditions surrounding social infrastructure development (in Japanese). https://www.mlit.go.jp/policy/shingikai/content/001342849.pdf. Accessed 15 Jan 2021

Davidson NC (2014) How much wetland has the world lost? Long-term and recent trends in global wetland area. Mar Freshw Res 65(10):934–941

Dottori F, Szewczyk W, Ciscar J-C, Zhao F, Alfieri L, Hirabayashi Y, Bianchi A, Mongelli I, Frieler K, Betts RA (2018) Increased human and economic losses from river flooding with anthropogenic warming. Nat Clim Chang 8(9):781–786

European Commission (2013) Green Infrastructure (GI) — Enhancing Europe's Natural Capital. https://eur-lex.europa.eu/legal-content/EN/TXT/?uri=CELEX:52013DC0249. Accessed 15 Jan 2021

GSI (2000) National survey of lakes and wetlands. Geospatial Information Authority of Japan (GSI). http://www.gsi.go.jp/kankyochiri/shicchimenseki2.html (in Japanese). Accessed 25 April 2018

Hermoso V, Morán-Ordóñez A, Lanzas M, Brotons L (2020) Designing a network of green infrastructure for the EU. Landscape Urban Plann 196:103732

Hirano T (2015) The effects of newly created ponds on hunting flights of eastern marsh harriers during the wintering season (in Japanese with English abstract). Bird Res 11:a21–a30

Hitachi River and National Highway Office (2020) Emergency flood control project in the Kuji River and the Naka River (in Japanese). https://www.ktr.mlit.go.jp/hitachi/hitachi00814.html. Accessed 26 Dec 2020

Hokkaido Regional Development Bureau (2018) Initiatives for river development: flood control measures in the Chitose River Basin. https://www.hkd.mlit.go.jp/sp/kasen_keikaku/kluhh40000001qfy.html. Accessed 25 April 2018

Ishii J, Hashimoto L, Washitani I (2011) Early vegetation growth in an experimental restoration site in the Watarase wetland. Jpn J Conserv Ecol 16:69–84

Ishiyama N, Akasaka T, Nakamura F (2014) Mobility-dependent response of aquatic animal species richness to a wetland network in an agricultural landscape (in Japanese with English abstract). Aquat Sci 76(3):437–449

Ishiyama N, Koizumi I, Yuta T, Nakamura F (2015a) Differential effects of spatial network structure and scale on population size and genetic diversity of the ninespine stickleback in a remnant wetland system. Freshwat Biol 60(4):733–744

Ishiyama N, Sueyoshi M, Nakamura F (2015b) To what extent do human-altered landscapes retain population connectivity? Historical changes in gene flow of wetland fish Pungitius pungitius. R Soc Open Sci 2(7):150033

Ishiyama N, Nagayama S, Iwase H, Nakamura F (2017) Restoration techniques for riverine aquatic connectivity: current trends and future challenges in Japan (in Japanese with English abstract). Ecol Civil Eng 19:143–164

Ishiyama N, Miura K, Inoue T, Sueyoshi M, Nakamura F (2020a) Geology-dependent impacts of forest conversion on stream fish diversity. Conserv Biol

Ishiyama N, Miura K, Yamanaka S, Negishi JN, Nakamura F (2020b) Contribution of small isolated habitats in creating refuges from biological invasions along a geomorphological gradient of floodplain waterbodies. J Appl Ecol 57(3):548–558

Jongman B, Ward PJ, Aerts JC (2012) Global exposure to river and coastal flooding: long term trends and changes. Glob Environ Chang 22(4):823–835

Knight AT, Cowling RM, Rouget M, Balmford A, Lombard AT, Campbell BM (2008) Knowing but not doing: selecting priority conservation areas and the research–implementation gap. Conserv Biol 22:610–617

Kobayashi Y, Higa M, Higashiyama K, Nakamura F (2020) Drivers of land-use changes in societies with decreasing populations: a comparison of the factors affecting farmland abandonment in a food production area in Japan. PLoS One 15(7):e0235846

Kumamoto Prefecture (2020) Pamphlet for special emergency project as the countermeasures against terrible disasters in the Shira river and the Kuro River (in Japanese). https://www.pref.kumamoto.jp/soshiki/105/5723.html. Accessed 26 Dec 2020

Lovell ST, Taylor JR (2013) Supplying urban ecosystem services through multifunctional green infrastructure in the United States. Landsc Ecol 28(8):1447–1463

Mallakpour I, Villarini G (2015) The changing nature of flooding across the Central United States. Nat Clim Chang 5(3):250–254

Ministry of Environment of Japan (2020a) The birth of a red-crowned crane chick at the Maizuru flood-control basin in Naganuma (in Japanese). http://hokkaido.env.go.jp/pre_2020/post_123.html. Accessed 26 Dec 2020

Ministry of Environment of Japan (2020b) Ramsar sites in Japan. https://www.env.go.jp/en/nature/npr/ramsar_wetland/pamph/index.html. Accessed 26 Dec 2020

Mitsch WJ, Gosselink JG (2000) The value of wetlands: importance of scale and landscape setting. Ecol Econ 35(1):25–33

MLIT Kantou Regional Development Bureau (2020a) The birth of a stork chick at the Watarase flood-control basin (in Japanese). https://www.ktr.mlit.go.jp/river/news/river_news00000009.pdf. Accessed 15 Jan 2021

MLIT Kantou Regional Development Bureau (2020b) Council for conservation and utilization of the Watarase flood-control basin (in Japanese). https://www.ktr.mlit.go.jp/tonejo/tonejo_index028.html. Accessed 15 Jan 2021

MLIT Kantou Regional Development Bureau (2020c) Information of the Watarase flood-control basin (in Japanese). https://www.ktr.mlit.go.jp/tonejo/tonejo00081.html. Accessed 15 Jan 2021

MLIT Kantou Regional Development Bureau (2020d) Wetland restoration plan in the Watarase flood-control basin (in Japanese). https://www.ktr.mlit.go.jp/tonejo/tonejo00661.html. Accessed 11 May 2021

Moilanen A, Nieminen M (2002) Simple connectivity measures in spatial ecology. Ecology 83:1131–1145

Nakamura F, Ishiyama N, Yamanaka S, Higa M, Akasaka T, Kobayashi Y, Ono S, Fuke N, Kitazawa M, Morimoto J (2020) Adaptation to climate change and conservation of biodiversity using green infrastructure. River Res Appl 36(6):921–933

Obata T, Ishii J, Kadoya T, Washitani I (2012) Effect of past topsoil removal on the current distribution of threatened plant species in a moist tall grassland of the Watarase wetland, Japan: mapping of selected sites for wetland restoration by topsoil removal. Jpn J Conserv Ecol 17:221–233

Osawa T, Ueno Y (2017) Research-implementation gap in Japanese ecology field. Jpn J Ecol 67:257–265

Rudnick D, Ryan SJ, Beier P, Cushman SA, Dieffenbach F, Epps C, Gerber LR, Hartter JN, Jenness JS, Kintsch J (2012) The role of landscape connectivity in planning and implementing conservation and restoration priorities. Issues Ecol 16:1–20

Saura S, Rubio L (2010) A common currency for the different ways in which patches and links can contribute to habitat availability and connectivity in the landscape. Ecography 33:523–537

Saura S, Torné J (2009) Conefor Sensinode 2.2: a software package for quantifying the importance of habitat patches for landscape connectivity. Environ Model Softw 24(1):135–139

Segawa A, Minato T, Yoshikawa K (2008) The history of flood plain development and the levee construction of the Ishikari River down stream (in Japanese with English abstract). J Construct Manage JSCE 15:429–440. https://doi.org/10.2208/procm.15.429

Shizuoka Prefecture (2017) Action plan for the conservation and utilization of the Asahata flood-control basin (in Japanese). http://asabata.org/wp-content/uploads/2017/04/4fad93fec4e8c2ca38736fb06cf4bb38.pdf. Accessed 15 Jan 2021

Soga M, Evans MJ, Tsuchiya K, Fukano Y (2020) A room with a green view: the importance of nearby nature for mental health during the COVID-19 pandemic. Ecol Appl e2248

Stevens CE, Gabor TS, Diamond AW (2003) Use of restored small wetlands by breeding waterfowl in Prince Edward Island, Canada. Restor Ecol 11:3–12

Suwa Y, Nishihiro J (2020) Distribution and location characteristics of flood-control reservoir in Japan (in Japanese with English abstract). Ecol Civ Eng 23:85–97

Tone River Upstream Office (2020) Report on the outflow of water in Typhoon No. 19 (in Japanese). https://www.ktr.mlit.go.jp/tonejo/tonejo00367.html. Accessed 26 Dec 2020

Wang J, Banzhaf E (2018) Towards a better understanding of green infrastructure: a critical review. Ecol Indicators 85:758–772

Yamanaka S, Ishiyama N, Senzaki M, Morimoto J, Kitazawa M, Fuke N, Nakamura F (2020) Role of flood-control basins as summer habitat for wetland species-a multiple-taxon approach. Ecol Eng 142:105617

Chapter 13
Natural Succession of Wetland Vegetation in a Flood-Control Pond Constructed on Abandoned Farmland

Junko Morimoto, Susumu Goto, Akito Kuroyanagi, Motoko Toyoshima, and Yuichiro Shida

Abstract Can a flood-control basin be considered a component of green infrastructure (GI) with not only disaster prevention functions but also biodiversity conservation functions? We studied the succession of wetland vegetation in a flood-control basin constructed in a floodplain. The number of species composing the buried seeds depended on the depth of the soil layer, with shallower (recent) layers having more species. Plants germinated from buried seeds in deep (ancient) soil layers were small and difficult to identify based on morphology alone, but DNA-based analyses made it possible to identify these species. The occurrence of three species of *Juncus* and *Cyperus* in the peat layer that developed between 840 and 1215 years ago indicates that it is possible to regenerate wetland plants from the past if the buried seeds are well preserved and in good condition. The excavated section of the experimental site, which was set in part of the flood-control basin, received dispersed seeds from the surrounding agricultural land and contained more species than did the soil layer itself. The shallower the water depth was (the shallower the excavation depth), the greater the number of species that occurred. In addition, species with different life forms occurred depending on the water depth. More species occurred in the entire flood-control basin than in the experimental site. The factors that determined the type of plant community were years since excavation, water depth, and water quality. Finally, we discussed the management practices concerning the supply of propagules and the management of suitable habitats as essential factors for flood-control basins constructed on fallow land to function as GI.

J. Morimoto (✉) · A. Kuroyanagi · M. Toyoshima
Graduate School of Agriculture, Hokkaido University, Sapporo, Hokkaido, Japan
e-mail: jmo1219@for.agr.hokudai.ac.jp

S. Goto
Graduate School of Agricultural and Life Sciences, The University of Tokyo, The University of Tokyo Forests, Tokyo, Japan

Y. Shida
Wildlife Research Institute Co. Ltd., Sapporo, Hokkaido, Japan

© The Author(s) 2022 209
F. Nakamura (ed.), *Green Infrastructure and Climate Change Adaptation*,
Ecological Research Monographs, https://doi.org/10.1007/978-981-16-6791-6_13

Keywords Buried seeds · Seed dispersal · Water depth · Succession · DNA barcoding · Field experiment

13.1 Introduction

The recent development of agricultural land has led to a global decline in wetland biodiversity (World Water Assessment Programme 2009; Zedler and Kercher 2005). This has led to concerns about the decline in biodiversity and ecosystem services in wetlands. On the other hand, in developed regions such as North America and the United Kingdom, the emergence of abandoned farmland has been reported since the beginning of the nineteenth century, and in Eastern Europe, Russia, and Latin America, this has occurred since the beginning of the twentieth century due to socioeconomic deterioration such as urbanization and industrialization and a decrease in the rural population (Baumann et al. 2011; Benayas and Bullock 2012; Kuemmerle et al. 2008). Agricultural land converted from wetlands by artificial draining is particularly prone to abandonment due to the high maintenance costs (Benjamin et al. 2005; Middleton 2003). Abandoned farmland that was once a wetland and adjacent to a river can be expected to serve as a flood-control basin and to restore wetland ecosystems (Morimoto and Shibata 2018). If these goals can be realized at the same time, it will be useful not only for disaster prevention but also for biodiversity conservation, making such sites beneficial green infrastructure (GI) from the perspective of improving the overall ecosystem services (Nakamura et al. 2020). In this chapter, we introduce a series of studies on the process of natural succession of wetland vegetation at a project site where a flood-control basin was established on abandoned farmland that had been a floodplain until it was developed into farmland in the twentieth century.

13.2 Materials and Methods

A flood-control basin (1150 ha) was constructed in a tributary of the Chitose River in northern Japan, Hokkaido ($42°55'$ N, $141°41'$ E, annual mean temperature of 7.2 °C, annual precipitation of 1005.6 mm), which was farmland that had been used as paddy and dryland fields for approximately 10 years and then left fallow for 7 years (refer to Ishiyama et al. of Chap. 12 for the positioning of the flood-control basin and overall flood management). We chose the flood-control basin named "*Maizuru Yusuichi*" in Japanese as our study site (hereafter referred to as Crane Pond). The oldest topographic map of 1920 and later maps confirm that Crane Pond had been at the bottom of a lake until 1960. Carbon dating analysis indicated that the humus layer from 0 m to 0.3 m below the surface was deposited after 1739; the clay layer from 0.3 m to 1.2 m below the surface, after between 1190 and 1275; and the peat layer from 1.2 m to 1.5 m below the surface, after between 775 and 790 or 800

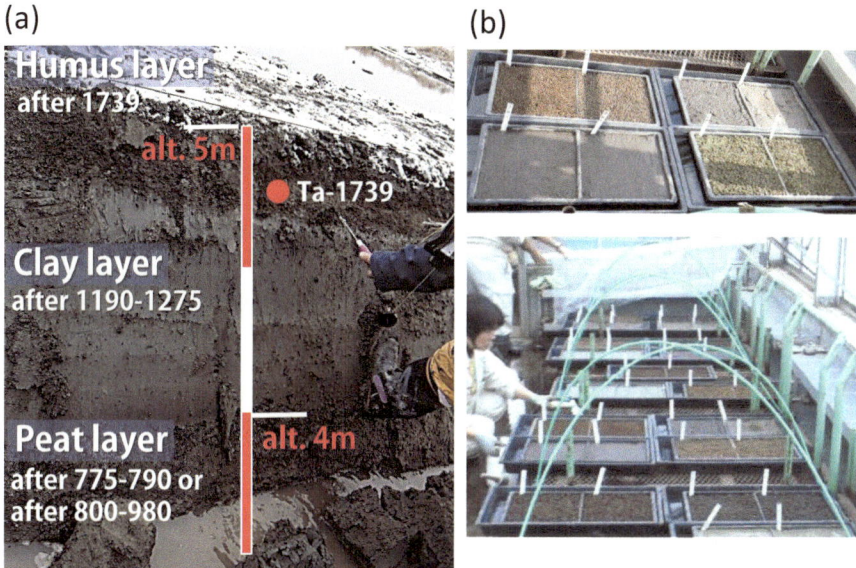

Fig. 13.1 Soil layers in Crane Pond (**a**) and assessment of the seed bank (**b**) Ta: volcanic deposit from Mt. Tarumae

and 980 (Fig. 13.1a). Therefore, it is assumed that this area, including Crane Pond, was a floodplain of a network of meandering rivers from the eighth or ninth century until, at the latest, the twentieth century, when the land was developed for farming. In this study area, before the construction of Crane Pond, an experiment was started to clarify the composition of the seed bank in the fallow farmland (2.2), and after the construction of Crane Pond, a survey of the naturally regenerated vegetation (2.3) was conducted.

13.2.1   Experiment Started Before the Construction of Crane Pond: Exploring the Seed Bank Species in Fallow Farmland

Soils from the humus, clay, and peat layers were collected in December 2014 and June 2015 from fallow farmland where Crane Pond was to later be constructed. The soil blocks were mixed by layer with 0.1 times the volume of water and loosened, and gravel and branches were removed. Two-centimeter-deep trays (0.0576 m^2) were filled with soil that was sterilized by drying, and each layer of soil was spread out at a thickness of 1 cm (Fig. 13.1b). The soil was submerged in water so that the water surface coincided with the soil surface. The soil was covered with mesh to prevent seed contamination from outside. As a control, we also prepared trays sown with sterilized soil at a depth of 1 cm. The germinating plants were observed until the development of new plants stopped (until the end of August for the soils collected in 2014 and until October for the soils collected in 2015). The plants were then grown until October 2016, and flowering and fruiting were confirmed. We identified the species by their morphological features except for three plants. These plants germinated from the peat layer as old as ca. 1000 years (hereafter, the ancient group). The size of the three plants was so small that we could not identify the species by morphological features alone, so chloroplast DNA information was used for species identification. First, we sampled collected leaves from the present group of *Juncus decipiens* (Buchenau) Nakai, *Juncus tenuis* Willd., *Juncus bufonius* L., and *Juncus wallichianus* Laharpe (which were candidate species based on their morphological features). Recently, DNA barcoding (barcode of life: BOL) has proven useful for species identification. For plants, a part of the sequence of the *rbc*L gene is effective for species identification. In this study, we extracted DNA from the leaves of ancient and present plants, amplified the BOL region of the collected samples, and identified species based on sequence data. After aligning the sequences using MEGA ver. 7, we checked the corresponding sequences against the NCBI database.

13.2.2 Experiment and Survey Started After the Construction of Crane Pond: Exploring the Vegetation that Regenerated Naturally

Crane Pond was constructed between 2009 and 2014 and was put into service in April 2015. In July 2017, we randomly settled 51 quadrats (2 m × 2 m) in Crane Pond, and the vegetation and aquatic environment was surveyed. Aquatic environmental parameters, including electrical conductivity (EC, μS/cm), dissolved oxygen (DO, %), pH, and depth (cm), of surface water were measured. Plant communities were identified by TWINSPAN (two-way indicator species analysis) from the vegetation dataset. The relationships between plant communities and the water environment were analyzed by canonical correspondence analysis (CCA).

We built an experimental site (0.17 ha) to observe the relationships between the regenerated species and water depth at the corner of Crane Pond (Fig. 13.2a). Excavation was carried out at three different depths (Fig. 13.2b, c). The water surface was at approximately the same level as the topmost ground surface (Fig. 13.2d). On the basis of the inaugural year of Crane Pond in 2015, we recorded the plant cover ratio of each species regenerated from 2015 to 2019. The succession over 5 years at three sections with different depths was monitored and recorded.

(a) Crane Pond, flood-control basin (ca. 200 ha)

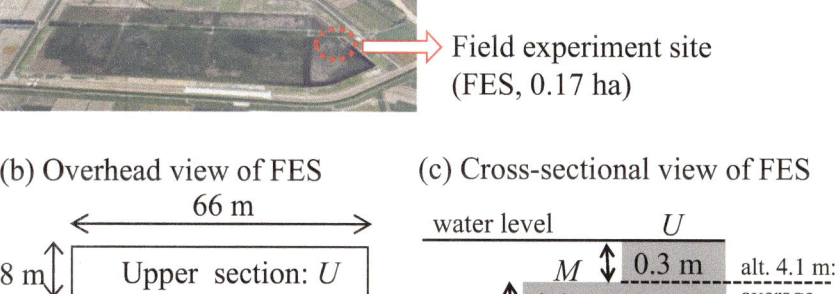

Field experiment site
(FES, 0.17 ha)

(b) Overhead view of FES

66 m

8 m	Upper section: U
8 m	Middle section: M
8 m	Lower section: L

(c) Cross-sectional view of FES

water level U

M 0.3 m   alt. 4.1 m:
L 0.3 m   average
                  altitude of
                  the pond

8 m   8 m   8 m

(d) Transition of the FES landscape over time

May, 2015   *July, 2017*

September, 2015   *July, 2018*

Fig. 13.2 Field experimental site built in a corner of a flood-control basin

13.3 Species Composition of Buried Seeds in Crane Pond

13.3.1 Species Identification by a Combination of Morphological and DNA Features

Three plants that germinated from the peat layer were too small to identify, but the first one (P1) was assumed to be *J. decipiens* based on its morphological features. *rbcL* DNA is used as a DNA barcoding region in plants (CBOL Plant Working Group 2009) because the diversity of *rbcL* is conserved within a species but is likely to vary between species. The *rbcL* DNA of many plant species is registered, which is advantageous in that a large amount of information is available but a disadvantage in that *rbcL* genes are often shared among closely related species, as shown by the results of our analysis. According to a phylogenetic tree based on the *rbcL* region (Tsubota et al. 2014), *Juncus* growing in wetlands in Hokkaido can be roughly classified according to the nucleotide sequence of this region. Therefore, we extracted DNA from leaves of P1 (Sample No. 4) and present plants of *J. decipiens* (Sample Nos. 1–3) and *J. tenuis* (Sample Nos. 5–6). After comparing the BOL sequences of the extracted DNA samples, at least five single-base substitutions occurred among *J. decipiens* and *J. tenuis*, and P1 had the same sequence as two *J. tenuis* plants collected in the southern part of Japan (Fig. 13.3). Then, we compared their sequences with those registered in the database of related species. We found that the sequences of Sample Nos. 1, 2, and 3 were identical to the sequence of *J. decipiens* (database ID number: KT695563) and that the sequences of Sample Nos. 4 (P1), 5, and 6 were identical to those of *J. tenuis* (KJ593488) and *J. bufonius* L. (KJ204357).

Since *J. decipiens* is abundant near the study site, we expected that P1 emerging from the peat layer might be this species before we performed the DNA analysis. However, the chloroplast DNA sequence clearly showed that P1 was not *J. decipiens* but rather *J. tenuis* or *J. bufonius*. *J. tenuis* and *J. bufonius* can be identified by the morphological features of their leaves (presence or absence of a membranous auricle in the leaf sheath). Finally, P1 was identified as *J. tenuis* because there was an auricle in the leaf sheath. Later, P2 and P3, which germinated from the

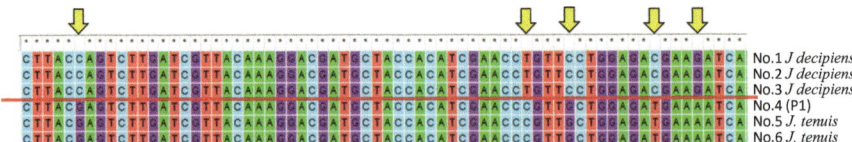

Fig. 13.3 Sequences of the *rbcL* (BOL) region of the chloroplast genome for six plants. P1 (No. 4) germinated from the peat layer and was initially assumed to be *Juncus decipiens*. However, the sequence was different from that of this species (Nos. 1, 2, and 3) at several sites (yellow arrows). We found that the sequence of P1 (No. 4) was completely consistent with that of *J. tenuis* (Nos. 5 and 6)

peat layer, were analyzed in the same way as P1, and the results were 100% consistent with *J. wallichianus* (registered as AB985730.1) collected in Hiroshima Prefecture. However, some of the sequences were different from those of the native *J. wallichianus* in Crane Pond, and further study is needed. Thus, the combination of morphological and DNA analyses showed the possibility of reliable species identification of small plants, which are difficult to identify based on morphological information alone.

13.3.2 Species Composition in Each Soil Layer from Different Depths

All plants except the three plants germinated from the peat layer were identified by morphological features alone. All the germinated species were present, and one of them, *Sparganium erectum* L. var. *coreanum* (H. Lév.) H. Hara, was a near-threatened species (Table 13.1). The number of species and plants was the highest in the latest layer of humus and was lower in older layers of clay and peat (Fig. 13.4). However, each layer produced unique species (Table 13.1). The planned elevation of Crane Pond (4.1 m above sea level) is comparable to the height of the clay layer, meaning that the humus layer containing the largest number of species has been stripped off and has spread out into the depression inside Crane Pond. Thus, the main contribution of buried seeds to natural succession in Crane Pond was assumed to be buried seeds from the humus and clay layers. In addition, notably, three species of *Juncus* and *Cyperus* germinated from the peat layer as old as ca. 1000 years old.

The result that the latest layer produced the most abundant species and plants coincides with the view that the germination ratio of seeds declines over time after production and that the seed longevity of most species is less than 5 years

Table 13.1 Number of plants from each soil layer

	Humus ($N = 16$)		Clay ($N = 14$)		Peat ($N = 32$) (m^{-2})	
	Mean	SD	Mean	SD	Mean	SD
Cyperus difformis	0.0	0.00	0.0	0.00	0.5	3.02
Juncus bufonius	1.1	4.1	0.0	0.00	0.0	0.00
Juncus decipiens	316.8	509.5	0.0	0.00	0.0	0.00
Juncus prismatocarpus	9.8	25.2	0.0	0.00	1.6	6.7
Juncus tenuis	0.0	0.00	0.0	0.00	0.5	3.02
Juncus wallichianus	10.9	20.5	0.0	0.00	0.0	0.00
Schoenoplectiella hondoensis	5.4	11.5	0.0	0.00	0.0	0.00
Schoenoplectiella triangulata	0.0	0.00	1.2	4.47	0.0	0.00
Schoenoplectus tabernaemontani	55.3	78.6	1.2	4.5	0.0	0.00
Sparganium erectum	2.2	8.2	0.0	0.00	0.0	0.00

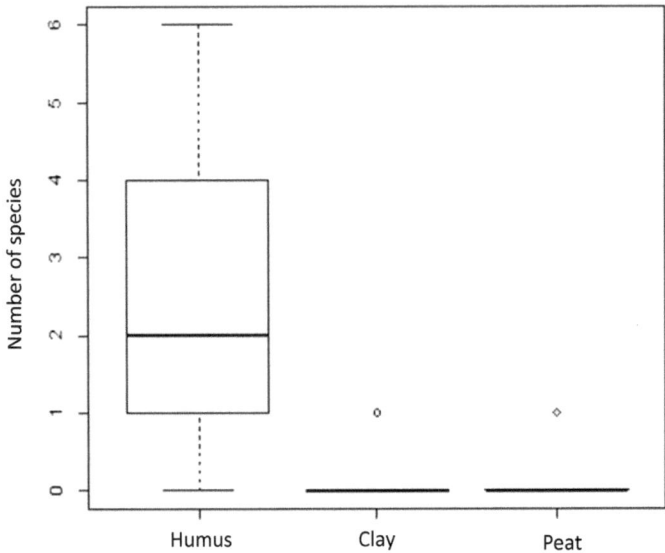

Fig. 13.4 Number of species ($/0.025$ m^2) from each soil layer. Open circles indicate outliers; the top, center, and bottom lines of each box indicate the third, second, and first quartiles, respectively. The horizontal lines above and below the box indicate the maximum and minimum values, respectively (refer to Fig. 13.1 for the position of each soil layer)

(Baskin and Baskin 2014). Long-lived seeds of more than 5 years have been reported occasionally (e.g., 100 years in Malvaceae, 80 years in Leguminosae, and 77 years in Lamiaceae), but reports of long-lived seeds exceeding 1000 years are very rare (Oga 1951). The germination of three species of *Juncus* and *Cyperus* from the ancient peat layer indicates their uniqueness. The conditions under which seeds can be stored for long periods of time are said to be cool, mildly alternating temperatures, and low oxygen (Baskin and Baskin 2014). These conditions might be easily possible in Crane Pond, which was the bottom of a lake until land reclamation for agriculture in the 1960s.

13.4 Natural Succession in Crane Pond

13.4.1 Features of Species Generated from Different Water Depths

Plant succession at the field experimental site is shown in Table 13.2. The number of species was higher than that contained in the soil layers (Table 13.1). This indicates that the number of samples was limited for the survey of buried seeds and that propagules dispersed by wind, water, and birds from surrounding agricultural land contributed to plant establishment at the experimental site. In addition, it was found that plants were first established in the shallow water stage and that plants with different life forms were established depending on the water depth (Table 13.2). Annual plants that emerged in the first year did not appear after the second year. The occurrence of alien species was negligible.

Hygrophytes and emergent macrophytes mainly emerged in the upper and middle sections half a year after the excavation; however, it took 3 years for plant establishment in the lower section, and floating-leaved macrophytes and submerged macrophytes emerged. Hygrophytes and emergent macrophytes are mainly dispersed by wind, whereas floating-leaved macrophytes and submerged macrophytes are mainly dispersed by water or birds (Kameyama and Ohara 2007; Mikulyuk and Nault 2009; Phartyal et al. 2018). It can be considered that the vegetation recovery in the lower section was slower than that in the upper and middle sections because the opportunities for dispersal by water inflow and bird visits to Crane Pond were more limited than those by wind.

Hygrophyte species that occurred in the first year were annual plants dependent upon disturbance (Euliss et al. 2004). Many of them were presumed to have not occurred after the second year because flood disturbance did not occur in the field experimental site. Human or natural disturbance is necessary to sustain disturbance-dependent species.

Mesophytes, including alien species, disappeared by the second year of the project, which means that there is no need to worry about alien species of mesophytic plants invading the water body as long as the wet environment is maintained.

13.4.2 Years from Excavation, Water Depth, and Water Quality Determine the Plant Community

From the field experiments, it was found that the native species occur in order from shallow to deep water, and their life forms depend on the water depth. Next, when we looked at plant establishment throughout the entire Crane Pond, 26 native species (including five rare species) appeared, which is more than that in the experimental site, and no exotic species appeared (Kuroyanagi et al. 2019). The 13 unique species that did not appear in the experimental site and in the seed bank appeared in Crane Pond (Fig. 13.5). Crane Pond is vast and includes areas where the water was deeper than that at the experimental site, and because of the large amount of incoming water and waterfowl flights, it was suitable for the establishment of floating-leaved, free-floating aquatic, and submerged macrophytes. Among the species that have contributed to vegetation recovery in Crane Pond, there are three species that are the same as those found in the buried seeds. Wetland vegetation in Crane Pond regenerated mainly from propagules supplied from outside Crane Pond, but buried

Table 13.2 Changes in plant species observed at the field experimental site

Site[a]	Life form[b]	Life history[c]	Red list or Blue list[d]	Species	The year of presence				
					2015	2016	2017	2018	2019
U	m	A		Persicaria longiseta	•				
		A,B	✓	Gnaphalium uliginosum	•				
	m-wet	P	✓	Trifolium spp.	•				
		P	✓	Rudbeckia laciniata		•			
		P		Juncus tenuis			•	•	•
	h	A		Persicaria sagittata var. sibirica	•	•	•	•	•
		A		Echinochloa crus-galli var. crus-galli	•	•			•
		A		Echinochloa crus-galli var. aristata	•		•		
		A		Cyperus difformis	•		•		
		A	***	Persicaria foliosa var. paludicola	•				•
		A		Persicaria lapathifolia	•				
		A		Juncus bufonius	•				
		A		Persicaria muricata	•				
		P		Lythrum salicaria	•	•			•
		P	✓	Phalaris arundinacea	•		•		•
		P	***	Carex capricornis		•	•	•	
		T		Salix udensis	•	•	•	•	•
		T		Salix schwerinii	•				•
		T		Salix triandra subsp. nipponica			•	•	•
		T		Salix integra				•	
	e,h	A	**	Monochoria korsakowii	•				
		P		Alisma plantago-aquatica var. orientale	•	•	•	•	•
		P		Juncus decipiens	•	•	•	•	•
		P		Scirpus radicans	•	•	•	•	•
		P		Juncus wallichianus	•	•	•	•	•
		P		Schoenoplectiella triangulata	•	•	•	•	
		P		Juncus prismatocarpus subsp. leschenaultii	•	•			
		P		Alisma canaliculatum	•		•		
		P		Cicuta virosa				•	•
	e	A		Schoenoplectus hotarui	•	•			
		P		Schoenoplectus tabernaemontani	•	•	•	•	•

(continued)

Table 13.2 (continued)

Site[a]	Life form[b]	Life history[c]	Red list or Blue list[d]	Species	The year of presence				
					2015	2016	2017	2018	2019
		P		*Typha latifolia*	•	•	•	•	•
		P		*Zizania latifolia*	•	•	•	•	
		P		*Bolboschoenus fluviatilis* subsp. *yagara*	•	•	•		
		P	*	*Sparganium erectum*	•	•			
		P		*Schoenoplectus triqueter*	•		•	•	
		P		*Eleocharis mamillata*		•	•	•	•
		P		Bolboschoenus koshevnikovii	•				
	fr	A		*Lemna aoukikusa* subsp. *aoukikusa*			•		
	fl	A		*Trapa japonica*					•
	s, fl	P		*Potamogeton octandrus* var. *octandrus*	•				
M	*e*	P		*Typha latifolia*				•	•
		P		*Zizania latifolia*					•
	fl	A		*Trapa japonica*				•	•
	s, fl	P		*Potamogeton octandrus* var. *octandrus*				•	•
	fr, s	P	*	*Utricularia japonica*					•
L	*fl*	A		*Trapa japonica*				•	•
	fr	A		*Lemna aoukikusa* subsp. *aoukikusa*					•
	s, fl	P		*Potamogeton octandrus* var. *octandrus*				•	•
	fr, s	P	*	*Utricularia japonica*					•

[a]*U* upper section, *M* middle section, *L* lower section

[b]Life form is identified for wetland plants and others based on the description from picture books. *h* hygrophyte, *e* emergent macrophyte, *fl* floating-leaved macrophyte, *fr* free-floating aquatic macrophyte, *s* submerged macrophyte, *m* mesophyte, *m-wet* methophyte that can live in a wet environment

[c]*A* annual plant, *B* biennial plant, *P* perennial plant, *T* tree

[d]We denoted the species categories of national (Japan) and regional (Hokkaido Prefecture) red lists according to Japanese Red lists (Ministry of Environment of Japan, 2017) and Hokkaido Red lists (Hokkaido Prefecture, 2001, 2017, 2018), respectively. The categories of Japanese Red lists (2017) are EN (Endangered), VU (Vulnerable), and NT (Near Threatened). The categories of Hokkaido Red lists (2001) are EN (Endangered), VU (Vulnerable), and R (Rare). We classified species as alien species according to Hokkaido Blue lists (Hokkaido Prefecture, 2010). ***, VU/R; **, NT/VU; *, NT/R; ✓, blue list

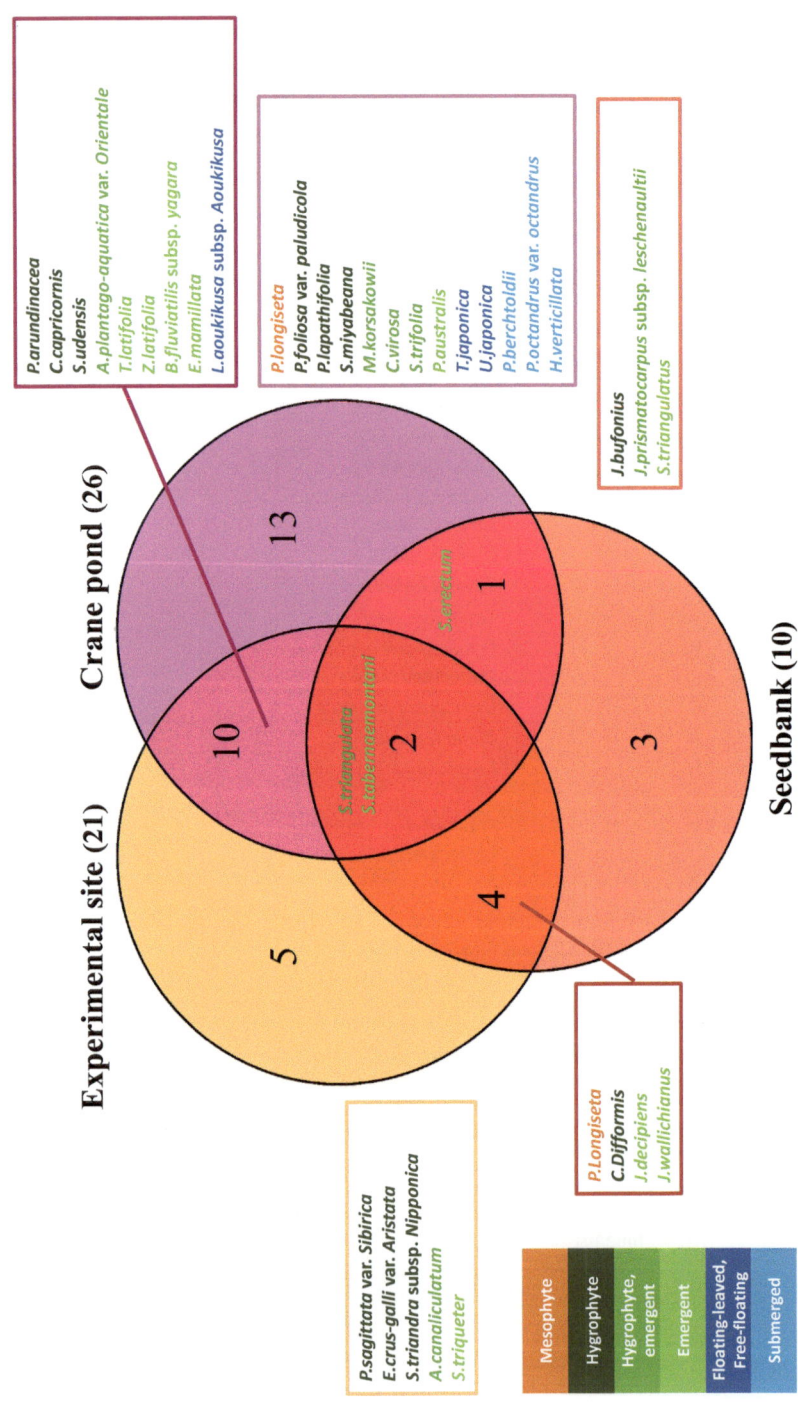

Fig. 13.5 Species in Crane Pond, the experimental site, and the seed bank. The numbers in the figure are the numbers of species

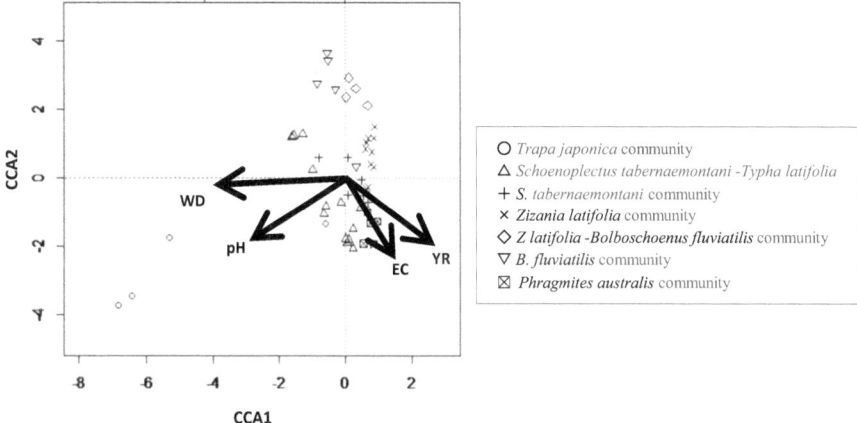

Fig. 13.6 Relationships between plant communities and environmental factors. WD, water depth; YR, years since excavation; EC, electrical conductivity

seeds also played an important role in maintaining the species diversity of wetland vegetation in Crane Pond.

Six types of plant species communities were present: the *Schoenoplectus tabernaemontani-Typha latifolia* community, *Schoenoplectus tabernaemontani* community, *Zizania latifolia* community, *Z. latifolia-Bolboschoenus fluviatilis* community, *B. fluviatilis* community, and *Phragmites australis* community. The factors characterizing the habitat of plant communities were the time since excavation, water depth, and water quality (Fig. 13.6). In Crane Pond, where 3–7 years have passed since excavation, floating-leaved *Trapa japonica* has developed in deeper water, and emergent macrophytes such as *Zizania latifolia* and *Phragmites australis* have developed in shallower water. These results coincide with those of the experimental site. Eutrophication, expressed as increasing EC, tended to increase with years since excavation, and the transition to a reed (*Trapa japonica*) community was observed (Fig. 13.6). These results suggest that creating various depths of water inside a flood-controlled basin and controlling the progress of eutrophication are effective in maintaining the various wetland species and plant communities.

13.5 Implications for Management of the Flood-Control Basin as GI

There are many cases in which past wetland vegetation did not recover on fallow land (e.g., González et al. 2017; Sheldon et al. 2016). The initial requirements for the recovery of wetland vegetation are the supply of propagules of wetland plants (Bischoff et al. 2009; Vécrin et al. 2002) and a suitable habitat for their growth and reproduction (Morimoto et al. 2017; Weiher and Keddy 1995). What kind of management is necessary for a flood-control basin built on fallow land to function as GI, conserve wetland ecosystems, and control floods? We will discuss the management of the supply of propagules and the management of suitable habitats.

13.5.1 Management of the Supply of Propagules

We can expect two types of propagules in the flood-control basin, buried and dispersed propagules. Our study showed that buried seeds were preserved even in the deep ancient soil layers that had accumulated at the time of the floodplain, and these seeds can be utilized for wetland restoration. A persistent seed bank functions as a gene pool (Bossuyt and Honnay 2008). Effective utilization of soil containing ancient seeds will enable wetland restoration considering genetic diversity. On the other hand, propagule dispersal from the surroundings is an essential requirement for the restoration of wetland vegetation with high diversity because the species that can be supplied from buried seeds are limited. Waterways running in an agricultural landscape function as refuges for wetland plants (Toyoshima et al. 2020). It is important to conserve these waterways as propagule sources for flood-control basins established in agricultural landscapes by organic farming or farming with low levels of agricultural chemicals.

13.5.2 Management for Suitable Habitat

Excavation is a general practice for the construction of flood-control basins, and the removal of surface soil is a general practice for the restoration of wetlands. Both practices have a similar effect on suitable habitat creation. Such practices increase the number of vacant sites, which permit the establishment of new plants (Klimkowska et al. 2010), decrease soil nutrients (Verhagen et al. 2001), increase the water level (Galatowitsch and van der Valk 1996; Klimkowska et al. 2010), and increase flood frequency (Hölzel and Otte 2004). However, both excavation and the removal of the surface soil will decrease the availability of seed banks in the soil, making these practices not a good idea if the propagule source is limited in the vicinity. Instead, it is recommended that drainage facilities be removed to raise the groundwater level and create a waterlogged environment to promote the effective use of seed banks (Morimoto and Shibata 2018). The reuse of topsoil removed during excavation can be an option, although keeping the waterlogged environment is a requirement to prevent the domination of alien species. In addition, a design that creates a variety of depths within the flood-control basin would contribute to the species diversity of plants, as it would promote the establishment of plants with different life forms suitable for different depths. After the construction of a flood-control basin, controlling the eutrophication of water from the surrounding farmland

and removing the sediment soils to maintain the water depth should be effective for the conservation of wetland vegetation. Ideally, river floods should be large enough to occasionally wash away the accumulated sediment in a flood-control basin.

One of the important attractions of GI is that it does not require or reduces maintenance costs. Even before the construction of the flood-control basin, it is essential to select a site and design a structure that aims to regenerate and maintain the wetland ecosystem by natural processes for the flood-control basin to function as GI.

Acknowledgments We appreciate Ms. Mitsuko Goto (The Asian Natural Environmental Science Center, The University of Tokyo) for carrying out the DNA analyses and Dr. Kazuo Yabe (Sapporo City University) for providing advice on the succession of wetland vegetation.

References

Baskin CC, Baskin JM (2014) Seeds: ecology, biogeography, and, evolution of dormancy and germination. Academic Press, San Diego, CA

Baumann M, Kuemmerle T, Elbakidze M, Ozdogan M, Radeloff VC, Keuler NS, Prishchepov AV, Kruhlov I, Hostert P (2011) Patterns and drivers of post-socialist farmland abandonment in Western Ukraine. Land Use Policy 28:552–562

Benayas JMR, Bullock JM (2012) Restoration of biodiversity and ecosystem services on agricultural land. Ecosystems 15:883–899

Benjamin K, Domon G, Bouchard A (2005) Vegetation composition and succession of abandoned farmland: effects of ecological, historical and spatial factors. Landsc Ecol 20:627–647

Bischoff A, Warthemann G, Klotz S (2009) Succession of floodplain grasslands following reduction in land use intensity: the importance of environmental conditions, management and dispersal. J Appl Ecol 46:241–249

Bossuyt B, Honnay O (2008) Can the seed bank be used for ecological restoration? An overview of seed bank characteristics in European communities. J Veg Sci 19:875–884

CBOL Plant Working Group (2009) A DNA barcode for land plants. Proc Natl Acad Sci U S A 106(31):12794–12797

Euliss NH, LaBaugh JW, Fredrickson LH, Mushet DM, Laubhan MK, Swanson GA, Winter TC, Rosenberry DO, Nelson RD (2004) The wetland continuum: a conceptual framework for interpreting biological studies. Wetlands 24:448–458

Galatowitsch SM, van der Valk AG (1996) The vegetation of restored and natural prairie wetlands. Ecol Appl 6:102–112

González E, Masip A, Tabacchi E, Poulin M (2017) Strategies to restore floodplain vegetation after abandonment of human activities. Restor Ecol 25:82–91

Hölzel N, Otte A (2004) Assessing soil seed bank persistence in flood-meadows: the search for reliable traits. J Veg Sci 15:93–100

Kameyama Y, Ohara M (2007) Population genetic consequences of predominant clonal reproduction in free-floating aquatic bladderworts. Jpn J Ecol 57:245–250

Klimkowska A, Kotowski W, Van Diggelen R, Grootjans AP, Dzierża P, Brzezińska K (2010) Vegetation re-development after fen meadow restoration by topsoil removal and hay transfer. Restor Ecol 18:924–933

Kuemmerle T, Hostert P, Radeloff VC, van der Linden S, Perzanowski K, Kruhlov I (2008) Cross-border comparison of post-socialist farmland abandonment in the carpathians. Ecosystems 11:614

Kuroyanagi A, Morimoto J, Shida Y, Shinsho H, Yabe K, Nakamura F (2019) Recovery of wetland vegetation in the Maizuru flood-retarding basin in the Chitose River watershed. J Jpn Soc Reveg Technol 45:45–50

Middleton BA (2003) Soil seed banks and the potential restoration of forested wetlands after farming. J Appl Ecol 40:1025–1034

Mikulyuk A, Nault ME (2009) Water chestnut (Trapa natans): a technical review of distribution, ecology, impacts, and management. Wisconsin Department of Natural Resources Bureau of Science Services, Madison, WI

Morimoto J, Shibata M (2018) Vegetation succession on fallow land. In: Nakamura F (ed) Biodiversity conservation using umbrella species: Blakiston's fish owl and the red-crowned crane. Springer Singapore, Singapore, pp 197–206

Morimoto J, Shibata M, Shida Y, Nakamura F (2017) Wetland restoration by natural succession in abandoned pastures with a degraded soil seed bank. Restor Ecol 25:1005–1014

Nakamura F, Ishiyama N, Yamanaka S, Higa M, Akasaka T, Kobayashi Y, Ono S, Fuke N, Kitazawa M, Morimoto J, Shoji Y (2020) Adaptation to climate change and conservation of biodiversity using green infrastructure. River Res Appl 36:921–933

Oga I (1951) The longevity of seeds – my dream – my longing – excavation of old lotus seeds in the Kemigawa Peatland, Chiba Prefecture. Collect Breed 13:206–207

Phartyal SS, Rosbakh S, Poschlod P (2018) Seed germination ecology in Trapa natans L., a widely distributed freshwater macrophyte. Aquat Bot 147:18–23

Sheldon CJ, Ficklin RL, Fawley KP, Fawley MW, Bataineh M, Nelson AS, Wilson S (2016) Vegetation diversity in natural and restored forested wetland sites in Southeast Arkansas. J Ark Acad Sci 70:37

Toyoshima G, Morimoto J, Nakamura F (2020) The role of waterways in agricultural land landscapes in preserving the diversity of aquatic and epiphytes. J Jpn Green Eng Soc 46:15–20

Tsubota H, Inoue Y, Nakahara-Tsubota M, Shimamoto T, Matsuda I, Uchida S, Mukai S (2014) Combination of DNA barcoding and molecular phylogenetic analysis as an identification tool for plant species: a case study using imperfect herbarium specimens. Hikobia 16:475–490

Vécrin MP, Diggelen R, Grévilliot F, Muller S (2002) Restoration of species-rich flood-plain meadows from abandoned arable fields in NE France. Appl Veg Sci 5:263–270

Verhagen R, Klooker J, Bakker JP, Diggelen R (2001) Restoration success of low-production plant communities on former agricultural soils after top-soil removal. Appl Veg Sci 4:75–82

Weiher E, Keddy PA (1995) The assembly of experimental wetland plant communities. Oikos 73:323–335

World Water Assessment Programme (2009) The United Nations World Water Development Report 3: Water in a Changing World. UNESCO, Paris; London: Earthscan

Zedler JB, Kercher S (2005) WETLAND RESOURCES: status, trends, ecosystem services, and restorability. Annu Rev Environ Resour 30:39–74

Chapter 14
Biodiversity Conservation through Various Citizen Activities in a Flood Control Basin

Jun Nishihiro, Shohei Koike, and Atsushi Ono

Abstract Flood control basins (FCBs) are artificially constructed facilities with the potential to conserve the biodiversity of floodplain ecosystems. However, the intensity of disturbance in FCBs is generally lower than that in natural floodplains; thus, habitats for plants with disturbance-dependent life histories and for animals that use temporary waters or mudflats can be lost in the years following FCB construction. Here, the Asahata FCB in Shizuoka city, Japan, was studied as an example, where the species diversity of wetland plants has been conserved as a result of diverse activities. Although most activities had objectives other than biodiversity conservation, they contributed to conserving plant diversity and providing habitats for endangered plants. The FCB is a green infrastructure that not only enables flood control but also supports activities for various purposes. Additionally, its proper use contributes to biodiversity conservation.

Keywords Biodiversity · Citizen participation · Education · Endangered plants · Welfare

14.1 Introduction: Floodplain and Flood Control Basin

14.1.1 Floodplain

In lowland plains, rivers that have not been artificially modified meander and change the channel morphology during flooding events (Bridge 2003). Such dynamics of

J. Nishihiro (✉)
National Institute for Environmental Studies, Ibaraki, Japan
e-mail: nishihiro.jun@nies.go.jp

S. Koike
Toho University, Tokyo, Japan

A. Ono
Showa Sekkei Co., Ltd., Osaka, Japan

the river channel result in a floodplain. Because the intensity of flood disturbances varies among locations within a floodplain, a mosaiclike ecosystem composed of heterogeneous elements is formed, providing a habitat for diverse flora and fauna (Junk et al. 1989; Tockner et al. 1998; Ward et al. 1999; Robinson et al. 2002; Tockner and Stanford 2002).

In recent years, the area and biodiversity of floodplains have decreased world-wide (Tockner and Stanford 2002; Ishiyama et al. 2017). In Japan, which is mostly mountainous, floodplain areas have been targeted for land use development. Many major Japanese cities are located on alluvial plains at low elevations, which constituted the seabed during the Holocene glacial retreat. Furthermore, these alluvial plains are suitable for modern rice cultivation; therefore, large areas of the plain have been converted into modernized agricultural land. In addition to land conversion, flow control by dams in the upper reaches of rivers causes biodiversity loss in floodplains. The reduction in flood disturbance in turn reduces the habitat areas of plants with life history characteristics that depend on disturbance and of animals that exploit temporary water (Nakamura et al. 2006).

14.1.2 Potential of Flood Control Basins

To conserve floodplain biodiversity, it is important to preserve areas affected by flood disturbances. In flood control basins (FCBs), the biodiversity of floodplain ecosystems can be conserved, even in Japan, where the population density is high and lowland plains are often subject to development. An FCB is an artificially constructed facility intended to reduce river flow due to rainfall occurring during peak floods (Fig. 14.1). Although FCBs are designed to reduce flood risk, they have similar hydrological characteristics to natural floodplains, i.e., they are temporarily inundated only during floods. Therefore, they can provide habitats for species with life history characteristics that depend on disturbances caused by floods.

There are 143 FCBs in Japan, with a total area of 15,104 ha (Suwa and Nishihiro 2020). However, not all of these factors may contribute to the conservation of floodplain ecosystems. There are many differences between FCBs and natural floodplains, a major difference being the intensity of the disturbance. In many FCBs, only water that exceeds the overflow dike, which is built as a relatively low part of the dike between the river channel and FCB, can enter the basin. Thus, there is insufficient energy to disturb vegetation and soil. In some areas of natural floodplains, the flood energy is sufficiently high to remove preexisting vegetation; however, identification of such areas in FCBs is difficult. In stable wetland environments, the biomass of aboveground vegetation commonly increases over time, and a transition to woodland occurs through vegetation succession (Salo et al. 1986). This change can be rapid because the nutrient input is high in FCBs as a consequence of sedimentation. Thus, habitats for plants with disturbance-dependent life histories and for animals that use temporary waters or mudflats can be lost within only a few years of FCB construction.

Fig. 14.1 The functioning Asahata flood control basin. (a) Floodwaters overflowing part of the levee (overflow dike), entering the FCB. (b) FCB during floods. (c) FCB under normal circumstances

14.1.3   Utilization of Land in a Flood Control Basin

FCB functions to reduce flood risk but can also be used for sports, recreational activities, environmental education, and paddy cultivation; additionally, their vegetation can be used as a source of materials for traditional thatched roofs. Appropriate planning of these activities is expected to contribute to biodiversity conservation. If so, disturbances caused by floods can be replaced by disturbances caused by human activities.

However, not all disturbances have a positive effect on biodiversity. Local biodiversity is high only under moderate disturbances ("intermediate disturbance hypothesis," Connell 1978). Excessive anthropogenic disturbances can lead to a decline in biodiversity. Moreover, the intermediate hypothesis is that species diversity within a spatial unit can be considered a homogeneous environment (α-diversity). Higher β-diversity, defined as a greater heterogeneity between different locations, is another characteristic of floodplain biodiversity. The effects of disturbance on FCB biodiversity should be evaluated in terms of both α- and β-diversity.

The Asahata FCB in Shizuoka city, Japan, was constructed to control floods in the Tomoe River. Many groups and individuals, including nature conservation groups, private companies, special schools for students with disabilities, and hospitals, participate in activities in the FCB (Nishihiro 2018). We studied the effects of these activities on the vegetation in the FCB. First, we provide background information and discuss the research results regarding the relationship between these activities and vegetation.

14.2   Asahata Flood Control Basin

14.2.1   Geomorphological Features and Changes in Land Use

The Tomoe River is 17.98 km long, and its watershed covers an area of 104.8 km^2. The watershed comprises approximately 7.6% of the total area of Shizuoka city but contains 47% of the total population (approximately 340,000 people). The lower parts of the river basin are urbanized areas.

The Tomoe River Basin was a bay during the Holocene glacial retreat, and most of the basin was flat. The FCB was built at the inflection point of the riverbed slope, where the river leaves the mountains and reaches the low land (Fig. 14.2). Until about 50 years ago, many swamps were scattered in this area (Yasumoto 1979). The paddy fields of this swampy area were inundated throughout the seasons, and rice productivity was low. Traditional fisheries for eels, loaches, shrimp, and carp were established in the Edo era (the late 1800s; Yasumoto 1979). From 1959 to 1973, a project was carried out to improve agricultural productivity, including wetland drainage and paddy field enlargement.

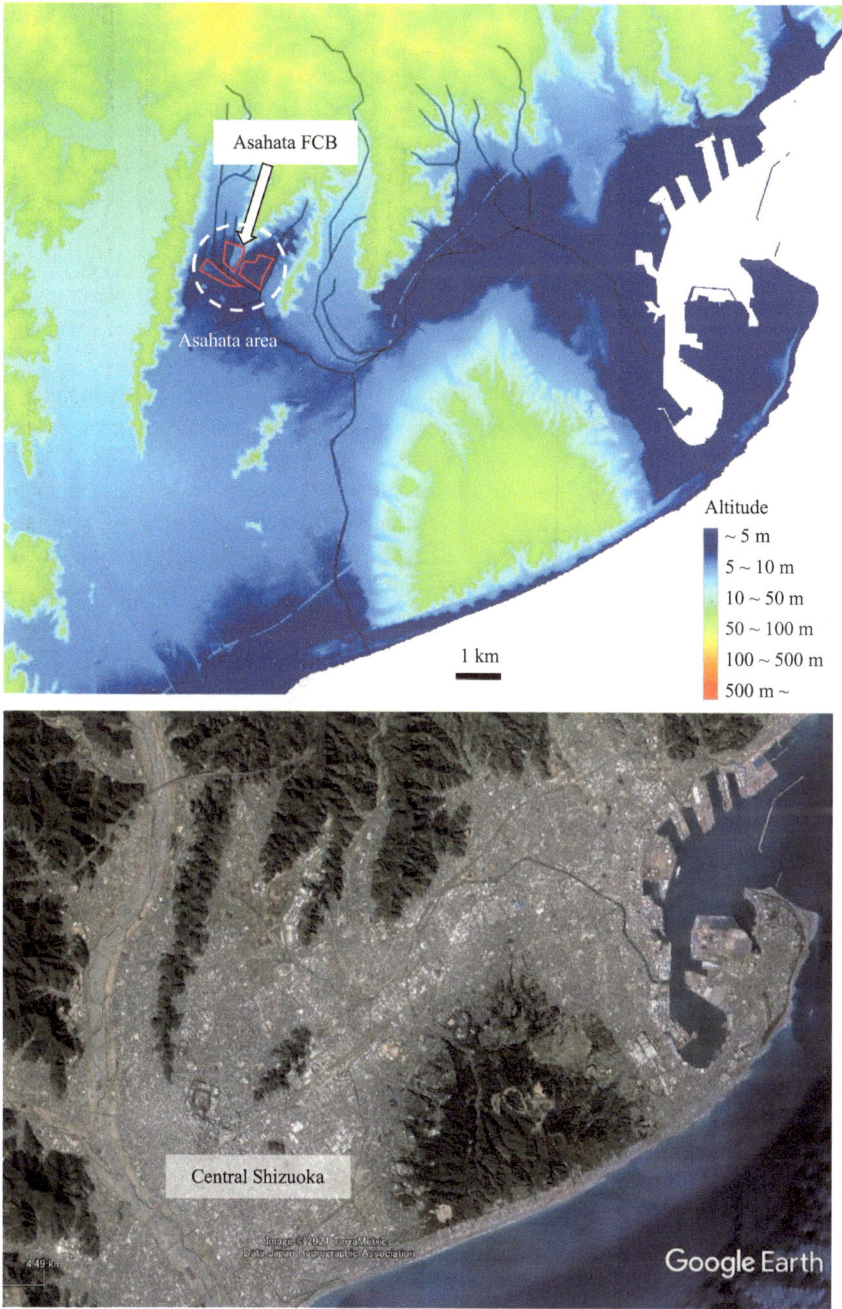

Fig. 14.2 Location of the Tomoe River (black line of upper panel) and location of the Asahata flood control basin. The lower panel shows Google Earth Image (March 2021 image)

Although this project greatly increased the rice productivity of agricultural land, the discharge capacity of the Tomoe River itself was low, and the risk of flooding in the watershed remained high. On July 7 and 8, 1974, Typhoon #8 caused heavy rain, leading to 41 deaths, the destruction of 32 houses, and above- and underfloor inundation of 11,981 and 14,143 houses, respectively. Because of this damage, a hydraulic control plan was established for the Tomoe River Basin in 1978, which included the construction of the Asahata FCB.

14.2.2 Activities in the Asahata Flood Control Basin

In the early 2000s, during the construction of the FCB, various plants, including the endangered disturbance-dependent species *Monochoria korsakowii*, grew extensively in the FCB due to disturbances caused by the construction work. The Nature Restoration Council, a local council for ecosystem management established under the Act to Promote Nature Restoration, was launched in 2003. The council initially focused on activities aimed at conserving biodiversity, of which the primary focus was the maintenance of the habitat of endangered plants. However, the range of activities was limited, and the number of participants did not increase over time.

In 2014, several new groups started to use the Asahata FCB, including people with disabilities attending special schools located near the FCB and staff from two local hospitals. New activities were encouraged by executives of the prefectural department and hospitals. Paddy fields were created to provide opportunities of job training to students with disabilities. With the provision of activities aimed mainly at welfare and education, local private companies and nongovernmental organizations also participated; thus, the total number of organizations participating in the Nature Restoration Council increased rapidly (Fig. 14.3). The number of participants in the

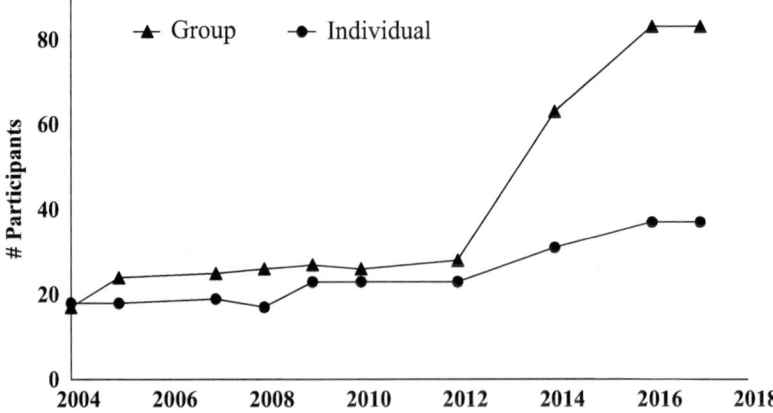

Fig. 14.3 Number of group and private participants of the Nature Restoration Council, Asahata flood control basin

council was about four times larger in the period when groups whose main purposes were welfare and education were included in the council than that in the period when it was composed of groups whose main purpose was biodiversity conservation.

14.3   Effect of Activities on Plant Diversity

14.3.1   Activities

We identified nine types of activities that caused anthropogenic disturbances in the Asahata FCB (Table 14.1, Fig. 14.4). Only one activity, management of *M. korsakowii* habitat, had biodiversity conservation as its primary purpose (Table 14.1). Other activities included the management of paddy fields for students with disabilities, as mentioned above, as well as conducting classes for students with disabilities; focusing on the study of soil seed banks; transplanting plants to other parts of the retarding basin; excavating a pond to increase the storage capacity of the FCB, collecting *Carex dispalata* and *Zizania latifolia* to make traditional grass hats and grass mats, respectively; controlled burning to prevent vegetation succession; and mowing to make firebreaks to facilitate controlled burning.

Some activities had biodiversity-related objectives. For example, an educational program for studying soil seed banks was conducted as follows:

1. The students attended a lecture on plant diversity and soil seed banks.
2. The students collected soil samples from the FCB and spread them in containers placed in the school garden.
3. The plants that emerged from the soil samples were checked and identified throughout the year, and the flora of the seed bank was recorded.
4. After the experiment, the soil samples containing high-density seed banks (newly dispersed seeds were added from the plants that emerged in the soil) were moved and spread in the FCB.

The main objective of this program was to educate students, but the content of the training program was closely related to the conservation of plant diversity. In another case, while the main purpose of excavating the pond was to increase the capacity for water retarding, biodiversity was also considered. In contrast, the activities involving collection of plant materials did not consider biodiversity conservation as an objective. Overall, there was a high diversity among activity objects.

14.3.2   Vegetation

The activities listed in Table 14.1 can have different impacts on the ecosystems of the FCB. Of the nine activities, four involved only mowing aboveground vegetation,

Table 14.1 Activities and their location in the Asahata flood control basin. Stars and circles represent the main actor and participants, respectively

Activities	Main objective	Stakeholder							Disturbance regime
		Nature conservation NGO	Welfare NGO	School for people with disabilities	Private company	Researchers	Local government	Citizens	
(a) Spreading soil containing seeds	Environmental education	o		★		o			Management of aboveground vegetation and soil disturbance
(b) Conservation of endangered plants	Biodiversity conservation	★				o			
(c) Transplant of *Zizania*	Public construction work						★		
(d) Paddy fields	Welfare and education		★	★	o	o			
(e) Pond construction	Flood control						★		
(f) *Carex* collection	Traditional hat making						★	★	Management of aboveground vegetation
(g) *Zizania* collection	Traditional mat making							★	
(h) Control burning	Vegetation management	o	o			★	o		
(i) Mowing	Fire stopping	o	o			★	o		

Fig. 14.4 Various activities in Asahata flood control basin. (a) Rice planting in a paddy field by students from a special-needs school. (b) Soil with high density of seed banks is being returned to the FCB by students from a special-needs school. (c) Collection of *Zizania* to make traditional rush mat. (d) Winter burning with the participation of many citizens

while five included both mowing and disturbing the soil. We conducted a vegetation survey to assess the impact of these activities on vegetation in the Asahata FCB. A 10 m² plot was set up in each area where the activities listed in Table 14.1 were conducted, and all vascular plants were recorded. A similar survey was conducted in a control plot in an area adjacent to each activity site. The control plots were similar in terms of altitude and soil origin.

In total, 97 plant species were recorded, of which 77 were native species (including 53 wetlands or aquatic species). Of these 97 species, 58 (60%) were observed only in plots that underwent activities (Fig. 14.5). In addition, five species on the Japanese Red List were recorded, all of which were found only in plots with activities. There were more species in the activity than in the control plots in areas where the soil was disturbed, with one exception (Fig. 14.6). Although the number of species was small in this exceptional plot, submerged plants were found only

Fig. 14.5 Endangered species observed in the sites with activities: *Monochoria korsakowii*, *Penthorum chinense*, and *Persicaria conspicua* (from left to right). The numbers of species (sp.) in sites with activities and/or without activities (control) are also shown

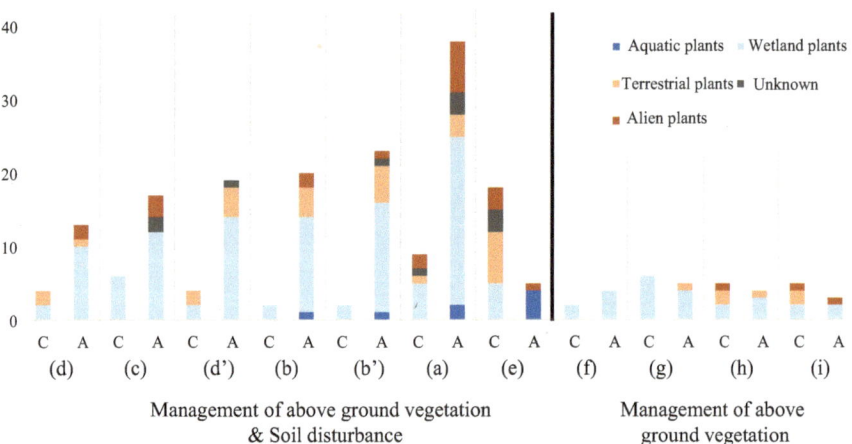

Fig. 14.6 Number of species observed in each site with activity (A) and control (C). The activities (a) to (i) correspond with these in Table 14.1. (b') and (d') represent the ridge of the paddy (b) and (d), respectively

in the plot. In other words, highly specific vegetation, which contributes to the β-diversity of flora in Asahata FCB, was established in this plot.

Activities in the Asahata FCB contributed to the maintenance of plant diversity. Most of the observed plants were wetlands and aquatic plants, suggesting that activities in the Asahata FCB facilitated the conservation of floodplain vegetation. Anthropogenic disturbances due to activity in the FCB might partly replace natural disturbances, although it will be necessary to compare the vegetation of FCB to that of a natural floodplain to confirm this hypothesis.

In many cases, the number of alien plants was also higher in the activity plots than that in the control plots, suggesting that disturbance can increase the opportunity for the invasion of many plants. Controlling alien species can be an issue in the management of biodiversity conservation.

14.4   Generality of the Results

This study examined activities in the Asahata FCB from the viewpoint of biodiversity conservation. However, these results cannot be applied to all FCBs. The floodplain-like vegetation observed here is thought to have been derived from the soil seed banks of plants that had grown in paddy fields and swamps before the FCB was built. The topsoil is removed in most FCBs to maintain the capacity for retarding floodwater. However, in the construction of the Asahata FCB, the topsoil from the paddy fields was first secured in another location and then returned to the surface of the FCB after construction. This was thought to have enabled the establishment of vegetation with a high species diversity. The use of topsoil is particularly important for the conservation of plant diversity in FCBs.

Our study showed that the utilization of wetlands usually increases α-diversity. However, not all activities in FCBs facilitate conservation, and some might significantly reduce biodiversity. For instance, under conditions of high invasion pressure from invasive alien species, disturbance can lead to a decrease in biodiversity. To prevent such situations, it is important to assess their impact before implementing activities. We consider that the Asahata FCB is being used appropriately, at least for the moment, owing to the consensus built among various stakeholders, including ecology experts.

14.5   Activity Redundancy and Conservation Sustainability

A shortage of participants may lead to discontinuation of the activities in the future, namely, for the "Use of *Zizania*" and "Use of *Carex*" activities listed in Table 14.1. In addition, the nature conservation group considers the aging of its members as problematic. Societal changes can make it difficult to continue an activity, the scale of which may have to be reduced. If an activity cannot be continued, the environment that depends on it will disappear, thus decreasing the biodiversity of the FCB. To maintain biodiversity under such uncertain circumstances, "activity redundancy," which refers to a variety of actors playing a similar role, is thought to be an important factor in long-term natural resource management. It will be important to allow activities with various purposes, not limited to biodiversity conservation, and to scientifically monitor the effects of these activities. It will also be important to have a mechanism to facilitate communication among participating groups and to help each other when necessary.

Acknowledgments We sincerely thank Ms. Kuriyama (Shizuoka city) for her cooperation. This study was partly supported by the Environment Research and Technology Development Fund JPMEERF20202001 of the ERCA of Japan.

References

Bridge JS (2003) Rivers and floodplains: forms, processes, and sedimentary record. Blackwell publishing, Oxford

Connell JH (1978) Diversity in tropical rain forests and coral reefs. Science 4335:1302–1310

Ishiyama N, Nagayama S, Iwase H, Akasaka T, Nakamura F (2017) Restoration techniques for riverine aquatic connectivity: current trends and future challenges in Japan. Ecol Civ Eng 19:143–164. (in Japanese)

Junk W, Bayley PB, Sparks RE (1989) The flood pulse concept in river-floodplain systems. In Dodge DP (ed). Proceedings of the international large river symposium (LARS). Canadian special publication of fisheries and aquatic sciences 106, pp. 110–127

Nakamura K, Tockner K, Amano K (2006) River and wetland restoration: lessons from Japan. Bioscience 56:19–429

Nishihiro J (2018) Efforts to balance the usage and conservation of biodiversity in Asahata retarding basin, Shizuoka. National Parks 763:16–19. (in Japanese)

Robinson CT, Tockner K, Ward JV (2002) The fauna of dynamic riverine landscapes. Freshw Biol 47:661–677

Salo J, Kalliola R, Hakkinen I, Makinen Y, Niemela P, Puhakka M, Coley PD (1986) River dynamics and the diversity of Amazon lowland forests. Nature 322:254–258

Suwa Y, Nishihiro J (2020) Distribution and location characteristics of flood-control reservoir in Japan. Ecol Civ Eng 23:85–97. (in Japanese)

Tockner K, Stanford JA (2002) Riverine flood plains: present state and future trends. Environmental Conservation 29:308–330

Tockner K, Shiemer F, Ward JV (1998) Conservation by restoration: the management concept for a river-floodplain system on the Danube River in Austria. Aquatic Conservation: Mar Freshw Ecosyst 8:71–86

Ward JV, Tockner K, Schiemer F (1999) Biodiversity of floodplain river ecosystems: ecotones and connectivity. Regul Rivers: Res Manage 15:125–139

Yasumoto H (1979) History of Asahata. Shizuoka city office. (in Japanese)

Part V
Urban and City Ecosystem

Chapter 15
Toward Holistic Urban Green Infrastructure Implementation

Takanori Fukuoka

Abstract This article aims to reexamine green infrastructure (GI) visions and frameworks and site-scale GI implementation toward holistic urban GI implementation. For GI visions and frameworks, both City of Philadelphia and Singapore have demonstrated progressive approaches to weave different divisions and planning, accomplishing holistic frameworks as well as enhancing GI implementation at site-scale level. Through three site-scale GI implementation case studies, the creative solutions have been demonstrated how streets, retention ponds, and the existing urban fabric were to transform into GI open spaces. This article suggests the way how GI could be interwoven into visions, planning, and design to create places for people.

Keywords Urban green infrastructure · Visions and framework · GI implementation · Open space

15.1 Toward Holistic Urban Green Infrastructure Implementation

Green infrastructure (GI) implementation needs to integrate visions and frameworks of the city scale with the ones of the site scale. In recent years, many practices have demonstrated good GI implementation; however, there are gaps between city-scale GI visions and site-scale GI implementation. Like many other concepts, the term GI often creates the confusion of its definition. As per Environmental Protection Agency (EPA) in the USA, GI is defined as "systems and practices that use mimic natural process such as infiltration and evapotranspiration."

T. Fukuoka (✉)
Faculty of Regional Environment Science, Department of Landscape Architecture Science, Laboratory of Landscape Design and Geoinformatics, Tokyo University of Agriculture, Tokyo, Japan
e-mail: tf206471@nodai.ac.jp

F. Nakamura (ed.), *Green Infrastructure and Climate Change Adaptation*,
Ecological Research Monographs, https://doi.org/10.1007/978-981-16-6791-6_15

Section 502 of the Clean Water Act defines green infrastructure as "the range of measures that use plant or soil systems, permeable pavement or other permeable surfaces or substrates, stormwater harvest and reuse, or landscaping to store, infiltrate, or evapotranspirate stormwater and reduce flows to sewer systems or to surface waters" (United States Environmental Protection Agency n.d.).

This sustainable stormwater management system has become a big trend since the early 2000s, and this system is introduced in the form of bioswales, green roofs, rain gardens, and such small implementation. On the other hand, the European Commission defines GI as "a strategically planned network of natural and semi-natural areas with other environmental features that was designed and managed to deliver a wide range of ecosystem services. It incorporates open spaces and other physical features in terrestrial and marine areas. On land, GI is present in rural and urban settings" (European Commission 2013). Though these two perspectives indicate principles of GI, it is not clear how both concepts can be applied for GI implementation. In other words, EPA expects GI put focus on solution, and performance of urban ecological system, especially for stormwater. EU covers much broader open spaces and natural systems as its framework. How can we implement these perspectives on the ground? This section aims to reexamine "visions and approaches" toward holistic urban green infrastructure implementation with reference to some project case studies.

15.2 GI Visions and Frameworks: "Green City, Clean Water" Citywide Green Infrastructure Implementation Frameworks in the City of Philadelphia

City of Philadelphia, USA, challenged on creating a new type of the holistic framework toward. strategic GI planning and implementation. Through a literature review and the interviews with the City of Philadelphia Water Department GI group, and by consulting "Green City, Clean Waters (GCCW)" to understand the development and framework of GI planning, three phases of GI planning and implementation were identified through the analysis of selected GI planning and policy (Fukuoka et al. (2020)). The first phase (1990s–2008) can be described as "Water quality control period." Combined sewer overflow (CSO) was a big problem in Philadelphia. Polluted water would be released directly to the river system after extensive storm event. CSO Control Policy was implemented in 1994, and the long term Control Plan by Water Department was released in 1997. In this phase, GI was implemented as a site-scale project within Watershed Planning (Office of Watershed). The other approach was a grassroot West Philadelphia project directed by Anne Spirn, professor at the University of Pennsylvania. West Philadelphia project demonstrates very progressive approaches which cover stormwater management for social issues and food production. Prof. Spirn collaborated with PWD (Philadelphia

Water Department) and advocated GI at early stage (West Philadelphia Landscape Project n.d.).

The second phase (2008–2012) can be described as "GI planning development period." Mayor Nutter (2008–2016) started it in 2008, and orchestrated very powerful leadership in creating sustainable cities. Multiple Policies and Guidelines such as Green Works (Sustainability Vision by Office of Sustainability), GCCW (GI Plan by PWD), and Green Philadelphia (Open Space Plan by Park and Recreation) are enforced from one after another.

The third phase (2012–) can be described as "GI implementation acceleration period." City of Philadelphia joined a partnership with EPA in 2012 to implement GI further. Contents of GI partnership include GI model project implementation, GI engineering development, and water quality research and communication. For organizational structure, GI Planning group was created in 2017.

City of Philadelphia shifted GI implementation unit from watershed-based implementation to district-scale implementation which is more effective to create synergy between city planning and GI implementation. In addition, Green Street Design Manual (PWD and Transportation), Stormwater Design Manual (PWD), and Planning and Design Manual (PWD) had been issued in 2018. Now, cities challenge to accelerate GI implementation further. GCCW performed key roles in setting GI goals, visions, and frameworks.

As a summary, Philadelphia developed the holistic framework toward strategic GI implementation. GI visions and frameworks set the big pictures and the essential structures for diverse governmental divisions in order to move forward with GI concepts. Those frameworks also help build up and activate a wide range of site-scale GI projects from green streets, open spaces to urban redevelopment. Especially, shifting GI framework scale from watershed-based one to district scale helped to create a synergetic effort with city planning work (Fig. 15.1). It was rather solo frame work before. The next phase for challenges aims to make this vision and framework much broader so that it will include new types of GI such as vacant lots or brownfields that are significantly increasing around the city periphery areas.

15.3   National-Scale Holistic GI Visions and Approaches: "ABC Water Design Guidelines in Singapore"

In contrast to GI implementation in Philadelphia, Singapore challenged to create national-scale holistic GI visions and frameworks with Design Guidelines. This case study is an excellent example of sustainable stormwater management at a national scale and may be applicable in other cities and nations in the similar monsoon climate. Singapore is located at the heart of Southeast Asia, and is comprised of 63 islands. Singapore Island is 42 km long from east to west and 23 km from north

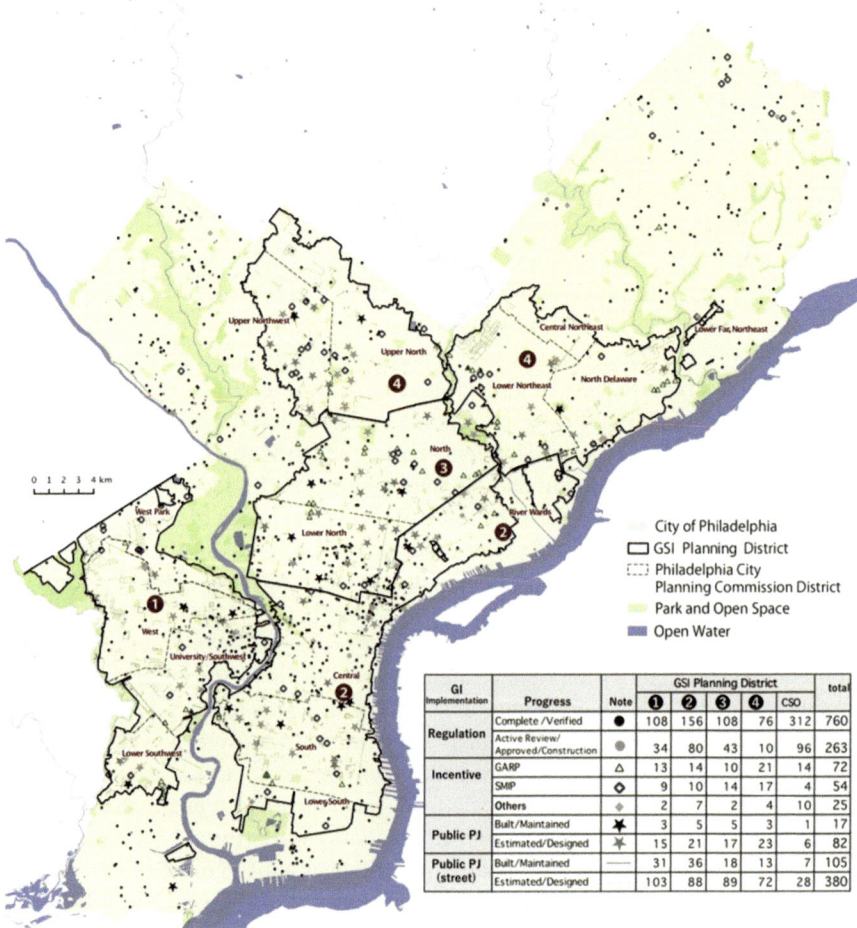

GI Implementation	Progress	Note	GSI Planning District					total
			①	**②**	**③**	**④**	CSO	
Regulation	Complete /Verified	●	108	156	108	76	312	760
	Active Review/ Approved/Construction	●	34	80	43	10	96	263
Incentive	GARP	△	13	14	10	21	14	72
	SMIP	◇	9	10	14	17	4	54
	Others	✦	2	7	2	4	10	25
Public PJ	Built/Maintained	✱	3	5	5	3	1	17
	Estimated/Designed	✳	15	21	17	23	6	82
Public PJ (street)	Built/Maintained	——	31	36	18	13	7	105
	Estimated/Designed		103	88	89	72	28	380

Fig. 15.1 GI implementation in the City of Philadelphia in 2019. *GSI indicates green stormwater infrastructure (Fukuoka et al. (2020) created it by using open Data PHL map Green Stormwater Infrastructure Public Projects Points/Street data set)

to south. Historically, Singapore has relied on Malaysia for 40% of its drinking water resource coming through pipelines. However, due to uncertain future with water problems, Singapore government made a decision to become self-sufficient of water by applying watershed-based management system. Especially, annual rainfall (around 2340 mm) is targeted as the major water resource to be wisely used together with gray water and the water stored in reservoirs. Public Utilities Board (PUB), Singapore's national water agency, has embarked on water management using the

3P (People, Public, and Private) approach to take joint ownership of Singapore's water resource management.

This is embodied in PUB's tagline – Water for All: Conserve, Value, Enjoy (Public Utilities Board 2016a).

The Active, Beautiful, Clean Waters (ABC Waters) Program, launched by PUB in 2006, is the cornerstone of the 3P approach. The program intends to transform Singapore's extensive network of reservoirs and water bodies into beautiful and clean streams, rivers, and lakes, creating a vibrant City of Gardens and Water. More than 100 potential locations will be identified as the site for the implementation of the program by 2030 (Public Utilities Board 2016b). As of June 2014, 23 projects had been completed (Public Utilities Board 2014).

The first edition of the ABC Waters Design Guidelines was launched in 2009 and the second edition in 2011. In June 2014, PUB upgraded the guidelines for locally built examples to the ones for the showcase developers, architects, and engineers who have incorporated the ABC Waters concept in their developments (Public Utilities Board 2016c). ABC Waters Program helps implement ABC Waters Design Guidelines and function based on GI as a hinge between visions and site-scale projects. This new set of reference material aims to meet the industry's needs better and continue building up technical expertise in the industry (Kato and Fukuoka 2016). First, the ABC Waters Design Guidelines provide actual stormwater design tools for three different stages which are applicable to various types of green infrastructure project scale. These stormwater design tools are categorized into (1) catchment elements, (2) treatment elements, and (3) conveyance and storage elements. First, catchment elements aim to collect water based on land use typologies such as road, canal, water bodies, pedestrian walkways, public open spaces, plazas, and buildings. For different surface conditions, appropriate stormwater planning and design methodologies are shown with a clear illustration. At the stage of project planning, research and concept were made based on the ABC Waters, and location and volume of buildings were to meet ABC Waters' goals, and applicable green infrastructure implementation methodologies were introduced at catchment stage. For example, building was divided into various catchment elements such as green roofs, terraced green balconies at multi levels, and ground level elements such as planted areas and water features. The ABC Waters Design Guidelines provided engineering and design procedures so that the basic knowledge and methodologies could be easily integrated into projects. Second, methodologies how to implement treatment elements such as swales, bioretention pond, detention pond, stormwater planter, rain garden, and cleaning biotope are explained through visual information. Water treatment part covers a wide range of methods, and this provides concept, benefits, and design methodologies for management issues. Third, conveyance and storage elements focus on large water bodies and provide methodologies such as bioengineering, erosion control, and water quality control.

Second, in order to realize ABC Waters concept, ABC Waters Certification and ABC Waters Professional Program were created to make loop of projects be recognized and adopted by people (Kato and Fukuoka 2016). The ABC Waters Certification launched by PUB on 1 July 2010 is a scheme to provide a recognition to public agencies and private developers who embrace the ABC Waters concept and incorporate ABC Waters Design Features in their developments (Public Utilities Board 2016d). Besides providing the recognition, the scheme also aims to ensure that the design features incorporated within the developments meet the minimum design standard. Since 2010, 62 projects have been certified by the ABC Waters Certification. Certification is judged in four categories: Active, Beautiful, Clean, and Innovative. Out of total 110 possible points, for a project to be certified, it needs to receive a minimum of 45 points with at least 5 points in each of the first three categories. ABC Waters Professional Program is supported by multiple institutions such as Institution of Engineers Singapore (IES) with the support of the Singapore Institute of Architects (SIA) and Singapore Institute of Landscape Architects (SILA) (Public Utilities Board 2016e).

As a summary, Singapore put force on setting nationwide Water Design Guidelines and provided both urban green infrastructure visions and frameworks at a national scale. Singapore's case study demonstrates "top-down" approaches and sets effective schemes to translate GI visions to real projects by creating multiple programs. Both Certificate and ABC Waters Professional Program help accelerate GI implementation. As described above, two cases demonstrated differently, but both are powerful in GI visions and frameworks toward implementation. In the next section, three cases of site-scale GI implementation in urban settings have been examined.

15.4 GI Approaches: Site-Scale GI Implementation

In this section, site-scale GI implementation project cases are introduced. GI visions and frameworks often express broader ideas but remain vague. At the same time, site-scale GI implementation tends to split into single bits of spaces such as bioswales, green roof, and rain garden. As shown in Fig. 15.2, toward urban GI implementation, each project still needs to contain a big picture about how each piece fits to create the whole GI. Site-scale GI's function is also difficult in setting a good balance among multiple choices such as stormwater management, biodiversity, heat mitigation, and others. Figure 15.3 illustrates GI goals with an envision. It varies from "disaster reduction," "healthy city" to "soft infrastructure, community." How can we set appropriate GI goals and clarify the function of GI required? In this section, three site-scale GI projects are introduced to describe its character and functionality.

Fig. 15.2 GI implementation in city scale

15.5 Street as GI: "City of Copenhagen's New GI Street Approaches"

City of Copenhagen, Denmark, illustrates an excellent GI street project adapting to the climate change. In July 2011, cloudburst, heavy stormwater event, caused a significant damage to the city. Extensive 150 mm of rainfall in 2 h left major part of the old city area under 1 m of water, and the damage caused by that was approximately 1 billion Euro worth (Fig. 15.4). Based on the research, the expected flood damage caused by both cloudburst and expected sea level rise forced City of Copenhagen to work on cloudburst master plan with blue-green infrastructure in order to reduce the future risk of disaster (ASLA 2016). Existing historic old city has no further spaces left without implementing GI to solve problems. In this master plan, strategic approach such as utilizing road as GI street was taken specifically for Copenhagen. Cloudburst master plan created frameworks to accelerate GI implementation and there are three essential points. First, precautions against cloudburst are developed based on "research." The City of Copenhagen worked on investigating multiple sets of data and identified high-risk areas. In addition, they model and map the large-scale catchment base of stormwater and

Fig. 15.3 GI goals with an envision

visualize vulnerable areas as well as their risks. Second, GI function and GI implemented places bridge over cloudburst by creating "Cloudburst Toolkit." This toolkit touches "design and quality" of spaces and aims to provide human-scale experiences to the place. Hot spots for future GI implementation projects are identified and rough images of designed places are illustrated according to the typologies based on the toolkit. Third, socioeconomic cost-benefit analysis was conducted, and it clarified that GI solutions have 50% of potential in saving over conventional gray infrastructure or piped solutions. As explained above, these three points helped utilize GI visions and frameworks for site-scale GI implementation.

Fig. 15.4 Cloudburst caused significant damages to Copenhagen (Ramboll Studio Dreiseitl)

Figure 15.5 illustrates a basic concept for transforming road into GI street. Two lanes 16 m wide road would be reduced into one lane road, and the rest of the space was converted into planted areas and pedestrian walkways. Inner 11 m space has a preventive function as a floodwater storage area to tackle once in 100-year storm event. Planted area has the function to store stormwater underground and to provide green and spaces for people. All GI streets had a plan to connect the stormwater flow and projected frameworks to reduce the flood risk. Similar to Singapore's ABC Water Design Guidelines, future site-scale GI projects were identified, and would be gradually implemented. As a summary, the cloudburst project suggests one possibility to implement GI into existing city fabric by transforming existing roads to GI. In 2018, City of Copenhagen selected 50 GI projects planned to be built, and 5 projects already completed. The speed of GI implementation is slower than expected due to the complicated coordination with existing underground infrastructure. The other raised issue is how cloudburst can be evaluated as a climate change adaptation project (Nakajima and Hoshino 2017). Especially, water quality control is yet unsolved to meet with EU water quality framework. Lessons learned from the cloudburst are how to set GI goals as well as evaluation framework to create multifunctional space in the urban settings.

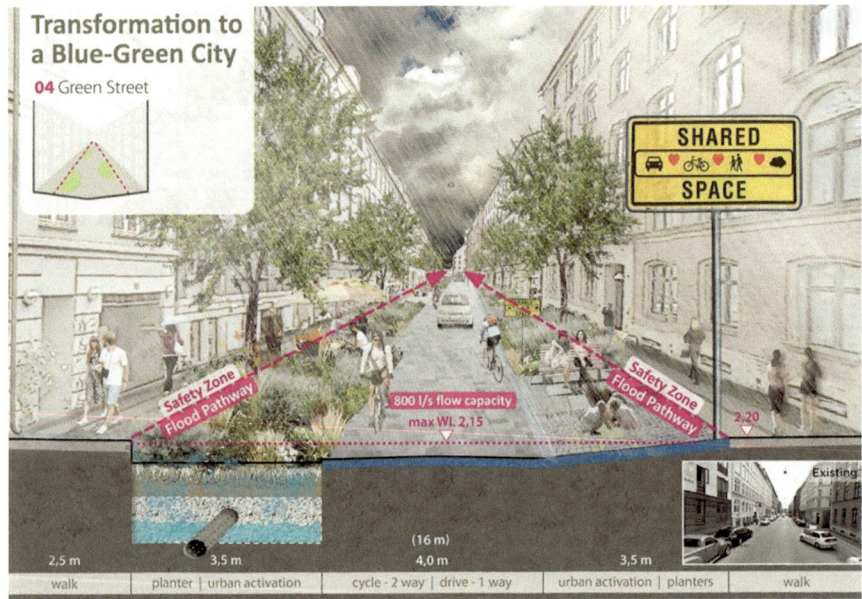

Fig. 15.5 GI implementation proposals in cloudburst master plan (Ramboll Studio Dreiseitl)

15.6 Kashiwanoha Aqua Terrace: "Closed Retention Pond to GI Open Space"

Kashiwanoha Aqua Terrace is the place that was redeveloped from flood control pond into multifunctional GI open space. It is located in Kashiwa, northeast of Tokyo, where 273 ha smart city is being developed. Aqua Terrace is a part of phase 2 developments with the concept of "Innovation Campus." The retention pond was formerly fenced around and had no access to the public and purely functioned as a flood control pond. Site area is about 2.4 ha big and retention capacity is approximately 73.720 m^3. Within 25 ha of phase 2 development, there was no central open space but the existing retention pond (Fig. 15.6). Thus, the idea of renovating closed retention pond to GI open space was strategically proposed.

To realize Kashiwanoha Aqua Terrace, the public-private partnership was formed by Kashiwa City, Kashiwanoha Urban Design Center (UDCK) and Mitsui Fudosan, a private developer. Prior to GI open space implementation, the master plan of vigorous "Innovation Campus" was issued in 2008 to illustrate "District of Mixed Usage" and "Walkable Street and Public Open Space Network" at the conceptual level. In 2015, the landscape design for Aqua Terrace started to connect the existing streets and the pond through applying seamless circulation loop, and multilevel seating on the retention pond slope. 800-m-long jogging loop around the pond provides a place for walking and jogging, while permeable paving provides cooling surface in a hot summer time. Multilevel seating on slope provides places of diverse

Fig. 15.6 Retention pond prior to redesign (Nikken Sekkei Ltd.)

activities for people and helps connect people and water. The challenge of Aqua Terrace design was to control the fluctuating water levels. The water level is usually kept at 40–80 cm; however, maximum of 4 m water level could be expected at once in 50-year storm event around this designed area. Trees and site elements were set carefully, and the lighting can endure against the floodwater. Instead of installing physical fences or gates against the dangerous water level after the storm event, lighting pole would blink red light when the water level has risen more than 1 m (Fig. 15.7). GI discussion tends to set aside the power of design; however, Kashiwanoha Aqua Terrace demonstrates an excellent design to weave GI's function and character into the place designed for people.

Regarding management of Kashiwanoha Aqua Terrace, UDCK plays a central role to bring all players and volunteers together. Volunteers help with planting management as well as cleaning up the walkway after the high water overflow. Events vary from "Outdoor Movie Night" to natural environment observation event for local children. Toward sustainable GI implementation, it is important to set organizational structures in order to manage GI properly. Lastly, Kashiwanoha Aqua Terrace represents a good GI open space model and illustrates the possibility of public open space in the era of mature society.

Fig. 15.7 Kashiwanoha Aqua Terrace as GI open space (Forward Stroke Inc.)

15.7 Minami-Machida Grandberry Park: Creating Livable, Sustainable City with Open Spaces

15.7.1 GI Visions and Frameworks

The Minami-machida Grandberry Park is a complex project in Machida city, southwest Tokyo. The train station, shopping mall, urban sports park, and Sakai River waterfront were integrated under the concept of "park life." The plan was to create a walkable community with seamless connections between the station, commercial facilities, parks, and the Sakai River. The inviting design starting from the station through the mall up to the sports park promotes "active lifestyle." This complex is located in the suburban residential area that was developed about 40 years ago with the station placed in the south and a semi-industrial area in the north. The urban redevelopment builds open spaces at its core. There are three major sites. Grandberry Park (shopping mall, privately owned, 8.3 ha), park life site (privately owned and operated facilities on former roads and public land, 0.5 ha), and Tsuruma Urban Sports Park (publicly owned, 7.1 ha), and a necklace of 14 open spaces in three large areas (Fig. 15.8). As illustrated, seven open spaces colored in blue are privately owned open spaces, and the other seven open spaces colored in green are publicly owned. In addition, yellow-green colored "park life site" is built by public-private partnership. Each of the 14 plazas has a variety of designs and unique atmosphere. For example, in the large Oasis Plaza, children can play on a

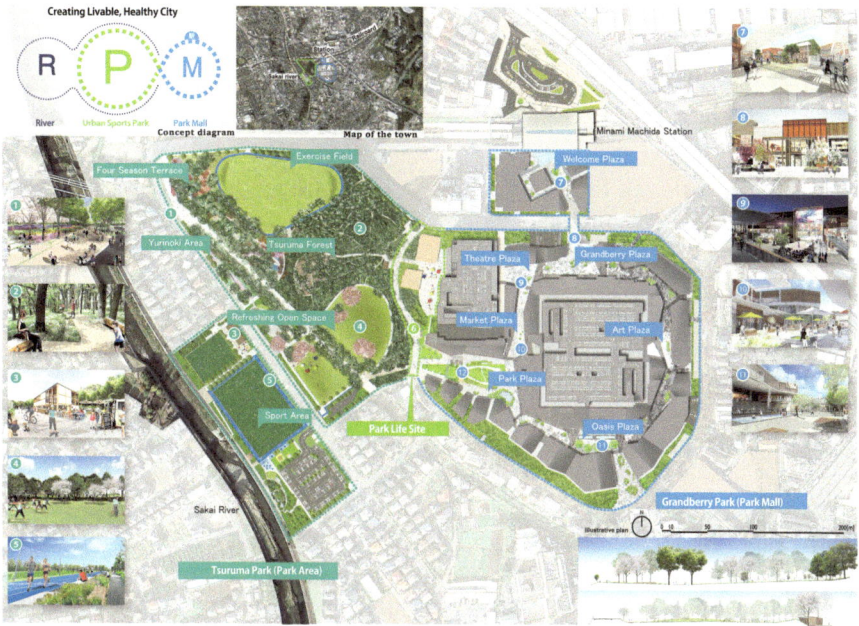

Fig. 15.8 Minami-machida Grandberry Park illustrative plan (Machida-Shi, Tokyo)

100-meter-long water carpet in summer, and in winter a skating rink is set up for multifunctional use.

How does this Minami-machida Grandberry Park function as GI? GI is strategically conceptualized in its vision and frameworks as networked open spaces. Formerly, the site was divided by the river, roads, urban sports park, shopping mall, and station. City of Machida (public) and Tokyu Corporation (private) formed public-private partnership in order to create a livable and sustainable city with open spaces. The following are the key frameworks related to GI. First, "Walkable City Network" is proposed from creating seamless connection of spaces through the district. A big decision was made to close the existing city road and create a 6-m-wide pedestrian street, which enhanced the physical integrity of Grandberry Park (shopping mall) and Tsuruma City Sports Park. In addition, fluid pedestrian circulation allows to connect the north and the south district that were formerly divided by railroad. Second, "Open Space Network" in both public and private land was created by proposing 14 open spaces. This necklace like open space structure allowed to promote not only walkability but lively urban lives. Third, "Sustainable Development Goals" raised by challenging to obtain "LEED-ND Gold" as common goals for all stakeholders. This vision helped implement sustainable stormwater management and other site-scale GI later on (Fukuoka 2020).

15.7.2 Design of Places and Public Engagement

By using visions and frameworks as guidelines, series of open spaces are designed to enhance its concept and character. "Walkable City" visions are developed for the broader "Active Design" concept. Active Design is a concept to support development or design for creating healthy cities (Osamura and Fukuoka 2020). In terms of health, when walkability improved, Tsuruma Park (Urban Sports Park) would function as the core site where people can participate in sports and become healthier. The 7.1 ha sports park provides diverse places for sports and the healthy activities. This park has tennis courts (three omni courts), artificial turf pitches (equivalent to three futsal pitches), athletic open space with playground (0.65 ha), clubhouse (studio, café, lockers, and showers), and parking lot for 133 cars. Active Design for promoting physical activities was implemented all over the park for the purpose of running, muscle training, jumping, etc. and yoga, dance, karate, etc., and classes are also open regularly at the studio (Fig. 15.9).

"Open Space Network" was implemented in various forms with unique character. For example, in the large Oasis Plaza, children can play on a 100-meter-long water carpet in summer, and in winter a skating rink is set up for multifunctional use. On the line of flow from the station to the park, plants of mainly native species are planted in connection with Tsuruma Park and Sakai River. Besides planting, there are many ideas worked out to allow comfortable usage of outdoor spaces. Various type of seating, shades, paving, and surrounding shops and façade create a setting for "park life" for all generations.

Regarding the aspect of sustainability, sustainable rainwater management was implemented as a part of Open Space Network. All rainwater at the station building is stored underground for reuse, and a permeable pavement has been installed on the site. The retention basin under the park is approximately 24,600 m^3. Sustainable stormwater management was implemented throughout the project by installing rain gardens and bioswales which promote temporary storage and infiltration.

Fig. 15.9 Tsuruma Urban Sports Park in Minami-machida Grandberry Park (left, before park renovation; right: after park renovation)

In the process of public engagement and participation, numerous workshops and activities were held for the citizen during the planning and design phase. Interestingly, citizen participation workshop was transformed into "Park School" where citizens themselves plan and implement temporal events working with other citizens of similar interests. In autumn, "Park School Festival" was successfully held a few times, and the participation of the citizen transformed it into more active, motivated activities in the park. In 2021, "Minami-machida-wo-minnna-no-machie foundation" (Creating Minami-machida Town for All) was founded by public-private partnership in order to proceed the city to the next stage. As a summary, the Minami-machida Grandberry Park demonstrates how big their visions and frameworks are and that the site-scale project was indeed integrated into a whole GI open space network. In this case, both city and developer envisioned big pictures such as "creating healthy, active cities" and "sustainable cities," and those visions realized the interconnected open spaces. In terms of public engagement, this case illustrates how participation and engagement process was integrated into the part of GI implementation in order to achieve a long term, sustainable management.

15.8   Toward Urban Green Infrastructure Implementation: Open Space as GI

Through examination of "visions and approaches" (15.1–3) and "site-scale GI project case studies" (15.4–7), the following are to recommend toward urban GI Implementation. Firstly, GI visions and frameworks need to provide broader perspectives from the heat mitigation in urban areas, reduction of water disaster to a healthy and walkable city. Frameworks also need to illustrate rough spatial images as well as functions of those spaces. City of Philadelphia demonstrated strategically implemented GI visions and frameworks and activated multiple planning and guidelines to accelerate GI implementation under very powerful leadership. On the other hand, Singapore developed nation-scale, top-down GI visions and frameworks in ABC Waters Design Guidelines. ABC Waters Design Guidelines set clear goals and provide typological GI approaches as well as demonstrate GI with actual model projects. Other good finding was circulative system of ABC Waters Design Guidelines with "ABC Waters Certification" and "ABC Waters Professional Program." This framework allows sustainable GI implementation over time.

Secondly, site-scale GI implementation needs to create places for multifunction and for people. The site-scale GI often tends to focus too much on monolithic function related to the stormwater management such as rain garden and bioswale. In Chap. 2, three GI implementation projects were introduced. Copenhagen's cloudburst demonstrates how "the strategies of climate change adaptation" were introduced into the current urban fabrics through transforming roads into multifunctional streets for all. Kashiwanoha Aqua Terrace developed interesting scheme to renovate closed, monofunctional retention pond into GI open space for diverse users.

At Minami-machida Grandberry Park, GI visions and frameworks can be seen in creating "Open Space Network," "Walkable City," and "Sustainable Development." At site scale, GI open space is created to achieve both GI function and its character as a place for people.

Lastly, this article tried to depict how GI visions and frameworks are translated into the site scale and physical GI implementation through various case studies. This article indicates how GI needs to be interwoven into vision, planning, and design, and creates places for people. As stated above, strategic GI visions and frameworks can catalyze a change to our city through physical manifestation of GI implementation.

References

ASLA Award 2016 Copenhagen Cloudburst (2016). https://www.asla.org/2016awards/171784.html

European Commission (2013) Green Infrastructure Enhancing Europe's National Capitol

Fukuoka T (2020) Catalyze changes to the City from open spaces. Nikkei BP Publishing, Green Infrastructure, pp 111–120

Fukuoka T, Katagiri Y, Kato S (2020) City of Philadelphia's framework towards strategic green infrastructure planning and implementation. J Jpn Inst Landsc Archit 83(5):673–678

Kato S, Fukuoka T (2016) Stormwater Management at a National Scale under the Asia Monsoon Climate: Case Study of Singapore's ABC Water Design Guidelines, International Symposium on Architectural Interchange in Asia (11th ISAIA) P.1905–1908

Nakajima N, Hoshino Y (2017) A study of the infrastructure development process and support system for climate change adaptation-in case of the cloudburst plan in municipality of Copenhagen, DE

Osamura Y, Fukuoka T (2020) Process of active design guidelines in New York City: focusing on ADG implementation. Reports of the City Planning Institute of Japan Nor.19. pp. 277–279

Public Utilities Board (2014) ABC waters design guidelines, 3rd edn. Public Utilities Board (PUB), Singapore

Public Utilities Board (2016a) Singapore Water Story. https://www.pub.gov.sg/watersupply/singaporewaterstory

Public Utilities Board (2016b) Active, beautiful, clean waters program. https://www.pub.gov.sg/abcwaters/about

Public Utilities Board (2016c) Design Guidelines. https://www.pub.gov.sg/abcwaters/designguidelines

Public Utilities Board (2016d) Certification. https://www.pub.gov.sg/abcwaters/certification

Public Utilities Board (2016e) ABC Waters Professionals. https://www.pub.gov.sg/abcwaters/abcwatersprofessionals

United States Environmental Protection Agency (n.d.) What is Green Infrastructure? https://www.epa.gov/green-infrastructure/what-green-infrastructure

West Philadelphia Landscape Project (n.d.). https://wplp.net

Chapter 16
Changes in the Use of Green Spaces by Citizens Before and During the First COVID-19 Pandemic: A Big Data Analysis Using Mobile-Tracking GPS Data in Kanazawa, Japan

Yusuke Ueno, Sadahisa Kato, Tomoka Mase, Yoji Funamoto, and Keiichi Hasegawa

Abstract To consider green spaces and parks as valuable green infrastructure that provides various socio-ecological benefits, including health, this study analyzed changes in the use of green spaces before and during the first COVID-19 pandemic in Japan, using mobile-tracking GPS data of Kanazawa citizens. The results showed that the declaration of a state of emergency in April–May 2020 changed the outing behavior of Kanazawa citizens, and there was a strong tendency for them to avoid going out, with a decrease in the number, time, and distance of outings. On the other hand, while citizens refrained from going out, the rate of decrease in the number of visits to green spaces was smaller compared to commercial areas, and the number of walks increased slightly from 2019. The distance traveled to green spaces is generally shorter in 2020, and the number of visits to green spaces near one's home increased in 2020. These findings suggest that those who had green spaces around their homes were able to use them for maintaining their health and refreshment during the COVID-19 pandemic, adding to the increasing evidence for the value of urban green spaces as part of nature-based solutions.

Y. Ueno (✉) · T. Mase
Faculty of Bioresources and Environmental Sciences, Ishikawa Prefectural University, Nonoichi, Ishikawa, Japan
e-mail: uenoyu@ishikawa-pu.ac.jp

S. Kato
Faculty of Environmental Studies, Tottori University of Environmental Studies, Tottori, Tottori, Japan

Y. Funamoto
Fukuyama Consultants Co., Ltd./Social Value Incubation Lab., Tokyo, Japan

K. Hasegawa
EY Strategy and Consulting Co., Ltd., Tokyo, Japan

F. Nakamura (ed.), *Green Infrastructure and Climate Change Adaptation*,
Ecological Research Monographs, https://doi.org/10.1007/978-981-16-6791-6_16

257

Keywords Behavioral change · COVID-19 · Green space · Public health · GPS data

16.1   Introduction

It is known that green spaces, essential element of green infrastructure, have functions such as improving people's health and providing recreational spaces. Previous studies reported that access and exposure to urban green spaces improves aspects of mental, physical, and social health of people (Zhou and Parves Rana 2012; Hartig et al. 2014; Fong et al. 2018; Kondo et al. 2018; Keeler et al. 2019; Bratman et al. 2019). For example, the statistical results from a Danish survey indicate that access to a garden or short distance to green areas from the dwelling is associated with less stress and a lower likelihood of obesity (Nielsen and Hansen 2007). These studies show that health is supported by a variety of mechanisms, including physical exercise, stress reduction, and social connections (Soga et al. 2021).

The COVID-19 pandemic triggered social reactions such as changes in people's behaviors including contact with nature. Lifestyle restrictions were forcefully set in place in response to the rapid spread of COVID-19 in early 2020. In many countries and cities, particularly in Europe and the USA, strict movement restrictions (e.g., stay-at-home order and restrictions on commuting and traveling) were placed, and the cities were "locked-down."

However, Japan's declaration of a state of emergency is different from the harsh measures taken by other countries. The prime minister at the time, Mr. Abe, declared a state of emergency in seven prefectures including Tokyo, Osaka, and Fukuoka on April 7, 2020, and expanded the scope of the state of emergency to the entire country on April 16 (NHK n.d.). In an emergency declaration, the prefectural governor can request residents to refrain from going out unnecessarily for a specified period and area, and request businesses to limit the hours of operation of their stores and facilities. This is different from other countries where cities are sealed off and severe penalties are imposed if the citizens do not comply. The restrictions on people's behavior under the emergency restrictions in Japan were voluntary (requested by the government), not mandatory, and there were no penalties and fines (NHK n.d.; NIKKEI 2020). Nevertheless, many people and businesses in Japan complied.

Regarding people's behaviors and the use of green spaces during the lockdown, there are recent research that showed how people's behaviors changed as to the use of green spaces before and during lockdown. For example, using STRAVA mobile application data, Venter et al. (2020) found a 291% increase in outdoor recreational activity over Oslo municipality, Norway, during March 2020 lockdown dates relative to the 3-yr baseline average. Both pedestrians (walking, running, hiking) and cyclists intensified activity on trails with greater greenery and in more remote areas. Pedestrian activity increased in city parks, peri-urban forest, as well as protected areas, highlighting the importance of access to green open spaces that are interwoven within the built-up matrix (Venter et al. 2020). Lu et al. (2020), by analyzing geotagged Instagram data of four Asian megacities (Hong Kong, Singapore, Tokyo, and Seoul), found that people were 5.3% more likely to use green spaces for every 100 additional weekly new infection cases during the COVID-19 pandemic in March 2020. They also showed that people preferred larger, nature parks that are close to city centers. Conducting an online survey, Lesser and Nienhuis (2020) found that activity restrictions during the COVID-19 pandemic differentially affected Canadians who were active and inactive, and that physical activity was strongly associated with well-being outcomes in inactive individuals. They also reported for active and inactive participants, those classified as "flourishing" indicated greater nature relatedness than those who scored lower on the mental health continuum. In sum, the three literature shows that during the lockdown those people who actively sought access to and used green spaces received health benefits.

Given the increasing evidence of overall health benefits of green spaces and based on the above literature review, the research hypothesis was that those who had green spaces around their homes under the declared state of emergency used them to maintain their health and refresh themselves, while those who did not have green spaces around their homes stayed at home without access to them. To consider greenery as an infrastructure that meets citizens' needs and solves societal challenges (i.e., green infrastructure), this study analyzed changes in the use of green spaces before and during the first COVID-19 pandemic in Japan, using mobile-tracking GPS data of Kanazawa citizens.

16.2   Materials and Methods

16.2.1   Study Area

The study area, Kanazawa City in Ishikawa Prefecture, is a region facing the Sea of Japan, with a total area of 468.79 km^2 and an estimated population of 463,387 (as of April 1, 2019). The urban area is surrounded by mountains, rivers, and the sea, and there are many green spaces in the city. There are 574 urban parks in Kanazawa City, with a total area of 308.20 ha (as of April 1, 2016) (Kanazawa City n.d.).

The first case of the COVID-19 infection was confirmed in Ishikawa Prefecture on February 21, 2020 (Ishikawa Prefecture 2020). Subsequently, as a measure to prevent the spread of the infection, all prefectural schools were temporarily closed, prefectural events were postponed or cancelled, prefectural facilities were closed, and residents were asked to refrain from going out unnecessarily and to refrain from traveling to and from other prefectures. On April 13, the prefecture issued its own emergency declaration, and on April 16, it was placed by the Japanese government as one of the "designated prefectures on alert," which means that the prefecture needs to take special measures to prevent the spread of COVID-19, and the residents of the prefecture were asked to refrain from going out. The alert in Ishikawa continued until May 14.

16.2.2 The GPS Data and Sample

The GPS data used in the analysis was the location data of cell phone terminals (mobile GPS data) collected by Blogwatcher, Inc. (location data service provider for smartphones) with the consent of the individual. It does not contain any personally identifiable information but is a combination of the latitude and longitude recorded by the GPS (with an error accuracy of about 50 to 100 m), the time recorded (every 5 minutes to several hours), and random alphanumeric characters used to identify the device.

The period covered by the data is 15 days each from May 1 to 15, 2019, and May 1 to 15, 2020. In Japan, there is a weeklong holiday in early May every year, and many people tend to enjoy leisure and travel during this period. Therefore, from the data collected during these periods, we extracted those records that continuously stayed in Kanazawa City at night as Kanazawa citizens. Furthermore, to enable comparison between 2019 and 2020, and to reduce the effect of variations in GPS data acquisition time between mobile devices, only devices with 2 years of records and data acquisition time of more than 20 hours per day were selected for analysis. As a result, the total number of data (terminals) used in this study was 13,710 for 2 years, and the total number of GPS logs was 3,542,408. Of these, 5482 terminals were able to record data in both 2019 and 2020, which is equivalent to about 1.18% of Kanazawa citizens. This study was approved by the Research Ethics Committee of Ishikawa Prefectural University, where the first author belongs, before being conducted.

16.2.3 Analysis Methods

16.2.3.1 Behavioral Changes of Kanazawa Citizens

To analyze changes in the behavior of Kanazawa citizens as a whole before and during the COVID-19 pandemic, comparisons were made between 2019 and 2020 with regard to the number of outings per day, time spent outside the home, distance outside the home (straight-line distance from the home), means of transportation, and types (large green areas, commercial areas) and locations of outings. The location of the home was presumed to be in the vicinity of the place where they mainly stay in the middle of the night. The means of transportation was determined based on the movement speed between GPS logs.

The number of outings was calculated based on the number of times people went out per day (the number of times they stayed outside their homes), and the time spent out was calculated based on the sum of the time spent outside their homes plus the time spent traveling (end of day – start of day – time spent at home). To determine whether a person was "staying" or not, data observed within a range of 100 m continuously for more than 5 minutes were judged to be "staying," and the coordinates of the center of gravity within that time were considered to be "staying position." As for the means of transportation, the number of outings by means (walk, walking; cycle, bicycling; car_other, car and others) was calculated separately for weekdays and holidays in 2019 and 2020.

16.2.3.2 Percentage Increase/Decrease in the Number of Visits to each Green Space and Distance Traveled

The number of visits for each large green space was divided into 2019 and 2020. We also calculated the percentage increase or decrease based on the number of visits in 2019 and 2020 for each green space. Moreover, we calculated the median access distance of visitors to each green space and examined the relationship between the rate of increase/decrease, the number of visits, and the distance traveled.

16.3 Results and Discussion

16.3.1 General Change of Activity Pattern

To describe the overall pattern of movement change, as a pre-analysis, we visualized the movement of people in Kanazawa on holidays in 2019 and 2020 from the collected GPS data (Fig. 16.1 and Fig. 16.2). Many people were seen visiting Kanazawa Station and Kenroku-en Garden area in 2019 (Fig. 16.1), but few people were seen in 2020 (Fig. 16.2).

16.3.2 Average Number of Outings and Total Time Spent out of the House per Day

Comparing 2019 and 2020, the average number of times people go out on weekdays and holidays decreased (Fig. 16.3). Similarly, comparing 2019 and 2020, the average total time spent out of the house decreased on both weekdays and holidays (Fig. 16.4). From these results, there was a tendency to avoid going out at all and for long periods of time in 2020.

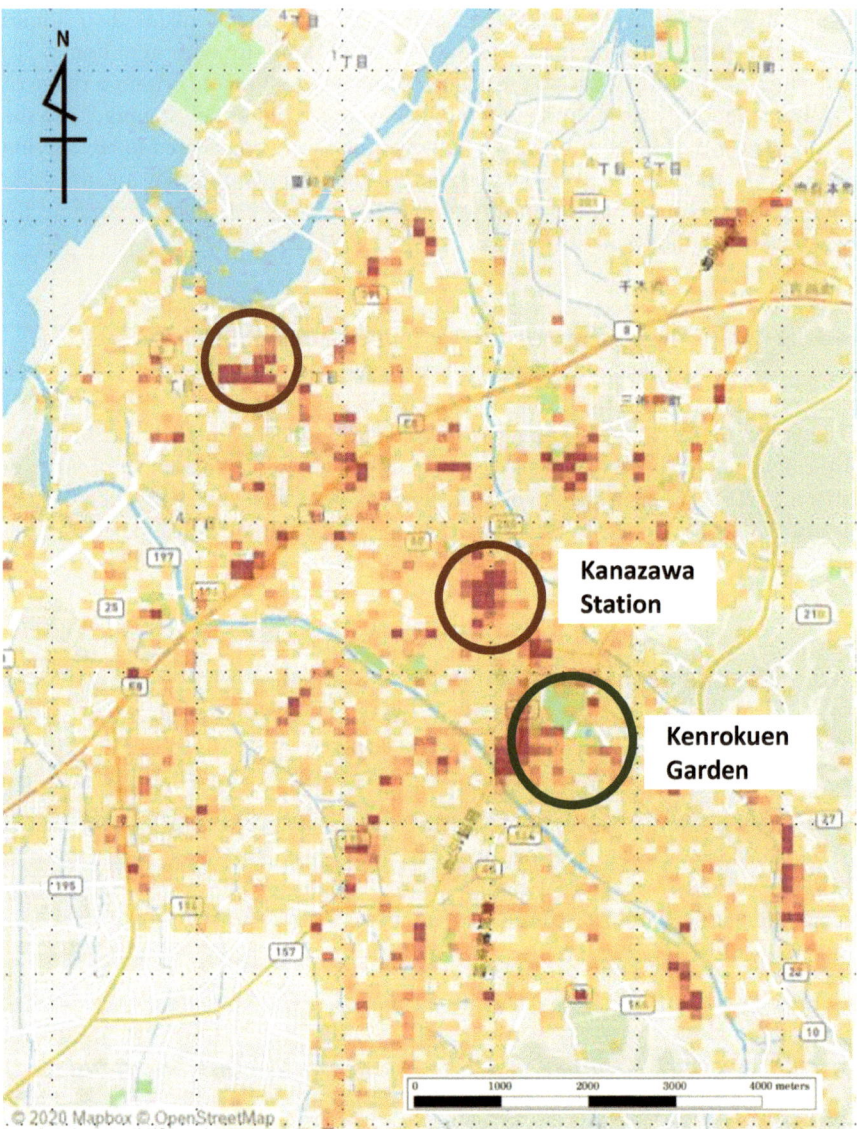

Fig. 16.1 Movement of people on holidays in 2019. The darker the color, the more people were concentrated in the area

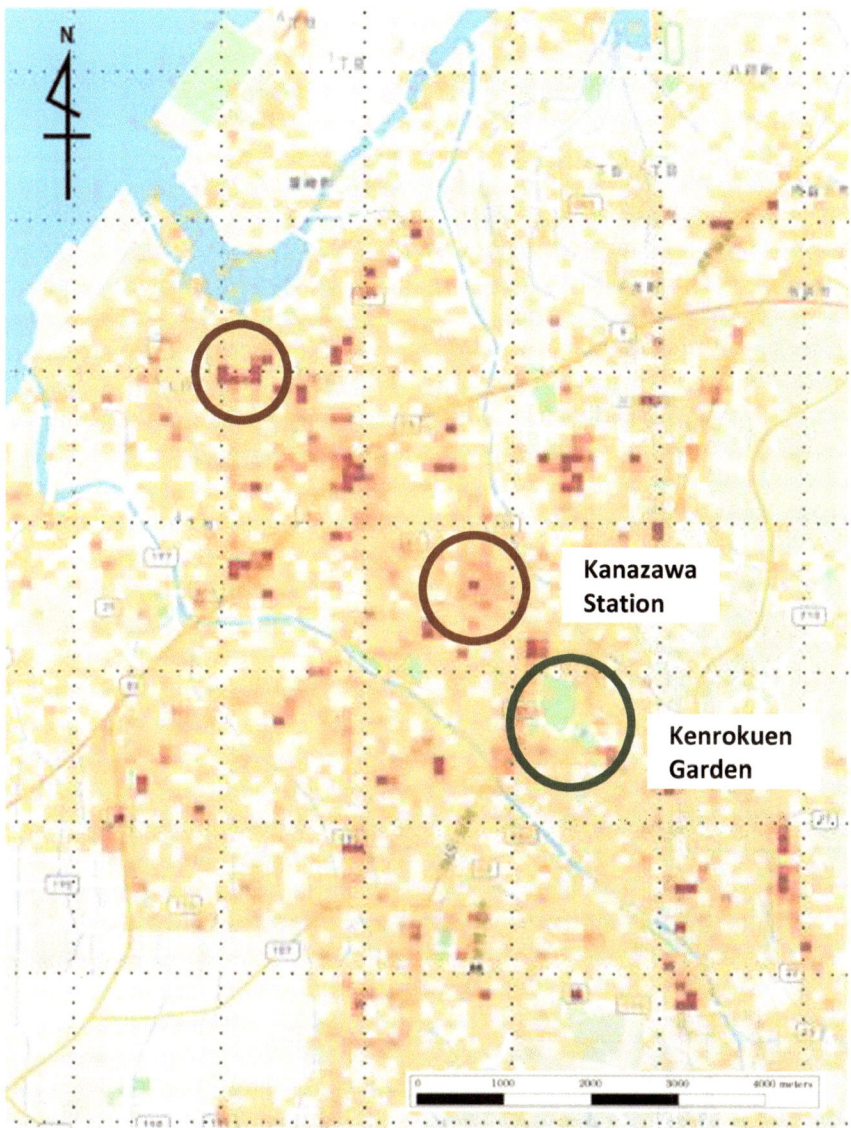

Fig. 16.2 Movement of people on holidays in 2020. The darker the color, the more people were concentrated in the area

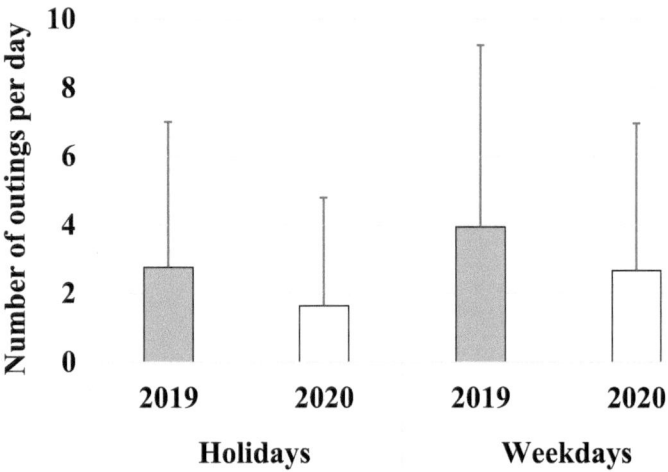

Fig. 16.3 Number of outings per day (mean ± S.D.)

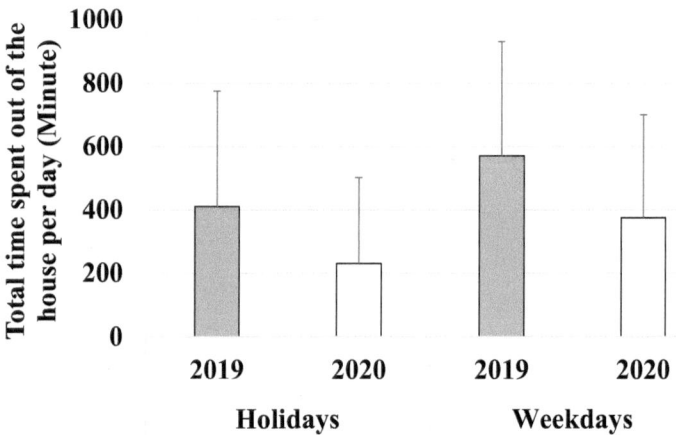

Fig. 16.4 Total time spent out of the house per day (mean ± S.D.)

16.3.3 Maximum Distance Traveled per Day

The most common distance traveled in both 2019 and 2020 was between 0 and 1000 m. In 2020, fewer people traveled more than 1000 m than in 2019. Furthermore, there were few trips of more than 4000 m during holidays in 2020 (Fig. 16.5). From these results, there was a tendency to avoid going out for long distances in 2020.

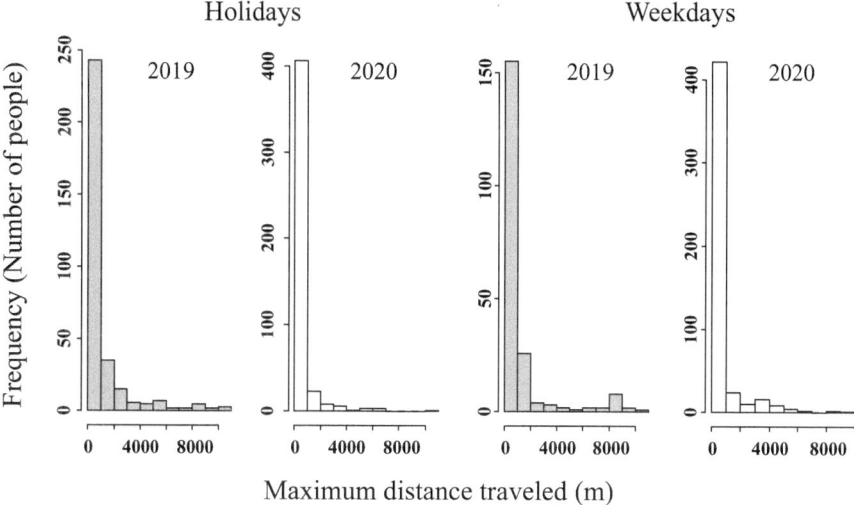

Fig. 16.5   Maximum distance traveled in a day

16.3.4   Number of Outings by Means of Transportation

The percentage of outings by car and other means of transportation (other than walking and bicycling) was the largest in both 2019 and 2020. 2020 saw an increase in the percentage of outings by walking and a relative decrease in the percentage of outings by bicycling and car and the others (Fig. 16.6). In 2020 compared to 2019, the percentage of outings by walking increased on holidays, with a change of 118.7% (Fig. 16.6).

16.3.5   Number of Visits by Destination

The percentage of visits by destination for green areas and commercial areas did not change significantly between 2019 and 2020 (Fig. 16.7). The change in the number of visits by destination in 2020 compared to 2019 showed that the number of visits decreased for all destinations in 2020, but the percentage decrease for green areas was smaller (Fig. 16.7), with a slight increase of 100.06% during the consecutive holidays in May.

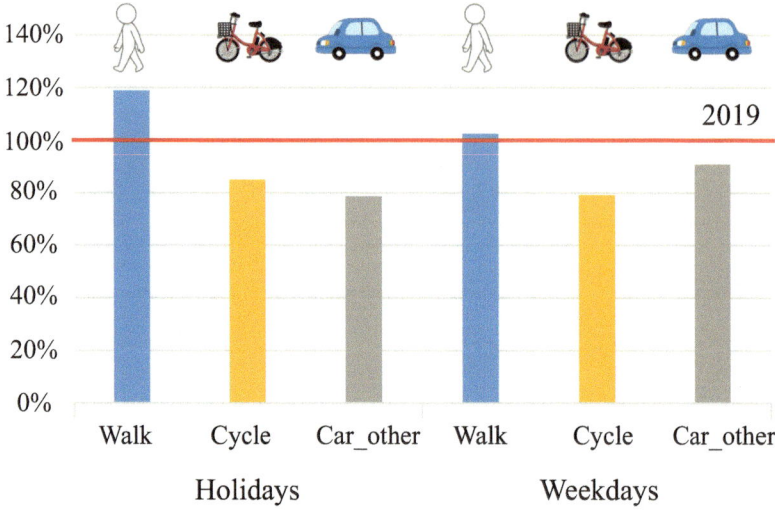

Fig. 16.6 Percentage change by means of travel in 2020 compared to 2019

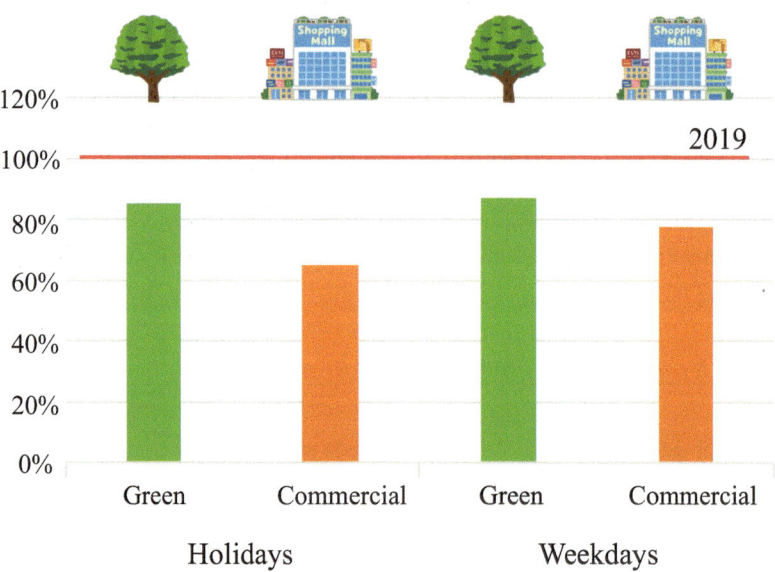

Fig. 16.7 Percentage change by destination in 2020 compared to 2019

16.3.6 Percentage Increase/Decrease in the Number of Visits to each Green Space and Distance Traveled

While the number of visits to large-scale green spaces in central Kanazawa (e.g., Kanazawa Castle Park, Ishikawa Shiko Memorial Park), which had many visitors from far away and were used more frequently in 2019, decreased significantly to half or less in 2020, the percentage of visitors from short distances within 1000 m increased in 2020 (Fig. 16.8). In Fig. 16.8, the horizontal axis is the percentage change in the number of visits at a given green space in 2020 compared to 2019. The median of travel distances in 2019 and 2020 for a given green space are plotted vertically at the corresponding percent change in the number of visits. The vertical axis is the access distance from the user's home, so a particular green space is plotted at the median distance in 2019 and 2020, respectively. Figure 16.8 shows that (1) compared to 2019, the distance visited is generally shorter in 2020 (2020 bubbles located below 2019 bubbles); (2) green spaces with a large increase in the number of visitors in 2020 (green spaces plotted on the right side of the graph) have a visiting distance of around 1000 m, which means that they are often visited by people from nearby areas. The results indicate that while the number of visits to green spaces and parks in the city center decreased, the number of visits to nearby parks and green spaces adjacent to residential areas increased. The results have implications for the importance of green spaces in and near residential areas during the periods of lifestyle restrictions.

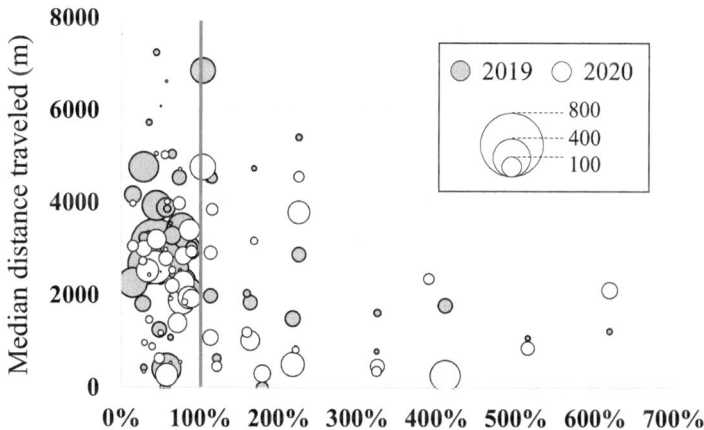

Fig. 16.8 Relationship between the rate of increase/decrease in the number of visits to each green space in Kanazawa City in 2020 compared to 2019 and the average distance traveled from home. Note that a bubble size indicates the number of visits

16.3.7 Green Space Planning in Kanazawa City in Relation to the Research Hypothesis

The results show that during the pandemic, people avoided going out at all, including visits to large green spaces in city center, popular destination in 2019, to avoid crowding and unnecessary contact with other people. However, Fig. 16.8 shows that visits within 1000 m of home increased under the COVID-19 pandemic, and visits to green spaces around residential areas within a short distance from home increased in particular. Therefore, we believe that nearby green spaces within a 1000 m radius provided people with opportunities to use them for health maintenance and refreshment, supporting the research hypothesis. Short distances to green areas from the dwelling are associated with less stress and a lower likelihood of obesity (Nielsen and Hansen 2007). Along with facilitating social cohesion, stress reduction is a strong pathway to health (De Vries et al. 2013). Also, Soga et al. (2021) found that the frequency of green space use and the existence of green window views from within the home were positively associated with five measures of mental health outcomes. Although Lu et al. (2020) found that in Tokyo during the pandemic people preferred large parks close to city center, this may be due to the Instagram data bias, where visual beauty and famous locations are favored over a local park near residential areas.

Our analysis also revealed that even if there is a green space within 1000 from home, many people do not use it. Although the parks closest to home are considered to be those that are close to daily life and can be used easily, they are not always the most frequently used (Shimomura et al. 1995). Also, Kaczynski et al. (2014) showed that distance to the closest park is not an important factor in park use, but diverse park facilities are associated with park use and physical activity. More broadly, it is generally believed that people's nature experiences are determined by "opportunity" and "orientation" (emotional affinity with nature) (Soga and Gaston 2016), and recent studies in particular have shown that the latter is actually more influential (Lin et al. 2014). In other words, these studies suggest that when planning green spaces in a city, it is important not only to consider park proximity but also to place green spaces including parks in a way that matches the needs of the users. Organizing activities and events, policies, and education to increase people's nature orientation is also important for encouraging people's use of and visit to green spaces.

16.4 Conclusions

The results of the analysis in this study showed that the declaration of a state of emergency under the COVID-19 pandemic in Japan in April–May 2020 changed the outing behavior of Kanazawa citizens, and there was a strong tendency for them to avoid going out, with a decrease in the number, time, and distance of outings. On the other hand, while citizens refrained from going out, the rate of decrease in the number of visits to green spaces was small, and the number of walks increased slightly from 2019. The number of visits to green spaces near one's home increased slightly compared to 2019, while there was a tendency to avoid using green spaces in city center. These results suggest that those who had green spaces around their homes were able to use them for maintaining their health, refreshment, and other purposes during the COVID-19 pandemic, according to the big data. On the other hand, those who did not have green spaces around their homes went out less frequently but did not show a trend of increasing the rate of staying at home.

Urban parks and green spaces are green infrastructure that provides a variety of benefits to urban residents, and by utilizing the functions that nature provides, they can also offer solutions to societal challenges in cities (i.e., nature-based solutions). Historically, urban parks were one of the urban facilities developed in the eighteenth and nineteenth centuries for the purpose of ensuring public health and reducing environmental pollution (Chadwick 1966). In recent years, it has been reported that green spaces have various functions such as disaster prevention (e.g., shelter and refuge in case of disasters), microclimate mitigation, formation of an aesthetically pleasing landscape, branding of the area, increase in land prices in the surrounding area, habitats for local organisms, and places for local community activities. However, the amount of green space around residential areas differs greatly among cities around the world, and the more affluent districts tend to have more developed green space. On the other hand, in the Sustainable Development Goals (SDGs) agreed upon by the United Nations to be achieved by 2030, Goal 11 "Make cities and human settlements inclusive, safe, resilient and sustainable" has a target 11.7: "By 2030, provide universal access to safe, inclusive and accessible, green and public spaces, in particular for women and children, older persons and persons with disabilities." This study, conducted under the COVID-19 pandemic, reiterates the need for green infrastructure that is closely related to people's lives, such as urban green spaces, and shows the importance of creating a society where all urban residents, regardless of wealth, can enjoy the benefits of green spaces through strategic urban planning.

References

Bratman GN, Anderson CB, Berman MG et al (2019) Nature and mental health: an ecosystem service perspective. Sci Adv 5:903–927

Chadwick GF (1966) The park and the town; public landscape in the 19th and 20th centuries. F.A. Praeger, New York

De Vries S, van Dillen SME, Groenewegen PP, Spreeuwenberg P (2013) Streetscape greenery and health: stress, social cohesion and physical activity as mediators. Soc Sci Med 94:26–33. https://doi.org/10.1016/j.socscimed.2013.06.030

Fong KC, Hart JE, James P (2018) A review of epidemiologic studies on greenness and health: updated literature through 2017. Curr Environ Heal reports 5:77–87

Hartig T, Mitchell R, de Vries S, Frumkin H (2014) Nature and health. Annu Rev Public Health 35:207–228. https://doi.org/10.1146/annurev-publhealth-032013-182443

Ishikawa Prefecture (2020) Circumstances of infected patients. https://www.pref.ishikawa.lg.jp/kansen/coronakennai2002.html. Accessed 16 Jan 2021

Kaczynski AT, Besenyi GM, Stanis SWA et al (2014) Are park proximity and park features related to park use and park-based physical activity among adults? Variations by multiple socio-demographic characteristics. Int J Behav Nutr Phys Act 11:114. https://doi.org/10.1186/s12966-014-0146-4

Kanazawa City (n.d.) Kanazawa City Types of Urban Parks https://www4.city.kanazawa.lg.jp/29004/park/syurui.html. Accessed 16 Mar 2021

Keeler BL, Hamel P, McPhearson T et al (2019) Social-ecological and technological factors moderate the value of urban nature. Nat Sustain 2:29–38

Kondo M, Fluehr J, McKeon T, Branas C (2018) Urban green space and its impact on human health. Int J Environ Res Public Health 15:445. https://doi.org/10.3390/ijerph15030445

Lesser IA, Nienhuis CP (2020) The impact of COVID-19 on physical activity behavior and Well-being of Canadians. Int J Environ Res Public Health 17:3899. https://doi.org/10.3390/ijerph17113899

Lin BB, Fuller RA, Bush R et al (2014) Opportunity or orientation? Who uses urban parks and why. PLoS One 9:87422. https://doi.org/10.1371/journal.pone.0087422

Lu Y, Zhao J, Wu X, Lo SM (2020) Escaping to nature in pandemic: a natural experiment of COVID-19 in Asian cities. doi:https://doi.org/10.31235/OSF.IO/RQ8SN

NHK Emergency declaration-1st situation (n.d.). https://www3.nhk.or.jp/news/special/coronavirus/emergency/. Accessed 22 Feb 2021

Nielsen TS, Hansen KB (2007) Do green areas affect health? Results from a Danish survey on the use of green areas and health indicators. Heal Place 13:839–850. https://doi.org/10.1016/j.healthplace.2007.02.001

NIKKEI (The Japan Economic Daily) (2020) In Europe and the United States, restrictions on private rights, penalties for going out, and enforcement differ from Japan. https://www.nikkei.com/article/DGXMZO57774880X00C20A4EA1000/. Accessed 22 Feb 2021

Shimomura Y, Masuda N, Abe D et al (1995) Study on Resident's behavior of Block Park in neighborhood. J Japanese Inst Landsc Archit 58:217–220. https://doi.org/10.5632/jila.58.5_217

Soga M, Gaston KJ (2016) Extinction of experience: the loss of human-nature interactions. Front Ecol Environ 14:94–101

Soga M, Evans MJ, Tsuchiya K, Fukano Y (2021) A room with a green view: the importance of nearby nature for mental health during the COVID-19 pandemic. Ecol Appl 31:e2248. https://doi.org/10.1002/eap.2248

Venter ZS, Barton DN, Gundersen V et al (2020) Urban nature in a time of crisis: recreational use of green space increases during the COVID-19 outbreak in Oslo, Norway. Environ Res Lett 15:104075. https://doi.org/10.1088/1748-9326/abb396

Zhou X, Parves Rana M (2012) Social benefits of urban green space. Mgmt Environ Qual Int J 23:173–189. https://doi.org/10.1108/14777831211204921

Chapter 17
Land Use Planning as a Green Infrastructure in a Rural Japanese Depopulated Town

K. Watanabe and K. Ishida

Abstract The purpose of this study is to show land use considering green infrastructure (GI) in a Japanese depopulated rural district. First, we classified six zones of the target area, depending on inundation risk, living environment, and area of each land use, by cluster analysis. As a result, we showed three characteristics of land use. These are the central built-up area with high disaster risk and high land price, the east side area with low disaster risk, and the paddy field area with high disaster risk. Next, we estimated the probability of vacant houses by logistic regression analysis. Based on the results and the six classified zones, we showed the area with high probability of vacant houses and high disaster risk. This area was observed at the central built-up area. Considering future depopulation, such a housing area in the central built-up area needs to shrink.

Keywords Inundation risk · Depopulated town · Cluster analysis · Land use planning

17.1 Introduction

Japanese population decline started from 2008 (Chino, 2009). After that, population has been continuously migrating from provincial cities to big cities. In particular, depopulation of small coastal towns and agricultural villages continues. Historically, most of Japanese built-up areas have been developed in coastal areas and plains. These are useful for the farming or water transportation, but these areas have much

K. Watanabe (✉)
Research Center for Management of Disaster and Environment/Department of Civil and Environmental Engineering, Graduate School of Technology, Industrial and Social Sciences, Tokushima University, Tokushima, Japan
e-mail: kojiro@tokushima-u.ac.jp

K. Ishida
Faculty of Business and Commerce, Kansai University, Osaka, Japan

© The Author(s) 2022

F. Nakamura (ed.), *Green Infrastructure and Climate Change Adaptation*,
Ecological Research Monographs, https://doi.org/10.1007/978-981-16-6791-6_17

271

inundation disaster risks. Since the advent of modern society, many structures made of concrete or other materials have been constructed to provide protection from a variety of disasters. In addition, recently, disaster mitigation for urban structures of land use control has been focused on. Green infrastructure (GI) is an important element of the land use plan.

In Japan, compact city policy is proceeding to make sustainable cities by utilizing limited resources. In many depopulated cities in Japan, it is difficult to arrange enough social infrastructure, due to financial problems. Therefore, it is necessary to relocate built-up areas in a more compact manner inside in a safe location. Since 2020, Act on Special Measures concerning Urban Reconstruction (Ministry of Land, Infrastructure, Transport and Tourism, Japan) has been promoted. As a foundation of the act, it is considered that natural disasters will become worse in Japan. The act includes land use control policy considering disaster risk.

Recently, natural disasters in Japan have become more severe; therefore, we need to change the urban structure utilizing GI also. Multifunction of GI is useful to solve some problems in depopulated areas. There are many places with GI functions, such as farmland, forest, vacant lot, etc., in depopulated localities. But recently, areas of unmanaged farmland or unmanaged forest have been increasing. Maintenance of such land use depends on the original character which is important to mitigate disaster risk. It can also contribute to a good natural landscape as regional resources. Thus, GI has an important role in sustainability of Japanese depopulated cities.

The purpose of the paper is to show land use policy with GI in a depopulated area. The target area is Zenzo River basin in Kaiyo town, Tokushima Prefecture, Japan. Kaiyo town is a Japanese depopulated municipality with a large forest area. Osato district is the central area in Kaiyo town. It is a plain area surrounded by mountain, river, and beach. Zenzo River is a tributary of Kaifu River; the basin overlapped Osato district. The target area has some disaster risks. These are building collapse and tsunami caused by the predicted Nankai Trough earthquake and inundation disaster from Kaifu River due to heavy rain. Tokushima Prefecture has assumed the seismic intensity level of 8 to 9 in Kaiyo town caused by Nankai Trough earthquake, and the probability of this earthquake is shown as a 70% within 30 years (Tokushima Prefecture Government 2013). Based on this assumption, many projects about disaster mitigation have been conducted.

Figure 17.1 shows land use of the target area. The target area is a suitable area utilized for GI of disaster mitigation because major land use is paddy fields and farmland. The built-up area is located along the roadside of national highway 55 and the east side of the target area. Population of the town has decreased from 8131 in 2005 to 6546 in 2015. Figure 17.2 shows that the population under 14 years old is decreasing, but population over 65 is increasing. The same trend is occurring in many Japanese provincial cities and towns.

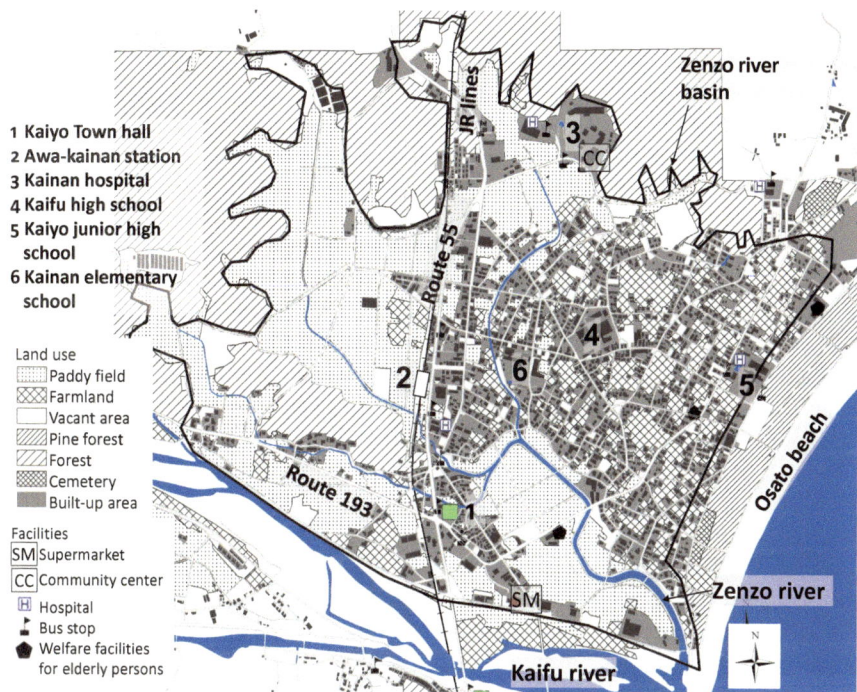

Fig. 17.1 Target area

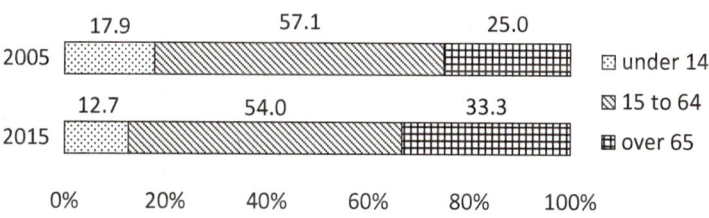

Fig. 17.2 Population ratio for each age group in the target area in 2005 and 2015

17.2   Disaster Risk and Living Environment

We classified the target area according to land characteristics, to show future land use trend. Elements of the classification are disaster risk, living environment, land use, and elevation. We used the housing damage level caused by inundation disaster as a disaster risk. Living environment index is used to show the potential of future residential location. Cluster analysis was conducted using these data, with a unit of 50 meters grid cell.

Table 17.1 Damage level for houses by inundation depth

	<0.5 m	0.5–0.9 m	1.0–1.8 m	> 1.8 m
Damage level	0.144	0.205	0.452	1.000

As disaster risk, we used two types of hazard, the river inundation (hazard 1) depth and the surface water inundation (hazard 2) depth. Hazard 1 shows the river inundation depth caused by inundation from Kaifu River by rainfall intensity having a recurrence interval of 50 years. This data is openly available at the website of National Land Numerical Information in Japan.[1] Hazard 2 shows the estimated surface water inundation depth caused by typhoon No. 14 in 2014 named "Fengshen" (Yokokawa et al. 2020).

Table 17.1 shows the damage level of houses according to inundation depth. We calculated the level for each grid cell, in Table 17.1 (Ministry of Land, Infrastructure, Transport and Tourism 2010; Disaster Management group in Cabinet office of Japan 2021). Hereafter, we show the damage level for houses by river inundation depth as risk 1 and damage level for houses by the surface water inundation depth as risk 2.

Figure 17.3 shows the inundation area caused by Kaifu River mainly distributes over the west side of target area. Over 3 m inundation area is observed at the south side area of Zenzo River. These areas are covered with paddy field. Inundation depth at the roadside of national highway 55 is estimated to be 0.5 to 2 m. Because the highway is a trunk road in the area, there are many commercial facilities in vicinity. It is important to manage the embankment of Kaifu River and to arrange evacuation facilities and routes. There is no inundation area on the east side. Elevation of this area is higher than others, so inundation from Kaifu River cannot reach the area.

Figure 17.4 shows that the surface water inundation area distributes for all areas diffusively. As is the same in Fig. 17.3, estimated surface water is observed at the paddy fields to the west side of the station and the south area of Zenzo River. If these paddy fields were to be changed to some other type of land use, the surface water will move to other areas. Such paddy fields are important to mitigate the flood disaster (Muto et al. 2018).

Table 17.2 shows the rate of the number of houses in the inundation area. In data of A in the table, it can be seen that 50% of houses are located in the non-inundation area, but 36% of houses are located in the 1.0–2.0 m and 2.0–3.0 m inundation area.

---

[1] https://nlftp.mlit.go.jp/ksj/

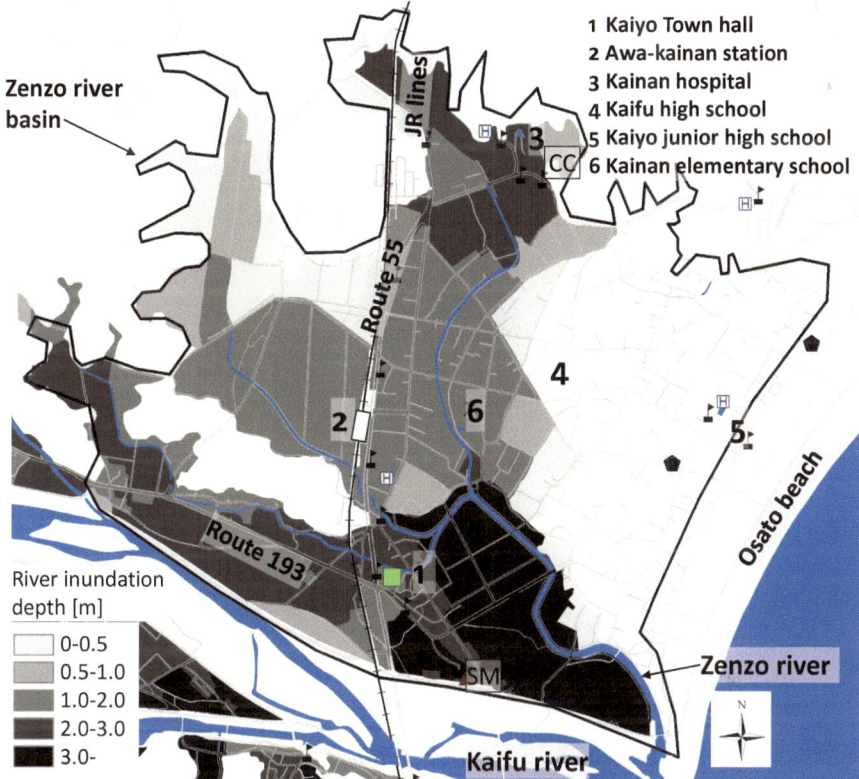

Fig. 17.3 River inundation depth (hazard 1)

Figure 17.3 shows the inundation area with 1.0–3.0 m includes a trunk road and town hall. Houses in this area need migration measures in order to protect them against inundation. In data of B in the table, it is evident that 78% of houses are located in the under 0.5 m inundation area. Although inundation depth is less than river inundation depth, Fig. 17.4 shows the inundation area is wide and diffusive. Therefore, it will result in surface water inundation over a large area.

Next, we calculated the living environment index. This index shows convenience level of living environment. We used the estimated land price as an alternative index of convenience level. According to the capitalization hypothesis, effect of infrastructure development is reflected as a land price increase. Based on the hypothesis, because it was assumed that areas with sufficient infrastructure have a good living environment, we estimated the land price to show convenience level.

Because there is no land price data for the all target areas, land price was estimated by multi-regression analysis. Objective variables were land prices of residential areas in depopulated areas of Shikoku Island region. This data shows the selected land prices used as an indicator, made by Japanese central and

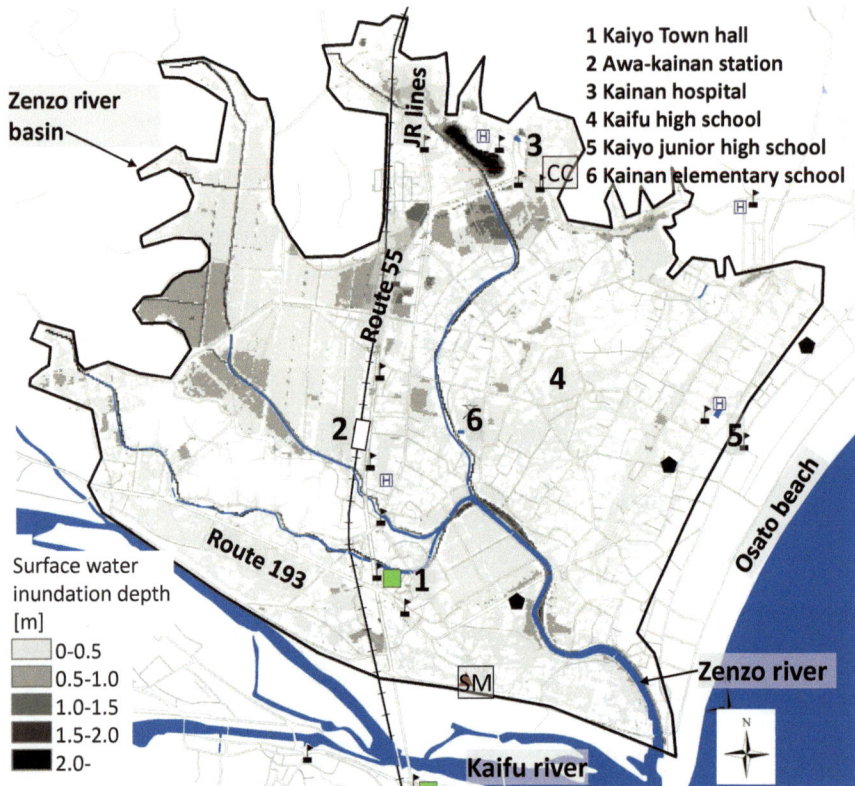

Fig. 17.4 Surface water inundation depth (hazard 2)

Table 17.2 Rate of number of houses in inundation area

		Inundation depth						Total
		no inundation	<0.5 m	0.5–1.0 m	1.0–2.0 m	2.0–3.0 m	>3.0 m	
A	2006	49.5	6.7	6.8	26.1	9.9	1.0	100
	2018	49.5	7.0	6.8	25.5	10.2	0.9	100
	2018–2006	0.0	0.3	0.0	−0.6	0.3	−0.1	
B	2006	20.2	78.1	0.7	0.7	0.2	0.0	100
	2018	20.3	78.0	0.8	0.8	0.2	0.0	100
	2018–2006	0.0	−0.1	0.0	0.1	0.0	0.0	

A: Rate of the number of houses in river inundation area [%].
B: Rate of the number of houses in surface water inundation area [%].

Table 17.3 Results of multi-regression analysis for land price estimation

		Estimate	Std. Error	t value	Pr(>|t|)
(intercept)		15550.69	1336.29	11.64	0.000
Population in 2010		27.30	1.83	14.91	0.000
Elevation		−18.84	4.86	−3.88	0.000
Distance from	Town hall and community center	−1.02	0.85	−1.19	0.234
	Elementary school	−0.23	0.56	−0.41	0.685
	Medical facilities	−1.13	0.47	−2.40	0.017
	Bus stop	−0.53	0.42	−1.24	0.215
	$R^2 = 0.5957$				

prefectural government. Such data was obtained from the website of the National Land Numerical Information[2] also. We assumed two factors of land price change as explanatory variables: population density and convenience of daily life. The population in 2010 is used as a population density factor. Elevation, distance from the town hall, community center, elementary school, hospitals, and bus stops are used as a convenience of daily life factor.

We used the equation to estimate a land price LP.

$$LP \sum_{j=1}^{6} \alpha_j \cdot f_j + \beta \tag{17.1}$$

where, f is each explanatory variables, α is the regression coefficient and β is the intercept.

Table 17.3 shows results of this analysis. The sign conditions of each of the coefficients are satisfied. Although the significance levels of three factors are not satisfied, coefficient of determination is 0.59; therefore, we used the model to estimate land price for all of the target areas. Figure 17.5 shows estimated land price.

Figure 17.5 shows two hot spots. These are located around Kainan Hospital and the town hall. This roadside area located between the town hall and station has many living facilities. The area is convenient to live. But most of the area is included in the river inundation area. Around Kainan Hospital, there is a community center and bus stops. The roadside area of national highway 55 near the hospital has some housing areas, but as with another hot spot, the area has river and surface water inundation risk also.

---

[2] https://nlftp.mlit.go.jp/ksj/

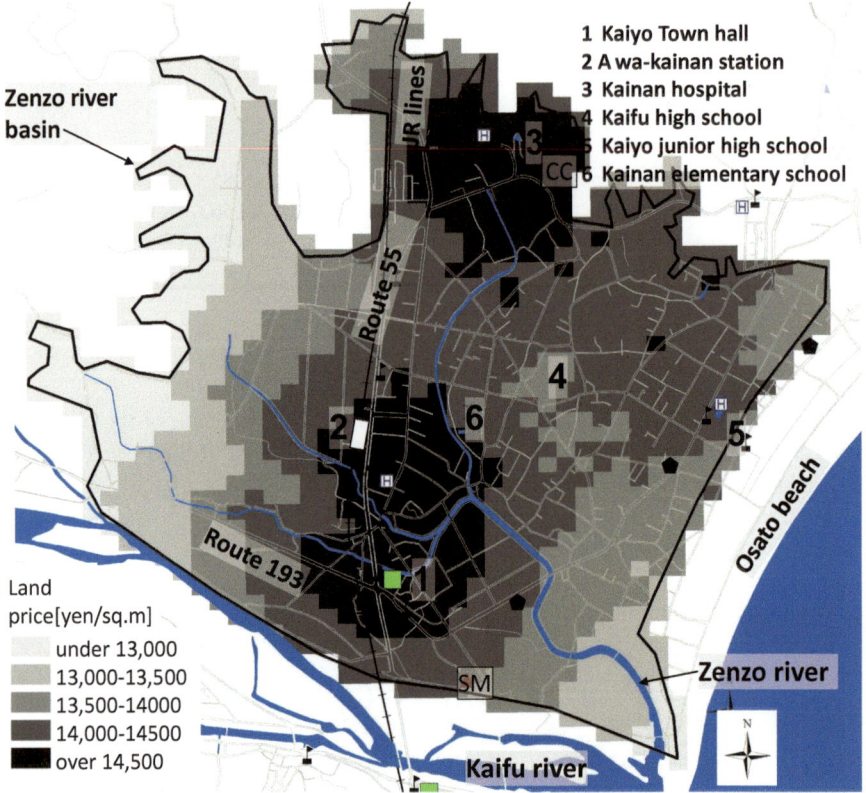

1 Kaiyo Town hall
2 A wa-kainan station
3 Kainan hospital
4 Kaifu high school
5 Kaiyo junior high school
6 Kainan elementary school

Zenzo river basin

JR lines

Route 55

Route 193

Osato beach

Zenzo river

Kaifu river

Land price[yen/sq.m]
- under 13,000
- 13,000-13,500
- 13,500-14000
- 14,000-14500
- over 14,500

Fig. 17.5 Distribution of convenience level as estimated land price

17.3 Classification of Target Area

We classified the target area for each zone using risk 1 and risk 2, convenience level, area of paddy fields, farmland, vacant areas, pine forests, other forests, and elevation, by Ward's method of hierarchical cluster analysis by statistics software R. Figure 17.6 shows the results of classification using risk 1, and Fig. 17.7 shows the results of classification using risk 2. Table 17.4 shows the average of data using cluster analysis for risk 1, and Table 17.5 shows the same data for risk 2.

Comparing Figs. 17.6 and 17.7, it can be seen that the trend of classification shows the same pattern. Cluster 4 of risk 1 shows the same pattern of cluster 5 of risk 2, and cluster 5 of risk 1 shows the same pattern of cluster 4 of risk 2.

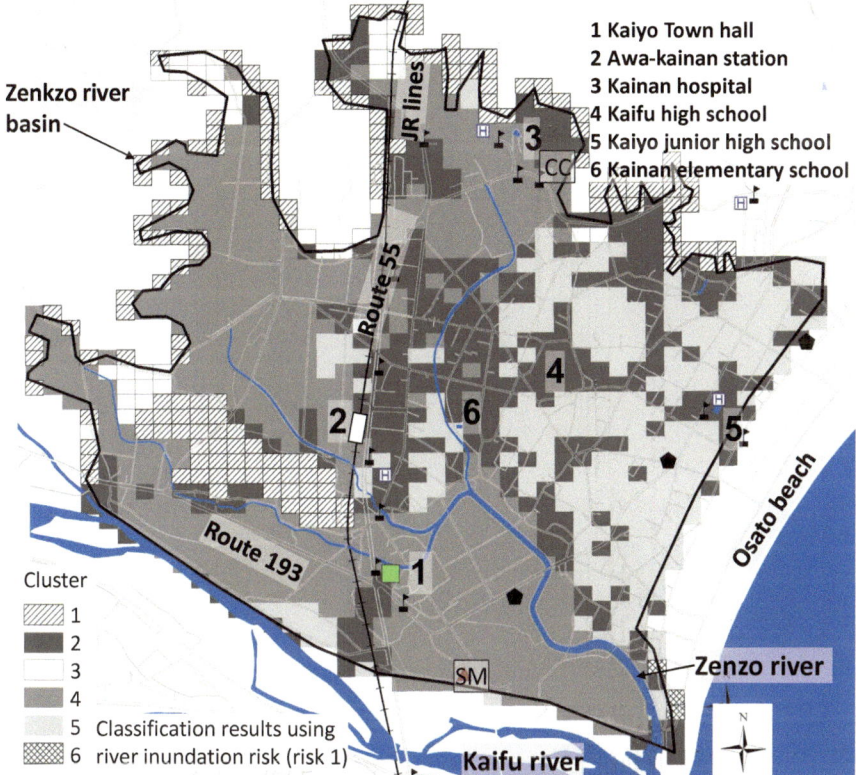

Fig. 17.6 Result of cluster analysis by risk 1

From these results, we can present the following land use trend:

1. Cluster 1 (zone 1).

    Zone 1, which is classified with cluster 1, includes the forest area mainly. This zone should be preserved as a preservation area.

2. Cluster 2 (zone 2).

    Zone 2, which is classified with cluster 2, includes the area with low inundation risk and high land price. Comparing Figs. 17.6 and 17.7, in the case of cluster 2 by risk 2, the area around the town hall is classified as cluster 2, but in the case of the same cluster for risk 1, the area is classified as cluster 4 mainly. For this reason, it is thought that there is a difference of risk levels 1 and 2. The area around the station leading to Kaifu High School is classified in this cluster. Because there are many houses and shops in the area, this area is suitable for town center and housing area, after having the arrangement of evacuation environment and reinforcement of buildings.

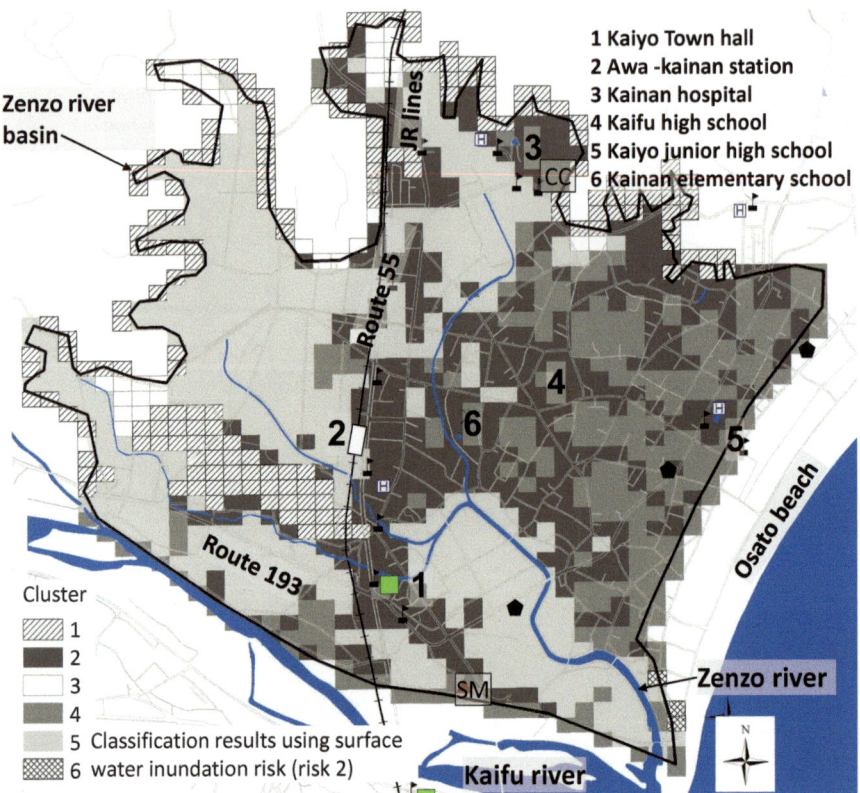

Fig. 17.7 Result of cluster analysis by risk 2

Table 17.4 Averages of the data using cluster analysis by risk 1

		Cluster					
		1	2	3	4	5	6
Damage level by river inundation		0.2	0.2	0.3	0.8	0.1	0.0
Elevation [m]		20.1	6.6	10.4	4.4	8.1	8.1
Land price [yen/sq.m]		13421.4	14214.5	13159.7	13908.0	14052.1	12876.7
Area of	Paddy field[sq.m]	170.8	204.4	234.2	1783.6	53.2	0.0
	Farmland[sq.m]	13.7	103.6	9.1	75.1	1401.8	46.7
	Vacant area[sq.m]	7.4	15.9	1417.6	7.5	2.2	0.0
	Pine forest[sq.m]	0.0	2.5	0.0	0.0	0.6	1366.7
	Forest[sq.m]	2271.8	105.1	593.3	63.3	12.2	0.0

Table 17.5 Averages of the data using cluster analysis by risk 2

		Cluster					
		1	2	3	4	5	6
Damage level by surface water inundation		0.1	0.1	0.1	0.1	0.2	0.0
Elevation [m]		19.4	6.4	10.4	7.8	4.4	8.1
Land price [yen/sq.m]		13440.6	14361.4	13159.7	13943.9	13818.9	12876.7
Area of	Paddy field[sq.m]	195.7	140.2	234.2	131.6	1973.7	0.0
	Farmland[sq.m]	12.7	250.6	9.1	1293.1	31.0	46.7
	Vacant area[sq.m]	17.7	15.4	1417.6	0.3	4.0	0.0
	Pine forest[sq.m]	0.0	0.0	0.0	1.1	1.5	1366.7
	Forest[sq.m]	2206.6	97.7	593.3	1.6	35.1	0.0

3. Cluster 3 (zone 3).

    Zone 3, which is classified as cluster 3, includes the vacant area and upland mainly. Because this zone is surrounded by a green area and paddy fields, new development in this zone should be prohibited, and current land use should be maintained.

4. Cluster 4 with risk 1 and cluster 5 with risk 2 (zone 4).

    Zone 4 which is classified with these clusters consists mainly of paddy fields. This cluster is overlapped with river and surface water inundation area. Because paddy fields have a storage and an infiltration function of rainwater, the cluster is important for mitigation of inundation disaster. Because it is easy to change paddy fields to the built-up area, as observed from past cases, new development in the zone should be prohibited.

5. Cluster 5 of risk 1 and cluster 4 of risk 2 (zone 5).

    Zone 5 which is classified with the cluster includes the farmland with diffusive built-up area. This area is located in the east side of Kaifu High School mainly. Because the area is slightly elevated, inundation risk is very low. Farmland has a disaster mitigation function as with paddy fields. Therefore, it is important to keep farmland to protect the nearby housing area.

6. Cluster 6 (zone 6).

    Zone 6, which is classified with the clusters, includes the pine forest mainly. Because pine trees are important to mitigate tsunami or wind disaster, it is important to maintain the present land use in zone 6.

17.4 Overlay Analysis

In the target area, depopulation continues and the number of vacant houses is increasing. Considering this point, the area needs to become smaller for a viable urban form in the future. Observing the distribution of vacant houses is useful information with regard to discussion about the land use plan. For this reason, we developed an estimation model to show a probability that changes from occupied to vacant houses by logistic regression analysis.

Objective variable is the ratio of vacant houses in 2018, per 50 meter cell. In Japan, it is difficult to obtain location data of vacant houses, due to this being private information. We arranged individual house data from a commercial GIS database which is named Z-Map town2 (ZENRIN Co., Ltd.). This data includes the location and nameplates for all houses. From the database, as a vacant house, we selected the house complying with the condition of having no nameplate in 2018, and we calculated the ratio of vacant houses.

As an explanatory variable, we assumed the following four factors: (1) inhabitant characteristics, (2) convenience level of living environment which is calculated in the model of Table 17.3, (3) area of houses, and (4) land use characteristics. Specifically, the number of inhabitants over 65 years of age is used to show the factor (1). The area of houses in 2006 is used to show the factor (3). The areas of paddy fields, farmland, and forest are used to show the factor (4). This estimation was carried out in the cell with houses in 2006.

We used the equation to estimate a probability Pv that changes from occupied to vacant houses.

$$Pv = 1/\left[1 + \exp\left\{-\left(\sum\nolimits_{j=1}^{7} \alpha_j \cdot f_j + \beta\right)\right\}\right] \qquad (17.2)$$

where, f is each explanatory variables, α is the regression coefficient and β is the intercept.

Table 17.6 shows the results of the analysis. The variables were selected by way of a stepwise method. As shown in Table 17.6, factors other than land use area and houses were not selected as significant variables. From this data, it is evident that with the smaller the area of houses, paddy fields, farmland, and green area, the probability of vacant houses will be larger. Because depopulation continues across the target area, the smaller house area the probability of vacant house will be high. Figure 17.8 shows the distribution of vacant houses which is estimated by the model. This figure shows that the probability in roadside area is about 0.4. These areas are residential areas but include diffusive paddy field and farmland also. Therefore, the

Table 17.6 Result of the logistic regression analysis

	Estimate (*10^{-3})	Std. Error	z values	Pr(>|z|)	
Intercept	391.7495	0.245	1.598	0.110	
Area of houses (2006) [sq.m]	−1.1701	0.000	−2.354	0.019	*
Area of paddy field (2006) [sq.m]	−0.4595	0.000	−2.420	0.016	*
Area of farmland (2006) [sq.m]	−0.4081	0.000	−2.828	0.005	**
Area of forest (2006) [sq.m]	−0.4374	0.000	−1.804	0.071	.
AIC = 758.46					

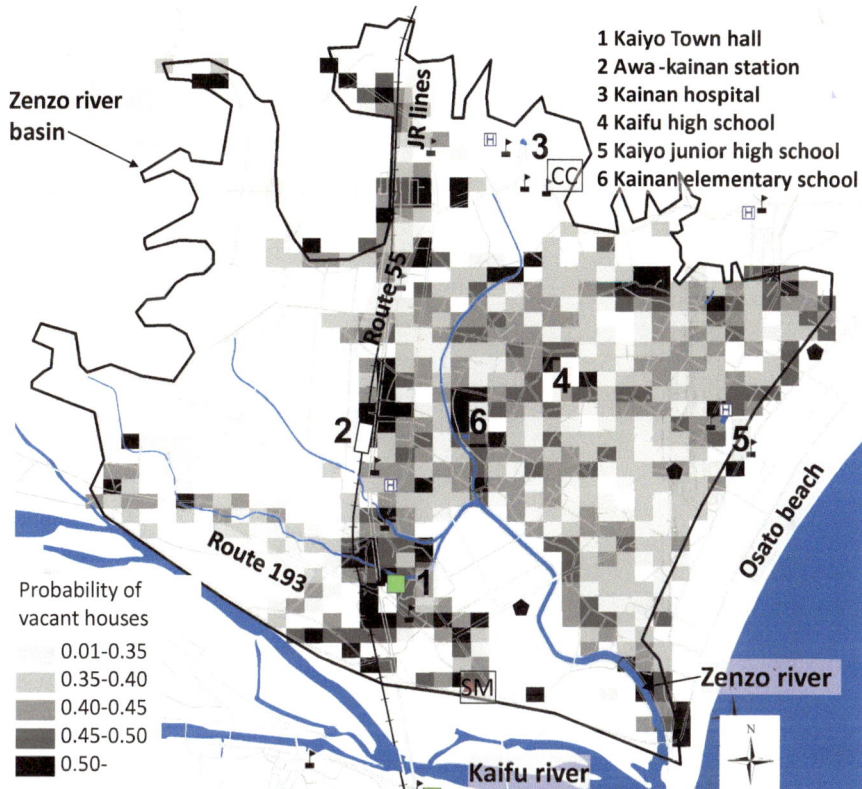

Fig. 17.8 Probability of vacant houses

probability is low. Because there are few houses in forest area, the probability is low. According to Fig. 17.8, a high probability area is observed mainly around the central built-up area and city hall. There are some areas of high probability near the elementary school also. These areas highlight the diffusive built-up area near the central area.

Figure 17.9 shows zone 2 with the probability of vacant houses over 0.5. The selected area has low inundation risk with a high convenience level, and high probability of vacant houses. Areas in Fig. 17.9 have a possibility of continuing the development of new houses because these areas are located in the central built-up area. The target area needs to be a smaller more compact built-up area due to

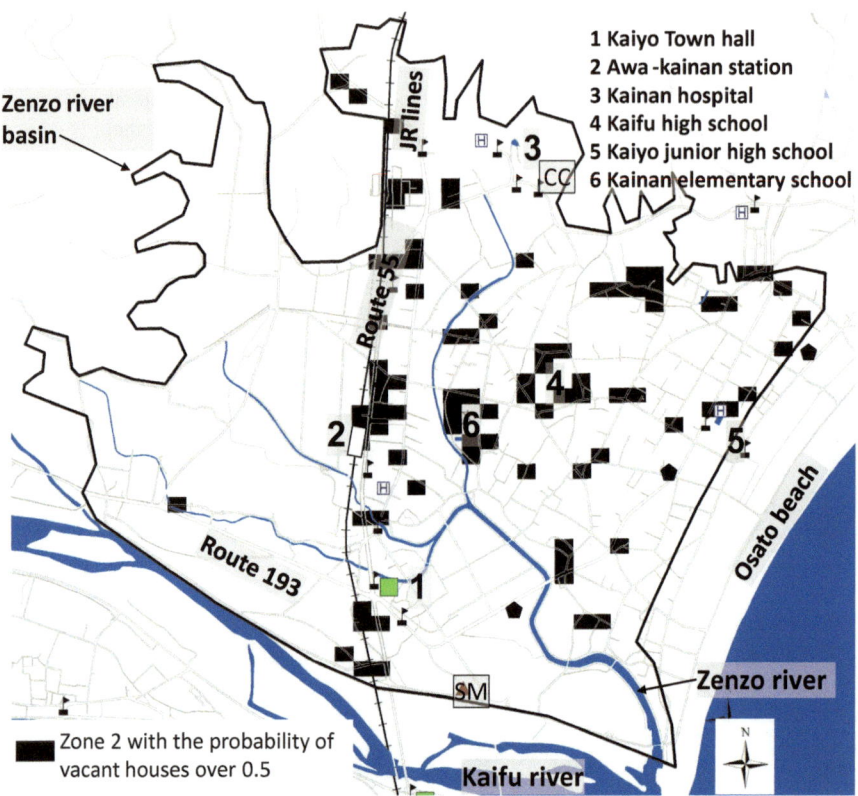

Fig. 17.9 Zone 2 with probability of vacant houses over 0.5

depopulation. Therefore, in the future, control of new development and change of vacant houses to natural land use like parks or farmland are necessary.

Figure 17.10 shows zone 4 with the probability of vacant houses over 0.5. Because zone 4 includes many paddy fields, if vacant houses in the area will appear, these are better to be changed to natural land use.

Figure 17.11 shows zone 5 with the probability of vacant houses under 0.5. The selected areas have low risk and low probability of vacant houses. Therefore, it is better to gather future housing mainly in this area.

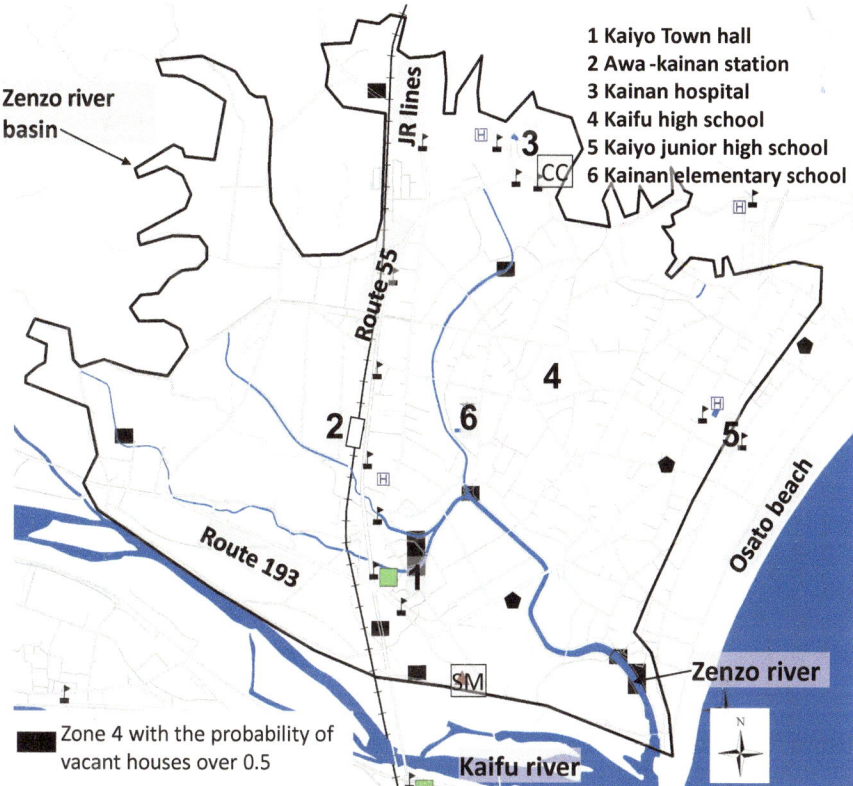

Fig. 17.10 Zone 4 with probability of vacant houses over 0.5

17.5   Land Use Regulation in the Target Area

Japanese national land is regulated by the National Land Use Planning Act. Five areas are specified in the act. These are urban area, agricultural area, forest area, nature park area, and nature conservation area. These areas are regulated by individual acts. There is no natural conservation area in the target area.

Land use in Japanese cities is regulated by the City Planning Act. In the act, designation of city planning area is required. This area is the same as in urban area. But the target area is not included in the designated city planning area, so it is difficult to control land use and buildings in the area.

Agricultural area is regulated by the Cropland Act. Because the target area is designated as an agricultural promotion area of the act, it is impossible for local government to control development in the area. But it is possible to change land use to other land use by satisfying some conditions. Figure 17.12 shows agricultural promotion area in the target area. This promotion area is overlapped by zone 4, so it is important to keep such an agricultural area.

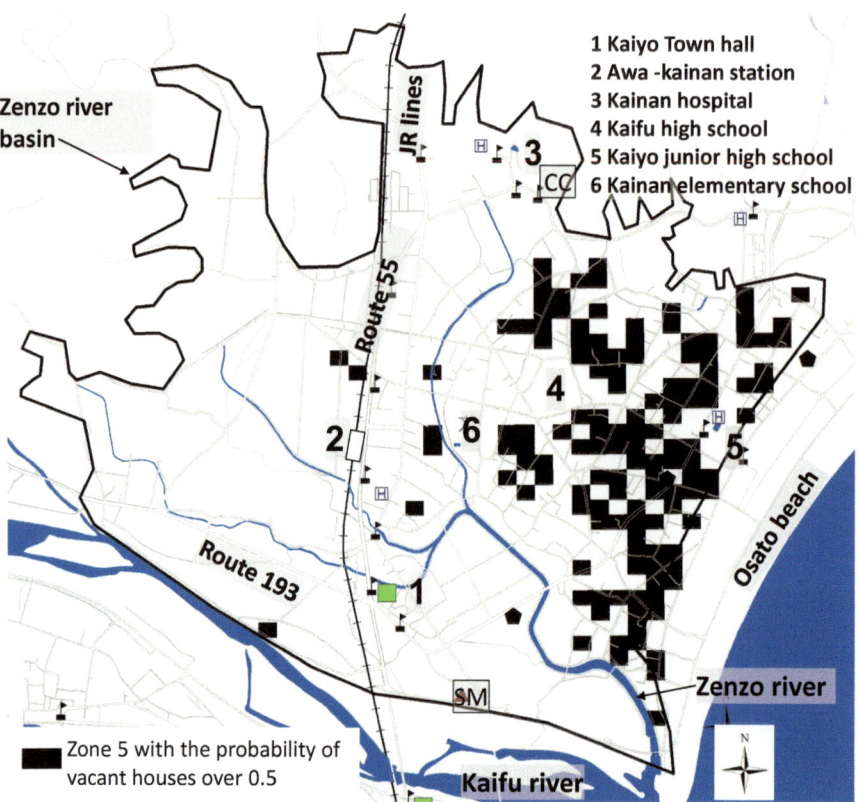

Fig. 17.11 Zone 5 with probability of vacant houses over 0.5

Forest area is regulated by the Forest Act. Around the mountainous area in the target area is designated as private forest of the regional forest plan, as shown in Fig. 17.12. Development in the forest requires permission from the local government. The target area is a depopulated town where there is little probability of any future large urban development. It is important to control small development in order to protect the forest area.

Nature Park is regulated by the Natural Parks Act. Osato beach in the target area is designated as a natural park area. This beach has an important role in GI in the target area, and continued management of this beach in the future is important.

Agricultural, forest, and beach areas are under legal control system, but there are no measures to manage them as built-up area or small agricultural area. The legal system is insufficient in the target area, and land use that should be included in the plans of local governments such as general master plan will be considered and the projects necessary for land use control will be promoted.

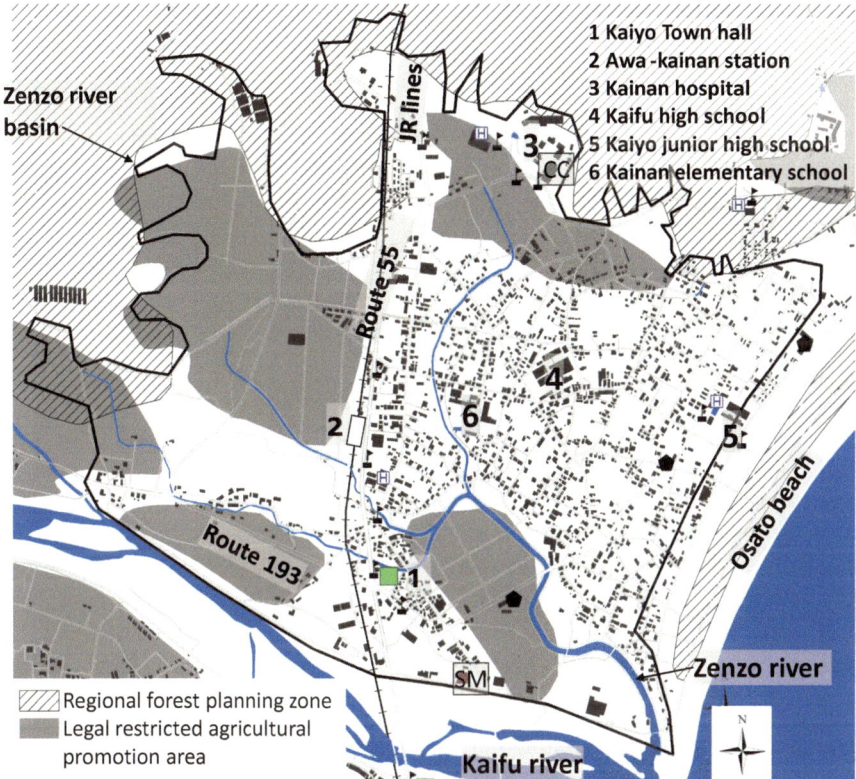

Fig. 17.12 Control area for farmland and forest

17.6 Conclusion

According to the second Japanese National Spatial Strategies of 2015, the concept of compact and resilient city is identified as an important strategy (Ministry of Land, Infrastructure, Transport and Tourism 2015). Recently, there have been many natural disasters in Japan, and rapid depopulation is continuing in many provincial towns and cities. In the future, such municipalities will be required to minimize the damage caused by disasters and reduce the scale to a suitable size in response to the declining population. In this study, we selected suitable areas using land characteristics classification and estimation of vacant houses probability as fundamental information for examining where and how to aggregate future, more compact built-up areas.

In the central area of the target area, although it is convenient, there is an inundation risk. Paddy fields on the west and south sides of the target area serve to help mitigate river and surface water inundation. Therefore, urban land use should be prohibited in such places. The eastern side near Osato beach is a lesser convenient place to live, but safer. Such a place should be a residential area by maintaining farmland and vacant areas as GI. Although the current central area is more convenient, there is a risk of river inundation; therefore, it is necessary to ensure safety by the condition of structures. Furthermore, from the viewpoint of increasing of living environment, it is also necessary to network the central area and surrounding residential areas by public transportation to eliminate movement restrictions. Important concepts for depopulated cities and towns in Japan are ensuring coexistence of disaster prevention using GI and retaining a comfortable living environment.

References

Chino M (2009) Tokei today no. 9 (short report by Statistics Bureau of Japan) (Japanese). https://www.stat.go.jp/info/today/009.html

Disaster Management group in Cabinet office of Japan (2021) Guidelines for the criteria for recognizing damage to dwellings disasters (Japanese)

Ministry of Land, Infrastructure, Transport and Tourism (2015) 2nd. National Spatial Strategies in 2015 (Japanese). https://www.mlit.go.jp/common/001100233.pdf

Ministry of Land, Infrastructure, Transport and Tourism (2010) Statistics report of inundation disaster in 2010 (Japanese)

Muto Y, Kotani S, Miyoshi M, Kamada M, Tamura T (2018) Retarding capability change of wetland paddy fields due to house land development -utilizing paddy fields as green infrastructure. Proceedings of the 21th IAHR-APD congress 2018, Vol 2, 1209–1218

Tokushima Prefecture Government (2013) Secondary damage estimation of Nankai Trough earthquake in Tokushima prefecture (Japanese). https://anshin.pref.tokushima.jp/docs/2013112100023/

Yokokawa R, Muto Y, Kamada M, Tamura T (2020) Flood control function of farmland as green infrastructure using inundation analysis model. J Jpn Soc Civil Eng Ser B1 76:463–468. https://repo.lib.tokushima-u.ac.jp/ja/115208

Chapter 18
Toward an Equitable Distribution of Urban Green Spaces for People and Landscapes: An Opportunity for Portland's Green Grid

Vivek Shandas and Dana Hellman

Abstract Portland, Oregon, has a relatively long history of progressive planning, environmental protection, and implementation of green infrastructure (GI) projects, enjoying associated benefits such as temperature reduction, energy savings, and aesthetic improvements. These efforts, while resulting in a robust suite of GI across the city, must be better coordinated to meet the challenges of the future. In this chapter, we provide an overview of GI implementation in Portland, emphasizing key projects and policies, as well as ongoing disparities in access to GI and its benefits. We then discuss the potential for activating a *green grid* in Portland by linking existing nodes of GI for maximum benefit. In addition to amplifying the positive biophysical effects of GI, we suggest that this approach can help alleviate ongoing socioeconomic disparities that make some communities more vulnerable to climate change and environmental degradation.

Keywords Green infrastructure · Socio-ecological resilience · Climate change · Environmental justice · Urban planning

18.1 Introduction

Green infrastructure (GI) is increasingly recognized as a crucial component of urban ecosystems, particularly as cities confront dual crises of climate change and degradation of environmental quality. In urban planning, discussions tend toward the immediate, biophysical or economic benefits of GI, including flood and heat control (Webber et al. 2020; Zölch et al. 2016), air quality improvements (Nowak et al. 2006), stormwater management (Copeland 2016; Eaton 2018), and carbon capture and climate change mitigation (Foster et al. 2011). This has been the case in Portland, Oregon, and in cities across the world (Meerow 2019; Naumann et al.

V. Shandas (✉) · D. Hellman
Portland State University, Portland, OR, USA

© The Author(s) 2022   289
F. Nakamura (ed.), *Green Infrastructure and Climate Change Adaptation*,
Ecological Research Monographs, https://doi.org/10.1007/978-981-16-6791-6_18

2011). However, recent scholarship suggests that GI can be more than a physical buffer between urban residents and looming environmental threats. It is a gateway through which urban planners and policymakers might respond to underlying disparities that create socio-ecological vulnerability and inhibit resilience (Bowen and Lynch 2017; Jennings et al. 2012; Zhu et al. 2019). If properly integrated, GI could be an essential tool enabling environmental justice and equity in the urban environment, alongside climate change mitigation, cost-effective risk reduction, and ecological protection.

Within the United States, Portland, Oregon offers a unique example of extensive municipal GI, which has been complemented by efforts in the private sector. Compared to other large cities on the US west coast and beyond, Portland features a relatively high density of urban canopy and other GI assets, as well as multiple GI-supportive programs and policies. Portland's progressive planning efforts have been well documented, including the establishment of an urban growth boundary in the 1970s and the earliest solid waste recycling program and the first climate action plan for a US city. Since the late1990s, Portland has been recognized for its pursuit of GI, decades before many other US cities. As such, it offers a compelling case through which to consider (1) the steps associated with successful GI implementation at a city scale and (2) opportunities to expand the positive effects of GI moving forward. Specifically, this chapter introduces the concept of a *green grid*.

The current state of GI in Portland is the result of a series of municipal efforts undertaken by various agencies. This work has largely been done ad hoc, implementing an uncoordinated array of GI features across the city. Such an approach was necessary in the early years of GI development, as municipal agencies experimented with different GI features, geographies, and linkages. While Portland now comprises a robust system of GI *nodes* (individual pieces of infrastructure), these are not well integrated. Completing the system by connecting existing nodes, and adding new ones where needed, is the next step toward effective, citywide GI. The green grid is a way of conceptually linking those disparate nodes to generate a protective grid of GI that encompasses the entire Portland area. While not yet a formal policy or project platform, this concept has been under consideration and development by cooperating agencies, including the Portland Bureau of Environmental Services, Bureau of Planning and Sustainability, Bureau of Transportation, and Department of Parks and Recreation.

In the subsequent sections, we will provide an overview of past GI projects and policies in Portland, illuminating the sociopolitical conditions that have allowed Portland to attain its current level of GI. We will then discuss divergent experiences of and access to GI and its effects on diverse human communities throughout the region. Finally, we will elaborate upon the need for a green grid model in planning and policymaking for the city of Portland. This includes conditions that must be in place to realize such an ideal and its potential to tackle a range of interrelated social and environmental issues through coordinated action.

18.2 An Abridged Overview of Green Infrastructure in Portland

With over 13 million large trees, Portland's urban forest provides innumerable services to its residents and businesses, including improvements in air quality, temperature reduction during heat events, stormwater mitigation, and energy conservation. While tree canopy has received significant attention both from members of the public and local decision-makers, it is supplemented by a range of additional GI assets, including bioswales (features that capture and filter stormwater), water retention and detention ponds, ecoroofs (also called green roofs), and vegetated green streets (Figs. 18.1 and 18.2). These features did not emerge by chance but represent the culmination of decades of community effort and political will. In this section, we offer an overview of a few salient GI projects and policies that have emerged over the past three decades. These examples offer a glimpse into the sociopolitical conditions that advanced Portland's current suite of GI, which remains unrivaled in most cities in the world.

Fig. 18.1 Vegetated ecoroofs, also called green roofs. These features reduce stormwater runoff and pollution and provide wildlife habitat (*Source: Portland State University*)

Fig. 18.2 Bioswales. These features capture stormwater and filter out pollutants (*Source: City of Portland*)

One of the earliest attempts to advance green infrastructure in Portland began in the early 1990s. At this time, a legal case against the Bureau of Environmental Services resulted in a 20-year project to update municipal sewer and stormwater infrastructure. Central to this project, called Tabor to the River (T2R), was the creation of thousands of bioswales that would capture rainwater and reduce combined sewer overflows, the original reason for the 1990s lawsuit. The installation of these facilities resulted in extraordinary insights into the engineering, ecological and social dimensions of such an ambitious project. To date, several publications have highlighted the outcomes from the T2R project, and once finished, the City of Portland expanded the project to install similar GI assets across the region (Shandas and Messer 2008; BenDor et al. 2018). The US Environmental Protection Agency has since identified these facilities as exemplars of similar ones built in other parts of the country.

In 2015, the City implemented a Tree Code, which describes specific actions that must be taken when managing the urban forest. The Tree Code, which is known to city planners as Title 11, addresses questions about the role of the urban forester and trees in development situations and identifies and creates a series of staff positions that are responsible for different aspects of urban forest management. Emergent from the Tree Code was the creation of a Heritage Tree program, which offers a means for any community member to nominate a tree, based on social and ecological significance, as one that can stand in perpetuity. The Tree Code also outlines specific responsibilities from a community-based advisory group called the Urban Forestry Commission. Around this time, the City also passed an ordinance requiring ecoroofs in all new buildings in Downtown Portland and providing incentives for developers throughout the region to consider them as well. Finally, these changes coincided with a zoning approach that addressed "environmental overlay zones." This approach identifies specific areas in the city in which to apply zoning codes aimed at providing additional environmental protections during building or infrastructure development.

One of the most recent green infrastructure adoptions by the City of Portland is a green roof program, a component of the Central City 2035 Plan (City of Portland 2021). The EcoRoof Requirement, as it is called, began in 2018 and requires vegetated roofs covering 100% for new buildings or retrofits in the Central City over net 20,000 square feet. While this provision only applies to the Downtown area, it sets a precedent and has accelerated research on, applications of, and corporate interest in green roofs. The expectation is that this program will encourage households, businesses, the educational sector, and other facilities to implement their own green roofs, thereby diverting stormwater from the City's combined sewer and water system, while providing habitat for urban wildlife, including beneficial insects and birds.

Efforts to advance GI have been championed and implemented by numerous politicians, community groups, and municipal departments. These entities exhibit the potential to complement one another in pursuit of a shared goal of GI, though their work to date has often been uncoordinated. This approach has not been ineffective; Portland has been and continues to be a leader in GI, both within the

United States and internationally. The culmination of 30 years of work provides a model for advancing GI elsewhere in the future. However, despite apparent progress, Portland is not without its environmental concerns. Green space is increasingly stressed by a growing urban population, the economic pressure for development, and conflicts over limited physical space. Additionally, tree loss has occurred across Portland owing both to climatic changes and pests that affect tree health. Even within the parameters of the protective Tree Code, developers are permitted to deviate from canopy cover requirements for a fee. The combined effects of human pressure and climate change pose a double threat to local communities and wildlife and a significant challenge for urban planners. Furthermore, impressive as Portland's current suite of GI may be, it is not evenly experienced by city residents. Rather, underlying socioeconomic inequities have meant that only segments of the population enjoy the benefits of GI, even as others are disproportionately exposed to environmental threats such as extreme heat, residential flooding, and air pollution.

18.3   Outcomes and Experiences of Green Infrastructure in Portland

The City of Portland has identified numerous, positive outcomes attributable to advancements in GI. These encompass physical and mental health, energy efficiency and decreased greenhouse gas emissions, aesthetic improvements, air quality improvements, and stormwater management (City of Portland 2010). Additional benefits have been well documented through case study research. For example, a study by Rao et al. (2014) found that Portland's current urban forest reduces annual healthcare costs by almost $five million by filtering nitrogen dioxide in the air, a byproduct of combustion that is a federally regulated pollutant. It has been further suggested that Portland's tree canopy can improve the birth outcomes of children (Donovan et al. 2011). Studies show that residential properties near green streets, large trees, and substantial canopy sell faster and for more money than similar homes without green features (Donovan and Butry 2011; Netusil et al. 2014); that communities living next to urban trees experience less isolation and fewer crimes, as well as greater engagement with neighbors and involvement with civic life (Shandas 2015); and that adding vegetation can decrease ambient temperatures in Portland's urban corridors and neighborhoods (Makido et al. 2019). These benefits are essential to social and economic well-being and can improve the quality of life for urban residents. However, local studies also reveal that communities across Portland do not share equally in these benefits. The map below shows the distribution of some notable GI elements, including tree canopy, ecoroofs, parks (open space), and bioswales, with notable differences in the concentration of these elements across the city (Fig. 18.3).

An examination of the distribution of Portland's urban forest indicates a sweeping range of canopy access, from 82% cover in some neighborhoods, to just 4%

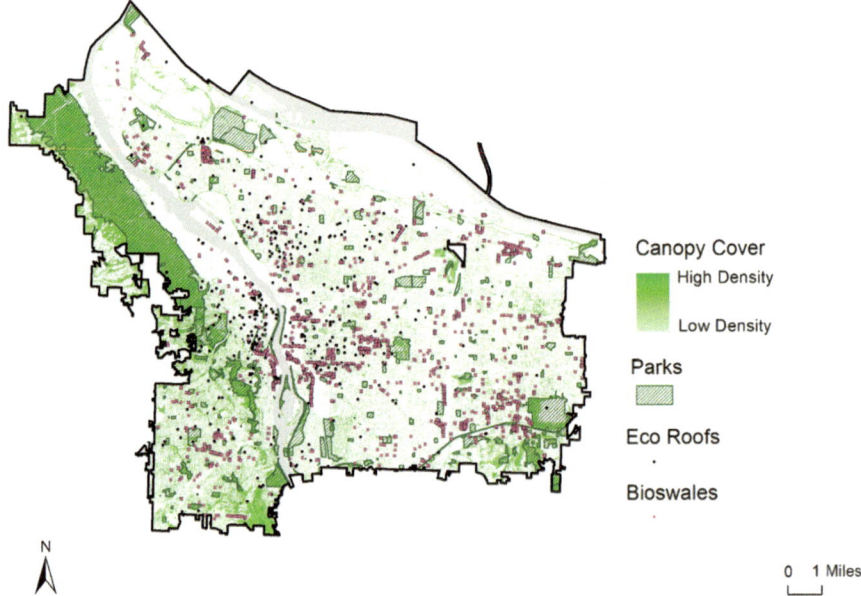

Fig. 18.3 Distribution of various GI elements in Portland, Oregon

in others (Fig. 18.4). While physical infrastructure and other urban features may preclude all areas of the city from having large tracts of forests, the current distribution suggests that some communities are disproportionately receiving benefits while others are not. A study by Voelkel et al. (2018) found that Portland's low-income residents and people of color (i.e., non-White), largely living on the city's east side, experienced higher summer temperatures than more affluent, White neighbors, largely concentrated on the west side. Additionally, Portland's history of redlining (that is, discrimination in housing sales based on race) can be understood as a facet of urban planning that has disproportionately situated people of color in higher-temperature neighborhoods (Hoffman et al. 2020). Locally, observed differences in socioeconomic status and exposure to urban heat correspond with differences in access to tree canopy. In fact, the City of Portland has recognized these inequities and is in the process of developing strategies to expand canopy into historically underserved neighborhoods. The city defines low-canopy as less than 25% canopy coverage and low-income as 50% or more of an area (e.g., neighborhood), making less than 80% of the median household income. These definitions provide a means for identifying those areas that meet both criteria and developing outreach strategies to expand tree plantings.

Portland is bisected by the Willamette River, which runs south to north and comprises tributaries across the city. Low-lying neighborhoods, particularly on the east side of the river, are prone to residential flooding when moderate or heavy precipitation events occur. As with redlining and urban heat, past development prac-

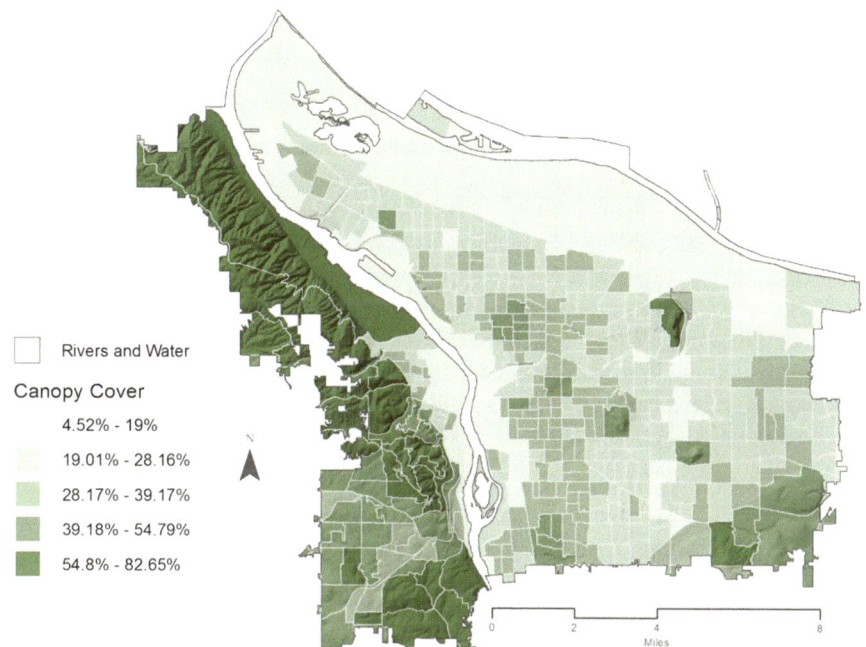

Fig. 18.4 Canopy cover in Portland, Oregon, visualized by census block group

tices have situated marginalized communities, especially those with low incomes, lower than average education, and people of color in floodplains, increasing their exposure to nuisance flooding (Kousky et al. 2020). A study by Fahy et al. (2019) found that flood exposure in Portland increased along spatial (east Portland) and socioeconomic lines. Unlike urban heat exposure and canopy access, Portland's GI assets related to urban flooding—specifically, green streets, bioswales, and ecoroofs—have already been concentrated in these high-risk, low-income areas (Chan and Hopkins 2017).

As planners and policymakers in Portland become more attuned to issues of equity and environmental justice, there is growing interest in targeting low-income or otherwise disinvested areas of the city with GI projects. The rationale is that all Portlanders should be able to enjoy green space and clean air and avoid flooding, extreme heat, and health assaults caused by poor environmental quality. However, this agenda is not without its complications. Sporadic GI investments may have the unintended consequence of making formerly disinvested neighborhoods more aesthetically pleasing, healthier or safer, raising property values and desirability (Ahmad et al. 2018). This in turn may lead to the displacement of area residents who were meant to be helped by GI upgrades, through a process sometimes referred to as "green gentrification" or "climate gentrification" (Anguelovski et al. 2019). This is a problem that may be exacerbated by the uncoordinated approach currently applied to GI planning in Portland. Unless all areas of Portland possess equitable access to

GI, it is unlikely that those residents most exposed to environmental harm and most in need of GI's benefits will be able to enjoy them. Rather, piecemeal approaches to GI encourage displacement and perpetuate disparities between socio-ecological haves and have-nots.

Historical development patterns, planning policies, and structural bias in decision-making systems have created social inequities that permeate the built environment. Solutions to common problems of climate change and environmental degradation amplify or overlook vulnerability for some while providing relief and rewards to others. This has been true even in Portland, which has made great strides both in GI and urban sustainability, as well as equity-focused environmental planning. GI is not a panacea for poverty, racism, and the many wicked problems facing urban communities. Yet, coordinated, integrated applications of GI across the city offer an opportunity to expand access to the ecosystem benefits and environmental protection it affords.

18.4   Next Steps: Activating the Green Grid

Portland is well-covered by GI assets. Even portions of the city that have been historically disinvested contain an array of bioswales, parks, and ecoroofs. To advance a green grid within the city, we consider each of these assets as a spatially explicit node of GI activity with associated benefits. While all GI is managed by a patchwork of private and public entities, none act alone and are rather part of a mosaic of efforts to enable the myriad ecosystem services that emerge from each node, large or small. The informal network of GI managers includes homeowners, business owners, developers, and residents who often participate in volunteer activities to improve the urban ecosystem. While informal, these entities are critical to support and occasionally supplant official entities who often lack the capacity and resources and reach to actively steward the GI. In addition, the vast majority of GI is located on private lands, which precludes official public organizations such as the Bureau of Planning and Sustainability, the Bureau of Environmental Services, Portland Parks and Recreation, and Oregon Metro from actively stewarding these GI systems. Whether the activation of Portland's green grid is accomplished through informal networks or formal ones, it will require a level of coordination and accountability for ensuring the grid's long-term viability. In this section, we identify a few of the essential components in developing a coordination and accountability system that can help maintain and enhance Portland's green grid of GI.

While broad legislation and programs, including the tree code and ecoroof ordinance, apply uniform standards across the city, to date, the authors know of no known attempts to integrate GI across sites or maximize benefits through coordinated actions. Characterizing a set of GI nodes—similar to those in standard planning practices that identify urban development centers (see, for example, the City of Portland's 2035 Comprehensive Plan)—offers a means for more effectively integrating the spatial and institutional aspects of this work. Here, we argue for an activation of Portland's green grid by connecting those many biophysical and spatially explicit nodes of existing GI, developing an intentional, holistic, well-coordinated strategy that aims to connect communities in need with beneficial GI and tackles environmental problems at the city scale. The green grid concept presents an opportunity to link up and integrate the many solitary components of Portland's GI landscape in response to the coordination and accountability challenges currently facing the region. We offer here a three-step approach for integrating the existing spatial nodes of GI and supporting a more coordinated effort that will ensure that urban ecosystems and communities will continue to receive essential ecosystem services from a green grid.

The first step is to describe the existing network of GI nodes across the city. This work has been underway by City officials but is not yet complete. While individual bureaus oversee specific aspects of each node—ecoroofs by Environmental Services, parks and trees by Parks and Recreation, etc.—developing a spatially explicit description of each of the green infrastructure features along with the entities that are responsible for their management can help to create an inventory for the current GI assets in the city. While definitions of GI vary from agency to community, we suggest maintaining a broad definition that encompasses the largest stretches of biophysical features that provide ecosystem services to areas of the city greater than a city block. Some attempts of such descriptions have emerged (Nascimento and Shandas 2021), though they are limited in terms of resolution and direct linkages to management entities. Of course, such descriptions will require highly granular data and the creation of a responsible entity that describes who might ensure the longevity of an individual node of GI. We have seen such efforts underway in several places around the world, including those mentioned in other parts of this GI compendium. We are not advocating for one entity to govern all the GI for an urban region but rather suggest an approach that identifies the specific GI nodes, functions, and expectations, which can translate into a co-management plan that ensures a coordinated response.

Second, we will need a means for understanding the distributional effects of GI in the city. As we mentioned earlier, the highly inequitable distribution of tree canopy provides communities with varying levels of ecosystem service, yet to date, we only have limited knowledge about the ways in which Portland's GI supports communities. Studies mentioned in the previous sections suggest potentially helpful ecosystem services associated with GI, though to date, the standards are largely based on its mere presence or absence. However, as we know from ecosystem studies, the health and overall quality of GI can vary depending on age, stress, maintenance, and other factors. More nuanced measures, variability, and qualitative outcomes associated with GI access and resulting ecosystem services are yet to be determined. A large tree, for example, provides greater amounts of ecosystem services to a community than a sapling. Assessing the quality of Portland's disparate GI is a herculean effort, requiring characterizations of the health and functioning of all features. Having recently completed the first exhaustive assessment of all public trees, which includes qualitative measurements of tree health, we know that the vast majority of green spaces are in private lands and will likely require a remotely sensed approach, such as Light Detection and Ranging (LiDAR) and hyperspectral assessments, to understand the health of GI. Efforts are currently underway to use remotely sensed data as a means for characterizing the quality of GI in Portland, though these will still require several years to complete. Coupling remotely sensed data with on-the-ground efforts to evaluate the ecosystem function of GI nodes will be an important second step to understanding the distributional effects of a green grid.

Coordination among formal and informal entities, taking inventory of GI, and characterizing the quality of ecosystem services can all be achieved through systematic and ongoing planning efforts. However, as a third step toward activating Portland's green grid, the City requires a collaborative approach that centers on values and ethics. This is essential for producing normative descriptions of GI and the contextual challenges facing it. Normative descriptions, in this case, will need to encompass a community conversation about preferable outcomes for Portland's green grid. What types of trees are preferable in which neighborhoods? How can we assign/adopt maintenance requirements? What forms of measurement will support an ecosystem function evaluation scheme? While calls for integrating community needs and perspectives into urban forestry efforts are gaining attention (Nascimento and Shandas 2021), still needed are approaches to articulate the extent to which diverse stakeholders view trees and GI as beneficial to community and individual health. If urban GI managers view its implementation simply as a matter of informing community members about the importance of trees, which has long been standard practice—such as one-way forms of communication—then opportunities to engage in broader discussions about equity will be lost. Arguably, much of the current emphasis on reconciling the disparate historical investments of GI within the Portland region has resulted in a prioritization of low-canopy and low-income neighborhoods (City of Portland 2018). However, we argue that without an explicit focus on the immediate needs of community members, effectively advancing concepts such as the green grid will fail.

One potential approach for fulfilling the promise of Portland's green grid is the use of scenario analysis techniques (Caughman et al. 2020). Scenario analysis, as applied in this context, offers a means for evaluating the outcomes of a specific set of green grid options through engagement with a diverse mix of stakeholders. With facilitation about specific themes (e.g., neighborhoods GI assets, maintenance options, performance outcomes, etc.), scenario analysis allows for the necessary reflectivity and integration of multiple disciplinary perspectives for evaluating outcomes. In this context, scenarios and models can play an important role in enabling community members and decision-makers to consider the myriad aspects of risk and opportunity associated with a green grid. Scenario analysis, as such, may enlighten participants as to how the future is likely to unfold, illuminate potential risks, help them assess alternative pathways to address risks, and smooth all phases of project implementation. Scenarios are scalable, often used in global, regional, and national environmental assessments carried out over the last decade, for example, the Intergovernmental Panel on Climate Change's Assessment Reports, the Millennium Ecosystem Assessment; Convention on Biological Diversity's Global Biodiversity

Outlook Reports, and United Nations Environment Programme's Environmental Outlook Reports. Nevertheless, we know of no uses of such an approach for advancing urban GI, which suggests a noteworthy oversight in the planning of our cities.

Integrating formal and informal organizational networks, describing and comprehensively documenting existing GI nodes, developing a functional basis for understanding and allocating ecosystems services, and developing normative approaches are fundamental pillars for achieving a robust green grid in Portland. If achieved, such a grid may offer protection against environmental threats while alleviating environmental disparities faced by Portland's most marginalized residents. A functional green grid means that the benefits of GI—clean air and water, cool temperatures, flood prevention, aesthetics—are enjoyed by all Portlanders, regardless of socioeconomic status or neighborhood of residence. Under the green grid, a robust system of GI assets covers the entire city, eliminating differences between those who do and do not have access. A green grid means that inequities associated with green gentrification may be avoided in the future because all parts of the city will be served. The implementation of GI in one neighborhood should not stimulate gentrification there if all neighborhoods are equally invested in. The coming decades will require even more care and intention to ensure that the existing charismatic mega-flora (i.e., old growth urban trees), along with the everyday shrubs and recently created bioswales, are enduring. If designed with local communities, and with clear understanding of goals, risks, and potential outcomes, Portland's GI can achieve more than climate change mitigation, economic development, or avoidance of physical threats. In fact, GI can improve human health, quality of life, environmental quality, and access to nature, while protecting treasured ecosystems and supporting a range of wildlife.

References

Ahmad M, Arkema C, Berman I, Brownwood B, Chan G, Zhu E (2018). Green infrastructure distribution in Portland, Oregon. Reed College Environmental Studies Junior Seminar. Retrieved from https://www.reed.edu/es/assets/ES300-2018-project-Green-Infrastructure-Distribution-FINAL.pdf

Anguelovski I, Connolly JJT, Pearsall H, Shokry G, Checker M, Maantay J et al (2019) Opinion: why green "climate gentrification" threatens poor and vulnerable populations. Proc Natl Acad Sci 116(52):26139–26143. https://doi.org/10.1073/pnas.1920490117

BenDor TK, Shandas V, Miles B, Belt K, Olander L (2018) Ecosystem services and U.S. stormwater planning: an approach for improving urban stormwater decisions. Environ Sci Pol 88:92–103. https://doi.org/10.1016/j.envsci.2018.06.006

Bowen KJ, Lynch Y (2017) The public health benefits of green infrastructure: the potential of economic framing for enhanced decision-making. Curr Opin Environ Sustain 25:90–95. https://doi.org/10.1016/j.cosust.2017.08.003

Caughman L, Plemmons N, Beaudoin F, Crim M, Shandas V (2020) The scenario Collaboratory: a framework for integrating environmental assessments and scenarios into municipal planning. In: Ninan K (ed) Environmental assessments. Edward Elgar Publishing, pp 215–230. https://doi.org/10.4337/9781788976879.00023

Chan AY, Hopkins KG (2017) Associations between Sociodemographics and green infrastructure placement in Portland, Oregon. J Sustain Water Built Environ 3(3):05017002. https://doi.org/10.1061/JSWBAY.0000827

City of Portland (2010) Portland's green infrastructure: quantifying the health, energy, and community livability benefits. City of Portland bureau of environmental services. Retrieved from https://www.portlandoregon.gov/bes/article/298042

City of Portland (2018) Growing a more equitable urban forest: Portland's Citywide tree planting strategy. Retrieved from: https://www.portland.gov/sites/default/files/2020-09/tree-planting-strategy-12.18.pdf

City of Portland (2021). The Central City 2035 Plan. Retrieved from: https://www.portland.gov/bps/cc2035/cc2035-documents

Copeland C (2016) Green infrastructure and issues in managing urban stormwater. Congressional Research Service Report 7:5700. Retrieved from https://fas.org/sgp/crs/misc/R43131.pdf

Donovan GH, Butry DT (2011) The effect of urban trees on the rental price of single-family homes in Portland. Oregon Urban Forestry & Urban Greening 10:163–168

Donovan GH, Michael YL, Butry DT, Sullivan AD, Chase JM (2011) Urban trees and the risk of poor birth outcomes. Health Place 17(1):390–393. https://doi.org/10.1016/j.healthplace.2010.11.004

Eaton TT (2018) Approach and case-study of green infrastructure screening analysis for urban stormwater control. J Environ Manag 209:495–504. https://doi.org/10.1016/j.jenvman.2017.12.068

Fahy B, Brenneman E, Chang H, Shandas V (2019) Spatial analysis of urban flooding and extreme heat hazard potential in Portland, OR. Int J Disast Risk Reduc 39:101117. https://doi.org/10.1016/j.ijdrr.2019.101117

Foster J, Lowe A, Winkelman S (2011) The value of green infrastructure for urban climate adaptation. The Center for Clean air Policy. Retrieved from http://www.ggi.dcp.ufl.edu/_library/reference/The%20value%20of%20green%20infrastructure%20for%20urban%20climate%20adaptation.pdf

Hoffman JS, Shandas V, Pendleton N (2020) The effects of historical housing policies on resident exposure to intra-urban heat: a study of 108 US urban areas. Climate 8(1):12. https://doi.org/10.3390/cli8010012

Jennings V, Johnson Gaither C, Gragg RS (2012) Promoting environmental justice through urban green space access: a synopsis. Environmental Justice 5(1):1–7. https://doi.org/10.1089/env.2011.0007

Kousky C, Netsil NR, Moldovan-Trujillo G (2020) The mispricing of flood insurance: a look at Portland, Oregon. University of Pennsylvania, Wharton Issue Brief. Retrieved from https://riskcenter.wharton.upenn.edu/wp-content/uploads/2021/01/The-Mispricing-of-Flood-Insurance-A-Look-at-Portland-Oregon.pdf

Makido Y, Hellman D, Shandas V (2019) Nature-based designs to mitigate urban heat: the efficacy of green infrastructure treatments in Portland, Oregon. Atmos 10(5):282. https://doi.org/10.3390/atmos10050282

Meerow S (2019) A green infrastructure spatial planning model for evaluating ecosystem service tradeoffs and synergies across three coastal megacities. Environ Res Lett 14(12):125011. https://doi.org/10.1088/1748-9326/ab502c

Nascimento LAC, Shandas V (2021) Integrating diverse perspectives for managing neighborhood trees and urban ecosystem Services in Portland, OR (US). Land 10(1):48. https://doi.org/10.3390/land10010048

Naumann S, Davis M, Kaphengst T, Pieterse M, Rayment M (2011) Design, implementation and cost elements of green infrastructure projects. Final report to the European Commission, DG Environment, contract no. 070307/2010/577182/ETU/F.1, ecologic institute and GHK consulting. Retrieved from https://www.ecologic.eu/sites/files/project/2014/documents/design-implementation-cost-elements-of-green-infrastructure-projects-2011-naumann_0.pdf

Netusil NR, Levin Z, Shandas V, Hart T (2014) Valuing green infrastructure in Portland, Oregon. Landsc Urban Plan 124:14–21. https://doi.org/10.1016/j.landurbplan.2014.01.002

Nowak DJ, Crane DE, Stevens JC (2006) Air pollution removal by urban trees and shrubs in the United States. Urban For Urban Green 4(3–4):115–123. https://doi.org/10.1016/j.ufug.2006.01.007

Rao M, George LA, Rosenstiel TN, Shandas V, Dinno A (2014) Assessing the relationship among urban trees, nitrogen dioxide, and respiratory health. Environ Pollut 194:96–104. https://doi.org/10.1016/j.envpol.2014.07.011

Shandas V (2015) Neighborhood change and the role of environmental stewardship: a case study of green stormwater infrastructure in the City of Portland (OR, USA). Ecol Soc 20(3):16. https://doi.org/10.5751/ES-07736-200316

Shandas V, Messer WB (2008) Fostering green communities through civic engagement: community-based environmental stewardship in the Portland area. J Am Plan Assoc 74(4):408–418. https://doi.org/10.1080/01944360802291265

Voelkel J, Hellman D, Sakuma R, Shandas V (2018) Assessing vulnerability to urban heat: a study of disproportionate heat exposure and access to refuge by socio-demographic status in Portland, Oregon. Int J Environ Res Public Health 15(4):640. https://doi.org/10.3390/ijerph15040640

Webber JL, Fletcher TD, Cunningham L, Fu G, Butler D, Burns MJ (2020) Is green infrastructure a viable strategy for managing urban surface water flooding? Urban Water J 17(7):598–608. https://doi.org/10.1080/1573062X.2019.1700286

Zhu Z, Ren J, Liu X (2019) Green infrastructure provision for environmental justice: application of the equity index in Guangzhou, China. Urban For Urban Green 46:126443. https://doi.org/10.1016/j.ufug.2019.126443

Zölch T, Maderspacher J, Wamsler C, Pauleit S (2016) Using green infrastructure for urban climate-proofing: an evaluation of heat mitigation measures at the micro-scale. Urban For Urban Green 20:305–316. https://doi.org/10.1016/j.ufug.2016.09.011

Part VI
Coast and Estuary Ecosystem

Chapter 19
Effectiveness and Sustainability of Coastal Hybrid Infrastructures for Low-Frequency Large-Scale Disasters: A Case Study of Coastal Disaster Assessment for a Complex Disaster

Ryoichi Yamanaka and Kosuke Nakagawa

Abstract This study aims to evaluate the functioning of hybrid infrastructures in coastal areas and identify the factors influencing their sustainability. The hybrid infrastructure targeted in this study is in the Osato area in southern Tokushima Prefecture, Japan, and consists of a seashore, coastal embankment, coastal forest, rice paddles, and dune. Numerical analysis and field observations of tsunamis and storm surges were conducted in this study. In the tsunami analysis, the response of the inundation area to the reference water level was evaluated assuming a complex disaster involving a tsunami, storm surge, and sea-level rise. In the storm surge analysis, the mechanism of beach deformation and damage in the coastal forest caused by Typhoon Hagibis (No. 19 in Japan) in 2019 was evaluated, and the historical dune formation process was discussed in this case study. Finally, the design strategy required for hybrid infrastructure is discussed.

Keywords Hybrid Infrastructure · Tsunami · Storm surge · Complex disaster · Sea-level rise

19.1 Current Status and Issues of Coastal Protection Measures in Japan

Japan has the sixth-longest coastline in the world. The total distance of the coastline is 35,295 km, and approximately 14,500 km of the coastline requires conservation

R. Yamanaka (✉)
Research Center for Management of Disaster and Environment, Tokushima University, Tokushima, Japan
e-mail: ryoichi_yamanaka@tokushima-u.ac.jp

K. Nakagawa
Kiso Kensetsu Consultant, Co., Ltd, Tokushima, Japan

measures. Coastal maintenance was conducted under the Coastal Act. The amendment of the Coastal Act in 1999 expanded the objectives of coastal development to include the environment, usage, and protection. Moreover, measures must cope with the increasing severity of disasters caused by climate change. However, the budget for coastal conservation is decreasing due to the continuing population decline, and the usage pattern of coastal areas is changing. Japan is, therefore, at a stage where new measures are needed for coastal conservation. The Japanese government's efforts to promote green infrastructure and ecosystem-based disaster risk reduction (Eco-DRR) are promoted by these circumstances. Furthermore, the reconstruction of infrastructure is expected to have a synergistic effect on disaster prevention, environmental conservation, and regional development, and it is necessary to clarify how this is to be achieved.

Hybrid infrastructure is considered a realistic strategy for applying green infrastructure to coastal areas. Hybrid infrastructures (Sutton-Grier et al. 2015) combine natural resources and artificial structures that are expected to result in benefits for green infrastructure. An example of a hybrid infrastructure in the coastal zone is a combination of coastal embankments as gray infrastructure and shallow beaches, coastal forests, rice paddies, and dunes as green infrastructure. Existing coastal embankments are designed based on past disaster records and, thus, effectively prevent the past maximum disaster level. In addition, the combination of gray and green infrastructure can be expected to be effective in preventing natural disasters beyond the expected scale. However, the design method of hybrid infrastructures in coastal areas has not yet been established and is still in the research stage; therefore, it is necessary to evaluate their effectiveness and explore management methods from different viewpoints.

A method similar to hybrid infrastructure has been used in the design of coastal protection facilities in Japan. It is called the "Integrated Shore Protection System." It is a method of protecting the coast from storm surges and coastal erosion by combining artificial coastal protection facilities such as levees, artificial beaches, and breakwaters. This system is superior in terms of access to the beach and scenery compared to the method that relies only on massive coastal levees. This system and that of hybrid infrastructure may look similar, but there is a difference in how the community interacts with the infrastructure. Because hybrid infrastructures are included in the category of green infrastructure, hybrid infrastructures also involve maintaining social infrastructure based on the coexistence of people and nature. In other words, hybrid infrastructures are not given unilaterally and managed by the government, like gray infrastructure, but are based on the actions of local people according to their history and intentions. Therefore, to realize hybrid infrastructures, the opinions of stakeholders and an understanding of the origins of the region based on a long-term perspective are indispensable; in other words, it is necessary to design infrastructure according to the characteristics of each region.

Based on the above background, this chapter discusses the effectiveness of hybrid infrastructures against natural disasters that may occur in the future and their maintenance and management measures including their origins and histories. A case study was conducted for a region with a hybrid infrastructure under the risk of tsunamis and storm surges. We assume a combined natural disaster caused by a tsunami, storm surge, and sea-level rise due to climate change. The occurrence of such combined disasters is infrequent and has not been a subject of serious discussion. Recently, however, the frequency of flood damage has prompted the Japanese government to approve disaster prevention plans based on maximum-expected levels rather than records. This study was conducted in response to such social demands.

19.2   Overview of the Study Area

The coastal area with the hybrid infrastructure targeted in this study is located in the Osato area of southern Tokushima Prefecture, Japan (Fig. 19.1). The Osato area has a semi-arc-shaped natural beach facing the Pacific Ocean, which is called the Osato coast. Behind the beach is a dune, on top of which there is a coastal embankment, pine forest, and village (Fig. 19.2). The length of the coastline is 2457 m, and the

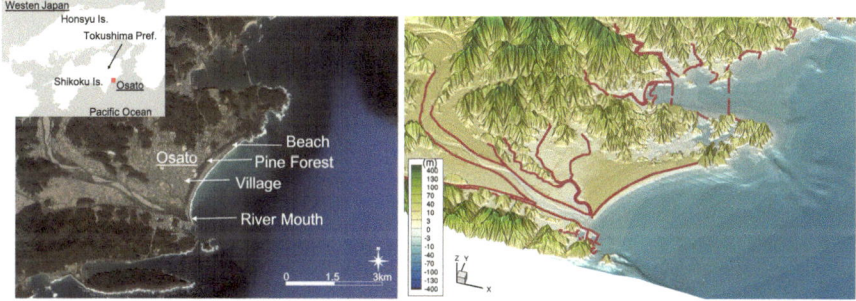

Fig. 19.1 Study area: (left figure) map of the surrounding area of Osato Coast and (right figure) elevation map; red line is the location of the embankment

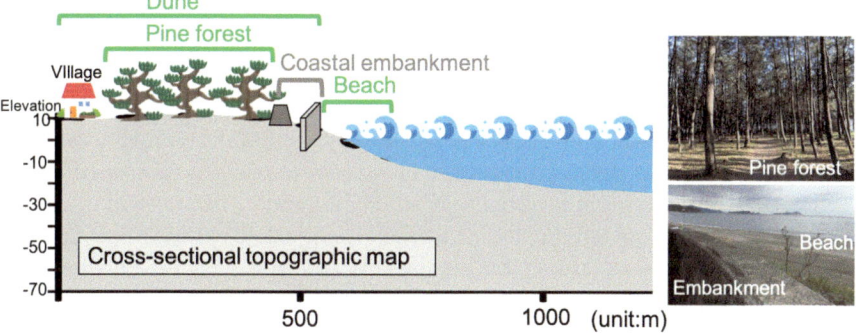

Fig. 19.2 Layout of hybrid infrastructure on the Osato Coast

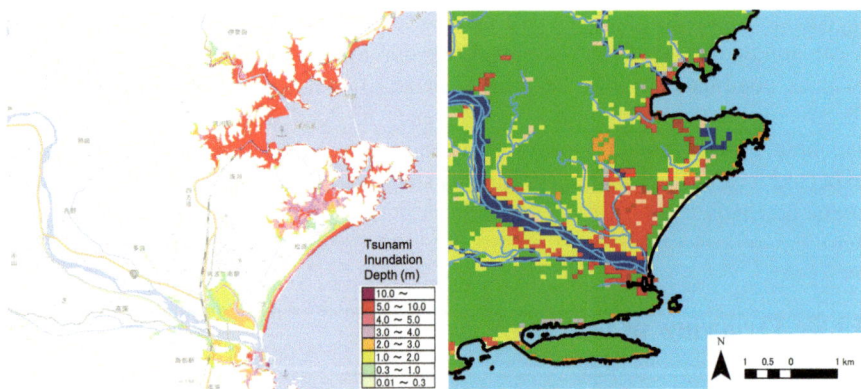

Fig. 19.3 Tsunami hazard map (inundation depth distribution map) published by Tokushima Prefecture (left figure) and land use map taken from the National Land Numerical Information Land Use Subdivision Mesh (right figure). In the right figure, red, yellow, green, blue, and orange indicate the residential area, rice paddies, forests, rivers, and wastelands, respectively

elevation of the dunes is approximately 10 m above sea level. It faces the Nankai Trough and has experienced periodic Nankai earthquake tsunamis, but there has been no significant damage to the village.

The Great East Japan Earthquake occurred in 2011, and the tsunami damage was more extensive than the prior hazard assessment. After the Great East Japan Earthquake in 2011, the Japanese government revised the hazard assessment of tsunamis originating from the Nankai Trough, which have an identical generation mechanism to the Great East Japan Earthquake. Figure 19.3 shows a tsunami hazard map of the area covered in this study. The Tokushima Prefecture made this hazard map public. This hazard map is for the case of the largest crustal movement estimated by the Japanese government, and the mean water level is set to the mean high tide, which can be regarded as a hazard indicator. Although the tsunami propagated from south to north, the Osato area was not inundated by the tsunami.

Table 19.1 shows the required functions of the hybrid infrastructure on the Osato coast. Beach, pine forest, and dune are examples of green infrastructure, whereas the coastal embankment is an example of gray infrastructure. The functions are divided into three categories, with reference to the Coastal Act: disaster prevention, environment, and regional development. Green infrastructure is expected to serve various functions unlike coastal embankments, which only prevent disaster. However, the performance of some green infrastructure is not easy to quantify. Considering that the evaluation of return on investment is strictly required when designing coastal protection facilities, quantifying these functions is critical. In hybrid infrastructures, where green and gray infrastructures are spatially arranged, the infrastructures may produce synergistic effects.

For example, the tsunami damping effect must be evaluated without dividing these infrastructures because the tsunami gradually dampens as it passes

Table 19.1 Expected functions of hybrid infrastructure on the Osato Coast

	Disaster prevention	Environmental functions	Regional development
Beach	Flood prevention Landslide prevention	Landscape conservation Ecosystem conservation Seawater purification Biological growth CO_2 absorption	Recreation Amenity improvement Fishery Local culture succession
Embankment	Flood prevention		
Pine forest	Brown sand prevention Flying salt prevention Windstorm protection Inundation reduction Space for evacuation	Landscape conservation Ecosystem conservation	Cultural heritage sites Place for local activities Forestry Tourist destination Religious sites Local symbols
Sand dune	Brown sand prevention Flood prevention Space for evacuation	Landscape conservation Ecosystem conservation Water purification Groundwater recharge	Environmental education Tourist destination Water supply

sequentially through the beach, embankment, and pine forest. For this reason, numerical simulations that can reproduce local conditions are effective, especially for performance related to physical processes.

Another advantage of using numerical simulations to assess the physical damage caused by tsunamis and storm surges is that the assumptions can be easily changed. For example, we can evaluate the effectiveness of the current infrastructure against a typhoon equivalent to the Second Muroto Typhoon (Typhoon 196118, NANCY), which caused extensive damage in the study area. Furthermore, we can predict what would happen if the second Muroto typhoon and the Nankai earthquake tsunami hit simultaneously. As mentioned above, the target area of this study has not been severely damaged by past Nankai earthquake tsunamis. However, if the scale and frequency of natural disasters increase due to climate change, predicting the damage caused by even such low-frequency disasters will provide helpful information for future disaster prevention measures. A possible scenario based on the hybrid infrastructure is that tsunamis and storm surges overtopping coastal levees and embankments will be attenuated by pine forests and dunes behind them, which will act as green infrastructure to prevent them from hitting the village. This study also focuses on the effects of such green infrastructure and draws a hybrid infrastructure function for consideration.

The disadvantage of simulations with virtual external forces is that it is difficult to assess the accuracy of the results. In projections of future climate change effects, multiple scenarios are set up, and policymakers are left to decide which simulation results to adopt. Because numerical simulations evolve as technology advances, policymakers should view the results of simulations as indicative of trends.

19.3 Scenario Analysis of Tsunami Hazard by Numerical Simulation

In the numerical simulation of tsunami hazards, it is necessary to simulate the physical processes from earthquake generation to tsunami propagation. Specifically, we estimated the crustal deformation associated with the earthquake, predicted the sea surface deformation associated with crustal deformation, calculated the tsunami propagation, and predicted the tsunami behavior in the run-up area. In this study, a community model was constructed to simulate physical processes related to tsunamis. JAGURS (Baba et al. 2015, https://github.com/jagurs-admin/jagurs), a tsunami simulator in a two-dimensional plane and high-performance tsunami calculation code, was used in this study. A two-dimensional model is advantageous because it has high prediction accuracy for long waves with the following characteristics: the horizontal behavior of water particles in the water does not depend on the water depth, and the velocity at the water surface and the bottom are almost the same. The tsunami calculation code is a community model that can handle nonlinear dispersive waves, developed in collaboration with the Japan Agency for Marine-Earth Science and Technology, the Meteorological Research Institute, and the Australian National University. JAGURS implements a nested algorithm to increase the resolution of specific regions and has a multi-scenario execution capability. JAGURS is based on a differential method using a staggered grid and the Leapfrog method and can be used in spherical or Cartesian coordinate systems. The code is written in Fotran90 and can be used for parallel and large-scale computations using OpenMP and MPI.

The semi-infinite homogeneous elastic body model (Okada 1985) was used to calculate the crustal movement in the fault model. The spatial distribution of crustal deformation can be obtained by assigning the reference position of the fault, depth from the surface, length of the fault, fault width, dip angle, strike angle, slip angle, and slip rate of the fault. Then, based on the distribution of crustal deformation, the elevation distribution can be modified, and sea-level deformation can be set.

The fault model used in this study was proposed by the Nankai Trough Mega Earthquake Model Study Group of the Cabinet Office of the Japanese government. In this study, we adopted scenario No. 3, in which a large slip zone is located in the south of the study area. Figure 19.4 shows the spatial distribution of the crustal movement. The maximum uplift is approximately 8 m in the southern part of the study area (top figure of Fig. 19.4), while the subsidence of 0.65 to 0.9 m was calculated in the target area (bottom figure of Fig. 19.4). In the target area, such a decrease in elevation over a wide area increases tsunami damage.

The topography for the numerical simulation was set up by combining different spatial resolutions for each area. This method is called "nesting," and it can improve computational efficiency. Figure 19.5 shows the mesh layout used in this study. Five mesh sizes ranged from 10 to 810 m. The influence of the geomaterials on the surface flow was expressed using Manning's roughness coefficient. Figure 19.6 shows the distribution of the roughness coefficients.

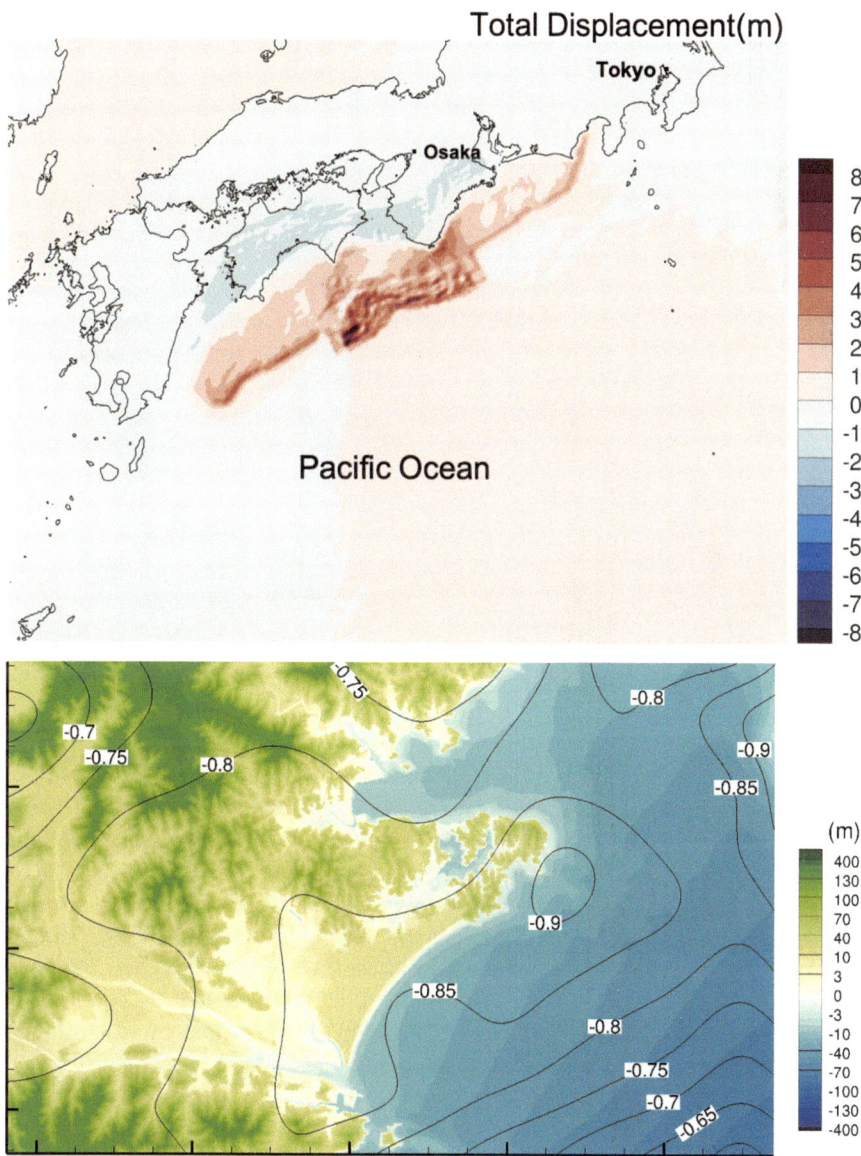

Fig. 19.4 Fault displacement distribution used for input (top figure) and magnified view in the area to be analyzed (bottom figure, coloring indicates elevation, contour lines indicate fault displacement distribution)

The scenarios assumed in this study are listed in Table 19.2. In all scenarios, the tsunami was input by the fault model described above, and the mean sea level varied for each scenario. In Case 1, as shown in Table 19.2, we set the mean high tide level

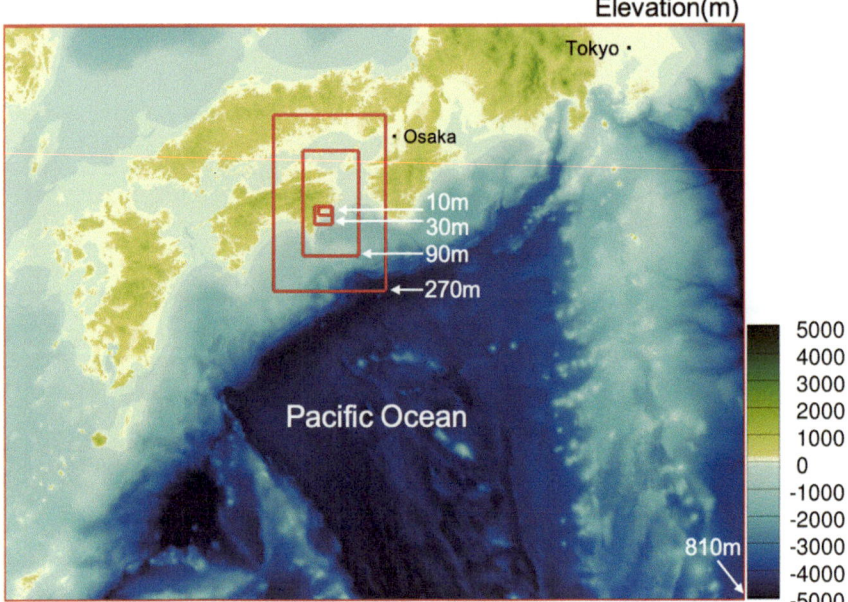

Fig. 19.5 Arrangement of computational grids and computational grids spacing (coloring indicates elevation)

as the mean water level. In Case 2, a typhoon of the same level as the second Muroto typhoon was assumed, and the sea level rise was set based on the records. In Case 3, we assumed that a typhoon larger than the Second Muroto Typhoon strikes with the same level of pressure and wind as the Isewan Typhoon (Typhoon 195915, VERA) and set the sea-level rise using storm surge predictions for Tokushima Prefecture. In Case 4, the IPCC (AR5, WG1) projections were used for sea-level rise, and in this case, the sea-level rise was set for the most severe warming between 2081 and 2100 (RCP8.5) in Case 2. In Case 5, we set the IPCC projection of sea-level rise in the case of the most severe warming (RCP8.5) in Case 3. Although it is difficult to estimate the probability of each scenario, the probability of periodic Nankai earthquake tsunamis is not considered here because it varies depending on when it is estimated. However, in this study, we interviewed experts in coastal engineering and obtained rough estimates of the probability of water level change, as shown in Table 19.2.

Figure 19.7 shows the results of the tsunami inundation simulations calculated based on the scenarios. The results are overlaid with satellite images and inundation depths with the indicated colors. The hazard map provided by the Tokushima Prefecture is also shown for comparison. The tsunami propagated from the south of the coast and moved to hit the coast and mouth of the river, then ran up to the land by flooding the coast or up the river. We compared the hazard maps and Case 1 in the inundation areas. The inundation areas were the same, but in the pine forests

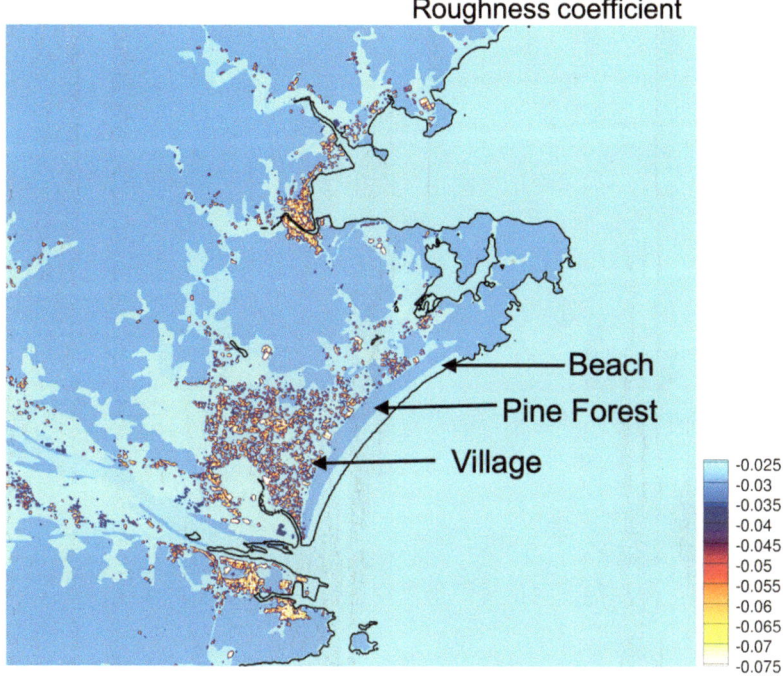

Fig. 19.6 Spatial distribution of roughness coefficients (negative values are due to model limitations, but are actually positive values)

Table 19.2 List of scenarios assumed in this study. The external tsunami force is the same for all scenarios, and only the reference water level is changed for each scenario

	Scenario assumptions	Mean sea level	Estimated probability of water level occurrence
Case1	Mean sea level of spring tide	T.P. + 0.92 m	Once a month
Case2	Storm surge at the level of the second Muroto typhoon	T.P. + 2.85 m	50 years
Case 3	A typhoon with the same pressure as the second Muroto typhoon and the same wind level as the Isewan typhoon	T.P. + 3.60 m	100 years
Case 4	Case 2 and the highest predicted average global mean sea level for 2081–2100 by IPCC (AR5 WG1)	T.P. + 3.67 m	150 years
Case 5	Case 3 and the highest predicted average global mean sea level for 2081–2100 by IPCC (AR5 WG1)	T.P. + 4.42 m	200 years

T.P. means Tokyo Peil

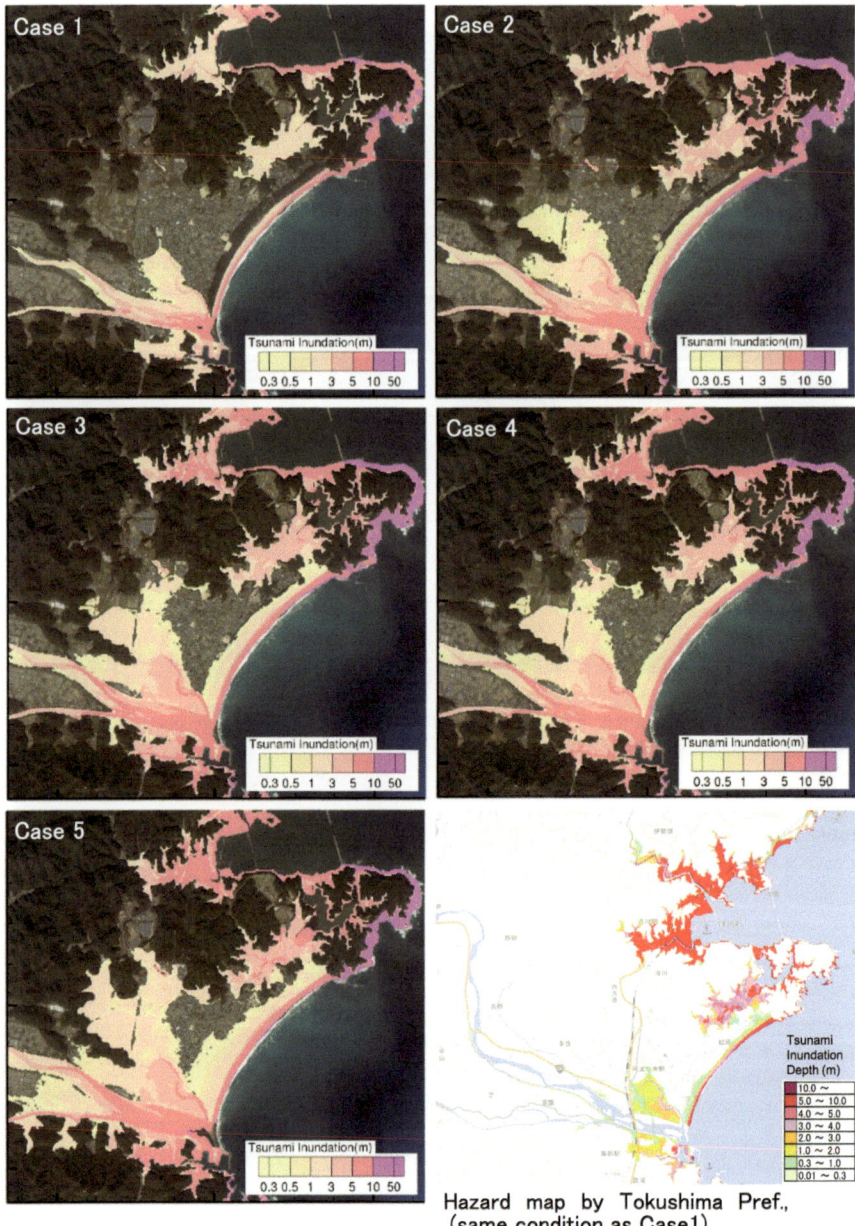

Hazard map by Tokushima Pref., (same condition as Case1)

Fig. 19.7 Distribution of inundation depths for each scenario calculated by numerical analysis and hazard map of inundation depth by Tokushima Prefecture

behind Osato Beach, the latter was more extensive near both ends of the beach. This is because the hazard map is a superposition of the maximum values obtained from tsunami inundation simulations using multiple fault models, which is different from the method used in this study. Therefore, the fault model in this study was considered appropriate for hazard assessment.

Next, we focused on a tsunami that ran from the beach to the pine forest. In Cases 1 and 2, the tsunami did not spread to the pine forest. However, in Cases 3 and 4, the tsunami inundation area extended to the entire pine forest area over the coastal embankment. In Case 5, the tsunami inundation area extended beyond the pine forest to the residential area. Similarly, when comparing the inundation areas spreading from the river mouth to the northwest, the inundation area in each case is different, and when the sea level rises, most of the area located behind residential areas will be inundated. Because rice paddies in the locations highlighted in yellow in the right figure of Fig. 19.3 are in the inundation area, the waterlogging effect of rice paddies was also included in the elevation distribution in this study. The above results can be summarized as follows. In Cases 1 and 2, the tsunami inundation was limited by coastal and river levees, and the scale of the external force was sufficient for the gray infrastructure to be effective. In Case 5, the tsunami reached the residential area beyond both gray and green infrastructures.

Figure 19.8 shows that the inundation area increases nonlinearly as the initial water level increases. Here, we consider the infrastructures that mainly contributed to reducing inundation in each scenario: Cases 1 and 2 can be regarded as the result of using gray infrastructures (coastal levees and river levees). Cases 3–5 are the result of using green infrastructure (dunes, pine forests, and rice paddies). The results imply that the effect of green infrastructure decreases nonlinearly as the external force increases. This is because not all levees break at once, and levees are installed in multiple layers at the water's edge and land.

The effects of gray and green infrastructure are interactive phenomena, and it is not easy to separate and discuss their physical disaster prevention effects. Understanding the effects of infrastructure holistically by changing the scale of external forces is important. It is possible to calculate the benefits of changing the conditions of infrastructure, such as the height of coastal embankments and vegetation density in pine forests, based on the assets in the flooded area. However, it should be noted that the assets in the flooded area cannot be fully expressed by only considering fixed assets such as houses and government facilities. It is essential to clarify what assets are needed to safeguard local peoples' lives and consider measures to increase infrastructure effectiveness.

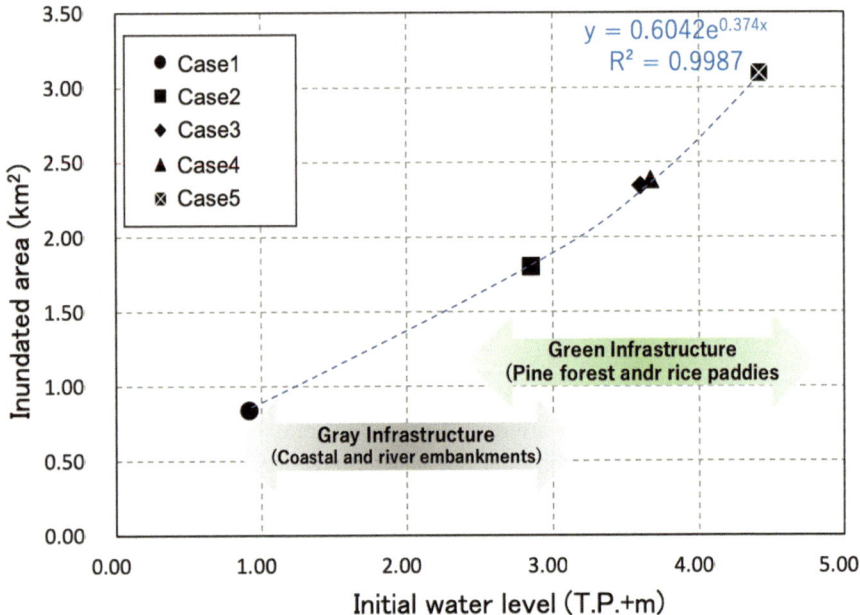

Fig. 19.8 Relationship between initial average water level and flooded area

19.4   Relationship Between Typhoons and the Formation Process of Hybrid Infrastructure

Typhoon 201919 (HAGIBIS) occurred in the Pacific Ocean on October 5, 2019. It gradually turned northward, approached the Tokai and Kanto regions of Japan from October 12 to 13, and disappeared east of the Kamchatka Peninsula on October 14. The pressure at the time of the approach was 945 hPa, and the maximum radius of the storm area was 370 km. The typhoon reached the Japanese archipelago with a very strong force, causing severe damage in western and eastern Japan. Although Shikoku Island, the target area in this study, was far from the typhoon's path, it caused severe damage to the Osato coast.

Figure 19.9 shows a summary of the information on typhoon-induced waves recorded at nearby observation sites. The blue squares show the significant wave height and the significant wave period at the time of the maximum wave height, and the yellow triangles show the significant wave period and the significant wave height during the maximum wave period. The solid and dashed lines represent the waveform gradient; the closer it is to the solid line, the larger is the swell size. The points in the red circle are the values for this typhoon, indicating that the swell reached its highest level in terms of both wavelength and period. This figure is based on the statistical results of the past 50 years. This means that the probability of the swell generated by this typhoon was more than 50 years.

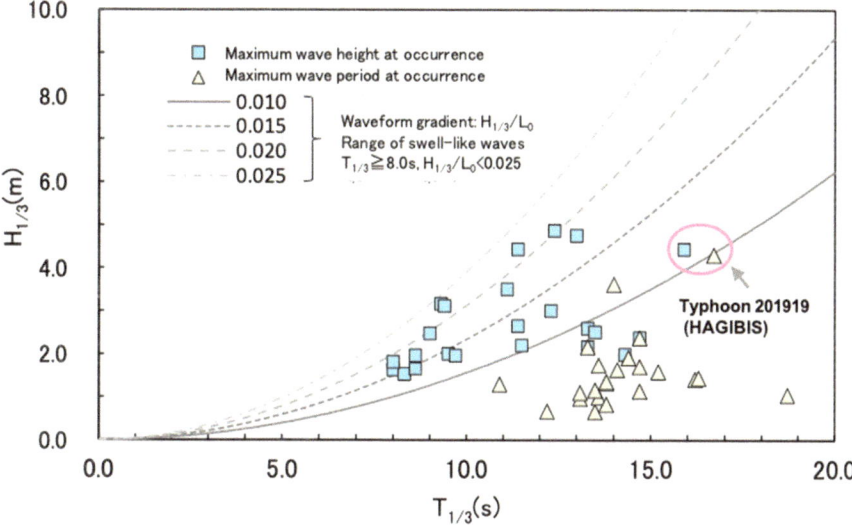

Fig. 19.9 Record of waves during past typhoons at the nearest observatory to the Osato Coast. The x-axis represents the significant wave period, and the y-axis represents the significant wave height

Photo 19.1 Aerial photo by drone (taken from the south side of the beach facing north, October 16, 2019) and a photo of the beach center near the shoreline

Photo 19.1 depicts a beach scene on October 16, 2019, 3 days after the typhoon passed. Beach deformation occurred even in the upper tidal zone, as well as large-scale sand movement. In addition, near the coastline in the center of the beach, fine sandy soil newly formed, which was not present before the passage of the typhoon. The gravel that was present before the typhoon hit was pushed out to the pine forest behind, and gravel and seawater invaded the pine forest. Such large-scale beach deformation has not been recorded in recent years. However, an 80-year-old local woman said that the beach returned to its shape when she was a child. This suggests that this type of beach deformation has occurred in the past.

Figure 19.10 shows the distribution of the change in ground height caused by the typhoon based on the observed data. This figure shows that the ground height

Fig. 19.10 Distribution map of the difference in ground elevation before and after the typhoon. Positive values indicate sedimentation, and negative values indicate erosion

increased at the beach center and decreased at both ends. According to published boring data (Tokushima Prefecture), the bottom sediment of the Osato Coast is sandy below the sea surface, whereas gravel spreads in the tidal zone, suggesting that the sandy ground seen in Photo 19.1 was formed because of the movement of sand from the seaside to the landside. The occurrence of beach deformation caused by the typhoon suggested the origin of the dune, which is a high ground where pine forests and villages are located. The mouth of the Kaifu River is located at the southern end of the Osato coast, which supplies sand to the sea area when the river flows out after heavy rains. Fine sand supplied by the river exists on the seabed of the Osato coast, but under normal wave conditions, this sand accumulates rather than moves. However, typhoons with billows pass through the area about once every 50 years, causing large-scale beach deformation, such as that seen here. As a result, sea sand was lifted inland, and dunes were formed over a long period. This means that the dune behind the hybrid infrastructure in the Osato area has a mechanism to enhance its disaster prevention function through natural processes.

After this large-scale beach deformation, residents raised the issue of seawater inundation into the pine forest, which prompted the government to conduct restoration work on beach deformation by moving the sand in the tidal zone offshore, but the same phenomenon is expected to occur if a similar typhoon passes through again. Therefore, this construction work is considered a countermeasure against high waves that occur frequently. Figure 19.11 shows the run-up area due to high

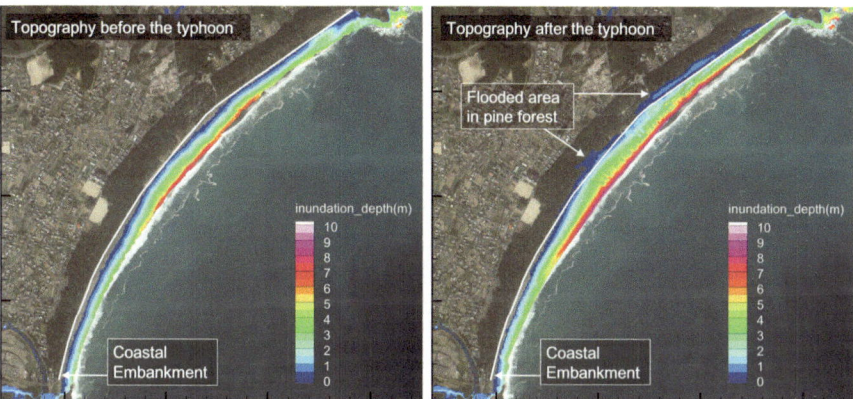

Fig. 19.11 Inundation area by high waves calculated by the wave model (comparison of different terrains applied)

waves calculated using the wave analysis model "Celeris" (Tavakkol and Lynett 2017). Celeris is a numerical wave analysis model based on the extended Boussinesq equations. A hybrid finite volume method was used to discretize the fundamental equations. In this study, boundary conditions based on the JONSWAP spectra were set up to reproduce the wave conditions during the typhoon. The directional function is assumed to follow the Mitsuyasu equation (Mitsuyasu et al. 1975). The boundary conditions of the wave spectra were obtained from the results of wide-area calculations based on the energy equilibrium equation, and the parameters at the boundary of the calculation domain were input. Three hundred wave inputs for the topography before and after the typhoon passage were used. When the topography before the typhoon was applied, the waves did not reach the pine forest area. In contrast, when the topography after the typhoon was applied, the waves reached the pine forest area. The white line in the figure is the boundary between the beach and pine forest, where the coastal embankment is located. In reality, beach deformation occurred in tandem with typhoons. However, as the beach deformation progressed, the situation changed, such that the waves could efficiently run up to the higher elevation pine forest area. This indicates that it is difficult to control seawater intrusion into the pine forest under the same scale of external force as the typhoon.

Knowing the origin of the natural environment in green infrastructure suggests how hybrid infrastructure can be maintained. Figure 19.12 shows a conceptual diagram of the effect of coastal embankments on the natural dune formation process. From this figure, to enhance the dunes as green infrastructure, it is better if there is no dike at the current location (bottom figure). However, if there is no embankment at the current location, preventing seawater from flooding the pine forest will be a challenge. Although this is not shown in detail here, many pine trees died due to seawater inflow during this time. This indicates the loss of ecosystem services provided by the pine forests. Thus, changing the position of the embankment causes a trade-off in ecosystem services. Such trade-offs of

Case with a coastal embankment

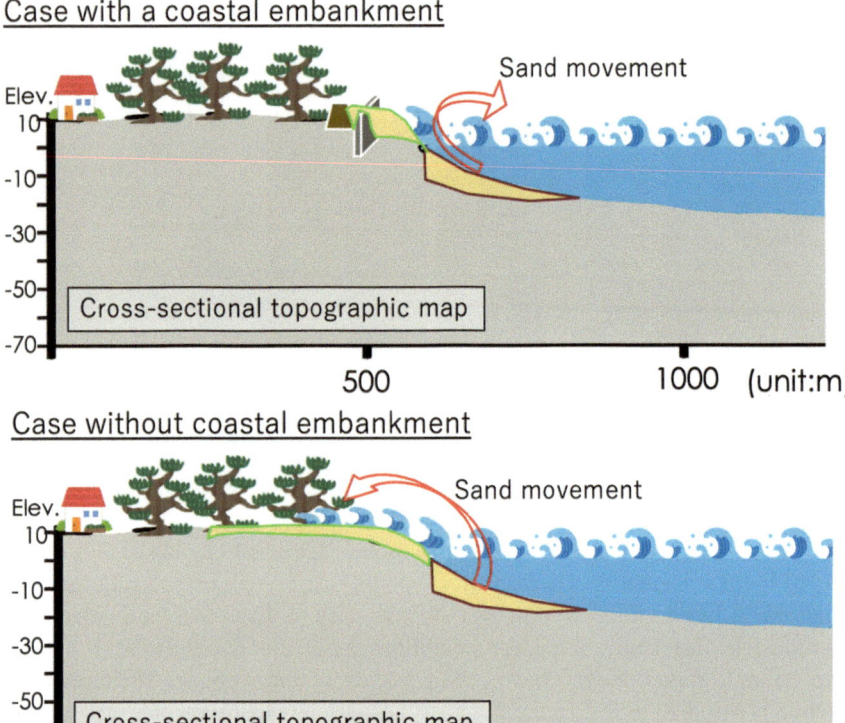

Case without coastal embankment

Fig. 19.12 Relationship between coastal embankments and sand movement due to meteorological events in hybrid infrastructure

ecosystem services brought by hybrid infrastructure should be considered during design. Even though various options have been proposed, residents' opinions are essential. In the Osato area, pine forests are essential for the sustainability of the community, and avoiding the loss of pine forests is the most important consideration. This area modified the coastal topography to increase the effectiveness of the coastal embankment, which is the gray infrastructure in the hybrid infrastructure. Waiting for nature to raise the elevation of the dunes and prevent the waves from inundating the pine forest is impractical. In other words, the time required for the formation of green infrastructure is another factor behind this choice. When designing hybrid infrastructures, it is also essential to understand how the functions of green infrastructure change over time.

Acknowledgments We sincerely thank Mr. Tadashi Saito for his assistance. We are grateful to Dr. Fumika Asanami, Prof. Yasunori Muto, Prof. Mahito Kamada, Prof. Futoshi Nakamura, Dr. Hajime Matsushima, and Dr. Yota Imai for helpful discussions. This research was performed by the Environment Research and Technology Development Fund (JPMEERF20184005) of the Environmental Restoration and Conservation Agency of Japan. This study was also supported by a Grant-in-Aid from the Ministry of Education, Science, Technology, and Sports of Japan (18K04659). Topographic data for tsunami analysis were provided by the Tokushima Prefecture and West Nippon Expressway Company Limited.

References

Baba T, Takahashi N, Kaneda Y, Ando K, Matsuoka D, Kato T (2015) Parallel implementation of dispersive tsunami wave modeling with a nesting algorithm for the 2011 Tohoku tsunami. Pure Appl Geophys 172:3455–3472

Mitsuyasu H, Tasai F, Suhara T, Mizuno S, Ohkusu M, Honda T, Rikiishi K (1975) Observation of the directional spectrum of ocean waves using a cloverleaf buoy. J Physical Oceanogr 5(4):750–760

Okada Y (1985) Surface deformation due to shear and tensile faults in a half-space. Bull Seismol Soc Am 75(4):1135–1154

Sutton-Grier AE, Wowk K, Bamford H (2015) Future of our coasts: the potential for natural and hybrid infrastructure to enhance the resilience of our coastal communities, economies and ecosystems. Environ Sci Pol 51:137–148

Tavakkol S, Lynett P (2017) Celeris: a GPU-accelerated open source software with a Boussinesq-type wave solver for real-time interactive simulation and visualization. Comput Phys Commun 217:117–127

Tokushima Prefecture: Awajiban. https://e-awajiban.pref.tokushima.lg.jp (in Japanese)

Chapter 20
Challenging a Hybrid Between Green and Gray Infrastructure: Coastal Sand-Covered Embankments

Hajime Matsushima and Xiangmei Zhong

Abstract While coastal areas are one of the most highly developed areas in the world, they have always been exposed to the danger of disaster. In response to the extreme events that increase every year owing to the effects of climate change, it is necessary to switch from coastal conservation that relies on existing infrastructure (gray infrastructure) to green infrastructure that utilizes ecotone. However, the reproduction of ecotones poses great challenges in many coastal areas. Therefore, we introduce a new approach to the complementary infrastructure development of gray and green functions, which involves conversion of the existing infrastructure, namely tide embankment, into dunes with sand-covering their slope, to be utilized as hybrid infrastructure. It was confirmed that the sand-covered embankments expanded the habitat of dune plants and protected the tide embankment body from high temperatures. This sand-covered embankment was also used as a place for environmental education for local elementary and junior high school students, and the multi-functionality of the hybrid infrastructure that grows over time was confirmed through the participation of various individuals.

Keywords Coast · Ecotone · Hybrid infrastructure · Eco-DRR · Sand-covered embankment · Environmental education · Disaster prevention education

20.1 Introduction

Millions of people rely on the proximity to the coast for their livelihoods because of their provision of opportunities for trade, economic activities, recreation, and livelihoods (Brown et al. 2014). It has been estimated that the percentage of the

H. Matsushima (✉)
Research Faculty of Agriculture, Hokkaido University, Sapporo, Japan
e-mail: matts@res.agr.hokudai.ac.jp

X. Zhong
Graduate School of Agriculture, Hokkaido University, Sapporo, Japan

world's population living in a near-coastal zone (below an elevation of 100 m, and closer than 100 km from the coast) was 28% (1.9 billion) in 2010 (Kummu et al. 2016).

Along with the population increase and rapid economic growth in coastal areas, there has been extensive conversion to urban and industrial zones, especially in the metropolitan inner bay areas, such as Tokyo Bay in Japan, where natural coastal lines that are mainly composed of tidal flats have disappeared due to development. Dense development along coastal areas has resulted in the loss of beaches and tidal flats as buffer zones between the sea and land, otherwise known as coastal ecotones. Furthermore, such developments are responsible for the deterioration of the inner bay environment, along with the loss of the water purification function of tidal flats as well as the functions of buffer zones to reduce the impact of natural disturbances such as strong wind, salt spray, and sand transportation.

This has resulted in a higher disaster risk due to the increased exposure of coastal communities to natural hazards, especially considering that storm surge disasters caused by extreme weather conditions and a rise in sea level are increasing every year. The IPCC estimated the global mean sea level (GMSL) rise in the Fifth Assessment Report (AR5). Under the most pessimistic scenario (RPC8.5), GMSL was likely to rise 0.45–0.82 m by 2081–2100. The impact of this GMSL rise seems small; however, it would lead to a loss of 83% of the Japanese coastal beach areas (Udo and Takeda 2017).

These predictions show that the developed coastal areas along the beach are exposed to high risks of disasters. To reduce disaster risks and exposure to natural hazards in developed coastal areas, it is necessary to construct larger and stronger structures, such as seawalls or tide embankments. These concrete structures have protected the coastal areas of Japan and supported their economic development throughout the latter half of the twentieth century.

The Ministry of Land, Infrastructure, Transport and Tourism in Japan reported that Japanese social capital stock management (allotted to build new infrastructure and/or maintain the existing ones) would not be sustainable beyond the year 2037. This alert was made based on a predictive budget shortage that resulted from increasing maintenance costs of concrete infrastructures as well as an aging and declining population. In particular, for shrinking cities (a designation for the urban areas becoming unpopulated due to population decline, a concerning phenomenon otherwise occurring in Japan), it is not feasible to construct larger and stronger concrete structures to deal with the situation. In other words, there is no choice besides using green infrastructure concepts to adapt to recent extreme weather events caused by climate change.

The concrete structures, otherwise known as "gray infrastructures" (for example, seawalls), have a single specific function (for example, coastal protection) and generally have superior effects as compared to natural ecosystems known as "green infrastructure (GI)" (for example, coastal sand dunes). However, many studies and reports have highlighted that multifunctional GI is often better than gray infrastructure in terms of overall capacity. GI helps protect and restore naturally functioning ecosystems and provides economic benefits such as facilitating an increase in property values and a decrease in the costs of public infrastructure and services. Investing in GI is often more cost effective than developing conventional public work projects such as flood control, water treatment systems, and storm water management (Benedict and McMahon 2002). However, the high degree of uncertainty in GI functions has been regarded as a problem when compared to those known for gray infrastructure (Onuma and Tsuge 2018). Since such discussions in the past were often unproductive and polarized between the gray and green approaches, it becomes necessary to consider a hybrid infrastructure that complementarily integrates both concepts (Nakamura et al. 2019).

This chapter outlines the functions of coastal ecotones as GI. We propose a hybrid infrastructure that functions complementarily to gray infrastructure, using seawalls on the Sendai coast, which was severely damaged by the tsunami disaster caused by the Great East Japan Earthquake on March 11, 2011, for instance.

20.2   Coastal Ecotones

Coastal landscapes, especially natural sandy coasts, are dynamic and ever-changing ecosystems in the vicinity of others such as shallow water bodies (coastal lagoons, estuaries, mangroves, fjords, etc.), sandy beaches, grasslands (dunes), shrubs, saltmarshes, forests, etc. (Fig. 20.1; Wootton et al. 2016). Such areas in which different biological communities, ecosystems, and biotic regions are in contact with each other are called ecotones. They are often steep transitional areas resulting

Fig. 20.1 Coastal ecotone of Ishikari coast, Hokkaido, Japan. Easily identifiable transition of ecosystem units by the distance from shoreline (differences in natural disturbance intensity)

from spatial gradients of abiotic and biotic environmental parameters (Kark 2013). In an ecotone, each ecosystem is maintained through a mutual relationship with the adjacent ecosystem. Therefore, it is important not only to protect individual ecosystems but also to maintain connectivity with the surrounding environment (Kark 2013).

Coastal ecotones have provided diverse ecosystem services. Liquete et al. (2013) defined the following services: provisioning services (fisheries, wood for buildings, and fuels), regulating and maintenance services (habitat for both fauna and flora, nutrient cycling, erosion prevention, waste-water treatment, moderation of extreme events), and cultural services (tourism, recreational, esthetic, and spiritual benefits), as listed in Table 20.1. These services also support the high resilience of coastal ecotones. Because sandy beaches are unstable landscapes, they are eroded by wind and waves, with the dunes covered with dune plants. These plants can cope with high sediment mobility by collecting airborne sand grains with their leaves and stems, protecting the sand from being blown away by the wind using their grass blades, and stabilizing the sand by fast-growing extensive and branched roots and rhizomes (Wootton et al. 2016). In addition, they can rapidly grow upward, surpassing sand burial and toward the sea, expanding the dune to zones of bare sand (Wootton et al. 2016). Therefore, these dune plants are responsible for dune formation and restoration after extreme weather events, coastal erosion, and human intervention (for example, off-road vehicles driven on coastal dunes) through autonomous resilience (Fig. 20.2). However, in order for dune plants to exert this autonomous resilience, it is necessary to not only protect the coastal ecotone itself but also manage the watershed sediment system in an integrated manner. In other words, it is important to maintain connectivity and sustainable dynamic stability, in which the ecosystem units that constitute the ecotone can function complementarily.

20.3   Recovery and Reconstruction from the Disaster of the Great East Japan Earthquake Using Gray Infrastructures

A catastrophic earthquake with Mw 9.0 occurred on March 11, 2011, at a depth of approximately 25 km and 130 km offshore of the northeastern part of Japan, and it was later named the Great East Japan Earthquake. This event had the largest magnitude ever recorded in Japan and continues to be the fourth-largest in the world since 1900. The earthquake also brought about a tsunami disaster, with an inundation area of approximately 500 km^2, registering 18,428 dead and missing people and 404,893 destroyed buildings (NPA 2020). Following the tsunami disaster in the coastal areas of eastern Japan, especially on the Pacific side of the Tohoku region, the Japanese government made a decision to construct larger and stronger concrete protection facilities to protect the country's inland areas from such natural hazards. Therefore, tide embankments were built in an extension of

Table 20.1 List and description of coastal ecosystem services

	Ecosystem services	Coastal-specific components
Provisioning services	Food provision	a. Fishing activities (including shellfishing) industrial or artisanal (either commercial or subsistence fishing). In general, fisheries are reported as total landings or catch per unit effort and, sometimes, corresponding jobs b. Aquaculture is the farming of aquatic organisms, including fish, crustaceans, mollusks, seaweeds, and algae
	Water storage and provision	a. Water abstraction in marine and coastal environments is mostly associated with coastal lakes, deltaic aquifers, or desalination plants b. Marine water may also be used for industrial cooling processes or coastal aquaculture in ponds and raceways
	Biotic materials and biofuels	a. This includes medicinal (e.g., drugs, cosmetics), ornamental (e.g., corals, shells), and other commercial or industrial resources (e.g., whale oil, fishmeal, seal leather, algal or plant fertilizers) b. Biomass to produce energy can have a solid form (like wood from mangroves), liquid (like fuels extracted from algal lipids or whale oil), or biogas (from decomposing material)
Regulating and maintenance services	Water purification	Treatment of human wastes (e.g., nitrogen retention); dilution; sedimentation, trapping, or sequestration (e.g., of pesticide residues or industrial pollution); bioremediation (e.g., bioaugmentation after marine oil spills); oxygenation of "dead zones"; filtration and absorption; remineralization; and decomposition
	Air quality regulation	Vegetation (e.g., in mangroves), soil (e.g., in wetlands), and water bodies (e.g., open ocean), due to their physical structure and microbiological composition, absorb air pollutants like particulate matter, ozone, or sulfur dioxide
	Coastal protection	Natural defense of the coastal zone against inundation and erosion from waves, storms, or sea-level rise. Biogenic and geologic structures that form the coastal habitats can disrupt the water movement and, thus, stabilize sediments or create buffering protective zones
	Weather regulation	For example, the influence of coastal vegetation and wetlands on air moisture and, eventually, on the saturation point and the formation of clouds

(continued)

H. Matsushima and X. Zhong

Table 20.1 (continued)

	Ecosystem services	Coastal-specific components
	Life cycle maintenance	The maintenance of key habitats that act as nurseries, spawning areas, or migratory routes (e.g., seagrasses, coastal wetlands, coral reefs, mangroves). These habitats and the connectivity among them are crucial for the successful life cycle of species. This also includes pollination (e.g., mangrove pollination) and seed and gamete dispersal by organisms. This service guarantees the maintenance of genetic diversity or gene pool protection
	Biological regulation	Control of fish pathogens especially in aquaculture installations; role of cleaner fishes in coral reefs; biological control on the spread of vector-borne human diseases; and control of potentially invasive species
Cultural services	Symbolic and aesthetic values	Coastal communities have always shown strong bonds to the sea due to the local identity. Natural and cultural sites linked to traditions and religion are numerous in the coastal zone. Both coastal and inland societies value the existence and beauty of charismatic habitats and species such as coral reefs or marine mammals
	Recreation and tourism	The appeal of marine ecosystems is usually linked to wilderness, sports, or iconic landscapes and species. It can be related to coastal activities (e.g., bathing, sunbathing, snorkeling, scuba diving) and offshore activities (e.g., sailing, recreational fishing, whale watching)

Modified from Liquete et al. (2013)

approximately 400 km along the coastlines, thereby replacing the dunes in the process of self-sustaining recovery (Fig. 20.3). The total cost of constructing these tide embankments was over $10 billion. Additionally, continuous maintenance costs are also required every year to ensure its integrity and functionality. Although these seawalls ensure a certain level of safety in coastal areas, it has been highlighted that they would prevent the connectivity of ecotones and degrade the resilience of dune ecosystems (Nishihiro et al. 2014).

To make things even worse, black pine trees were replanted in the hinterlands of the seawalls, with an embankment of mountain soil. Coastal forests have been planted in this area for a long time, but most of them were uprooted and/or washed away by the tsunami (Sakamoto 2012; Tanaka et al. 2013). The damage incurred in the trees was a result of the high groundwater levels. This prevented the growth of the plants' root system, thereby restraining the taproot (vertical root) from burrowing deep and developing plate roots (horizontal roots) in thin soil layers (Oda 2001; Hirano et al. 2018). Moreover, the plate root system exhibits lower resistance to uprooting (Dupuy et al. 2005), due to which the black pine trees could not endure the tsunami forces. On account of this, 2-m-high embankments with mountain soils were placed over the sand dunes before replanting new trees to ensure sufficient soil

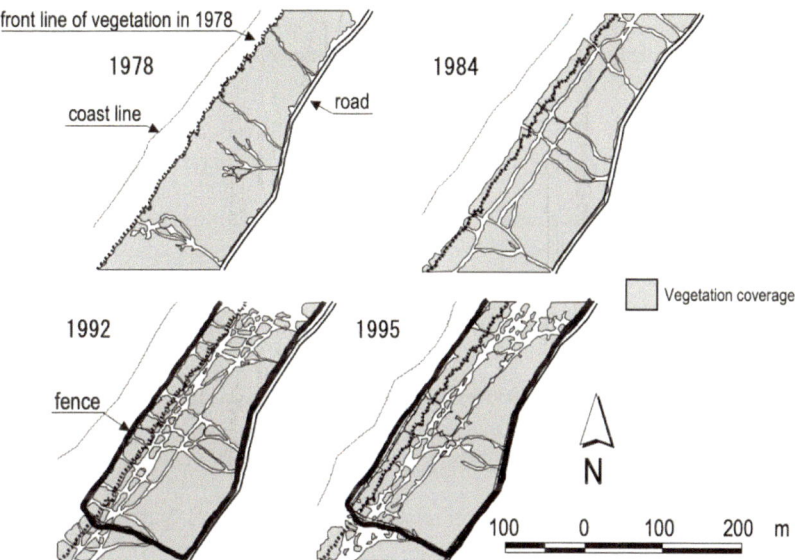

Fig. 20.2 Resilience of coastal dunes on Ishikari coast, Hokkaido, Japan. (**a**) Foredunes were eroded by winter storm in 2010; (**b**) dune plants recovered on the beach and regrowth foredunes within several years; (**c**) impact of an off-road vehicle driven on the coastal dune vegetation. The track of off-road vehicles increased each year until 1992. Once the coastal manager protected these dunes using a fence, dune vegetations recovered 1% year^{-1} autonomously (Matsushima et al. 2000)

layers for the forest beds. This strategy aimed to encourage the growth of deeper and more vast root systems for holding up a tree and soil particles together.

Adding mountain soil to dune systems causes drastic environmental changes. For example, hardening and densification of mountain soil embankments reduce water permeability and promote the formation of puddles on the surface, which ironically results in a significant inhibition of black pine growth (Ono et al. 2016). Moreover, the seed bank contained in the mountain soil increases the number of inland exotic plants in coastal areas. As a result, natural habitats in coastal ecosystems have been

Fig. 20.3 Coastal dunes and seawalls in Yuriage coast, Miyagi, Japan. (**a**) Autonomously recovered dunes with coastal plants (*Calystegia soldanella, Leymus mollis, Carex kobomugi*, etc.), June 17, 2011; (**b**) The dunes were replaced by seawalls at the same place as (**a**), August 18, 2019

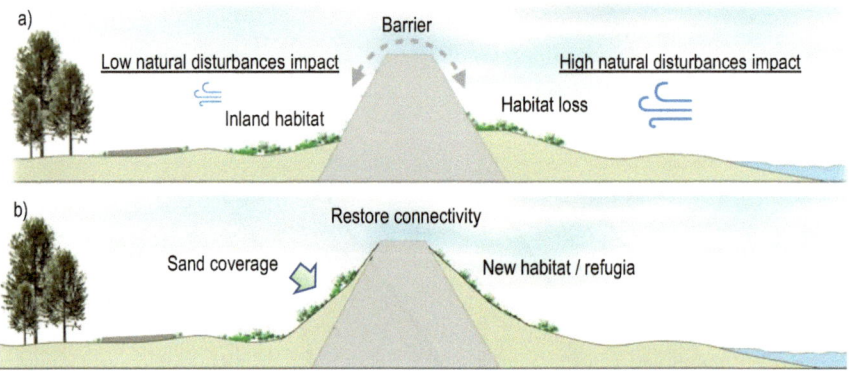

Fig. 20.4 Images of interrupted and connected coastal ecotones. (**a**) Embankment-interrupted seaside and hinterland environment; (**b**) sand coverage on the embankment slopes will restore the connectivity and create new habitats and refugia

lost or deteriorated, and their biological communities have suffered changes in the composition of species.

As a result of these recovery and reconstruction processes, the connectivity of coastal ecotones was interrupted by seawalls, and each ecosystem unit was drastically changed by embankment, as shown in Fig. 20.4a.

20.4 Sand-Covered Embankments as Hybrid Infrastructure for Coastal Community Resilience

To rehabilitate coastal ecotones and reduce disaster risk and exposure to natural hazards, it is necessary to reduce urban land-use along the coastlines and maintain buffer zones in their natural condition (Matsushima and Ferreira 2022). However, it may be difficult to maintain or gain enough space to restore dunes in highly urbanized coastal areas. For instance, residence relocation is not easy, but vacant houses are increasing in urban areas due to a population decline. Moreover, fallow lands are also increasing each year. Therefore, coastal protection and management solutions must be considered thoroughly.

On the other hand, it is relevant to consider what can be done in areas such as the Tohoku region, where seawalls have already been constructed. In Miyagi Prefecture, Japan, a pilot project is being conducted by citizens and a group of researchers, covering the seawalls by sand and turning it into a dune, since 2019 (Matsushima et al. 2019). Conventional studies have shown that coastal plants are required to be moderately disturbed by wind, waves, salt spray, and sand transportation (Moreno-Casasola 1986; Oka 2010). Therefore, the effect of wind reduction by seawalls leads to a decrease in natural disturbance from the sea to the hinterland, and as a result, promotes the invasion and settlement of inland (noncoastal) plants and woods. In addition, there was a high risk of habitat shrinkage or loss due to the severe impact of natural disasters at the beach located on the seaside of the seawalls. Therefore, it was considered necessary to restore the connectivity with sand over the tide embankments between the seaside and hinterland ecosystems and to create refugia for biological communities, as shown in Fig. 20.4b.

Traditionally, sand dune systems have been used to protect hinterlands in Japan, as in other countries. In some cases, sand dunes have been artificially formed and used as natural dikes (Perk et al. 2019). In the same line of reasoning, there also exists a conceptual hybrid plan that combines the strength of a seawall with the multiple functions of a sand dune (Almarshed et al. 2020).

The pilot project in Miyagi prefecture began on June 5, 2019, with the cooperation of the Sendai Civil Engineering Office, in Miyagi Prefecture. Using the dredged sand from the Nanakita River estuary sandbar, the sand was covered over an area of 150 m, with a thick layer of 30 cm, on both slopes of the tide embankments (T.P. 7.2 m, Slope gradients 1:2). The Nanakita River constitutes the same sediment transport system as the pilot site. It was also confirmed in advance that the sediment used had the same particle size composition and pH value as the pilot site beach. To reduce the mixing of mud as much as possible, the surface sand of the estuary sandbar was used as the dredged sand. Since it was collected from a place strongly affected by waves, the EC value showed a considerably high average of ten times, and more compared to the sand around the pilot site. However, 1 year later, it indicated the same value as the surrounding area.

As a result, the vegetation coverage in the sand-covered area increased fivefolds during 2019–2020, as shown in Fig. 20.5. It displayed the efficiency to promote vegetation coverage to cover the seawall with sand. This hybridization of green and gray infrastructure is expected to ensure the connectivity of ecotones and generate new habitats, in addition to the disaster protection function. The seawall will also be protected from time deterioration effects, namely those caused by wave attacks, sand erosion, ultraviolet radiation, corrosion due to sea breeze, and sun heating. In fact, when comparing the surface temperature of the seawall with and without sand

Fig. 20.5 Vegetation coverage area after 1 year (2020). (**a**) Hinterland side slope; (**b**) seaside slope

cover, the daily range sometimes became 30 °C or more without sand cover, but remained within 5 °C under sand cover. A moisturizing effect was also expected because the plants were settled in. Consequently, the lifespan of this seawall will be extended by hybridization.

This pilot project was conducted with citizen group "Kita-no-satohama Hana-no-kakehashi network." This group involved local residents, the elementary school, and junior high school students as a segment of their educational program. It is imperative to connect local residents to the coast. This is because many coastal areas have been designated as disaster risk areas and are being separated from daily life due to them being dangerous places, especially for young children. In this program, students went to the beach to learn coastal environments and collected the seeds of coastal plants to grow seedlings in their schools. Furthermore, the next students came to transplant grown seedlings on these sand-covered embankments to promote "hybrid infrastructure." This project was called "raise the tide embankment" by local residents and students. They learnt more about coastal environments while also inculcating disaster prevention education and learning from local history, culture, and identities. From an infrastructure maintenance aspect, citizens would be an accessible and feasible source of plant maintenance. In addition, because elementary and junior high schools enroll new students every year, it can be stated that this is a developmental project that guarantees sustainability and involvement of a greater number of people each year.

20.5 Conclusion

Needless to say, it is important to preserve coastal ecotones and utilize their functions. On the other hand, the scope of hybrid infrastructures such as sand coverage on the tide embankments that complementarily utilize the functions of green and gray infrastructures has also become apparent. Therefore, by covering the tide embankments with sand, it is possible to nourish the ecotones while maintaining the required disaster prevention and reduction functions.

In addition, the sand-covered embankments allow the participation of various people. In the case of conventional gray infrastructure, it was difficult for local citizens to participate in its maintenance. However, in the case of these sand-covered embankments, even elementary school students can participate in the regeneration process of the native dune plant community. Currently, it is difficult for children to approach these coastal areas designated as disaster risk areas, but through such maintenance of the sand-covered embankments, it is possible to touch and learn about the natural environment of these areas. Furthermore, they will be able to learn not only about the natural environment but also about the local history and culture that has lived and grown with nature. It will also be used as a place for disaster prevention and reduction education.

Sand-covered embankments, a hybrid infrastructure aimed at restoring ecosystems, could contribute to the establishment of new, resilient coastal communities. This effort has just begun. We would like to collect more data and establish a new infrastructure that is multifunctional, allows various people to participate, and grows over time.

References

Almarshed B, Figlus J, Miller J, Verhagen HJ (2020) Innovative coastal risk reduction through hybrid design: combining sand cover and structural defenses. J Coastal Res 36(1):174–188

Benedict MA, McMahon ET (2002) Green infrastructure: smart conservation for the 21st century. Renew Resour J 20(3):12–17

Brown S, Nicholls R, Hanson S et al (2014) Shifting perspectives on coastal impacts and adaptation. Nat Clim Chang 4:752–755. https://doi.org/10.1038/nclimate2344

Dupuy L, Fourcaud T, Stokes A (2005) A numerical investigation into the influence of soil type and root architecture on tree anchorage. Plant Soil 278:119–134

Hirano Y, Todo C, Yamase K, Tanikawa T, Dannoura M, Ohashi M et al (2018) Quantification of the contrasting root systems of *Pinus thunbergii* in soils with different groundwater levels in a coastal forest in Japan. Plant Soil 426:327–337

Kark S (2013) Effects of ecotones on biodiversity. In: Levin SA (ed) Encyclopedia of biodiversity, vol 3, 2nd edn. Academic Press, Waltham, pp 142–148

Kummu M, Moel H, Salvucci G, Viviroli D, Ward PJ, Varis O (2016) Over the hills and further away from coast: global geospatial patterns of human and environment over the 20th–21st centuries. Environ Res Lett 11:034010

Liquete C, Piroddi C, Drakou EG, Gurney L, Katsanevakis S, Charef A et al (2013) Current status and future prospects for the assessment of marine and coastal ecosystem services: a systematic review. PLoS One 8(7):e67737. https://doi.org/10.1371/journal.pone.0067737

Matsushima H, Ferreira SMF (2022) Urbanization of Coastal Areas: loss of coastal dune ecosystems in Japan. In: Gonçalves SC, Ferreira SMF (eds) Sandy beaches as endangered ecosystems: environmental problems and possible assessment and management solutions. CRC Press. (in printing)

Matsushima H, Aikoh T, Kondo T, Asakawa S (2000) A study on the change of beach plants cover area on Ishikari coast. Papers on Environmental Information Science 14:295–300. In Japanese with English summary

Matsushima H, Suzuki A, Kimura K, Zhong X, Hirabuki Y (2019) Greening on the seawall: challenge to conversion from gray to green infrastructure. In: Proceedings of the JpGU meeting 2019: H-DS10-03

Moreno-Casasola P (1986) Sand movement as a factor in the distribution of plant communities in a coastal dune system. Vegetation 65:67–76

Nakamura F, Ishiyama N, Yamanaka S, Higa M, Akasaka T, Kobayashi Y et al (2019) Adaptation to climate change and conservation of biodiversity using green infrastructure. River Res Applic 36:921–933

Nishihiro J, Hara K, Hirabuki Y (2014) Biodiversity conservation and infrastructure reconstruction after a large-scale disaster: lessons from the coastal regions of southern Sendai Bay. Jpn J Conserv Ecol 19:221–226. (in Japanese with English summary)

NPA (National Police Agency) (2020) Police measures and damage situation of the Tohoku-Pacific Ocean Earthquake in 2011. Press release on June 10th, 2020

Oda T (2001) Study on the reaction of planted tree root systems to water-logging and its application to developing forests in damp lowlands of coastal sand dunes. Spec Bull Chiba Pref For Res Center 3:1–78. (in Japanese with English summary)

Oka K (2010) Coastal environment and biodiversity : a case study of conservation and restoration of coastal sand dune vegetation. J Japan Soc Reveget Technol 35(4):503–507. (in Japanese)

Ono K, Imaya A, Takahashi K, Sakamoto T (2016) Evaluation of the berms built on the Restoration of the Mega-Tsunami-Damaged Coastal Forests—Comparison with the effects of soil-scratching as a soil physical correction method among the various types of machinery. Bull Forestry Forest Prod Res Inst 15:65–78. (in Japanese with English summary)

Onuma A, Tsuge T (2018) Comparing green infrastructure as ecosystem-based disaster risk reduction with gray infrastructure in terms of costs and benefits under uncertainty: a theoretical approach. Int J Disaster Risk Reduct 32:22–28

Perk L, van Rijn L, Koudstaal K, Fordeyn J (2019) A rational method for the design of sand dike/dune systems at sheltered sites; Wadden Sea coast of Texel, The Netherlands. J Mar Sci Eng 7:324

Sakamoto T (2012) Regeneration of coastal forests affected by tsunami. Forestry and Forest Products Research Institute

Tanaka N, Yagisawa J, Yasuda S (2013) Breaking pattern and critical breaking conditions of Japanese pine trees on coastal sand dunes in huge tsunami caused by great East Japan earthquake. Nat Hazards 65:423–442

Udo K, Takeda Y (2017) Projections of future beach loss in Japan due to sea-level rise and uncertainties in projected beach loss. Coast Eng J 59(2):1740006-1–1740006-16. https://doi.org/10.1142/S057856341740006X

Wootton L, Miller J, Christopher MMS, Peek M, Williams A, Rowe P (2016) Dune manual. Sea Grant Consortium, New Jersey

Chapter 21
Green Infrastructures in Megacity Jakarta: Current Status and Possibilities of Mangroves for Flood Damage Mitigation

Yukichika Kawata

Abstract Jakarta is Southeast Asia's most disaster-vulnerable city and recently has suffered from severe floods almost every year. Flood events have worsened mainly because of ground subsidence, reduction of green spaces, and littering into streams. The mangrove forests remaining in coastal urban areas of northern Jakarta can be utilized for mitigating flood damage. In the first two sections of this chapter, an overview of flood events is presented and the status of mangroves in Jakarta Bay is introduced. Mangroves may serve as hard and soft measures for disaster reduction. Because enlargement of the area of mangroves in urban Jakarta is difficult, in the last section, soft measures, especially the utilization of mangroves as an icon for disaster reduction, are proposed. Activities such as the incorporation of a mangrove design as a logo in documents of disaster prevention education may enhance citizens' recognition of disaster risks because they will see both logos and mangroves repeatedly in real life.

Keywords Floods · Icon for disaster reduction · Jakarta · Mangroves · Vulnerability

21.1 Introduction

Despite the high risk of natural disasters, people prefer to reside in coastal regions globally. Approximately 40% of the world's population is estimated to live within 100 km of the coast (United Nations 2017). In Indonesia, 65% of people live in coastal areas (Ilman et al. 2011), where rapid population increase and economic development have been observed. The special capital region of Jakarta (hereafter referred to as Jakarta) is Southeast Asia's most populous city, and its population

Y. Kawata (✉)
Faculty of Economics, Kindai University, Osaka, Japan
e-mail: ykawata@kindai.ac.jp

© The Author(s) 2022
F. Nakamura (ed.), *Green Infrastructure and Climate Change Adaptation*,
Ecological Research Monographs, https://doi.org/10.1007/978-981-16-6791-6_21

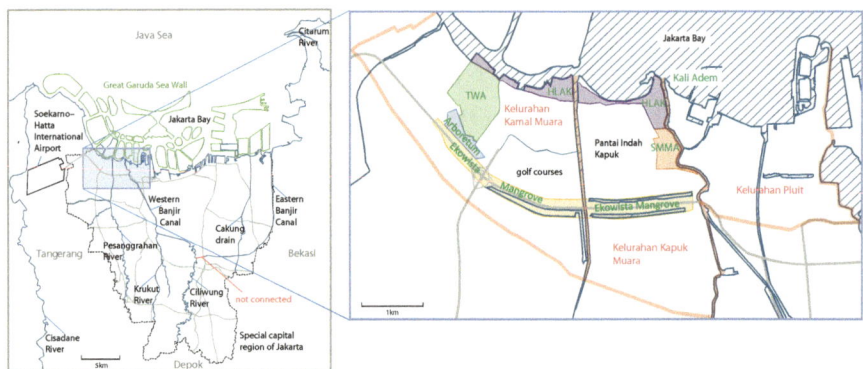

Fig. 21.1 Maps of Jakarta (left) and mangrove areas in Angke Kapuk (right)

size tripled in the past 50 years (Fig. 21.1). According to the Indonesian Disaster Database (DIBI), the majority of natural disasters are hydro-meteorological events (Badan Nasional Penanggulangan Bencana [BNPB] 2021). Based on the DIBI, flooding accounts for a large proportion of all disaster-related deaths (28.0%), injuries (37.6%), and houses destroyed (35.2%) and evacuated (95.5%) in Jakarta after 2000.

Natural disaster risk is the product of natural hazards, exposure, and vulnerability (United Nations Development Programme 2004). While the occurrence of natural hazards such as tsunami is uncontrollable, exposure is controllable but may not be in the short-term. A large number of poor and marginalized people settle in coastal areas of Asian countries (Pomeroy et al. 2006; Winsemius et al. 2018). In Jakarta, the poorest and most vulnerable people live in coastal areas illegally, and those under the most severe conditions suffer tidal flooding every day. They are reluctant to change settlements because of high land prices in safer places and good accessibility to their workplaces; they frequently return to these locales if they are forced into involuntary resettlement (Marschiavelli 2008; Padawangi 2012; Marfai et al. 2015). Vulnerability is the most controllable factor in present-day Jakarta, and hard and soft measures are applicable. Hard measures comprise artificial measures, including seawalls (gray infrastructures), and natural measures, including mangroves, which possess inherent coastal-disaster prevention functions (green infrastructures, GI) (Tanaka 2009; Nurhidayah and McIlgorm 2019), while soft measures minimize vulnerability through, for example, disaster prevention education.

Because many developing countries are located in tropical/subtropical regions, it is feasible to utilize GI, such as mangroves, coral reefs, and seagrass meadows, as ecological measures for disaster mitigation; this is referred to as Ecosystem-based Disaster Risk Reduction (Eco-DRR). By so doing, developing countries can ease the financial difficulty of constructing gray infrastructures (Esteban et al. 2017; Takagi 2019). Other advantages of GI include self-maintenance (Gedan et al. 2011) and self-repairing functions when damaged (Ferrario et al. 2014). Mangroves exert and enhance ecological resilience in coastal areas (Alongi 2008; Gilman et al. 2008;

Takagi 2019). For example, mangroves can migrate in response to sea-level rise (SLR) and sustain coastal ecological resilience (Nyström and Folke 2001; Alongi 2008).

This chapter selects Jakarta as a coastal, highly-populated city with high vulnerability and reviews the current status and possibilities of urban mangroves in Jakarta Bay for natural disaster reduction. Currently, the range of mangrove habitat in urban Jakarta Bay is limited, and enlargement of habitat may be difficult. Therefore, this chapter also briefly discusses the potential role of mangroves as an icon in disaster prevention, that is, a kind of soft measure, utilizing mangroves as a visual cue for reminding the population of natural hazard risks.

21.2 Floods in Jakarta

21.2.1 Most Vulnerable People

Jakarta is located in the northwest of Java Island (Fig. 21.1). Population size has exceeded ten million. Thirteen rivers discharge into Jakarta Bay. The biggest are the Ciliwung and Cisadane rivers, and nine rivers are considered to be associated with floods (Marfai et al. 2015). Yusuf and Francisco (2009) provide a list of 132 vulnerable provinces in Southeast Asian countries, of which part of Jakarta's municipalities are ranked 1st (Central Jakarta), 2nd (North Jakarta), 3rd (West Jakarta), 5th (East Jakarta), and 8th (South Jakarta).

Globally, the poorest people are the most vulnerable to natural hazards (Esteban et al. 2013; Tauhid and Zawani 2018). In Jakarta, the poor gather in informal urban settlements called kampung,[1] developed along, for example, riverbanks (Padawangi 2012), where people suffer from higher vulnerability than other urban residents (Firman et al. 2011; Arkema et al. 2013; Colven 2020). Jakarta's poor population was 0.373 million (3.57%), and the percentage is the highest in North Jakarta (5.35%) (Badan Pusat Statistik [BPS] 2018).

21.2.2 Causes of Floods

The frequency and magnitude of floods have been higher since the 2000s, and the 2007 flood was the worst in the past three centuries (Marfai et al. 2015; Colven 2020) (Fig. 21.2). The factors influencing the severity of floods are (1) heavy precipitation in the wet season, (2) SLR, (3) El Niño–Southern Oscillation (ENSO), (4) urbanization, and (5) accumulation of garbage in streams (Steinberg 2007;

---

[1] The term kampung is not a synonym for slum (Irawaty 2018) and does not necessarily consist of only poor people. However, kampung indicates the settlement of the poor in this chapter.

Fig. 21.2 Submerged roads near the Citraland Mall by the 2020 Jakarta Floods (taken by the author on January 1, 2020)

Marfai et al. 2015; Budiyono et al. 2016; Nurhidayah and McIlgorm 2019). The SLR in Jakarta coasts is 0.57 cm/year (Firman et al. 2011), which exacerbates flood damage in association with subsidence. The ENSO may also influence precipitation and inter-annual oscillations in sea surface height, causing more severe floods (Budiyono et al. 2016; Esteban et al. 2017; Takagi 2017).

Urbanization includes (4a) losses and degradations of urban lakes, called setu or situ, in the Ciliwung and Cisadane river watersheds (Henny and Meutia 2014), (4b) decrease of green areas and agricultural lands (Permatasari et al. 2016; Purwoko et al. 2016), and (4c) land subsidence. Approximately 10–20% of urban lakes have been lost (Henny and Meutia 2014), and the green space ratio reduced from 40% in 1985 to 9% in 2002 (Steinberg 2007), which may increase surface runoff, facilitate land subsidence, and cause downstream inundation (Firman 2009). Subsidence is mainly attributed to groundwater extraction and is considered to be the most important factor affecting future flood events (Budiyono et al. 2016; Takagi et al. 2016b). The highest subsidence has observed near the coast at a rate up to 15–20 cm/year between 1995 and 2005 (Chaussard et al. 2013). Moreover, (4d) sedimentation of silt in the downstream has worsened the flood events (Colven 2020).

Fig. 21.3 Accumulation of garbage in Waduk Pluit (Pluit Reservoir, downstream of Ciliwung river) (taken by Dr. Hajime Matsushima on March 17, 2015)

Finally, (5) Jakarta has a long history of garbage dumping in canals and rivers from the age of Batavia (1619–1942) (Steinberg 2007), which reduces river flows (Fig. 21.3). Currently, 57.0–85.4%, 8.1–31.4%, and 0.4–19.9% of wastes are from houses, industry, and livestock, respectively (Purwoko et al. 2016). The total volume of garbage discharged is 23,400 m^3/day, of which 14,700 are disposed by the City Sanitation Office, but some portion of the rest may be thrown into streams (Steinberg 2007). Furthermore, many do not receive garbage collection services (Colven 2020). Because riverbank settlers throw garbage into the river, the width of the Angke river was diminished from 40–60 m to 5–10 m (Steinberg 2007). A large proportion of rainwater from upstream contains garbage and may worsen downstream floods (Purwoko et al. 2016).

21.2.3 Current Countermeasures

Some countermeasures have been implemented. First, river dredging and expansion have been conducted as river normalization projects (Colven 2017, 2020). A flood canal, called Western Banjir Canal (WBC), was constructed and connected to the downstream of the Ciliwung river in the early twentieth century. The Eastern Banjir Canal (EBC) was developed to further mitigate floods and was utilized after 2011, but the plan to connect the WBC and EBC remains unrealized (the part indicated

Fig. 21.4 River cleaning activities by local government in Western Banjir Canal (downstream of the Ciliwung river) near the Seasons City Trade Mall (taken by the author on November 30, 2019)

by a red line in the left-hand chart of Fig. 21.1). Second, the central government plans to make Ciawi and Sukamahi retention basins in Bogor district to reduce the water volume of the Ciliwung river. Third, the National Capital Integrated Coastal Development program is ongoing, which includes the construction of the Great Garuda Sea Wall (GGSW, 32 km) in the gulf of Jakarta and the construction of artificial islands, including one with the shape of Garuda[2] (more than 1000 ha) (Colven 2017; Nurhidayah and McIlgorm 2019). The construction of GGSW is controversial because it may cause damage to coastal fisheries and mangroves and does not mitigate subsidence (Colven 2017).

Other activities include the removal of waste and trash from rivers and canals by Jakarta's Public Facility Maintenance Agency (Figs. 21.4 and 21.5). Local communities also conduct river and canal cleaning voluntarily (Marschiavelli 2008; Marfai et al. 2015; Purwoko et al. 2016). In coastal and riverside kampung, people living in single-storied houses add a second floor as a place of refuge (Marschiavelli 2008; Marfai et al. 2015; Budiyono et al. 2016; Wijayanti et al. 2017). Takagi et al. (2016a) recommend escaping to the second floor rather than moving long distances to higher places when dykes break. Some people in Kali Adem, a coastal

---

[2] A legendary bird used in the Indonesian national emblem.

Fig. 21.5 River cleaning activities by pasukan orange (orange troops) of the Jakarta's Public Facility Maintenance Agency at Saluran Cideng (Cideng waterway) (taken by the author on October 4, 2019)

village, elect to live in fishing boats to avoid floods (Padawangi 2012). Some local communities utilize mosques as evacuation sites (Marfai et al. 2015).

21.3   Mangroves in the Gulf of Jakarta

21.3.1   Recognition of Mangroves as GI in Indonesia

The advantage of GI is gradually being acknowledged in Indonesia. The 1992 Flores Island Earthquake Tsunami highlighted the bio-shield function of mangroves in Indonesia (Karminarsih 2007). After the 2004 Aceh Tsunami, Indonesians recognized the term tsunami much more (Baeda et al. 2015) and became more aware of mangrove's Eco-DRR function (Karminarsih 2007). The Indonesian government tried to create a green belt on the Banda Aceh coast after the 2004 Aceh Tsunami, although it remains incomplete (Dalimunthe 2018). Simultaneously, the Minister of Forestry launched a plan to reforest 600 thousand hectares of depleted mangroves (Barbier 2008). The government proclaimed the National Strategy for Mangrove Ecosystem Management (Indonesian Government 2012). Mangrove rehabilitation program Ayo Tanam Mangrove (Let's Plant Mangrove) has been implemented nationwide by the Ministry of Fisheries and Marine Affairs for more than a decade, and mangrove afforestation is regarded as a prospective way of improving Eco-DRR in Indonesia (Dalimunthe 2018).

21.3.2 Functions and Status of Mangroves

Mangroves are a group of halophytic plants inhabiting the intertidal zone of tropical and subtropical regions. Giri et al. (2011) estimate the world's mangrove area is 137,760 km^2 in 2000, of which 22.6% (31,130 km^2) is in Indonesia. Indonesian mangroves are distinguished by having the largest number of species and the largest area (Ilman et al. 2011), but about 30% of their area has been lost since 1980 (Giesen et al. 2006; Murdiyarso et al. 2015) mainly because of transformation to shrimp ponds (63%) and agricultural lands (32%) (Giri et al. 2008).

Mangroves may protect coastal areas from tsunamis, tidal surges, floods, and cyclones (or typhoons, hurricanes) by mitigating waves and winds and minimize damage to property and human life (e.g., Danielsen et al. 2005; Kathiresan 2012; Unnikrishnan et al. 2013; Sandilyan and Kathiresan 2015). Some types of mangroves have prop roots or pneumatophores, which strengthen their durability against winds and facilitate the accretion of sediment, resulting in wave attenuation (Alongi 2008; Ostling et al. 2009). Accumulation of sediment preserves or even expands the coastline (Marois and Mitsch 2015). The sturdy trunks and extensive root complexes of mangroves, along with accumulated sediments, mitigate the speed of flood, and sediments also absorb floodwater (Ilman et al. 2011). Moreover, mangroves mitigate the impact of dyke-breaking tsunamis, a kind of flooding "caused by the rupture of a coastal dyke" (Takagi et al. 2016a, p. 1630). People may save their lives from tidal waves by climbing, clinging to, or hanging in mangrove trees (Forbes and Broadhead 2007; Sandilyan and Kathiresan 2015).

21.3.3 Mangroves in North Jakarta

The most famous and extensively examined mangroves in Jakarta Bay are Mangrove Angke Kapuk (MAK) in Kelurahan Kamal Muara (Kamal Muara village) and Kelurahan Kapuk Muara (Kapuk Muara village) of Kecamatan Penjaringan (Penjaringan district) (Fig. 21.6; right-hand chart of Fig. 21.1). The MAK consists of mangrove forests in (1) Hutan Lindung Angke Kapuk (HLAK, Angke Kapuk protected forest, 44.76 ha), (2) Taman Wisata Alam (TWA, Angke Kapuk national park, 99.82 ha), (3) Arboretum (10.51 ha), (4) Suaka Margasatwa Muara Angke (SMMA, Muara Angke Wildlife Reserve, 25.02 ha), and (5) Ekowista Mangrove (95.50 ha) (Kusmana et al. 2013). There are 11 true (exclusive) mangroves in MAK, and the dominant species are *Avicennia* spp. and *Rhizophora* spp. (Pambudi et al.

Fig. 21.6 Riparian forests in HLAK (taken by Dr. Emma Colven on May 25, 2019)

2018). Some residential areas including Pantai Indah Kapuk (PIK), a relatively wealthy area, and golf courses are also located near the MAK. The east side of the MAK is Kelurahan Pluit (Pluit village), which is another village of Penjaringan district and contains mangrove forest (6) Kali Adem.

The total mangrove area in Jakarta has decreased from 1165.33 ha in 1980 to 165.28 ha in 2016 (Hilmi et al. 2017). Mangroves in Angke Kapuk and adjacent areas have been transformed into residential and industrial areas, highways, an airport, and fish ponds since the late 1970s (Ambinari et al. 2016; Kusmana 2017). Wastes and pollutants from upstream and residential areas in MAK are also significant factors of mangrove degradation (Hilmi et al. 2017; Sofian et al. 2020). One of the main wastes is plastic debris in MAK (Cordova and Nurhati 2019; Sofian et al. 2020), mostly polystyrene (44.62%), a polymer used for food packaging (Purwoko et al. 2016; Cordova et al. 2021). Plastic debris prevents air uptake by pneumatophores (Purwoko et al. 2016) and reduces the strength of mangrove roots (Sofian et al. 2020) and interferes with mangrove regeneration by inhibiting the growth of seeds (Sasongko et al. 2014; Putri et al. 2015). Other pollutants, including heavy metals, nitrogen, and phosphorus, affect some sorts of mangroves (Hamzah and Setiawan 2010; Ambinari et al. 2016; Purwoko et al. 2016; Pambudi et al. 2018). Furthermore, groundwater extraction, land subsidence, and associated changes of hydrological conditions may result in the loss of mangroves (Gilman et al. 2008; Ilman et al. 2016). Because of waste accumulation and pollution, mangrove diversity diminishes (Sasongko et al. 2014; Putri et al. 2015), and those tolerant to pollution such as *Avicennia marina* and *Rhizophora mucronata* dominate in the MAK (Wibowo 2006; Pambudi et al. 2018).

Several mangrove rehabilitation projects have been implemented in HLAK (Penyusun 2009; Sasongko et al. 2014), Jakarta Fishing Port in Pluit village (Takagi et al. 2016a), and Kali Adem (Rahadian et al. 2019). The rehabilitation at HLAK is conducted by a group Sahabat Bakau (Friends of Mangrove). They have constructed 1780-m-long stone breakwater, 100 m from the coast, and 8000 mangroves were planted (Penyusun 2009).

Mangrove degradation causes significant CO_2 emissions (Kathiresan 2012; Murdiyarso et al. 2015; Sandilyan and Kathiresan 2015) and the loss of natural disaster mitigation (Gilman et al. 2008; Giri et al. 2008). For example, the PIK was formerly a mangrove forest able to store excess rainfall and high tides, but this water storage function was lost, and the number of floods increased; thus, the toll road surface (roads go through the Ekowista Mangrove in Fig. 21.1) was raised by 1.2 m to prevent flood impacts (Ilman et al. 2011). Removal of mangroves also caused the loss of coastal sediments (Penyusun 2009).

21.4 Possibilities of Mangroves for Flood Damage Mitigation

To address Jakarta's current issues, including floods, subsidence, and loss of greenery (Sasongko et al. 2014), rehabilitation of mangroves is recommended, but suitable areas are limited. The most realistic option might be to conserve current mangroves in an appropriate state while emphasizing soft measures more. In this section, requirements and limitations of rehabilitation are summarized, and potential mangrove functions as soft measures are discussed.

21.4.1 Hard Measures: Rehabilitation of Mangroves for GI

Several points should be noted in promoting the restoration of mangroves. A minimum width or area is required to sustain mangroves, to maintain species richness, and to support the provision of ecosystem services (Duke et al. 2007; Ilman et al. 2011; McIvor et al. 2012). There is controversy as to whether urban and aquacultural areas can be restored to mangroves (Giri et al. 2008; Biswas et al. 2009; van Oudenhoven et al. 2015), and if a suitable area is not fully secured, mangroves may not exert tidal attenuation function (Takagi 2017). Locations of restored mangroves are critical for disaster reduction. Mangroves reduce the rate of fatalities of a tsunami when they are located in front of residential areas (Danielsen et al. 2005) but escalate damage when settled areas are located between the coast and mangrove forest (Bayas et al. 2011). Another restriction is coastal squeeze, which happens because of SLR and urbanization. Belize's example demonstrates that mangroves need to move landward 500 m for 1 m SLR (Guannel et al. 2016), but expansion inland is limited by many factors, including the presence of a developed land surface (Gilman et al. 2008; Pontee 2013; Marois and Mitsch 2015). Other conditions, including water depth and contamination level, may also be restrictive (Sasongko et al. 2014; Putri et al. 2015; Takagi et al. 2016a; Takagi 2019). Mangroves located in creeks, estuaries, and inlets may increase the height of a tsunami and accelerate the flow, resulting in more severe damage (Forbes and Broadhead 2007; Tanaka 2009; Power 2013).

Existing studies highlight the importance of considering biological diversity (Tanaka 2009; Takagi 2019), original vegetation (Karminarsih 2007; Purwoko et al. 2016), species that have a high tolerance of floods (Alongi 2008) and tsunamis (Kathiresan and Rajendran 2005; Tanaka 2009; Yanagisawa et al. 2009), and a mangrove landscaping strategy that effectively reduces the impacts of tsunamis and high waves (Hilmi 2018). Because polluted sediments and floating garbage may hinder the growth of mangroves, species with high tolerance to these factors are recommended (Putri et al. 2015; Pambudi et al. 2018). The most promising species for disaster prevention are *Avicennia marina* for floods (Alongi 2008) and *Rhizophora* spp. for tsunami (Kathiresan and Rajendran 2005; Yanagisawa et al. 2009), which dominate the MAK.

The current status of contamination and urbanization may be restrictive to any substantial improvement in mangrove diversity and area. A more promising idea may be to combine mangrove improvements with other remedies. Groundwater regulations contribute to diminishing subsidence and flood events (Budiyono et al. 2016; Esteban et al. 2017), although similar regulations have already been implemented in Jakarta, but in an unsatisfactory manner. Multiple lines of defense (Lopez 2009), whereby the coast is protected by a combination of seascape elements, including coral reefs, seagrass meadows, and mangroves (Guannel et al. 2016), attracts the attention of practitioners (Arkema et al. 2017) because multiple lines of defense may exert more disaster prevention functions (Barbier 2016; Guannel et al. 2016), along with the provision of varied ecological services, and establish a solid environmental complex.

21.4.2 Soft Measures: Mangroves as a Disaster Prevention Icon

Mangroves can be utilized more when applying soft measures for disaster reduction. The MAK may serve as an icon, similar to the panda used by the World Wide Fund for Nature as its icon or logo. In multiple activities involving citizens, authorities can utilize mangroves as an iconic cue to make citizens think of vulnerability to natural disasters and the necessity to stop littering.

First, risk-mitigating actions, including early-warning systems and disaster prevention education, should be emphasized more (Bayas et al. 2011; Marois and Mitsch 2015), and designs with mangroves could be adopted in documents of such disaster prevention initiatives. Some victims of the 2006 Java Tsunami did not flee to safer places when they saw the telltale drawdown because of a lack of disaster prevention education and early-warning systems (Kerr and Andrew 2007). Regardless of the existence of gray infrastructures, including concrete dikes, too much expectation for GI may lead to loss of life (Dahdouh-Guebas et al. 2005). Multiple malls scattered in Jakarta (Steinberg 2007) may serve as evacuation areas. Special consideration for those in kampung is necessary because they suffer floods every week (Marfai et al. 2015; Esteban et al. 2017; Colven 2020) and tend to underestimate their vulnerability. One feasible way of enhancing kampung residents' disaster prevention activities includes participation in mangrove rehabilitation programs, which may also improve their knowledge of disaster prevention.[3] Tanaka (2009) states that vegetated areas such as mangroves can be used in disaster prevention education by local temples or local authorities. Because Indonesians regularly hold religious meetings, reduction of littering and improvement of disaster prevention knowledge through education are feasible. The iconic mangroves, printed in educational materials, may enhance participants' recognition when used repeatedly.

Second, because deforestation of mangroves causes sedimentation of creeks, canals, and estuaries (Wolanski et al. 1992) and increases flood events, mangroves can be planted along upstream riverbanks. Riparian mangrove forests can also be utilized as an icon of disaster prevention for local residents.

Acknowledgments The author is grateful for constructive comments on an earlier version of the manuscript, especially from Futoshi Nakamura and Hajime Matsushima. The author also thanks Emma Colven and H. Matsushima for providing pictures.

References

Alongi DM (2008) Mangrove forests: resilience, protection from tsunamis, and responses to global climate change. Estuar Coast Shelf Sci 76:1–13. https://10.1016/j.ecss.2007.08.024

Ambinari M, Darusman D, Alikodra HS, Santoso N (2016) Penataan peran para pihak dalam pengelolaan hutan mangrove di perkotaan: Studi kasus pengelolaan hutan mangrove di teluk Jakarta. Jurnal Analisis Kebijakan Kehutanan 13(1):29–40. https://doi.org/10.20886/jakk.2016.13.1.29-40

Arkema KK, Guannel G, Verutes G, Wood SA, Guerry A, Ruckelshaus M, Kareiva P, Lacayo M, Silver JM (2013) Coastal habitats shield people and property from sea-level rise and storms. Nature Clim Change 3:913–918. https://doi.org/10.1038/nclimate1944

---

[3] To promote their participation, benefits, including the assignment of some rights, such as mangrove ownership (Biswas et al. 2009), financial payment for participation (Karminarsih 2007; Sasongko et al. 2014), and reflection of their opinion in rehabilitation programs (Padawangi 2012; Dalimunthe 2018; Martínez-Espinosa et al. 2020), should be emphasized more.

Arkema KK, Griffin R, Maldonado S, Silver J, Suckale J, Guerry AD (2017) Linking social, ecological, and physical science to advance natural and nature-based protection for coastal communities. Ann N Y Acad Sci 1399:5–26. https://doi.org/10.1111/nyas.13322

Baeda AY, Suriamihardja DA, Umar H, Rachman T (2015) Tsunami mitigation plan for Manakarra Beach of West Sulawesi Province, Indonesia. Procedia Engineering 116:134–140. https://doi.org/10.1016/j.proeng.2015.08.274

Barbier EB (2008) In the wake of tsunami: lessons learned from the household decision to replant mangroves in Thailand. Resour Energy Econ 30:229–249. https://doi.org/10.1016/j.reseneeco.2007.08.002

Barbier EB (2016) The protective service of mangrove ecosystems: a review of valuation methods. Mar Pollut Bull 109:676–681. https://doi.org/10.1016/j.marpolbul.2016.01.033

Bayas JCL, Marohn C, Dercon G, Dewi S, Piepho HP, Joshi L, van Noordwijk M, Cadisch G (2011) Influence of coastal vegetation on the 2004 tsunami wave impact in West Aceh. Proc Natl Acad Sci 108(46):18612–18617. https://doi.org/10.1073/pnas.1013516108

Biswas SR, Mallik AU, Choudhury JK, Nishat A (2009) A unified framework for the restoration of southeast Asian mangroves—bridging ecology, society and economics. Wetl Ecol Manag 17:365–383. https://doi.org/10.1007/s11273-008-9113-7

Badan Nasional Penanggulangan Bencana [BNPB] (2021) Data dan Informasi Bencana Indonesia. Retrieved March 2, 2021, from http://dibi.bnpb.go.id/

Badan Pusat Statistik [BPS] (2018) Data dan informasi kemiskinan kabupaten/kota tahun 2018. Retrieved March 8, 2021, from https://www.bps.go.id/publication/

Kabupaten/Kota Tahun (2018). Retrieved March 8, 2021, from https://www.bps.go.id/publication/

Budiyono Y, Aerts JCJH, Tollenaar D, Ward PJ (2016) River flood risk in Jakarta under scenarios of future change. Nat Hazards Earth Syst Sci 16:757–774. https://doi.org/10.5194/nhess-16-757-2016

Chaussard E, Amelung F, Abidin H, Hong S-H (2013) Sinking cities in Indonesia: ALOS PALSAR detects rapid subsidence due to groundwater and gas extraction. Remote Sens Environ 128:150–161. https://doi.org/10.1016/j.rse.2012.10.015

Colven E (2017) Understanding the allure of big infrastructure: Jakarta's great Garuda Sea wall project. Water Alternatives 10(2):250–264

Colven E (2020) Subterranean infrastructures in a sinking city: the politics of visibility in Jakarta. Crit Asian Stud 52(3):311–331. https://doi.org/10.1080/14672715.2020.1793210

Cordova MR, Nurhati IS (2019) Major sources and monthly variations in the release of land-derived marine debris from the greater Jakarta area, Indonesia. Sci Rep 9:18730. https://doi.org/10.1038/s41598-019-55065-2

Cordova MR, Ulumuddin YI, Purbonegoro T, Shiomoto A (2021) Characterization of microplastics in mangrove sediment of Muara Angke wildlife reserve, Indonesia. Mar Pollut Bull 163:112012. https://doi.org/10.1016/j.marpolbul.2021.112012

Dahdouh-Guebas F, Jayatissa LP, Di Nitto D, Bosire JO, Lo Seen D, Koedam N (2005) How effective were mangroves as a defence against the recent tsunami? Curr Biol 15(12):R443–R447. https://doi.org/10.1016/j.cub.2005.06.008

Dalimunthe SA (2018) Who manages space? Eco-DRR and the local community. Sustainability 10:1705. https://doi.org/10.3390/su10061705

Danielsen F, Sorensen MK, Olwig MF, Selvam V, Parish F, Burgess ND, Hiraishi T, Karunagaran VM, Rasmussen MS, Hansen LB, Quarto A, Suryadiputra N (2005) The Asian tsunami: a protective role for coastal vegetation. Science 320(5748):643. https://doi.org/10.1126/science.1118387

Duke NC, Meynecke J-O, Dittmann S, Ellison AM, Anger K, Berger U, Cannicci S, Diele K, Ewel KC, Field CD, Koedam N, Lee SY, Marchand C, Nordhaus I, Dahdouh-Guebas F (2007) A world without mangroves? Science 317(5834):41–42. https://10.1126/science.317.5834.41b

Esteban M, Tsimopoulou V, Mikami T, Yun NY, Suppasri A, Shibayama T (2013) Recent tsunamis events and preparedness: development of tsunami awareness in Indonesia, Chile and Japan. International Journal of Disaster Risk Reduction 5:84–97. https://doi.org/10.1016/j.ijdrr.2013.07.002

Guannel G, Arkema K, Ruggiero P, Verutes G (2016) The power of three: coral reefs, seagrasses and mangroves protect coastal regions and increase their resilience. PLoS One 11(7):e0158094. https://doi.org/10.1371/journal.pone.0158094

Esteban M, Takagi H, Mikami T, Aprilia A, Fujii D, Kurobe S, Utama NA (2017) Awareness of coastal floods in impoverished subsiding coastal communities in Jakarta: tsunamis, typhoon storm surges and dyke-induced tsunamis. International Journal of Disaster Risk Reduction 23:70–79. https://doi.org/10.1016/j.ijdrr.2017.04.007

Ferrario F, Beck MW, Storlazzi CD, Micheli F, Shepard CC, Airoldi L (2014) The effectiveness of coral reefs for coastal hazard risk reduction and adaptation. Nat Commun 5:3794. https://doi.org/10.1038/ncomms4794

Firman T (2009) The continuity and change in mega-urbanization in Indonesia: a survey of Jakarta–Bandung region (JBR) development. Habitat Int 33(4):327–339. https://doi.org/10.1016/j.habitatint.2008.08.005

Firman T, Surbakti IM, Idroes IC, Simarmata HA (2011) Potential climate-change related vulnerabilities in Jakarta: challenges and current status. Habitat Int 35(2):372–378. https://doi.org/10.1016/j.habitatint.2010.11.011

Forbes, K. & Broadhead, J. (2007). The role of coastal forests in the mitigation of tsunami impacts. RAP PUBLICATION 2007/1. Food and Agriculture Organization of the United Nations Regional Office for Asia and the Pacific. Bangkok. Retrieved September 4, 2020, from http://www.fao.org/forestry/14561-09bf06569b748c827dddf4003076c480c.pdf

Gedan KB, Kirwan ML, Wolanski E, Barbier EB, Silliman BR (2011) The present and future role of coastal wetland vegetation in protecting shorelines: answering recent challenges to the paradigm. Clim Chang 106:7–29. https://doi.org/10.1007/s10584-010-0003-7

Giesen W, Wulffraat S, Zieren M, Scholten L (2006) Mangrove guide book for Southeast Asia. FAO and Wetlands International

Gilman EL, Ellison J, Duke NC, Field C (2008) Threats to mangroves from climate change and adaptation options: a review. Aquat Bot 89(2):237–250. https://doi.org/10.1016/j.aquabot.2007.12.009

Giri C, Zhu Z, Tieszen LL, Singh A, Gillette S, Kelmelis JA (2008) Mangrove forest distributions and dynamics (1975–2005) of the tsunami-affected region of Asia. J Biogeogr 35:519–528. https://doi.org/10.1111/j.1365-2699.2007.01806.x

Giri C, Ochieng E, Tieszen LL, Zhu Z, Singh A, Loveland T, Masek J, Duke N (2011) Status and distribution of mangrove forests of the world using earth observation satellite data. Glob Ecol Biogeogr 20:154–159. https://doi.org/10.1111/j.1466-8238.2010.00584.x

Hamzah F, Setiawan A (2010) Akumulasi logam berat Pb, Cu, dan Zn di hutan mangrove Muara Angke, Jakarta Utara. Jurnal Ilmu dan Teknologi Kelautan Tropis 2(2):41–52

Henny C, Meutia AA (2014) Urban Lakes in megacity Jakarta: risk and management plan for future sustainability. Procedia Environ Sci 20:737–746. https://doi.org/10.1016/j.proenv.2014.03.088

Hilmi E, Kusmana C, Suhendang E, Iskandar (2017) Correlation analysis between seawater intrusion and mangrove greenbelt. Indonesian Journal of Forestry Research 4(2):151–168. https://doi.org/10.20886/ijfr.2017.4.2

Hilmi E (2018) Mangrove landscaping using the modulus of elasticity and rupture properties to reduce coastal disaster risk. Ocean & Coastal Management 165:71–79. https://doi.org/10.1016/j.ocecoaman.2018.08.002

Ilman M, Wibisono ITC, Suryadiputra INN (2011) State of the art information on mangrove ecosystems in Indonesia. (translated by Suryadiputra, W.) technical report. https://doi.org/10.13140/RG.2.1.3967.9120

Ilman M, Dargusch P, Dart P, Onrizal (2016) A historical analysis of the drivers of loss and degradation of Indonesia's mangroves. Land Use Policy 54:448–459. https://doi.org/10.1016/j.landusepol.2016.03.010

Indonesian Government (2012) Peraturan Presiden Republik Indonesia Nomor 73 Tahun 2012 tentang Strategi Nasional Pengelolaan Ekosistem Mangrove. Retrieved March 2, 2021, from http://dishut.jabarprov.go.id/data/menu/PP73tahun2012.pdf

Irawaty DT (2018) Jakarta's Kampungs: their history and contested future. Retrieved June 14, 2021, from https://escholarship.org/uc/item/55w9b9g

Karminarsih E (2007) Pemanfaatan Ekosistem Mangrove bagi Minimasi Dampak Bencana di Wilayah Pesisir. Jurnal Manajemen Hutan Tropika 13(3):182–187

Kathiresan K (2012) Importance of mangrove ecosystem. International Journal of Marine Science 2(10):70–89. https://doi.org/10.5376/ijms.2012.02.0010

Kathiresan K, Rajendran N (2005) Coastal mangrove forests mitigated tsunami. Estuar Coast Shelf Sci 65(3):601–606. https://doi.org/10.1016/j.ecss.2005.06.022

Kerr AM, Andrew H (2007) Natural barriers to natural disasters. Bioscience 57(2):102–103. https://doi.org/10.1641/B570202

Kusmana C, Valentino N, Mulyana D (2013) Ensiklopedia Flora Mangrove di Kawasan Hutan Angke Kapuk. Sahabat Bakau

Kusmana C (2017) Lesson learned from mangrove rehabilitation program in Indonesia. Jurnal Pengelolaan Sumberdaya Alam dan Lingkungan 7(1):89–97. https://doi.org/10.19081/jpsl.2017.7.1.89

Lopez JA (2009) The multiple lines of defense strategy to sustain coastal Louisiana. J Coast Res 54:186–197. http://www.jstor.org/stable/25737479

Marfai MA, Sekaranom AB, Ward P (2015) Community responses and adaptation strategies toward flood hazard in Jakarta, Indonesia. Nat Hazards 75:1127–1144. https://doi.org/10.1007/s11069-014-1365-3

Marois DE, Mitsch WJ (2015) Coastal protection from tsunamis and cyclones provided by mangrove wetlands – a review. International Journal of Biodiversity Science, Ecosystem Services & Management 11(1):71–83. https://doi.org/10.1080/21513732.2014.997292

Marschiavelli MIC (2008) Vulnerability assessment and coping mechanism related to floods in urban areas: a community-based case study in kampung Melayu. Gadjah Mada University, Indonesia. Retrieved January 19, 2021, from https://webapps.itc.utwente.nl/librarywww/papers_2008/msc/ugm/marschiave.pdf

Martínez-Espinosa C, Wolfs P, Velde KV, Satyanarayana B, Dahdouh-Guebasa F, Hugé J (2020) Call for a collaborative management at Matang mangrove Forest reserve, Malaysia: an assessment from local stakeholders' view point. For Ecol Manag 458:117741. https://doi.org/10.1016/j.foreco.2019.117741

McIvor AL, Möller I, Spencer T, Spalding M (2012) Reduction of wind and swell waves by mangroves. Natural coastal protection series: report 1. Cambridge coastal research unit working paper 40. Retrieved March 12, 2021, from http://www.naturalcoastalprotection.org/documents/reduction-of-wind-and-swell-waves-by-mangroves

Murdiyarso D, Purbopuspito J, Kauffman JB, Warren MW, Sasmito SD, Donato DC, Manuri S, Krisnawati H, Taberima S, Kurnianto S (2015) The potential of Indonesian mangrove forests for global climate change mitigation. Nat Clim Chang 5(12):1089–1092. https://doi.org/10.1038/NCLIMATE2734

Nurhidayah L, McIlgorm A (2019) Coastal adaptation laws and the social justice of policies to address sea level rise: an Indonesian insight. Ocean & Coastal Management 171:11–18. https://doi.org/10.1016/j.ocecoaman.2019.01.011

Nyström M, Folke C (2001) Spatial resilience of coral reefs. Ecosystems 4(5):406–417. http://www.jstor.org/stable/3658801

Ostling JL, Butler DR, Dixon RW (2009) The biogeomorphology of mangroves and their role in natural hazards mitigation. Geogr Compass 3(5):1607–1624. https://doi.org/10.1111/j.1749-8198.2009.00265.x

van Oudenhoven APE, Siahainenia AJ, Sualia I, Tonneijck FH, van der Ploeg S, de Groot RS, Alkemade R, Leemans R (2015) Effects of different management regimes on mangrove ecosystem services in Java, Indonesia. Ocean & Coastal Management 116:353–367. https://doi.org/10.1016/j.ocecoaman.2015.08.003

Padawangi R (2012) Climate change and the north coast of Jakarta: environmental justice and the social construction of space in urban poor communities. In: Holt WG (ed) Urban areas and global climate change. Emerald Group Publishing Limited, Bingley, pp 321–339. https://doi.org/10.1108/S1047-0042(2012)0000012016

Pambudi AC, Gusviga BH, Fahrezi ZA (2018) Analysis of mangrove forest change in Muara Angke Jakarta by using geographical information system and remote sensing. Prosiding Seminar Nasional Penginderaan Jauh 2018:192–198

Penyusun T (2009) Bersahabat dengan Hutan Bakau: Restorasi Ekologis Hutan Mangrove, Hutan Lindung Angke Kapuk. Pantai Indah Kapuk, Sahabat Bakau

Permatasari PA, Setiawan Y, Khairiah RN, Effendi H (2016) The effect of land use change on water quality: a case study in Ciliwung watershed. IOP Conference Series: Earth and Environmental Science 54:012026. https://doi.org/10.1088/1755-1315/54/1/012026

Pomeroy RS, Ratner BD, Hall SJ, Pimoljinda J, Vivekanandan V (2006) Coping with disaster: rehabilitating coastal livelihoods and communities. Mar Policy 30:786–793. https://doi.org/10.1016/j.marpol.2006.02.003

Pontee N (2013) Defining coastal squeeze: a discussion. Ocean & Coastal Management 84:204–207. https://doi.org/10.1016/j.ocecoaman.2013.07.010

Power WL (2013) Review of Tsunami Hazard in New Zealand (2013 Update), GNS Science Consultancy Report 2013/131. Retrieved March 13, 2021, from https://www.wremo.nz/assets/Publications/Review-NZ-Tsunami-Hazard-2013.pdf

Purwoko PF, Wulandari AA, Benariva AP, Tiara A, Sabiel MQT, Risaandi R, Jannati A, Nugraha A, Noriko N, Priambodo TW (2016) Ketahanan Vegetasi Wilayah Mangrove Suaka Margasatwa Muara Angke. DKI Jakarta terhadap Sampah dari Aliran Sungai Seminar Nasional Perhimpunan Biologi Indonesia 2016:140

Putri L, Yulianda F, Wardiatno Y (2015) Pola zonasi mangrove dan asosiasi makrozoobentos di wilayah Pantai Indah Kapuk, Jakarta. Bonorowo Wetlands 5(1):29–43. https://doi.org/10.13057/bonorowo/w050104

Rahadian A, Leilan F, Arafat IN, Lestari TA (2019) Ecosystem mangrove management in urban area: case study mangrove Kali Adem Jakarta Indonesia. IOP Conference Series: Earth and Environmental Science 399:012008. https://doi.org/10.1088/1755-1315/399/1/012008

Sandilyan S, Kathiresan K (2015) Mangroves as bioshield: an undisputable fact. Ocean & Coastal Management 103:94–96. https://doi.org/10.1016/j.ocecoaman.2014.11.011

Sasongko DA, Kusmana C, Ramadan H (2014) Strategi pengelolaan hutan lindung Angke Kapuk. Jurnal Pengelolaan Sumberdaya Alam dan Lingkungan 4(1):35–42. https://doi.org/10.19081/jpsl.2014.4.1.35

Sofian A, Kusmana C, Fauzi A, Rusdiana O (2020) Evaluasi Kondisi Ekosistem Mangrove Angke Kapuk Teluk Jakarta dan Konsekuensinya Terhadap Jasa Ekosistem. Jurnal Kelautan Nasional 15(1):1–12. https://doi.org/10.15578/jkn.v15i1.7722

Steinberg, F. (2007). Jakarta: environmental problems and sustainability. Habitat Int 31(3–4), 354–365. doi:https://doi.org/10.1016/j.habitatint.2007.06.002

Takagi H, Mikami T, Fujii D, Esteban M, Kurobe S (2016a) Mangrove forest against dyke-break-induced tsunami on rapidly subsiding coasts. Nat Hazards Earth Syst Sci 16:1629–1638. https://doi.org/10.5194/nhess-16-1629-2016

Takagi H, Esteban M, Mikami T, Fujii D (2016b) Projection of coastal floods in 2050 Jakarta. Urban Clim 17:135–145. https://doi.org/10.1016/j.uclim.2016.05.003

Takagi H (2017) Design considerations of artificial mangrove embankments for mitigating coastal floods – adapting to sea-level rise and long-term subsidence. Nat Hazards Earth Syst Sci. https://doi.org/10.5194/nhess-2017-61

Takagi H (2019) "Adapted mangrove on hybrid platform" – coupling of ecological and engineering principles against coastal hazards. Results in Engineering 4:100067. https://doi.org/10.1016/j.rineng.2019.100067

Tanaka N (2009) Vegetation bioshields for tsunami mitigation: review of effectiveness, limitations, construction, and sustainable management. Landsc Ecol Eng 5:71–79. https://doi.org/10.1007/s11355-008-0058-z

Tauhid FA, Zawani H (2018) Mitigating climate change related floods in urban poor areas: green infrastructure approach. Journal of Regional and City Planning 29(2):98–112. https://doi.org/10.5614/jrcp.2018.29.2.2

United Nations (2017) Factsheet: People and Oceans. Retrieved March 2, 2021, from https://www.un.org/sustainabledevelopment/wp-content/uploads/2017/05/Ocean-fact-sheet-package.pdf

United Nations Development Programme (2004) Reducing disaster risk: a challenge for development. A global report. United Nations Development Programme, Bureau for Crisis Prevention and Recovery, New York. Retrieved March 1, 2021, from https://www.diplomacy.edu/reducing-disaster-risk-challenge-development

Unnikrishnan S, Singh A, Kharat MG (2013) The role of mangroves in disaster mitigation: a review. In: Filho WL (ed) Climate change and disaster risk management. Springer, Cham. https://doi.org/10.1007/978-3-642-31110-9_16

Wibowo K (2006) Pelestarian hutan mangrove melalui pendekatan mina hutan (silvofishery). Jurnal Teknologi Lingkungan 7(3):227–233. https://doi.org/10.29122/jtl.v7i3.386

Wijayanti P, Zhu X, Hellegers P, Budiyono Y, van Ierland EC (2017) Estimation of river flood damages in Jakarta, Indonesia. Nat Hazards 86:1059–1079. https://doi.org/10.1007/s11069-016-2730-1

Winsemius HC, Jongman B, Veldkamp TIE, Hallegatte S, Bangalore M, Ward PJ (2018) Disaster risk, climate change, and poverty: assessing the global exposure of poor people to floods and droughts. Environ Dev Econ 23(3):328–348. https://doi.org/10.1017/S1355770X17000444

Wolanski E, Mazda Y, Ridd P (1992) Mangrove hydrodynamics. In: Robertson AI, Alongi DM (eds) Tropical mangrove ecosystems. American Geophysical Union, New York, pp 43–62. https://doi.org/10.1029/CE041p0043

Yanagisawa H, Koshimura S, Goto K, Miyagi T, Imamura F, Ruangrassamee A, Tanavud C (2009) The reduction effects of mangrove forest on a tsunami based on field surveys at Pakarang cape, Thailand and numerical analysis. Estuar Coast Shelf Sci 81:27–37. https://doi.org/10.1016/j.ecss.2008.10.001

Yusuf AA, Francisco H (2009) Climate change vulnerability mapping for Southeast Asia. In: Economy and environment program for Southeast Asia. Retrieved March 8, 2021, from https://www.preventionweb.net/files/7865_12324196651MappingReport1.pdf

Chapter 22
Implementation of Japanese Blue Carbon Offset Crediting Projects

Tomohiro Kuwae, Satoru Yoshihara, Fujiyo Suehiro, and Yoshihisa Sugimura

Abstract The term "blue carbon" is still rather new, having been coined in 2009. However, the blue carbon concept and the role of blue carbon stored in shallow coastal ecosystems, as part of nature-based or green infrastructure, in mitigating climate change and providing other ecosystems services, such as disaster risk reduction, infrastructure resilience, erosion control, and land formation, have attracted the interest of many people worldwide. In this chapter, we first summarize the current status of blue carbon initiatives, including for carbon offsetting, worldwide. Then, we review three blue carbon offset credit projects that have already been implemented in Japan: (1) the blue carbon offset crediting projects of (1) Yokohama City, the first in the world; (2) Fukuoka City, the second such project in Japan; and (3) the first Japanese national governmental demonstration project. Finally, we discuss the need to accelerate the development of blue carbon offset credit projects and related initiatives in the future.

Keywords Climate change · Blue economy · Blue finance · Nature-based solutions (NbS) · J-Blue Credit

T. Kuwae (✉)
Coastal and Estuarine Environment Research Group, Port and Airport Research Institute, Yokosuka, Japan

Japan Blue Economy Association (JBE), Yokosuka, Japan
e-mail: kuwae@p.mpat.go.jp

S. Yoshihara · F. Suehiro
Environmental Planning Department, Yachiyo Engineering Co., Ltd, Tokyo, Japan

Y. Sugimura
Department of Joint Research on Environment & Disaster in Coastal and Port Areas, Faculty of Engineering, Kyushu University, Fukuoka, Japan

F. Nakamura (ed.), *Green Infrastructure and Climate Change Adaptation*,
Ecological Research Monographs, https://doi.org/10.1007/978-981-16-6791-6_22

353

22.1 Introduction

Nature-based or green-gray infrastructure combines conservation or restoration of ecosystems with the selective use of conventional engineering approaches to deliver climate change resilience and adaptation benefits. Green-gray approaches strategically blend "green" natural and working landscapes and other open spaces that conserve ecosystem functions and values with "gray" human-engineered technology. The utilization of coastal green-gray (also termed "blue" or "natural") infrastructure is attracting increasing attention from a policy perspective (Sutton-Grier et al. 2015; World Bank 2017). Shallow coastal ecosystems (SCEs), for example, mangroves, tidal marshes, seagrass meadows, and macroalgal beds, through their role as part of the global carbon cycle and as a natural defense against external climate factors such as sea-level rise and storm surges, are a good example of green-gray infrastructure (Kuwae and Crooks 2021).

In a report jointly published in 2009 by the United Nations Environment Programme Planning Unit (UNEP), the United Nations Food and Agriculture Organization (FAO), and the United Nations Educational, Scientific, and Cultural Organization (UNESCO) (Nellemann et al. 2009), "blue carbon" is defined as carbon captured by marine organisms. The ocean is a particularly important carbon reservoir because blue carbon stored in seafloor sediments can remain undecomposed and unmineralized for long periods of time (up to several thousand years). It is estimated that 186 million tons (Kuwae and Hori 2019) (Fig. 22.1) to 238 million tons (Nellemann et al. 2009) of carbon are annually buried beneath the seafloor, and SCEs account for about 73–79% of this carbon. This role of blue carbon in SCEs as a climate mitigation measure has attracted the interest of many people worldwide. Typical SCEs, including mangroves, tidal marshes, and seagrass meadows, are now being called "blue carbon ecosystems" (Macreadie et al. 2019).

Blue carbon initiatives are currently moving from the advocacy stage to social penetration, policy-making, and implementation stages (International Partnership for Blue Carbon; https://bluecarbonpartnership.org/). About 20% of the countries that have joined the Paris Agreement have pledged to use SCEs as a climate change mitigation option in their nationally determined contributions (NDCs), and these countries are moving toward measuring national blue carbon amounts and accounting for them in their greenhouse gas inventories. About 40% of those countries have pledged to use SCEs to adapt to climate change as part of conservation, protection, and reforestation initiatives and through planning efforts such as Integrated Coastal Zone Management and fisheries management (Herr and Landis 2016; Martin et al. 2016). Australia (Kelleway et al. 2017) and the United States (Crooks et al. 2018) have also begun including blue carbon in their numerical emissions reduction targets and have started to calculate blue carbon in accordance with the 2013 Supplement to the 2006 United Nations Intergovernmental Panel on Climate Change (IPCC) Guidelines for National Greenhouse Gas Inventories: Wetlands (Wetlands Supplement) (IPCC 2014). The Conference of the Parties (COP) 25 to the United Nations Framework Convention on Climate Change (UNFCCC), held in Spain in

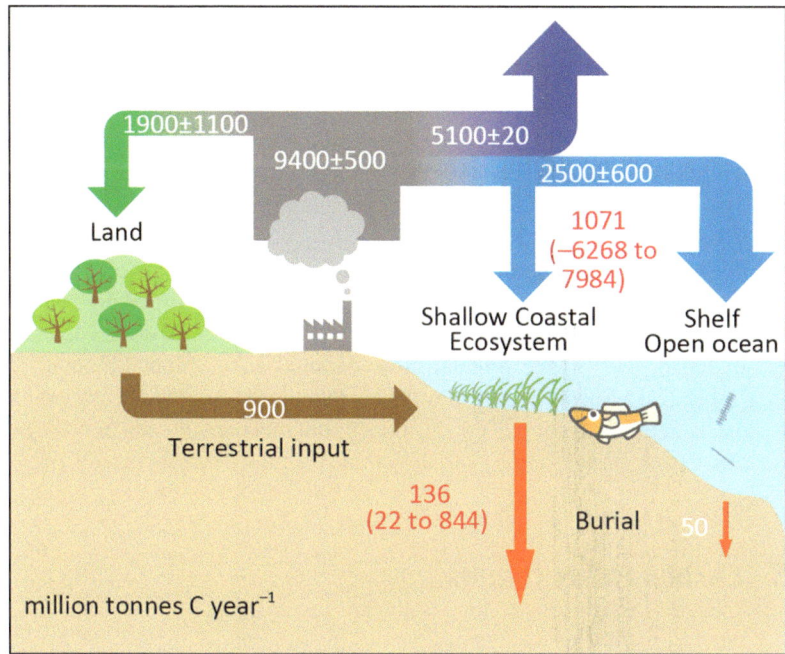

Fig. 22.1 Global carbon cycling (modified from Kuwae and Hori 2019). Data for the blue carbon sink potential (red type, showing mean and range) are from Kuwae and Hori (2019); atmospheric data (mean ± SD for 2010–2019) are from Friedlingstein et al. (2020); and terrestrial input data are from IPCC (2013)

2019, was positioned as a "Blue COP," and the importance of the ocean as part of the global climate system was mentioned for the first time in the COP-adopted document.

In conjunction with these international efforts, local communities have begun to recognize the importance of blue carbon and efforts toward its social implementation. Private companies, regional administrative organizations, and individual managers and engineers have expressed a strong interest in future socioeconomic development that incorporates the conservation or restoration of blue carbon ecosystems as new business opportunities (Thomas 2014; Nobutoki et al. 2019). In this context, Japan has clearly stated its intention to pursue the potential of blue carbon as a carbon dioxide (CO_2) sink in the "Long-Term Strategy as a Growth Strategy Based on the Paris Agreement," approved by the Cabinet in 2019. Other countries have also developed policy frameworks and conducted case studies related to blue carbon (e.g., Laffoley and Grimsditch 2009; Crooks et al. 2011; Lutz et al. 2014; Wylie et al. 2016; EX-ACT 2017; Howard et al. 2017; Villa and Bernal 2018; Wyndham-Meyers et al. 2018).

One of the most efficient and effective ways of promoting climate change mitigation is to implement emissions trading systems (ETS). The two main types of

ETS are "cap-and-trade" and "baseline-and-credit." Cap-and-trade systems impose a cap on aggregate emissions levels but allow trades of allowances between covered entities. Baseline-and-credit systems define an emission baseline and reward verified emission reductions beyond that baseline with tradable offset credits. Offset credits, as an incentive for climate change mitigation, are well established for forest and agricultural ecosystems, and several voluntary carbon markets have certified blue carbon offset methodologies and implementation protocols (Kelleway et al. 2020; Sapkota and White 2020).

In this chapter, we first outline three blue carbon offset credit projects that have already been socially implemented in Japan: the blue carbon offset credit projects of (1) Yokohama City, the world's first, and (2) of Fukuoka City, the second such project in Japan, and (3) the first Japanese national governmental demonstration project (Table 22.1). Second, we review the challenges encountered in implementing these projects in terms of people, goods, money, and mechanisms and how they were solved. Finally, we discuss issues and directions for future project expansion.

22.2 Carbon Offset Credits

The socioeconomic aims of blue carbon initiatives include improving the capital value and economic benefits of SCEs and their cost-effectiveness as public works, as well as promoting local business (Thomas 2014). These benefits referred to here include economic incentives, including carbon offset credits (carbon trading), payments for ecosystem services, and income from funds (Murray et al. 2011; Herr et al. 2015).

There are at present two types of carbon trading markets: voluntary markets and compliance markets. Voluntary markets have been developed to credit actions taken to reduce greenhouse gas emissions, primarily by private sector companies, to reduce a company's environmental footprint, to demonstrate corporate social responsibility, and to enhance public relations. Compliance markets, which deal with mandatory emission reductions imposed by regulations, are driven by the demand for allowances and offsets by regulated greenhouse gas emitters.

Historically, carbon offset credits have been implemented by a top-down approach, in which international markets are first established and credit markets at the national and local government levels are subsequently created. However, in the new framework adopted at COP 21 in 2015, which is legally binding after 2020 as part of the Paris Agreement, mitigation measures are undertaken in a unique way by each country, and a mutual verification mechanism (i.e., the pledge and review approach) is the basic policy. Thus, to implement the new policy of the Paris Agreement, both global and local climate change countermeasures will be promoted. Also, the use of monetary incentives to appeal to the private sector requires a bottom-up approach, in which markets are newly established at the local governmental level, as we describe in this chapter, and privately led projects are developed.

Table 22.1 Current blue carbon (BC) credit schemes in Japan

		Yokohama BC Credit	Fukuoka BC Credit	J-Blue Credit
Year established		2015	2019	2021
Carbon market		Voluntary	Voluntary	Voluntary, but also compliance targeted
Developer and secretariat		Yokohama city	Fukuoka city	Japan Blue Economy Association (JBE) approved by the Japanese government
Validation and verification body (VVB)		Not established	Not established	Established with members independent of the secretariat (JBE)
Approver		Yokohama City	Fukuoka City	JBE
Spatial coverage		Within Yokohama City and some collaborating local governments	Within Fukuoka City	Nationwide
Inclusion into national accounting		Not yet	Not yet	Targeted by 2023
Inclusion into national emission trading systems (ETS)		Not yet	Not yet	Targeted by 2025
Project activities	BC sink creation/ restoration/ conservation (IPCC methodology applied)[a]	Seagrass (Tier 1)	Seagrass (Tier 3)	Seagrass (Tier 3)
		Macroalgal beds (N/A)		Macroalgal beds (N/A)
				Salt marshes and tidal flats (Tier 1)
	CO_2 emission reduction	Yes[b]	Not yet	Not yet
Trading price		8000 yen/ton CO_2	8000 yen/ton CO_2	Negotiated transaction

[a]See IPCC (2014)
[b]See Nobutoki et al. (2019)

For the social implementation of carbon credit schemes, independent methods for the measurement, reporting, and verification (MRV) of credits are needed. These methods involve accurate, objective, and quantitative measurement of carbon based on scientific and technological knowledge, transparent reporting, and verification. The submission of greenhouse gas inventories to the UNFCCC Convention Secretariat is based on the MRV principle.

Mitigation of climate change by storing atmospheric CO_2 in the sea via natural systems can be achieved by three approaches: (1) newly creating target ecosystems (i.e., carbon reservoirs and atmospheric CO_2 sinks), (2) reducing the decline of target ecosystems through restoration and conservation, and (3) improving the management of target environments and ecosystems (i.e., improving carbon storage rates and the CO_2 uptake rate per unit area).

Various guidelines for measuring carbon storage and CO_2 uptake by blue carbon ecosystems and for creating credits for blue carbon have been developed. Australia has included blue carbon ecosystems in its national greenhouse gas accounts, and the Australian Government's Emissions Reduction Fund has developed comprehensive guidelines (Kelleway et al. 2017). Other organizations that have produced guidelines include the IPCC (IPCC 2014), Conservation International, UNESCO, the International Union for Conservation of Nature (Howard et al. 2014), UNEP and the Center for International Forestry Research (Crooks et al. 2014), and the Verified Carbon Standard (2015), which is an independent carbon trading certification body in the United States. In Japan, guidance documents describing measurement methods for seagrass meadows, tidal flats, embayments, and port facilities have been prepared (Tokoro et al. 2015).

22.3 Offset Crediting by the Yokohama Blue Carbon Project

22.3.1 Background

In the Yokohama City Action Plan for Global Warming Countermeasures, the city of Yokohama, Japan, has set a target of reducing greenhouse gas emissions by 7% by 2021 and by 30% by 2030, compared to 2013 levels. Yokohama City is a member of the C40 Cities Climate Leadership Group (C40; https://www.c40.org/) and the Local Governments for Sustainability (ICLEI; https://www.iclei.org), which are international networks of cities that are actively working to combat climate change. In addition, it was selected as the only Asian member of the Carbon Neutral Cities Alliance (CNCA; https//carbonneutralcities.org) in 2015.

The Yokohama Blue Carbon Project, started in 2014, aims to create a variety of synergistic effects between the environment (e.g., through water purification, biodiversity conservation), society (e.g., through enhancement of amenities and the Yokohama brand), and the economy (e.g., by increasing supplies of resources and food, and increasing tourism), by implementing global warming countermeasures

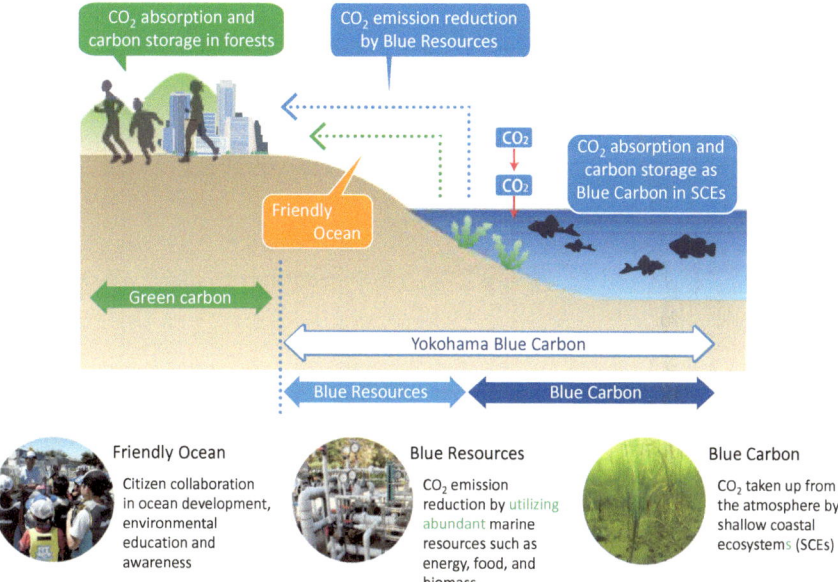

Fig. 22.2 Framework of the Yokohama Blue Carbon Project (modified from Suehiro et al. 2020)

(Nobutoki et al. 2019). In addition to "blue carbon," which refers to the utilization of SCEs as CO_2 sinks, this project introduced the concept of "blue resources," which are resources tailored for effective utilization of the abundant marine energy, food, and biomass resources for CO_2 emission reduction (Fig. 22.2). Furthermore, the project employs the "Friendly Ocean" concept to promote citizen collaboration in ocean development, environmental education, and environmental awareness.

The Yokohama Blue Carbon Project certifies the amounts of greenhouse gas absorption and reduction by blue carbon or blue resources as blue carbon credits and promotes carbon offsetting by trading these credits. Since the project's initiation in 2014, the amounts of credits created (Table 22.2) and used (Table 22.3), as well as the number of users, have been increasing each year (Suehiro et al. 2020). Although social penetration by this scheme has been gradually increasing, all of these credits were from blue resources until 2018. Therefore, in 2019, Yokohama City introduced the certification and offsetting of blue carbon credits to revitalize the project.

22.3.2 Calculation Methodology

The blue carbon offset credit scheme is based on both the IPCC Guidelines (IPCC 2014), which outline methodologies for calculating the CO_2 sink capacity of blue carbon ecosystems (mangroves, tidal marshes, and seagrass meadows), and on methodologies described by Kuwae et al. (2019). The methodologies of Kuwae et

Table 22.2 Amounts of carbon offset credit created and certified by the Yokohama Blue Carbon Project, as of March 2020 (modified from Suehiro et al. 2020)

| | Blue Carbon | | | | Blue Resources | | | | Amount of carbon credit offset created (tonnes CO_2) | | | | | |
	Eelgrass bed maintenance	Kelp forest maintenance	Seaweed farming — Wakame seaweed	Seaweed farming — Kelp	Local production for local consumption of wakame	Introduced seawater heat pump	Alternative use of LNG-fueled tugboats	Alternative use of hybrid tugboats	2019	2018	2017	2016	2015	2014
Yokohama City	●								12.3					
Fisheries Cooperative of			●						0.2					
Yokohama City				●					5.4					
NGO "SATOUMI initiative"		●							0.2					
Hyuuga City			●						0.5					
Fudai Village			●	●					58.0					
Hannan City				●					2.3					
Subtotal of Blue Carbon									**78.9**					
Yokohama Hakkeijima Inc.					○				1.0	0.7	1.7	0.7	0.7	2.1
						○			1.4	0.8	1.0	1.1	0.9	0.9
Fisheries Cooperative of Yokohama City					○				4.9	18.4	19.5	22.3	16.0	13.7
NPO "Association for Shore Environment Creation"					○				1.0	1.0	2.0	1.0	1.0	
Shin-Nippon Kaiyosha Corporation							○		143.3	164.6	31.5			
								○	89.4	112.6				
Subtotal of Blue Resources									**241.0**	**298.1**	**55.7**	**25.1**	**18.6**	**16.7**
Total									**319.9**	**298.1**	**55.7**	**25.1**	**18.6**	**16.7**

Table 22.3 Amounts of carbon offset credit used by the Yokohama Blue Carbon Project, as of March 2020 (modified from Suehiro et al. 2020)

	Users		Offset emissions	Amount of carbon offset credit used (tonnes CO_2)					
				2019	2018	2017	2016	2015	2014
Events	Marine sports events	A	Use of transportation by participants and staff Energy consumption at the venue	28.4	33.8	36.4	18.8	14.5	
		B	Use of transportation by participants and staff Energy consumption at the venue	5.7	5.5	6.2	7.3	8.3	3.1
		C	Use of transportation by staff Disposal of locally generated garbage Paper usage for the guidebook			1.8	2.1		
	Sports events	D	Use of transportation by visitors	27.8	3.6	5.4			
	Environmental seminars	E	Use of fuel for generator Power consumption Waste disposal	2.2	2.0				
Corporate activities	Construction	F	Power and water consumption in the building Vehicle operation	16.8	10.0	4.0	1.0		
		G	Vehicle operation		86.0				
	Energy	H	Power consumption	6.1					
	Food	I	Power consumption	6.2					
		J	Vehicle operation	1.0					
	Printing	K	Use of power, natural gas, water, and fuel	1.0	1.0	1.0			
	Retail	L	Product manufacturing Waste disposal		1.0	4.0			
	Manufacturing	M	Product manufacturing Waste disposal	1.0	1.0				
		N	Product manufacturing and transportation		2.2				
		O	Product manufacturing		16.0				
		P	Use of transportation by visitors	0.7	2.3				
	Service	Q	Use of transportation by employees	0.9					
	Waste disposal	R	Vehicle operation	22.0					
Individuals	–	S	Use of aircraft for traveling	0.5					
			Total	**120.3**	**164.4**	**58.8**	**29.2**	**22.8**	**3.1**

al. (2019) are based on compilations of domestic and international data, and while in line with the IPCC Guidelines (IPCC 2014), they can be used to estimate the CO_2 sink capacities of macroalgal beds and tidal flats as well as those of other blue carbon ecosystems throughout Japan. Although this scheme is, in a sense, a social experiment and unique to Yokohama City, to the best of our knowledge, it is the first project in the world that credits the CO_2 sink capacities of seagrass meadows, macroalgal beds, and seaweed aquaculture.

Both the IPCC Guidelines (IPCC 2014) and Kuwae et al. (2019) calculate the CO_2 sink capacity of an SCE as the product of the area (ha) of the target ecosystem (referred to as "activity data") and the amount of CO_2 absorbed per unit area (removal coefficient):

$$\text{Annual } CO_2 \text{ sink capacity (tons } CO_2/\text{year)}$$
$$= \text{activity data (ha)} \times \text{removal coefficient (tons } CO_2/\text{ha/year)} \tag{1}$$

The default (Tier 1) value (1.58 tons CO_2/ha/year) in the IPCC Guidelines (IPCC 2014) was used as the removal coefficient for eelgrass beds (*Zostera marina*), and removal coefficients estimated by Kuwae et al. (2019) were used for macroalgal beds [*Sargassum*, 2.7 tons CO_2/ha/year; wakame and arame kelps (*Undaria, Ecklonia*, and *Eisenia*), 4.2 tons CO_2/ha/year]. The CO_2 sink capacity of seaweed farms was calculated as the product of the net primary production of the seaweeds and the residual rate (the percentage of net primary production that is stored for a long period in the ocean carbon pool without reverting to CO_2) (Kuwae et al. 2019):

$$\begin{aligned}&\text{Annual } CO_2 \text{ sink capacity (tons } CO_2/\text{year)}\\&\quad = \text{net primary production (tons } CO_2/\text{year)} \times \text{residual rate (7.7\%)}\end{aligned} \quad (2)$$

22.3.3 Implementation

Equation (1) was used to calculate the CO_2 sink capacity in 2019 of the eelgrass beds maintained and managed at Sea Park Yokohama (35.3394°N, 139.6360°E). The distribution area of eelgrass was determined by conducting a GPS logger survey during the eelgrass blooming season in June 2019 and using location information provided by park managers. The recorded area was 7.8 ha, and the estimated CO_2 sink capacity of the eelgrass beds was 12.3 tons CO_2/year, and the certified credit amount was 12.3 tons CO_2 (Table 22.2).

Equation (2) was used to calculate the CO_2 sink capacity of the wakame (*Undaria*) cultivated (22.8 tons wet weight) by the Kanazawa Branch of the Yokohama City Fisheries Cooperative Association (35.3312°N, 139.6375°E) in 2019. The estimated CO_2 sink capacity was 0.2 tons CO_2/year, and the certified credit amount was also 0.2 tons CO_2.

The scheme initially targeted blue carbon and blue resources within the Yokohama City area. However, increased awareness of blue carbon initiatives in Japan has led to the expansion of this scheme to other municipalities, which are collaborating with Yokohama City to further develop blue carbon offsets (Table 22.2).

The number of credit applicants and the total amount of certified credits for blue resources have been increasing each year. In 2017, Yokohama City established a methodology for calculating reductions in CO_2 emissions achieved by replacing tugboats fueled by heavy oil first with LNG-fueled tugboats and then, in 2018, with hybrid tugboats. For details of the methodologies of other blue resources, see Nobutoki et al. (2019).

At the beginning of the project, the main use of the credits was to offset CO_2 emissions generated by short-term events; however, by the third year (2016) of the project, credits had begun to be used to offset CO_2 emissions from ongoing corporate activities (Table 22.3). Credits were used by individuals for the first time in 2019.

22.4   Offset Credit System of Fukuoka City

22.4.1   Background and Framework

In Fukuoka City, the "Hakata Bay NEXT Conference" was established in 2018 to promote collaborations among citizens, fisheries, businesses, educators, and government in carrying out environmental, economic, and societal improvements to the rich environment of Hakata Bay in Fukuoka City to pass on to future generations (Sugimura et al. 2021). At present, the conference is working on the conservation, restoration, and utilization of the Hakata Bay environment with a focus on eelgrass bed creation. As part of these activities, a blue carbon offset credit scheme and a funding scheme that utilizes a part of ship entry fees and donations from companies as financial resources for environmental conservation and restoration activities were established (Fig. 22.3). The offset credit scheme is the second for blue carbon in

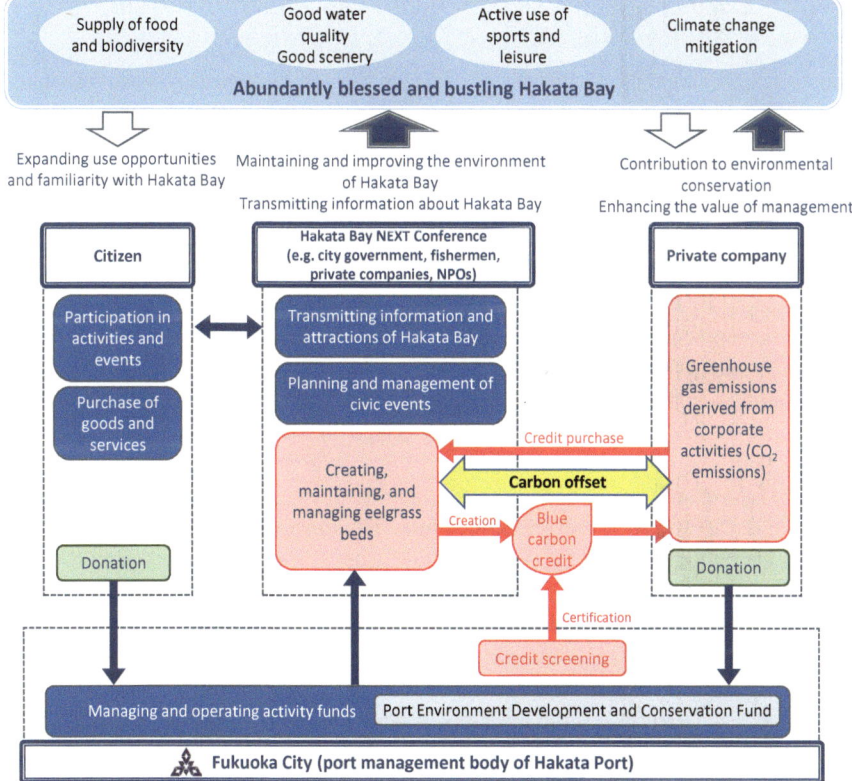

Fig. 22.3 Framework of the Fukuoka City blue carbon offset credit system (modified from Sugimura et al. 2021)

Japan after Yokohama City, and the funding scheme is the first such scheme in the country.

In the Fukuoka City blue carbon offset credit scheme, the amount of CO_2 absorbed as a result of the creation, maintenance, and management of eelgrass beds and macroalgal beds in Hakata Bay is designated as a blue carbon credit. Proceeds from the sale of credits by Fukuoka City, the project implementer, are returned to the Hakata Bay NEXT Conference to fund its environmental conservation activities, including eelgrass bed-related activities (Fig. 22.3). The funding scheme utilizes income from port operations and donations from companies and individuals for the conservation and creation of Hakata Bay environments. According to this scheme, 2.5% of port charges collected from ships, as well as donations from individuals and corporations, are used for Hakata Bay conservation and creation projects in the current fiscal year, and another 2.5% of the port charges are accumulated in the Port Environment Improvement and Conservation Fund for future projects.

22.4.2 Carbon Offset Credit Scheme and Implementation

The CO_2 sink capacity of the target eelgrass beds at Hakata Bay was calculated as described in Sect. 22.3.2 for the Yokohama City scheme, using Eq. (1) and the methodologies in the IPCC Guidelines (IPCC 2014) and Kuwae et al. (2019). However, the Fukuoka City scheme used the modeled (Tier 3) value (2.7 tons CO_2/ha/year) for the removal coefficient of eelgrass beds. Modeled values were also used for the removal coefficients of macroalgal beds [*Sargassum*, 1.09 tons CO_2/ha/year; wakame kelp (*Undaria*), 0.45 tons CO_2/ha/year) (Sugimura et al. 2021). Both the eelgrass bed and macroalgal bed projects are considered to be eligible targets for offset credits because the former are managed beds, and intensive efforts are underway to establish the latter on new seawalls.

In an area survey conducted in May 2019, when the turbidity of seawater was relatively low in Hakata Bay (33.6136°N, 130.3125°E), the total area of the eelgrass beds was estimated by generating a composite image based on aerial images obtained by a multicopter (aerial drone), visually classifying the cover classes, and then calculating the area of each cover class. The distribution area and cover class of eelgrass beds in the offshore area, which were difficult to determine from the aerial images, were corrected by using visual surveys by divers. An area survey of macroalgal beds was conducted visually by divers.

The estimated areas of eelgrass beds and macroalgal beds were 15.6 and 2.9 ha, respectively, the resulting estimated CO_2 sink capacities were 42.1 and 1.3 tons CO_2/year, respectively, and the amounts of certified credits were 42.1 and 1.3 tons CO_2, respectively (Sugimura et al. 2021).

22.4.3 Funding Scheme and Implementation

In 2020, Fukuoka City's general budget of 38,533,000 yen included 100,000 yen for the Hakata Bay Environment Conservation and Creation Project and 3,625,000 yen (equivalent to 2.5% of port charges) for the Port Environment Improvement and Conservation Fund Reserve to be used for the Hakata Bay Environmental Conservation and Creation Project. In addition, 3,625,000 yen in port charges (equivalent to the reserved fund) and 100,000 yen in anticipated donations were budgeted for the Hakata Bay Environmental Conservation and Creation Project. Thus, 2.5% of port charges would be accumulated in the Port Environment Improvement and Conservation Fund, and another 2.5% would be used for the Hakata Bay Environmental Conservation and Creation Project during the current fiscal year. All donations would also be used for the project during the current fiscal year.

22.5 Offset Crediting Demonstration by the Japanese National Government

22.5.1 Background

In 2020, the Japanese Ministry of Land, Infrastructure, Transport and Tourism (MLIT 2020) approved the foundation of the Japan Blue Economy Association (JBE; https://www.blueeconomy.jp/en/) as a Collaborative Innovation Partnership (CIP), a type of national government-approved private cooperative. JBE is the first CIP for the marine environment in Japan. The ministry also decided to implement an offset credit demonstration project in collaboration with JBE.

A variety of the initiatives in JBE are implemented by researchers, engineers, and practitioners in different fields and positions, all of which are included on an equal footing under the supervision of the government, to promote collaboration among various partners such as private companies, municipalities, NGOs, and NPOs. JBE is developing detailed methodologies for blue economy projects based on scientific and technological evidence and quantitative evaluations of ecosystem services and their economic value. The anticipated contributions of JBE to partners include the following:

1. NPOs/NGOs: Raising national awareness of coastal environmental initiatives implemented at local sites and obtaining funding for those initiatives.
2. Private companies: Quantifying their environmental, social, and corporate governance (ESG) efforts as key performance indicators (KPIs) and, in particular, contributing to society through Sustainable Development Goal (SDG) activities, including SDG 14 (Life Below Water), SDG 13 (Climate Action), and SDG 6 (Clean Water and Sanitation).

3. Citizens and educators: Developing and identifying methods for measuring how much CO_2 is taken up by vegetation growing on shorelines in their own community.

22.5.2 Characteristics of the Carbon Offset Scheme

The national demonstration project and the carbon offset scheme "J-Blue Credit" are currently set up for the voluntary market, but the compliance market will also be targeted after blue carbon becomes included, similar to the "J-Credit," in the national baseline-and-credit system of Japan. JBE, as secretary of the national demonstration project, is not dependent on subsidies but is managed as an independent corporation. This makes the scheme different from those operated by local governments, as detailed below.

22.5.2.1 Multiple Methodological Options

The first difference is that JBE prepares multiple methodological options for determining removal coefficients and activity data (areas). As incentives for options that are technically difficult but result in more certain (more accurate and reliable) estimates, the amount of the credits certified is increased and transaction fees are decreased. In this way, a project as a whole can be expected to automatically improve the accuracy and reliability of the estimation methodology and achieve facility development. In contrast, negative incentives such as high fees and a small amount of credits certified are provided for options that are technically easy and thus likely to be implemented in many situations, at least in the near future.

22.5.2.2 Continuous Review and Revision

Second, established methodologies are subject to continuous review and revision by JBE, whereas the methodologies of many domestic and international credit schemes are more or less fixed. A flexible scheme in which methodologies are reviewed and revised in light of the rapid progress in science and technology (e.g., area determination using both aerial and above-water sailing drones and remote sensing) should improve certainty and reduce costs, although project management may become more burdensome.

22.5.2.3 Variable Transaction Prices

Third, dynamic pricing such as direct trading and auctions is introduced. The unit price of credit transactions is often fixed when local governments are in charge of the carbon offset scheme secretariat because the fixed price is apt to the secretariat to minimize the risk of failure. However, such fixed prices may not reflect the balance of supply and demand. In interviews about their reasons for purchasing blue carbon credits, buyers indicated that they appreciated or sympathized with not only climate change measures but also other co-benefits aligned with the buyer's goals and branding messages. By allowing the price to vary, these co-benefits can be reflected in the transaction price, thus increasing the unit price.

22.5.3 Case Study

A case study was conducted in the Yokohama Bay Side Marina (35.3837°N, 139.6506°E), where both eelgrass and *Sargassum* beds were newly created by transplantation. The offset credit scheme adopted methodologies in both the IPCC Guidelines (IPCC 2014) and Kuwae et al. (2019); thus, the CO_2 sink capacity of SCEs was calculated by using Eq. (1) (see Sect. 22.3.2). From the candidate options prepared for the JBE scheme, based on the multiple option policy described in Sect. 22.5.2.1, the modeled value of 4.9 tons CO_2/ha/year was selected for the removal coefficient of the eelgrass beds and the modeled value of 2.7 tons CO_2/ha/year was selected for the *Sargassum* beds (Kuwae et al. 2019). An area survey of the eelgrass and *Sargassum* beds was conducted in March 2017. The total area of these beds was estimated by generating a composite image based on aerial images taken by a multicopter (drone), visually classifying the cover classes, and then calculating the area of each cover class.

The estimated areas were 10.3 ha for the eelgrass beds and 0.3 ha for the *Sargassum* beds, the resulting CO_2 sink capacities were 50.2 tons CO_2/year and 0.6 tons CO_2 /year, respectively, and the certified credit amounts were 22.8 tons CO_2 and 22.8 tons CO_2, respectively, based on the credit reduction policy that considered the uncertainty, as described in Sect. 22.5.2.1.

22.6 Keys for Successful Implementation

Highlights and breakthroughs in terms of mechanisms, people, goods, and money associated with the project implementations described in Sects. 22.3, 22.4, and 22.5 are summarized below.

22.6.1 Yokohama City

1. Yokohama City has already implemented a number of initiatives to combat global warming in coastal areas through industry–government–academia–private partnerships.
2. Yokohama City was able to decide to establish its own scheme ahead of the rest of the world despite the incomplete scientific knowledge of blue carbon and the lack of social implementations.
3. The project was promoted in collaboration with stakeholders (scientists, engineers, consultants, and many other experts), who willingly and quickly provided the know-how for resolving the conflict between "as easy as possible" and "high certainty" schemes.
4. The use of the sea area within an aquarium as a field for a demonstration experiment was graciously agreed upon.
5. The project generated enthusiasm and energy among municipal officials.
6. The very positive attitude of the credit creators and credit users toward marine environments fitted with the objectives of this project.
7. An international fund was utilized, and the project was included in the long-term city budget.
8. The sale of credits generated a profit.

22.6.2 Fukuoka City

1. Fukuoka City established the Hakata Bay Environmental Conservation Plan.
2. Various entities, including private citizens and citizens' groups, fishermen, businesses, educators, and the government, have successfully worked together toward the conservation and creation of marine ecosystems.
3. A foundation has been laid for industry–government–academia–private sector cooperation.
4. The Hakata Bay NEXT Conference was established as the project's main entity.
5. New financial resources were secured through the creation of a new funding scheme utilizing port charges.

22.6.3 National Demonstration Project

1. Japan already had a framework of government-approved private corporations and related laws to promote industry–government–academia–private partnerships.
2. The Japanese national government has been hosting blue carbon study groups and discussion committees for several years.
3. Successful efforts by local governments have been recognized in Japan.

4. The JBE CIP was founded, and a national demonstration project was led by people from JBE who support local government initiatives.
5. Costs for the foundation of JBE and the national demonstration project were minimized.

22.7 Challenges for Future Blue Carbon Offset Schemes

22.7.1 Quantification and Reducing Uncertainty

The accumulation of scientific and technological knowledge is still very important for connecting blue carbon initiatives to policy-making and implementation (Macreadie et al. 2019). Here, we propose four areas of blue carbon research, based on Kuwae and Hori (2019), that require further quantification relevant to carbon offset credit schemes:

1. Key data on carbon stocks and flows (primarily CO_2 uptake and blue carbon storage) for various SCEs still need to be gathered by conventional methods (Kuwae et al. 2016; Crosswell et al. 2017). In particular, more continuous long-term, large-scale in situ observation data, as well as monitoring data obtained by remote sensing technologies, are needed to measure changes in the distribution areas of target ecosystems.
2. New measurement techniques need to be established for carbon stocks and flows that researchers have not been able to quantify by using conventional methods, especially seasonal fluctuation and the drifting amount of the macrophyte biomass, as well as the formation of refractory dissolved organic matter in SCEs (Wada et al. 2008; Orr 2014; Hill et al. 2015; Krause-Jensen and Duarte 2016; Duarte and Krause-Jensen 2017; Hamaguchi et al. 2018; Jiao et al. 2018; Abo et al. 2019; Ortega et al. 2019). Underlying processes and mechanisms also need to be elucidated by using the new measurement techniques.
3. The spatiotemporal variability of measured carbon stocks and flows, particularly in response to long-term disturbances such as climate change (Arias-Ortiz et al. 2018; Watanabe et al. 2019) and altered food web structures (Atwood et al. 2015), as well as short-term disturbances such as storm and tsunami events (Cahoon et al. 2003), needs to be estimated.
4. The temporal scale required for a functional response from restored SCEs needs to be evaluated (O'Connor et al. 2020; Kuwae and Crooks 2021). The importance of the temporal scale used is clear: for seagrass meadows, 10–20 years might be needed for the soil organic carbon pool or carbon accumulation rate of restored and managed sites to achieve the levels natural sites take in the case of seagrass meadows (Duarte et al. 2013; Greiner et al. 2013; Marbà et al. 2015), at least 25–100 years are required but for tidal marshes (Craft et al. 2003; Burden et al. 2013), 20–25 years for mangrove wetlands (Osland et al. 2012; Salmo et al. 2013), and 7–17 years for blue carbon ecosystems generally (mangroves, tidal marshes,

seagrass meadows) (reviewed by O'Connor et al. 2020). These temporal scales are comparable to those required for the recovery of coastal habitats (Duarte et al. 2020).

22.7.2 Considering Emissions from SCEs

In this chapter, we have focused only on the CO_2 absorption and climate change mitigation effects of SCEs. However, CO_2 emissions from SCEs are also possible, as a result of development projects and human activities such as dredging, excavation, and aquaculture (e.g., Macreadie et al. 2015; Serrano et al. 2016; Atwood et al. 2017; Kauffman et al. 2017; Lovelock et al. 2017). Thus, it is of equal importance to measure, report, and verify the reduction of carbon stocks and increase in CO_2 emissions from SCEs caused by human activities for the future establishment of disincentives.

22.7.3 Expanding the Scope of Carbon Offsetting

Blue carbon ecosystems such as mangroves, tidal marshes, and seagrass meadows have been shown to have climate change mitigation functions and benefits (Duarte et al. 2013), but we should not limit ourselves to these ecosystems when considering ways to mitigate climate change mitigation and provide other co-benefits. The scope of blue carbon offset schemes should be broadened to include other ecosystems, known as "potential blue carbon ecosystems," which can also play important roles in climate change mitigation (Crooks et al. 2019; Hoegh-Guldberg et al. 2019; Kuwae and Hori 2019; Kuwae et al. 2019; Lovelock and Duarte 2019). For instance, tidal mudflats can be viewed, along with tidal marshes and mangroves, as a type of intertidal blue carbon ecosystem; although they lack large vegetation, their microphytobenthos can take up atmospheric CO_2 and their soils can store the captured carbon. Moreover, similar coastal ecosystems and carbon storage mechanisms can be found in arid regions (e.g., microbial mat systems and coastal sabkhas) (Schile-Beers et al. 2019). Among potential blue carbon ecosystems, macroalgal beds are gaining recognition (e.g., Froehlich et al. 2019; Krause-Jensen and Duarte 2016; Krause-Jensen et al. 2018; Lovelock and Duarte 2019; Ortega et al. 2019; Queirós et al. 2019; Watanabe et al. 2020). Although global estimates of atmospheric CO_2 uptake rates by macroalgal beds are few and extremely uncertain, estimates of the global annual CO_2 gas exchange rate indicate that among global shallow coastal ecosystems macroalgal beds may be the largest contributor to the net CO_2 uptake rate (Fig. 22.4).

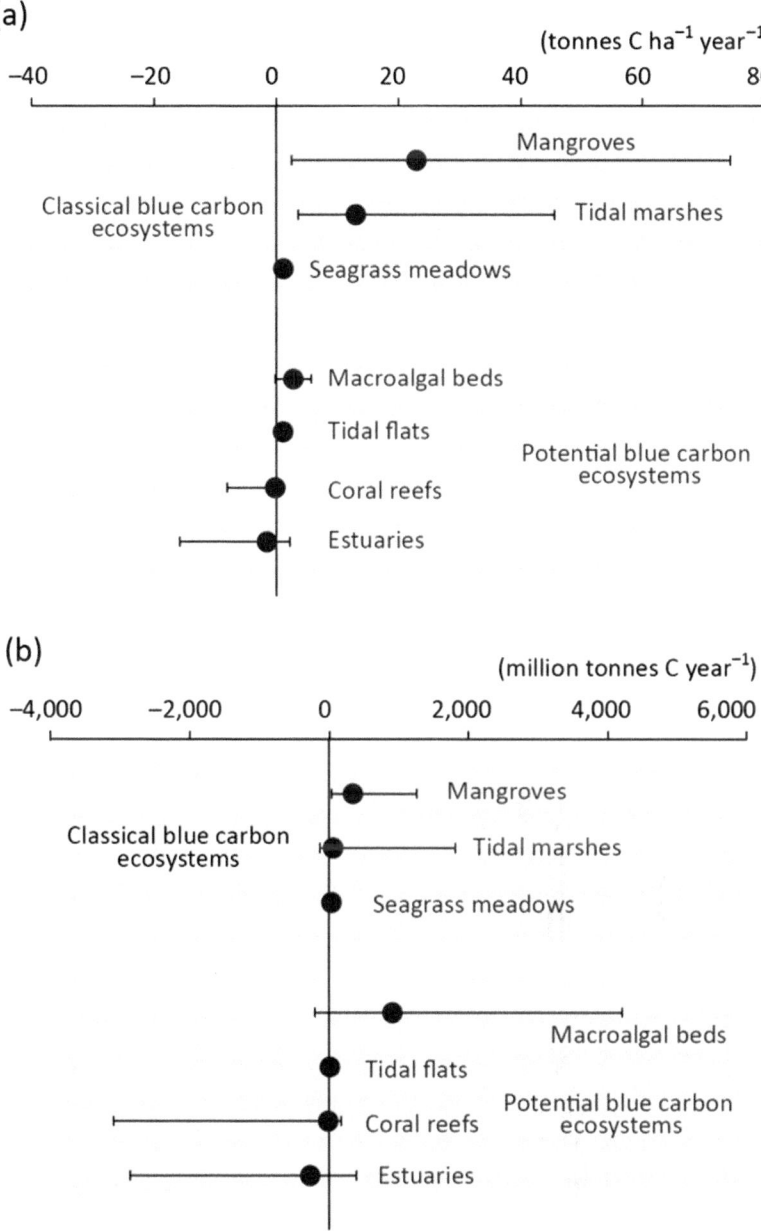

Fig. 22.4 Global annual CO_2 gas exchange rates (a) per unit area and (b) on a global basis (modified from Kuwae and Hori 2019). Positive values indicate the uptake of atmospheric CO_2. Bars indicate maximum and minimum values

22.7.4 Considering Co-benefits

Although this chapter focused on the specific ecosystem service of climate regulation, SCEs provide various ecosystem services and thus represent natural capital. It is important to manage this natural capital to take advantage of co-benefits such as food provision, recreation, environmental purification, health, and employment creation and to contribute to food security, ecosystem integrity, and biodiversity (Okada et al. 2019). However, highlighting climate change countermeasures, in which society is becoming increasingly interested and concerned, can help to initiate or accelerate the conservation and restoration of SCEs.

22.7.5 Monetizing and Crediting Co-benefits

Ensuring that the effectiveness and importance of SCEs are widely perceived and understood by the various coastal stakeholders, researchers, engineers, and economists is of first importance when quantifying SCE functions and monetizing, and ideally crediting, the full range of the provided benefits (Okada et al. 2021). Even if practitioners perceive SCEs to be cost-effective and to have an acceptable internal rate of return, the lack of quantitative evidence of the benefits of SCEs may discourage planners, policy-makers, and investors from deciding to implement or invest in blue carbon projects (Emerton 2013).

When considering the cost-effectiveness of SCEs, it is preferable to first quantify all of their functions and to consider trade-offs among them and then to monetize them according to the results they produce, rather than to base the monetization of the benefits of SCEs on willingness to pay as determined by questionnaire surveys. Function-based monetization, which involves the direct evaluation of multifunctionality supported by numerical evidence, is more likely to satisfy coastal stakeholders and to help secure public financing or attract funds from private companies and investors (Vanderklift et al. 2019).

For most projects, there is strong pressure not to waste money, and projects that are highly uncertain or have a high risk of failure from the perspective of cost-effectiveness are unlikely to be approved by decision-makers (Hinkel et al. 2014, 2018). Risk-taking decision-making is difficult not only in the case of public works projects but also in the case of investor-financed pro-environmental activities undertaken by private companies and through private donations. Hence, risk exposure must be reduced by improving the cost-effectiveness of projects.

22.7.6 Challenges to Expanding Credit Trading

Although the quantitative social impact of blue carbon offset credits is minimal at this time, given the potential of blue carbon for mitigating climate change, expanding the volume and enhancing the social impact of credit trading are important (Fig. 22.1). Some future challenges to be addressed are listed below.

First, it is necessary to further improve the motivation of both credit creators and buyers. For example, the Japanese government has set a goal to account for blue carbon in the national inventory by 2024. Making blue carbon offset credit schemes that contribute directly to such a national indicator is expected to improve the motivation of credit creators. In addition, for credit buyers, being able to reflect the blue carbon credits they offset in the NDCs of the Paris Agreement will enable them to contribute directly to the international community's goals and motivate them to contribute to not only their own corporate social responsibility (CSR) but also their ESG indicators.

Second, it is necessary to enhance offset credit transaction products to generate interest by more participants. One idea to increase the number of participants is to present an array of trading products by assessing the economic value of co-benefits and allowing them to be traded together with carbon. By doing so, credit creators can expect to increase their sales proceeds by increasing both the unit price and transaction volume. In turn, credit buyers would be motivated by the ability to choose trading products that better fit their goals and branding messages.

Third, increasing the number of demonstration projects, such as the project promoted by JBE in collaboration with the Japanese government and the collaboration with local government promoted by the City of Yokohama, and accumulating good practices will contribute to raising awareness and interest in the blue carbon offsetting, which will in turn motivate participants.

Finally, we should not forget the role of the credit secretariat, which mediates transactions. At present, because the amount of blue carbon credits being traded worldwide is low, it is difficult to maintain the system in a stable and sustainable manner with the income from intermediary fees. Increasing the trading volume and the unit price by increasing the number of participants and motivating them to participate in the credit system is thus important for the smooth operation of the secretariat. In addition, a system such as a validation and verification body (VVB), which is independent of the credit secretariat as established by JBE, is important to ensure the credibility of the system through enhanced validation and verification. This credibility is essential to increase the number of participants in the credit system.

References

Abo K, Sugimatsu K, Hori M, Yoshida G, Shimabukuro H, Yagi H, Nakayama A, Tarutani K (2019) Quantifying the fate of captured carbon: from seagrass meadow to deep sea. In: Kuwae T, Hori M (eds) Blue carbon in shallow coastal ecosystems: carbon dynamics, policy, and implementation. Springer, Singapore, pp 251–271

Arias-Ortiz A, Serrano O, Masqué P et al (2018) A marine heatwave drives massive losses from the world's largest seagrass carbon stocks. Nat Clim Chang 8:338–344

Atwood TB, Connolly RM, Ritchie EG et al (2015) Predators help protect carbon stocks in blue carbon ecosystems. Nat Clim Chang 5:1038–1045

Atwood TB, Connolly RM, Almahasheer H et al (2017) Global patterns in mangrove soil carbon stocks and losses. Nat Clim Chang 7:523–528

Burden A, Garbutt RA, Evans CD et al (2013) Carbon sequestration and biogeochemical cycling in a saltmarsh subject to coastal managed realignment. Estuar Coast Shelf Sci 120:12–20

Cahoon DR, Hensel P, Rybczyk J et al (2003) Mass tree mortality leads to mangrove peat collapse at Bay Islands, Honduras, after Hurricane Mitch. J Ecol 91:1093–1105

Craft C, Megonigal P, Broome S et al (2003) The pace of ecosystem development of constructed *Spartina alterniflora* marshes. Ecol Appl 13:1417–1432

Crooks S, Herr D, Tamelander J, Laffoley D, Vandever J (2011) Mitigating climate change through restoration and management of coastal wetlands and near-shore marine ecosystems: challenges and opportunities. https://www.uncclearn.org/sites/default/files/inventory/wb87.pdf

Crooks S, Emmer I, Murdiyarso D, Brown B (2014) Guiding principles for delivering coastal wetland carbon projects. Report to the United Nations Environment Program and the Center for International Forestry Research. http://www.cifor.org/publications/pdf_files/Books/BMurdiyarso1402.pdf

Crooks S, Sutton-Grier AE, Troxler TG, Herold N, Bernal B, Schile-Beers L, Wirth T (2018) Coastal wetland management as a contribution to the US National Greenhouse Gas Inventory. Nat Clim Chang 8:1109–1112

Crooks S, Windham-Myers L, Troxler TG (2019) Defining blue carbon: the emergence of climate context for coastal carbon dynamics. In: Windham-Myers L, Crooks S, Troxler TG (eds) A blue carbon primer: the state of coastal wetland carbon science, practice and policy. CRC Press, Boca Raton, pp 1–8

Crosswell JR, Anderson IC, Stanhope JW et al (2017) Carbon budget of a shallow, lagoonal estuary: transformations and source-sink dynamics along the river-estuary-ocean continuum. Limnol Oceanogr 62:S29–S45

Duarte CM, Krause-Jensen D (2017) Export from seagrass meadows contributes to marine carbon sequestration. Front Mar Sci 4:13

Duarte CM, Losada IJ, Hendriks IE, Mazarrasa I, Marbà N (2013) The role of coastal plant communities for climate change mitigation and adaptation. Nat Clim Chang 3:961–968

Duarte CM, Agusti S, Barbier E et al (2020) Rebuilding marine life. Nature 580:39–51

Emerton L (2013) Using valuation to make the case for economic incentives: promoting investments in marine and coastal ecosystems as development infrastructure. In: Essam M (ed) Economic incentives for marine and coastal conservation: prospects, challenges and policy implications. Earthscan Press, London

EX-ACT (2017) EX-ante carbon balance tool, blue carbon and fisheries. Food and Agriculture Organization. http://www.fao.org/3/a-i8342e.pdf

Friedlingstein P et al (2020) Global Carbon Budget 2020. Earth Syst Sci Data 12:3269–3340

Froehlich HE, Afflerbach JC, Frazier M, Halpern BS (2019) Blue growth potential to mitigate climate change through seaweed offsetting. Curr Biol 29:3087–3093

Greiner JT, McGlathery KJ, Gunnell J, McKee BA (2013) Seagrass restoration enhances "blue carbon" sequestration in coastal waters. PLoS One 8:e72469

Hamaguchi M, Shimabukuro H, Hori M, Yoshida G, Terada T, Miyajima T (2018) Quantitative real-time polymerase chain reaction (PCR) and droplet digital PCR duplex assays for detecting *Zostera marina* DNA in coastal sediments. Limnol Oceanogr Method 16:253–264

Herr D, Landis E (2016) Coastal blue carbon ecosystems. Opportunities for nationally determined contributions: policy brief. IUCN, Gland; The Nature Conservancy, Washington, DC

Herr DT, Agardy D, Benzaken F et al (2015) Coastal "blue" carbon: a revised guide to supporting coastal wetland programs and projects using climate finance and other financial mechanisms. IUCN, Gland. https://doi.org/10.2305/IUCN.CH.2015.10.en

Hill R, Bellgrove A, Macreadie PI et al (2015) Can macroalgae contribute to blue carbon? An Australian perspective. Limnol Oceanogr 60:1689–1706

Hinkel J, Lincke D, Vafeidis AT et al (2014) Coastal flood damage and adaptation costs under 21st century sea-level rise. Proc Natl Acad Sci U S A 111:3292–3297

Hinkel J et al (2018) The ability of societies to adapt to twenty-first-century sea-level rise. Nat Clim Chang 8:570–578

Hoegh-Guldberg O, Northrop E, Lubchenco J (2019) The ocean is key to achieving climate and societal goals. Science 365:1372–1374

Howard J, Hoyt S, Isensee K, Pidgeon E, Telszewski M (2014) Coastal blue carbon: methods for assessing carbon stocks and emissions factors in mangroves, tidal salt marshes, and seagrass meadows. Conservation International, Intergovernmental Oceanographic Commission of UNESCO, International Union for Conservation of Nature, Arlington, Virginia

Howard J, McLeod E, Thomas S et al (2017) The potential to integrate blue carbon into MPA design and management. Aquat Conserv Mar Freshwat Ecosyst 27(S1):100–115

IPCC (2013) Fifth assessment report of the intergovernmental panel on climate change. IPCC, Geneva

IPCC (2014) In: Hiraishi T, Krug T, Tanabe K et al (eds) 2013 supplement to the 2006 IPCC guidelines for national greenhouse gas inventories: wetlands. IPCC, Geneva

Jiao N, Wang H, Xu G, Aricò S (2018) Blue carbon on the rise: challenges and opportunities. Natl Sci Rev 5:464–468

Kauffman BJ, Arifanti VB, Trejo HH et al (2017) The jumbo carbon footprint of a shrimp: carbon losses from mangrove deforestation. Front Ecol Environ 15:183–188

Kelleway J, Serrano O, Cannard T et al (2017) Technical review of opportunities for including blue carbon in the Australian Government's emissions reduction fund. CSIRO, Canberra

Kelleway JJ et al (2020) A national approach to greenhouse gas abatement through blue carbon management. Glob Environ Chang 63:102083

Krause-Jensen D, Duarte CM (2016) Substantial role of macroalgae in marine carbon sequestration. Nat Geosci 9:737–742

Krause-Jensen D, Lavery P, Serrano O, Marbà N, Masque P, Duarte CM (2018) Sequestration of macroalgal carbon: the elephant in the blue carbon room. Biol Lett 14:20180236

Kuwae T, Crooks S (2021) Linking climate change mitigation and adaptation through coastal green–gray infrastructure: a perspective. Coast Eng J 63:188–199

Kuwae T, Hori M (2019) Blue carbon in shallow coastal ecosystems. Springer, Singapore, 373 p

Kuwae T, Kanda J, Kubo A et al (2016) Blue carbon in human-dominated estuarine and shallow coastal systems. Ambio 45:290–301

Kuwae T, Yoshida G, Hori M, Watanabe K, Tanaya T, Okada T, Umezawa Y, Sasaki J (2019) Nationwide estimation of the annual uptake of atmospheric carbon dioxide by shallow coastal ecosystems in Japan. J Jpn Soc Civil Eng B2 75:10–20

Laffoley DA, Grimsditch G (2009) The management of natural coastal carbon sinks. IUCN, Gland, 53 pp. https://portals.iucn.org/library/sites/library/files/documents/2009-038.pdf

Lovelock CE, Duarte CM (2019) Dimensions of blue carbon and emerging perspectives. Biol Lett 15:20180781

Lovelock CE, Atwood T, Baldock J et al (2017) Assessing the risk of carbon dioxide emissions from blue carbon ecosystems. Front Ecol Environ 15:257–265

Lutz SJ, Neumann C, Bredbenner A (2014) Building blue carbon projects: an introductory guide. https://gridarendal-website-live.s3.amazonaws.com/production/documents/:s_document/317/original/building_blue_carbon_projectsLowRes.pdf?1489067674

Macreadie PI, Trevathan-Tackett SM, Skilbeck CG et al (2015) Losses and recovery of organic carbon from a seagrass ecosystem following disturbance. Proc R Soc B Biol Sci 282:1–6

Macreadie PI et al (2019) The future of blue carbon science. Nat Commun 10:3998

Marbà N, Arias-Ortiz A, Masqué P, Kendrick GA, Mazarrasa I, Bastyan GR, Garcia-Orellana J, Duarte CM (2015) Impact of seagrass loss and subsequent revegetation on carbon sequestration and stocks. J Ecol 103:296–302

Martin A, Landis E, Bryson C, Lynaugh S, Mongeau A, Lutz S (2016) Blue carbon: nationally determined contributions inventory. Appendix to: coastal blue carbon ecosystems: opportunities for nationally determined contributions. GRID, Arendal, Norway

MLIT (2020) Ministry of land, infrastructure, transport and tourism. https://www.mlit.go.jp/report/press/port06_hh_000216.html

Murray BC, Pendleton L, Jenkins WA, Sifleet S (2011) Green payments for blue carbon: economic incentives for protecting threatened coastal habitats. Nicholas Institute for Environmental Policy Solutions, Report NI, 11(04). https://nicholasinstitute.duke.edu/sites/default/files/publications/blue-carbon-report-paper.pdf

Nellemann C, Corcoran E, Duarte CM, et al (2009) Blue Carbon: a rapid response assessment. United Nations Environmental Programme, GRID-Arendal, Birkeland Trykkeri AS, Birkeland

Nobutoki M, Yoshihara S, Kuwae T (2019) Carbon offset utilizing coastal waters: Yokohama blue carbon project. In: Kuwae T, Hori M (eds) Blue carbon in shallow coastal ecosystems. Springer, Singapore

O'Connor JJ, Fest BJ, Sievers M, Swearer SE (2020) Impacts of land management practices on blue carbon stocks and greenhouse gas fluxes in coastal ecosystems- a meta-analysis. Glob Chang Biol 26:1354–1366

Okada T et al (2019) Method for the quantitative evaluation of ecosystem services in coastal regions. PeerJ 6:e6234

Okada T, Mito Y, Akiyama YB et al (2021) Green port structures and their ecosystem services in highly urbanized bay. Jpn Cost Eng J. https://doi.org/10.1080/21664250.2021.1911194

Orr KK (2014) Floating seaweed (Sargassum). In: Laffoley D, Baxter JM, Thevenon F, Oliver J (eds) The significance and management of natural carbon stores in the open ocean. Full report. IUCN, Gland, pp 55–67

Ortega A, Geraldi NR, Alam I et al (2019) Important contribution of macroalgae to oceanic carbon sequestration. Nat Geosci 12:748–754

Osland MJ, Spivak AC, Nestlerode J et al (2012) Ecosystem development after mangrove wetland creation: plant–soil change across a 20-year chronosequence. Ecosystems 15:848–866

Queirós AM et al (2019) Connected macroalgal-sediment systems: blue carbon and food webs in the deep coastal ocean. Ecol Monogr 89:e01366

Salmo SG, Lovelock CE, Duke NC (2013) Vegetation and soil characteristics as indicators of restoration trajectories in restored mangroves. Hydrobiology 720:1–18

Sapkota Y, White JR (2020) Carbon offset market methodologies applicable for coastal wetland restoration and conservation in the United States: a review. Sci Total Environ 701:134497

Schile-Beers L et al (2019) Carbon sequestration in arid blue carbon ecosystems: a case study from the United Arab Emirates. In: Windham-Myers L, Crooks S, Troxler TG (eds) A blue carbon primer: the state of coastal wetland carbon science, practice and policy. CRC Press, Boca Raton, pp 327–339

Serrano O, Ruhon R, Lavery PS et al (2016) Impact of mooring activities on carbon stocks in seagrass meadows. Sci Rep 6:23193

Suehiro F, Suzuki H, Yoshihara S, Shuji Okazaki S (2020) Study on the world's first credit certification for bule carbon in eelgrass fields in Yokohama City. J Jpn Soc Civil Eng G3:49–53

Sugimura Y, Kobayashi T, Mito Y, Yoshihara S, Okada T, Kuwae T (2021) The establishment and future prospects of blue carbon offset system at Hakata Port. J Jpn Soc Civil Eng 77:31–48

Sutton-Grier AE, Wowk K, Bamford H (2015) Future of our coasts: the potential for natural and hybrid infrastructure to enhance the resilience of our coastal communities, economies and ecosystems. Environ Sci Pol 51:137–148

Thomas S (2014) Blue carbon: knowledge gaps, critical issues, and novel approaches. Ecol Econ 107:22–38

Tokoro T, Watanabe K, Tada K, Kuwae T (2015) Guideline of blue carbon (CO_2 absorption and carbon sequestration) measurement methodology in port areas. Technical note of the port and airport research institute no 1309

Vanderklift MA et al (2019) Constraints and opportunities for market-based finance for the restoration and protection of blue carbon ecosystems. Mar Policy 107:103429

Verified Carbon Standard (2015) Methodology for tidal wetland and seagrass restoration. http://www.v-c-s.org/methodologies/methodology-tidal-wetland-and-seagrass-restoration-v10

Villa JA, Bernal B (2018) Carbon sequestration in wetlands, from science to practice: an overview of the biogeochemical process, measurement methods, and policy framework. Ecol Eng 114:115–128

Wada S, Aoki MN, Mikami A et al (2008) Bioavailability of macroalgal dissolved organic matter in seawater. Mar Ecol Prog Ser 370:33–44

Watanabe K, Seike K, Kajihara R, Montani S, Kuwae T (2019) Relative sea-level change regulates organic carbon accumulation in coastal habitats. Glob Chang Biol 25:1063–1077

Watanabe K, Yoshida G, Hori M, Umezawa Y, Moki H, Kuwae T (2020) Macroalgal metabolism and lateral carbon flows can create significant carbon sinks. Biogeosciences 17:2425–2440

World Bank (2017) Implementing nature based flood protection: principles and implementation guidance. http://documents.worldbank.org/curated/en/739421509427698706/Implementing-nature-based-flood-protection-principles-and-implementation-guidance

Wylie L, Sutton-Grier AE, Moore A (2016) Keys to successful blue carbon projects: lessons learned from global case studies. Mar Policy 65:76–84

Wyndham-Meyers L, Crooks S, Troxler T (2018) A blue carbon primer: the state of coastal wetland carbon science, practice and policy. CRC Taylor and Francis, Boca Raton, 352 p

Part VII
Economic Evaluation

Chapter 23
Understanding Preference Differences Among Individuals for the Reduction in Flood Risk by Green Infrastructure

Yasushi Shoji, Takahiro Tsuge, and Ayumi Onuma

Abstract The purpose of this study is to grasp, through a discrete choice experiment, the general public's preferences regarding green infrastructure that provides flood-control services. Green infrastructure, unlike artificial structures (gray infrastructure) such as continuous artificial levees, can potentially handle floods that exceed what is envisioned at the planning stages. However, there is also the possibility that they may not be able to handle the expected floods. People's preferences could be heterogeneous when it comes to an infrastructure that has such a risk. The results of the latent class model indicated that people's preferences regarding green infrastructure were heterogeneous. Respondents who regard green infrastructure as not contributing to nature conservation and as an excuse to carry out unnecessary river-management projects evaluated gray infrastructure more favorably. It was also revealed that the more confident respondents were in providing their answers, the more likely they were to support green infrastructure. These results may suggest that more understanding will be required for a consensus to be formed regarding the use of green infrastructure.

Keywords Flood risk · Green infrastructure · Preference heterogeneity · Discrete choice experiment · Latent class model

Y. Shoji (✉)
Research Faculty of Agriculture, Hokkaido University, Sapporo, Japan
e-mail: yshoji@for.agr.hokudai.ac.jp

T. Tsuge
Graduate School of Global Environmental Studies, Sophia University, Tokyo, Japan

A. Onuma
Faculty of Economics, Graduate School of Economics, Keio University, Tokyo, Japan

23.1 Introduction

23.1.1 Background

Floods affect more people globally each year than any other disaster (IFRC 2020). In 2019, 127 floods affected 69 countries, killed 1586 people, and displaced ten million more (IFRC 2020; IDMC 2019). Japan, which has an annual rainfall almost twice the world's average, also suffered disasters, including floods (IFRC 2020). At least 140,000 people in Japan were affected by floods (including tidal waves and debris flows) during the 60 years between 1950 and 2010 (MLIT 2019). In response, the country has built artificial structures (gray infrastructure), such as continuous artificial levees and large dams to protect people's lives and assets from floods (Nakamura et al. 2020). These efforts have achieved some successes; for example, the flooded area in urban districts decreased significantly in the latter half of the 1980s (MLIT 2019). However, the flooded area in urban districts has not declined any further since 1990, and the number of flood victims began to rise in the 2010s (MLIT 2019).

This is partly due to fluctuations in rainfall patterns resulting from climate change. Since the 1970s, fluctuations in annual rainfall have been increasing (JMA 2021a). The number of "extremely heavy" rainfalls (50–80 mm per hour) and "intense" rainfalls (more than 80 mm per hour) is on the rise (JMA 2021b). The continuous artificial levees and large dams that have been constructed so far do not function adequately to control megafloods caused by such extreme rainfall. This is because, while many of these structures are designed to withstand, for example, a flood that may be caused by the rainfall of once-in-100-years intensity, megafloods are caused by rainfall that exceeds that level.

In response to this situation, expanding the size of gray infrastructure is being considered. However, in addition to fiscal problems, a major concern is that gray infrastructure may drastically transform the natural environment. This is because such facilities are changing flow, sediment, and large wood regimes (Lytle and Poff 2004; Nakamura et al. 2017) and exert a great negative impact on the biodiversity of aquatic and riparian organisms (Nakamura et al. 2020). The general public is becoming reluctant to accept a major expansion of gray infrastructure in part because of this scientific insight. For example, a plan was created to artificially construct a waterway stretching 40 km in the Chitose River watershed in Hokkaido, Japan, to prevent flooding. However, the plan raised concerns about its potential impact on wetlands and coastal ecosystems. After years of debate, the plan was scrapped in 1999, and a new river management project featuring a flood control basin was implemented (Yamanaka et al. 2020; Nakamura et al. 2020; Kim et al. 2021).

Against this backdrop, expectations are growing for green infrastructure. Green infrastructure can be broadly defined as a strategically planned network of high-quality natural and semi-natural areas with other environmental features and is designed and managed to deliver a wide range of ecosystem services and protect biodiversity in both rural and urban settings (European Commission 2013). Green infrastructure in urban and surrounding rural areas can complement large-scale gray infrastructure in areas in the form of flood protection (IPBES 2019). The term "green infrastructure" is also used in the context of rainwater treatment in urban areas. However, the meaning of the term used in this study is more or less the same as this definition: ecosystem-based disaster risk reduction (Eco-DRR), which refers to "the suitable management, conservation, and restoration of ecosystems to reduce disaster risk" (Renaud et al. 2013). In the case of the Chitose River mentioned earlier, the inside of the constructed flood control basin has been restored from farmland to wetlands, resulting in the return of the red-crowned crane (*Grus japonensis*), a rare species and the national bird of Japan, to the area. While the embankment of the flood control basin is a gray infrastructure facility, the flood control basin as a whole is designed as a green infrastructure facility. In addition to the flood control basin, the formation of forests in the upper reaches of the river and the maintenance of wetlands in the river basin could become part of the green infrastructure. The above-mentioned flood control basin at the Chitose River will require maintenance and renovation expenses in the future. However, a great advantage of green infrastructures, such as forests and wetlands, is that they do not incur huge maintenance costs (Nakamura et al. 2020).

There is a major difference in the disaster-prevention functions between gray and green infrastructure. Gray infrastructure can cope with expected floods without failure (for example, a flood caused by the rainfall of once-in-100-years intensity), but it cannot handle any flooding that exceeds that level. Meanwhile, many green infrastructures could potentially handle floods that exceed those envisioned at the planning stages. However, there is also a possibility that they may not be able to deal with the expected floods. For example, although Nakamura et al. (2020) pointed out that wetlands could reduce peak river flows, there is a possibility that the capacity of wetlands to retain rainwater may be affected by seasonal factors and the cumulative rainfall leading up to the rainfall immediately before the flood. If a gray infrastructure facility (such as a dam) is compared to a glass, a green infrastructure facility may be compared to a sponge. A dry sponge can adequately absorb a glass full of water (flood), but a sponge that retains water will not be able to absorb water even if it is less than a glass full. Because green infrastructure uses the natural environment, its disaster-prevention function always involves certain risks. Gray infrastructure also has a limited capacity to absorb water, but it can certainly withstand flooding in situations with lower than an expected flood flow level.

The adoption of green infrastructure for flooding requires land-utilization planning and consensus-building, and various ecosystem services provided by green infrastructure must be considered in the process (Goldstein et al. 2012; Guerrero et al. 2017). Among these, flood control constitutes the core of ecosystem services, and people's preferences regarding these services provide important insights into building a consensus. Meanwhile, people's preferences regarding the flood-control services of green infrastructure could be heterogeneous. Some people may prefer expanding the size of gray infrastructure that reliably serves their functions, considering that gray infrastructure has already achieved certain results. Others may consider the notion that rainfall has intensified and put more emphasis on green infrastructure's potential capability to handle floods that exceed those envisioned levels at the planning stages. Still, others may call for a drastic policy shift toward the adoption of green infrastructure by emphasizing its capability to preserve biodiversity and provide ecosystem services. Understanding such preferences may provide important insights into forming a consensus regarding the adoption of green infrastructure.

23.1.2 The Purpose of the Study

The purpose of this study is to grasp, through a discrete choice experiment, the preferences of the general public regarding flood-control services of green infrastructure that involve certain risks. The insights gained from this study can be used in consensus building toward the adoption of green infrastructure. A discrete choice experiment is a method of grasping the preferences of individual survey respondents by asking them to choose an option that they prefer from choice sets that combine multiple alternatives (Louviere et al. 2000). The discrete choice experiment in this study assumes that gray infrastructure can handle a flood that may be caused by the rainfall of once-in-100-years intensity. The respondents were presented with a hypothetical scenario under which a measure was implemented to address a flood that may be caused by the rainfall of once-in-150-years intensity. As a means of bolstering the flood-control services so that they could handle a flood caused by the rainfall of once-in-150-years intensity, four alternatives were created: (1) bolstering gray infrastructure, (2) bolstering green infrastructure, (3) bolstering both gray and green infrastructure at a 1:2 ratio, and (4) bolstering both gray and green infrastructure at a 2:1 ratio. Six choice sets that combine these four alternatives were created ($_4C_2 = 6$), and the respondents were asked to choose the preferred alternative among each choice set. As mentioned earlier, it is assumed that respondents' preferences vary with respect to green infrastructure. This study, in order to grasp such heterogeneity in preferences, uses a latent class model to analyze the respondents' choices.

23.2 Methods

23.2.1 Literature Review

The discrete choice experiment was developed by Louviere and Hensher (1982) and Louviere and Woodworth (1983) as part of the stated preference approach. Discrete choice experiments enable the assessment of individual preferences by asking respondents to choose among various multi-attribute scenarios (Louviere et al. 2000). It was initially used in marketing and transportation (e.g., Hensher 1994; Louviere 1994); however, it is now applied to a wide variety of fields, including environmental valuation and healthcare (e.g., Ryan et al. 2008; Hoyos 2010). While this study focuses on the single attribute of flood-control services, people's preferences regarding this matter can be understood in the same way.

To understand the preferences of survey respondents, we carried out a discrete choice experiment by giving the respondents a batch of hypothetical alternatives and asking them to choose what they would prefer the most. While there are various approaches for modeling choice results (e.g., a mixed logit model, see Train (2009) for more details.), this study used a latent class model. The latent class model was proposed by McFaden (1986) and subsequently applied to empirical studies by Swait (1994). The latent class model, which assumes that respondents are divided into several segments, estimates the preference parameters of each segment and the probability of each respondent belonging to a certain segment. In contrast to the mixed logit model (e.g., Train 1998), the advantage of the latent class model is that, by adopting a membership function for the probability of each respondent belonging to a certain segment, it can grasp not only the heterogeneity of preferences but also the basis of such heterogeneity (Boxall and Adamowicz 2002). Details pertaining to the methodological aspect of the latent class model are stated in the Appendix.

Analysis using a discrete choice experiment or a latent class model is widely practiced in the field of environmental valuation. Such analysis is also used to evaluate green infrastructure and the restoration of nature (e.g., Milon and Scrogin 2006; Birol et al. 2009; Kim et al. 2021). Birol et al. (2009) used a discrete choice experiment to evaluate wetland management methods regarding flood risk, biodiversity, and recreation. Milon and Scrogin (2006) also used a discrete choice experiment to assess people's preferences regarding ecosystem services of wetlands. They also attempted to grasp the heterogeneity of preferences using a latent class model. Kim et al. (2021) targeted the green infrastructure of the Chitose River to evaluate people's preferences regarding its ecosystem services using a latent class model.

In the context of flood risk management, the study of risk perception and risk communication has gained increasing interest (Kellens et al. 2013). There exist many case studies on perception and social behavior dealing with floods (Wachinger et al. 2013; Lechowska 2018). However, our study does not focus on risk perception of flood per se, but rather on the supply uncertainty of the flood-control services. In the literature of environmental valuation, several studies using the stated preference approach try to understand preferences over uncertain outcomes (Roberts et al. 2008). There are several studies that use discrete choice experiments in this context (Glenck and Colombo 2013; Imamura et al. 2016; Glatte et al. 2019). Imamura et al. (2016) examined people's attitudes toward disaster-prevention risk in coastal areas using a discrete choice experiment. They found that coastal citizens did not prefer the excessive degradation of ecosystems due to constructing coastal structures and argued that the introduction of green infrastructure could be a solution. This study can be positioned as one such study.

23.2.2   Survey Design

The discrete choice experiment was conducted using an online questionnaire survey. The survey respondents were provided with graphs and texts that explained the scenarios in detail. Since the purpose of this study is to evaluate the risk of flood-control services, a scenario was established under which flood-control services would be bolstered as an additional measure against intensifying rainfalls. Japan already has a gray infrastructure. Thus, a scenario for replacing it with green infrastructure would be unrealistic. A scenario under which green infrastructure is considered an addition to existing functions would be more realistic. In the survey questionnaire, Fig. 23.1 was presented to the respondents, which explained the characteristics of the existing gray infrastructure facilities.

Fig. 23.1 indicates that gray infrastructure can withstand a flood that may be caused by the rainfall of once-in-100-years intensity (safety: 100%; the probability of flooding: 0%) but cannot withstand any flood caused by rainfall that exceeds the once-in-100-years intensity (safety: 0%; flooding will certainly occur). Next, the respondents were presented with a scenario where flood-control services would be bolstered as an additional measure (Fig. 23.2). The respondents were explained that this additional measure would be implemented with the use of gray infrastructure. To advance the understanding of the survey respondents, gray infrastructure was compared to glass and green infrastructure to a sponge.

The meaning of green infrastructure was explained to the survey respondents using the example of wetlands shown in Nakamura et al. (2020). Then, a situation where flood-control services would be bolstered through green infrastructure was

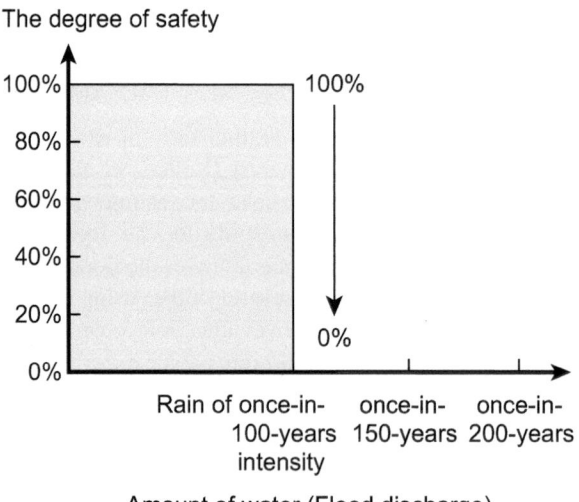

Fig. 23.1 The characteristics of the existing gray infrastructure facility

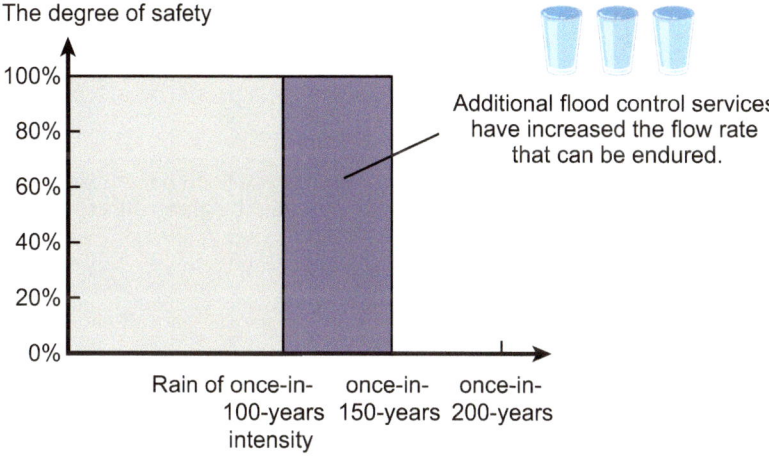

Fig. 23.2 Additional flood-control services are provided by gray infrastructure

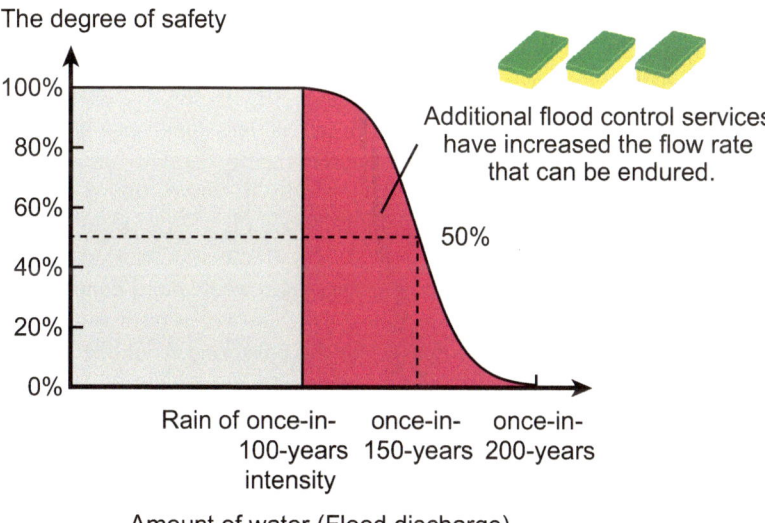

Fig. 23.3 Additional flood-control services are provided through green infrastructure only

explained (Fig. 23.3). This study assumes that there is a 50% chance that the flood-control service, if bolstered by green infrastructure, would withstand a flood caused by rainfall of once-in-150-years intensity. It would, however, not withstand a flood caused by the rainfall of once-in-200-years intensity, but there is a likelihood that it would withstand a flood caused by the rainfall of once-in-150-years or more and less than once-in-200-years intensity. It is also possible that it may not

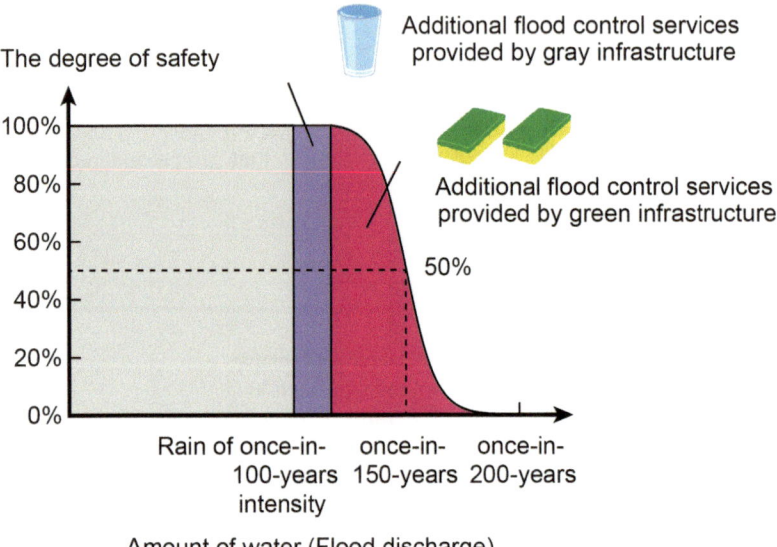

Fig. 23.4 Gray infrastructure provides 2/3, and green infrastructure 1/3, of the additional flood-control services

be able to withstand a flood caused by rainfall of less than once-in-150-years intensity. However, the existing gray infrastructure facility certainly prevents floods caused by rainfall of once-in-100-years intensity. In the survey questionnaire, the respondents were asked whether they would choose Figs. 23.2 or 23.3 after the above explanations were provided.

Then, explanations were given regarding a situation where flood-control services would be bolstered through a combination of gray and green infrastructure (Figs. 23.4 and 23.5). Figure 23.4 shows that 1/3 of the bolstering would be discharged through gray infrastructure and 2/3 through green infrastructure (one glass and two sponges). The facility would certainly withstand a flood that may occur as a result of the heaviest rainfall of approximately 117 years. However, the probability that it will withstand a flood caused by the heaviest rainfall of 150 years or longer is lower than a situation where the bolstering would be discharged by green infrastructure alone. Figure 23.5 assumes that 2/3 of the bolstering would be discharged through gray infrastructure and 1/3 through green infrastructure (two glasses and one sponge). The facility would certainly withstand a flood that may occur as a result of the heaviest rainfall in approximately 133 years, but the probability that it will withstand a flood caused by the heaviest rainfall of 150 years or longer is lower than a situation where 2/3 of the bolstering is discharged by green infrastructure alone.

Regarding the bolstering of flood-control services, the four alternative plans have differing ratios of green infrastructure (0%, 33%, 67%, and 100%). Thus, there are six combinations that contain two alternatives each ($_4C_2 = 6$). A choice had already been made regarding a pair that consists of an alternative plan under which

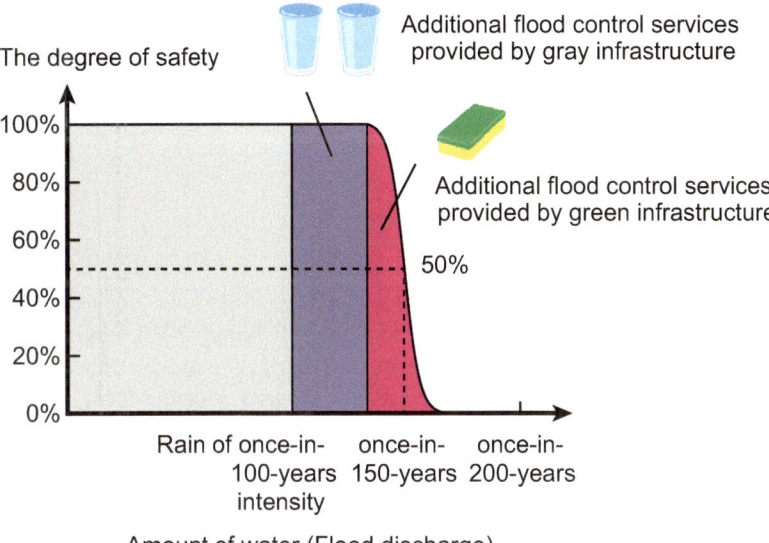

Fig. 23.5 Gray infrastructure provides 1/3, and green infrastructure 2/3, of the additional flood-control services

the bolstering would be discharged 100% by gray infrastructure and an alternative plan under which the bolstering would be discharged 100% by green infrastructure. Therefore, the respondents were asked to choose their preferred option regarding each of the five remaining pairs (choice sets). Since all combinations were presented, the number of times that each alternative appeared in the series of six choice tasks was the same.

In the online questionnaire survey, in addition to the questions for the discrete choice experiment, the respondents were also asked about their attributes (age, gender, income, etc.), whether they knew about green infrastructure, and their attitude toward green infrastructure. Regarding their attitudes toward green infrastructure, the following statements were evaluated on a seven-point Likert scale (from strongly agree to strongly disagree) (e.g. Jamieson 2004).

- Green infrastructure is necessary
- Green infrastructure will contribute to the conservation and restoration of the natural environment
- Green infrastructure should be increased in the future
- In a river management project, the method that offers the best flood prevention should be chosen regardless of whether it is related to green infrastructure.
- Green infrastructure is an excuse to carry out unnecessary river management projects

In addition, since the content of this questionnaire survey may have been difficult for the general public to understand, the respondents were also asked whether they provided their answers with confidence. It is assumed that questions other than those for the discrete choice experiment are incorporated into the membership function for the model shown in the Appendix.

23.3 Results

23.3.1 Surveys

The online questionnaire survey was conducted in February 2019, targeting members of the general public in Hokkaido. A research firm conducted the survey. A survey request was sent to ordinary citizens registered with the firm. They were asked to provide answers on the website. In conducting the survey, care was taken to ensure that the composition of the respondents reflected that of the people of Hokkaido with respect to gender and age. A total of 1054 responses were obtained from the website. Of these responses, 914 were chosen for analysis. These were from respondents who answered all questions and did not provide answers too quickly (those who answered questions too quickly may not have read the questions well).

23.3.2 Descriptive Statistics

A total of 46.8% of the respondents were men, while 53.2% were women. People in their 20s comprised 13.6% of all respondents. Those in their 30s, 40s, 50s, and 60s were 16.3%, 21.8%, 20.4%, and 28.0% of the total respondents, respectively. According to the 2015 national census for Hokkaido, men comprised 48.4%, while women comprised 51.6% of the overall population in the prefecture (Statistics Bureau of Japan 2015). Those in their 20s, 30s, 40s, 50s, and 60s were 14.2%, 18.4%, 21.8%, 20.3%, and 25.3%, respectively (Statistics Bureau of Japan 2015). Thus, the composition of the respondents is very similar to that of Hokkaido residents with respect to gender and age.

The results showed that 3.6% of the respondents responded that they knew the meaning of the term "green infrastructure," while 16.6% responded that they had heard the term. A total of 79.8% of the respondents responded that they did not know the meaning of the term. Table 23.1 indicates respondents' attitudes toward green infrastructure. Their responses were classified into "positive" (strongly agree, agree, somewhat agree), "neutral" (cannot say either way), and "negative" (strongly disagree, disagree, somewhat disagree). Asked if they provided their answers with confidence, 23.6% gave affirmative answers, while 34.5% responded that they could not say either way. A total of 41.9% responded that they were not confident.

Table 23.1 Attitude toward Green Infrastructure

	Positive (%)	Neutral (%)	Negative (%)
Green infrastructure is necessary	57.5	38.7	3.7
Green infrastructure will contribute to the conservation and restoration of the natural environment	62.1	33.4	4.5
Green infrastructure should be increased in the future	54.9	40.4	4.7
In a river management project, the method that offers the best flood prevention should be chosen regardless of whether it is related to green infrastructure	52.5	41.1	6.3
Green infrastructure is an excuse to carry out unnecessary river management projects	15.3	60.2	24.5

Table 23.2 Estimation results by a conditional logit model

Green infrastructure ratio	Coefficient
0% (gray infrastructure)	−2.5115
33.3%	0.9196[***]
66.6%	1.1232[***]
100%	0.4687[***]
Number of observations	5484
Loglikelihood	−2834.06
Log-likelihood ratio index	0.23

[***]$p < 0.01$, [**]$p < 0.05$

23.3.3 The Conditional Logit Result

The estimation results of the conditional logit model are presented in Table 23.2. The alternative-specific constant, which is a constant term, was excluded from the estimation results because it was not statistically significant. The preference parameters are all estimated as effect-coded dummy variables (Bech and Gyrd-Hanse 2005), with a situation where every facility is a gray infrastructure facility as the base (0%). Dummy variables, in general, have a base value of zero, but effect-coded dummy variables have a base value of −1. For effect-coded dummy variables, the coefficient for the gray infrastructure level can be calculated as the sum of the other levels multiplied by −1 (−{0.9196 + 1.1232 + 0.4687} = −2.5115). We assume that preferences are heterogeneous and that it would be unrealistic to assign 0 to a situation where every facility is a gray infrastructure facility. For this reason, we used effect-coded dummy variables. If a preference parameter is positive and statistically significant, it has a positive impact on utility. As a result, the probability increases that an alternative at such a level will be chosen. In contrast, if a preference parameter is negative and statistically significant, the probability declines when an alternative at such a level is chosen. However, the estimation results of the effect-coded dummy variables can change depending on the base value. Thus, it was considered whether a preference parameter is positive, negative, or an absolute value. Also, we considered the difference between preference parameters. The likelihood ratio index, which indicates the appropriateness of the model, was 0.23. LRI is regarded as favorable if it is between 0.2 and 0.4 (McFadden 1978).

23.3.4 The Latent Class Result

The estimation results of the latent class model are presented in Table 23.3. The number of segments is specified in the table. There are three segments. Segments are often determined by considering the interpretability of the results, in addition to statistical indicators (Swait 1994; Scarpa and Thiene 2005; Hynes et al. 2008). AIC, AIC3, and BIC indicated that a six-segment model, a five-segment model, and a three-segment model would be the most appropriate. Thus, the most appropriate number of segments differed depending on the statistical indicators. If the number

Table 23.3 The estimation results by the latent class model

	Coefficient		
	Segment 1: Green infrastructure oriented (24.2%)	Segment 2: Gray infrastructure oriented (13.4%)	Segment 3: Intermediate (62.5%)
Green infrastructure ratio			
0% (gray infrastructure)	-7.6866^{***}	4.3394^{***}	-5.2154
33.3%	1.3699^{***}	-0.4523^{***}	2.1526^{***}
66.6%	2.8622^{***}	-1.3531^{***}	2.4041^{***}
100%	3.4545^{***}	-2.5340^{***}	0.6587^{***}
Membership function			
Constant	-2.6608^{***}	-1.5545^{**}	0 (fixed)
Green infrastructure will contribute to the preservation and restoration of the natural environment (seven-point Likert scale)	0.0345	-0.3737^{***}	0 (fixed)
Green infrastructure is an excuse to carry out unnecessary river management projects (seven-point Likert scale)	0.0323	0.3605^{***}	0 (fixed)
I answered this survey confidently (dummy variable: Yes = 1)	0.4265^{***}	0.1175	0 (fixed)
Number of observations			5484
Number of respondents			914
Loglikelihood			-2088.30
Log-likelihood ratio index		0.45	

$^{***}p < 0.01, ^{**}p < 0.05$

of segments is higher than four, segments will emerge that are difficult to interpret. Thus, we adopted a three-segment model.

The estimation by the latent class model included gender, age, income, and so on as part of the membership function. However, none of the differences were statistically significant. Regarding the variables related to attitude, the statements "Green infrastructure is necessary" and "Green infrastructure should be increased in the future" were excluded because they overlap with discrete choice experiment questions. However, the statement "In a river management project, the method that offers the best flood prevention should be chosen regardless of whether it is related to green infrastructure" was not statistically significant. Whether respondents provided their answers confidently was also included in the membership function because it influenced segmentation. Table 23.3 shows a model that retains only the statistically significant variables.

As the table shows, the classification is as follows: segment 1, 37.0%; segment 2, 38.0%; and segment 3, 24.9%. Regarding the estimated coefficients of the membership function, it should be noted that the parameter for Segment 3 was standardized as 0 for standardization. Therefore, the interpretation of the parameters of Segments 1 and 2 is that their values are relative to Segment 3. As in the case of the preference parameter, if the membership parameter is positive and statistically significant, the evaluation rating is high regarding the applicable questions, and respondents who answered "yes" are more likely to be classified as part of a segment. If the parameter is negative and statistically significant, the likelihood of being classified becomes lower.

23.4 Discussion

23.4.1 The General Attitude toward Green Infrastructure

While only 3.6% of the survey respondents responded that they knew the meaning of the term "green infrastructure," 79.8% said that they did not know the term. Thus, green infrastructure may not be familiar to most members of the general public. After the meaning of green infrastructure was explained, the respondents were asked whether they thought green infrastructure would be necessary. In response, 57.5% gave positive answers. Although 38.7% of the respondents were neutral regarding this issue, only 3.7% provided negative answers. They provided similar responses to the statement "Green infrastructure will contribute to the preservation and restoration of the natural environment" and the statement "Green infrastructure should be increased in the future." Regarding the statement "In a river management project, the method that offers the best flood prevention should be chosen regardless of whether it is related to green infrastructure," 57.5% gave positive answers. As expected at the outset, the provision of flood prevention services is regarded as the priority in a river management project. In Japan, bid-rigging was sometimes practiced for many large public works projects, including river management projects. For this reason, there was a certain number of people who chose the statement "Green infrastructure is an excuse to carry out unnecessary river management projects." Overall, most people were unfamiliar with the term "green infrastructure." Thus, even though people's attitudes toward green infrastructure are favorable in general, they have not been solidified at this time.

23.4.2 Interpretation of the Results of the Discrete Choice Experiment

The results of the conditional logit model shown in Table 23.2 are the average evaluation results of the respondents. The preference parameter for gray infrastructure was low: -2.5115. Meanwhile, the rating rises when green infrastructure is mixed. However, the rating declines once again when green infrastructure becomes 100%. These results may indicate that the respondents regarded green infrastructure as desirable and provided high evaluations, but they still hesitated to support a complete switch to green infrastructure, which is a concept that they learned for the first time.

Fig. 23.6 shows a diagram that makes the results of the latent class model easier to understand. Segment 3 comprised 62.5% of all respondents. The changes in the value of the preference parameter are also similar to those of the conditional logit model. For this reason, the results of this group are reflected strongly in the results of the conditional logit model, which reflects average answers.

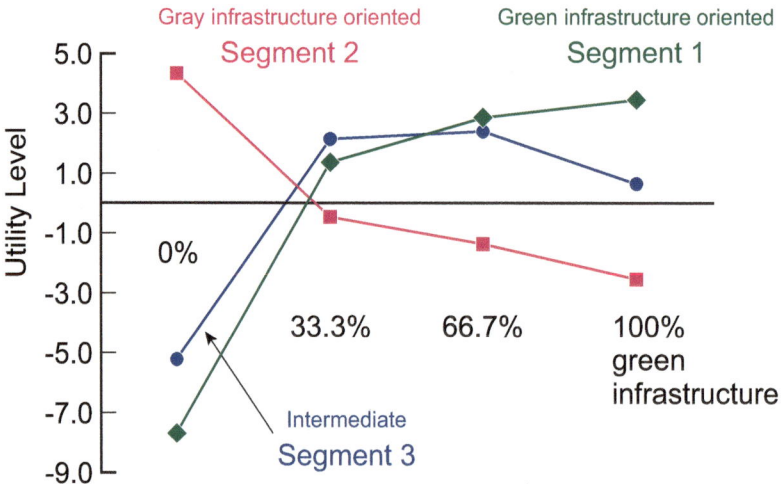

Fig. 23.6 The Estimation Results by the Latent Class Model

Next, Segment 2 is highly likely to include people who responded negatively to the statement "Green infrastructure will contribute to the preservation and restoration of the natural environment" and affirmatively to the statement "Green infrastructure is an excuse to carry out unnecessary river management projects." As for the preference parameter, the evaluation rating for gray infrastructure is the highest, and the rating declines as the ratio of green infrastructure increases. Thus, respondents who find problems with green infrastructure tend to support existing gray infrastructure. Segment 2 can be named "gray infrastructure oriented" segment.

Meanwhile, Segment 1 is not different from Segment 3 regarding the attitude toward green infrastructure. However, Segment 1 included more people who provided their responses confidently in the discrete choice experiment. Fig. 23.6 shows that the value of the preference parameter for Segment 1 is similar to that of Segment 3 and that the two differ only with respect to the evaluation of "100% gray infrastructure" and "100% green infrastructure." Segments 1 and 3 can be named "green infrastructure oriented" and "intermediate" segment, respectively. This may be interpreted to mean that people, as they increase their understanding of green infrastructure, provide a low evaluation of "100% gray infrastructure" and a high evaluation of "100% green infrastructure." Whether the respondents knew about green infrastructure was also included in the membership function for analysis, but the coefficient was not statistically significant. What mattered was not whether the respondents knew about green infrastructure, but whether they provided their responses confidently. This is an intriguing point.

23.4.3 Several Perspectives toward Consensus Building

This study seeks to use the obtained insights to help build a consensus regarding the adoption of green infrastructure. Several issues can be raised from this perspective. First, the green infrastructure is unfamiliar to most people. For this reason, awareness should be raised. The issue was explained only briefly in the survey questionnaire, but the overall evaluation of green infrastructure was high, and there were relatively few negative responses. This is a significant advantage when considering the use of green infrastructure.

On the other hand, 41.9% of the respondents did not respond to this questionnaire with confidence, while 23.6% said they provided their answers confidently. Those who were not confident greatly exceeded those who were, indicating that the content of the survey was difficult for many people. A total of 52.5% of the respondents provided affirmative answers to the statement "In a river management project, the method that offers the best flood prevention should be chosen regardless of whether it is related to green infrastructure," meaning that they were highly interested in flood-prevention services. However, the general public who is familiar with conventional river management projects (i.e., people who think that river management projects are based on risk-free gray infrastructure) may have found it difficult to understand immediately that green infrastructure carries certain risks.

Still, those who provided their responses with confidence are more likely to be included in Segment 1, a group of people who highly evaluated a plan to bolster flood-control services through green infrastructure alone. This may indicate that the better the people understand the issue, the more likely they are to support green infrastructure. Conversely, this means that those who insist on gray infrastructure may do so because of a lack of understanding. This does not argue that green infrastructure should be forced by people. However, if many people oppose green infrastructure, it may be necessary to consider whether they do so because green infrastructure is not adequately understood. It is also important to note that Segment 2 tended to think that green infrastructure is an excuse to carry out unnecessary river management projects. As shown in Fig. 23.6, the preference for Segment 2 has an opposite trend to that of Segments 1 and 3. Clearing the doubts of people, including Segment 2, is also important in building consensus for introducing green infrastructure.

23.5   Concluding Remarks

We conducted a discrete choice experiment to grasp the preferences of the general public regarding the bolstering of flood-control services using green infrastructure that carries certain risks. While people's awareness of green infrastructure was low, they generally provided a favorable evaluation after the issue was explained. Green infrastructure, a new concept for many respondents, is a preferable option. However, on average, their evaluation of a plan to rely entirely on green infrastructure was not particularly high in part because they had not solidified their stance regarding this issue. Nevertheless, their preferences regarding green infrastructure are heterogeneous. Those who found problems with green infrastructure provided a high evaluation of gray infrastructure. It was also revealed that those who provided their responses confidently tended to have a high evaluation of green infrastructure. These results may indicate that the general public's understanding of the mechanism of green infrastructure is inadequate at this time; also, some members of the general public are suspicious of green infrastructure. Therefore, more understanding as well as removing doubts about green infrastructure is required when it comes to building a consensus on the adoption of green infrastructure.

Acknowledgments We Gratefully Acknowledge Financial Support from the Environment Research and Technology Development Fund [Number 4-1504 and 4-1805] from the Japanese Ministry of the Environment

Appendix

Data from the choice tasks for the discrete choice experiment, performed six times per respondent, were quantified using a random utility model. In the random utility model, it is assumed that the utility function is determined by the sum of the factors that influence the utility (the ratio of green infrastructure in this study) and the random term that is unobservable and stochastic. The utility function for alternative plan i is defined as $U_i = V_i + \varepsilon_i$. U_i is the total utility for alternative plan i, and V_i is its observable deterministic term (the ratio of green infrastructure). Meanwhile, ε_i is an unobservable random term. The probability that alternative plan i will be chosen from combination C is the same as the probability that U_i will be larger than U_j. Thus, it can be described as follows:

$$\Pr(i) = \Pr\left[U_i > U_j\right] = \Pr\left[V_i - V_j > \varepsilon_j - \varepsilon_i\right] \forall j \neq i, \forall j \in C.$$

It is assumed that, in the most basic conditional logit model that does not presuppose classification (i.e., respondents are not divided into several segments with homogeneous preferences like the latent class model) and presupposes that the utility of the respondents is homogeneous (McFadden 1973), the error term is distributed as type-I extreme values. In this case, the probability that alternative plan i will be selected can be expressed as follows:

$$\Pr(i) = \exp\left(\mu\beta x_i\right) / \sum_{j \in C} \exp\left(\mu\beta x_j\right).$$

Here, x_i is a vector of the observed variables, and β is a vector of utility or preference parameters. β can be estimated by the maximum likelihood method, and the scale parameter μ is assumed to be 1 in a single sample (Ben-Akiva and Lerman 1985). The conditional logit model does not presuppose classification. Thus, it is equivalent to evaluating the average preferences of respondents.

In contrast, the latent class model assumes that respondent n belongs to latent class s, which cannot be observed in advance. Swait (1994) assumes a latent membership likelihood function $M^*_{n|s} = \gamma'_s z_n + \zeta_{n|s}$. Here, z_n represents the socioeconomic and/or attitude characteristics of respondent n (in this study, the respondents' attributes and attitude toward green infrastructure), γ'_s is a vector of membership parameters, and $\zeta_{n|s}$ is an unobservable random term. For the error terms, a type-I extreme value distribution, which is independently distributed, is assumed for individuals and segments. The probability that respondent n is classified as part of segment s is as follows:

$$P_{n|s} = \exp\left(\lambda\gamma_s z_n\right) / \sum_s \exp\left(\lambda\gamma_s z_n\right),$$

where λ is a scale parameter. The joint choice probability of a respondent belonging to segment s regarding the six choice sets can be expressed as follows:

$$P_{n|s}\left(i_1, i_2, \ldots, i_6|s\right) = \prod_{t=1}^{6} P_{n|s}\left(i_t|s\right) = \prod_{t=1}^{6} \exp\left(\mu_s \beta_s x_{i,t}\right) / \sum_{j \in C} \exp\left(\mu_s \beta_s x_{j,t}\right),$$

where β_s and μ_s are the utility and scale parameters specific to segment s, respectively, and t is the number of choice sets presented to each respondent. Ultimately, the unconditional probability of respondent n belonging to segment s regarding the series of alternatives can be obtained by the following formula:

$$P_n\left(i_1, i_2, \ldots, i_6\right) = \sum_s \left[\exp\left(\lambda \gamma_s z_n\right) / \sum_s \exp\left(\lambda \gamma_s z_n\right)\right] \bullet \prod_{t=1}^{6} \left[\exp\left(\mu_s \beta_s x_{i,t}\right) / \sum_{j \in C} \exp\left(\mu_s \beta_s x_{j,t}\right)\right].$$

References

Bech M, Gyrd-Hanse D (2005) Effects coding in discrete choice experiments. Health Econ 14:1079–1083

Ben-Akiva ME, Lerman SR (1985) Discrete choice analysis: theory and application to travel demand. The MIT Press, Cambridge

Birol E, Hanley N, Koundouri P, Kountouris Y (2009) Optimal management of wetlands: quantifying trade-offs between flood risks, recreation, and biodiversity conservation. Water Resour Res 45:1–11

Boxall PC, Adamowicz WL (2002) Understanding heterogeneous preferences in random utility models: a latent class approach. Environ Resour Econ 23:421–446

European Commission (2013) Communication from the commission to the European parliament, the council, the European economic and social committee and the committee of the regions green infrastructure (GI)—enhancing Europe's natural capital. Brussels, SWD(2019) 184. https://ec.europa.eu/transparency/regdoc/rep/1/2013/EN/1-2013-249-EN-F1-1.Pdf

Glatte M, Brouwer R, Logar I (2019) Combining risk attitudes in a lottery game and flood risk protection decisions in a discrete choice experiment. Environ Resour Econ 74:1533–1562

Glenck K, Colombo S (2013) Modelling outcome-related risk in choice experiments. Aust J Agric Resour Econ 57:559–578

Goldstein JH, Caldarone G, Duarte TK, Ennaanay D, Hannahs N, Mendoza G, Polasky S, Wolny S, Daily GC (2012) Integrating ecosystem-service tradeoffs into land-use decisions. Proc Natl Acad Sci U S A 109:7565–7570

Guerrero AM, Shoo L, Iacona G, Standish RJ, Catterall CP, Rumpff L, de Bie K, White Z, Matzek V, Wilson KA (2017) Using structured decision-making to set restoration objectives when multiple values and preferences exist. Restor Ecol 25:858–865

Hensher DA (1994) Stated preference analysis of travel choice: the state of practice. Transportation 21:107–133

Hoyos D (2010) The state of the art of environmental valuation with discrete choice experiments. Ecol Econ 69:1595–1603

Hynes S, Hanley N, Scarpa R (2008) Effects on welfare measures of alternative means of accounting for preference heterogeneity in recreational demand models. Am J Agric Econ 90:1011–1027

IDMC (Internal Displacement Monitoring Centre) (2019) Global report on internal displacement. http://www.internal-displacement.org/sites/default/files/publications/documents/2019-IDMC-GRID.pdf

IFRC (International Federation of Red Cross and Red Crescent) (2020) World Disasters Report 2020. https://media.ifrc.org/ifrc/wp-content/uploads/2020/11/20201116_WorldDisasters_Full.pdf

Imamura K, Takano KT, Mori N, Nakasizuka T, Managi S (2016) Attitudes toward disaster-prevention risk in Japanese coastal areas: analysis of civil preference. Nat Hazards 82:209–226

Intergovernmental Science-Policy Platform on Biodiversity and Ecosystem Services (IPBES) (2019) Summary for policymakers of the global assessment report on biodiversity and ecosystem services–unedited advance version, Bonn, Germany. https://ipbes.net/global-assessment

Jamieson S (2004) Likert scales: how to (ab)use them. Med Educ 38:1212–1218

JMA (Japan Meteorological Agency) (2021a) Annual precipitation in the world. https://www.data.jma.go.jp/cpdinfo/temp/an_wld_r.html

JMA (Japan Meteorological Agency) (2021b) Changes in the number of days of heavy rainfall or extremely hot days. https://www.data.jma.go.jp/cpdinfo/extreme/extreme_p.html

Kellens W, Terpstra T, de Maeyer P (2013) Perception and communication of flood risks: a systematic review of empirical research. Risk Anal 33:24–49

Kim H, Shoji Y, Tsuge T, Kubo T, Nakamura F (2021) Relational values help explain green infrastructure preferences: the case of managing crane habitat in Hokkaido, Japan. People Nat 3:861–871

Lechowska E (2018) What determines flood risk perception? A review of factors of flood risk perception and relations between its basic elements. Nat Hazards 94:1341–1366

Louviere JJ (1994) Conjoint analysis. In: Bagozzi R (ed) Advances in marketing research. Blackwell Publishers, Hoboken, pp 223–259

Louviere JJ, Hensher DA (1982) Design and analysis of simulated choice or allocation experiments in travel choice modeling. In: Transportation Research Record. Transportation Research Board, Commission on Sociotechnical Systems, National Research Council, National Academy of Sciences, Washington, p 890

Louviere JJ, Hensher DA, Swait JD (2000) Stated choice methods: analysis and applications. Cambridge University Press, Cambridge

Louviere JJ, Woodworth G (1983) Design and analysis of simulated consumer choice or allocation experiments: an approach based on aggregate data. J Mark Res 20:350–367

Lytle DA, Poff NL (2004) Adaptation to natural flow regimes. Trends Ecol Evol 19:94–100

McFadden D (1973) Conditional logit analysis of qualitative choice behavior. In: Zarembka P (ed) Frontiers in econometrics. Academic Press, New York, pp 105–142

McFadden D (1978) Quantitative method for analyzing travel behaviour of individuals: some recent developments. In: Hensher DA, Stopher PR (eds) Behavioural travel modelling. Groom Helm, London, pp 279–318

McFaden D (1986) The choice theory approach to market research. Mark Sci 5:275–297

Milon JW, Scrogin D (2006) Latent preferences and valuation of wetland ecosystem restoration. Ecol Econ 56:162–175

MLIT (The Ministry of Land, Infrastructure, Transportation and Tourism) (2019) Statistical Survey on Flood Damage. https://www.e-stat.go.jp/dbview?sid=0003161327

Nakamura F, Ishiyama N, Yamanak S, Higa M, Akasaka T, Kobayashi Y, Ono S, Fuke N, Kitazawa M, Morimoto J, Shoji Y (2020) Adaptation to climate change and conservation of biodiversity using green infrastructure. River Res Appl 36:921–933

Nakamura F, Seo JI, Akasaka T, Swanson FJ (2017) Large wood, sediment, and flow regimes: their interactions and temporal changes caused by human impacts in Japan. Geomorphology 279:176–187

Renaud FG, Sudmeier-Rieux K, Estrella M (eds) (2013) The role of ecosystems in disaster risk reduction. United Nations University Press, Tokyo

Roberts DC, Boyer TA, Lusk JL (2008) Preferences for environmental quality under uncertainty. Ecol Econ 66:584–593

Ryan M, Gerard K, Amaya-Amaya M (2008) Using discrete choice experiments to value health and health care. Springer, Dordrecht

Scarpa R, Thiene M (2005) Destination choice models for rock climbing in the northeastern alps: a latent-class approach based on intensity of preferences. Land Econ 81:426–444

Statistics Bureau of Japan (2015) Population census. https://www.stat.go.jp/data/kokusei/2015/

Swait J (1994) A structural equation model of latent segmentation and product choice for cross-sectional revealed preference choice data. J Retailing Consum Serv 1:77–89

Train KE (1998) Recreation demand models with taste variation over people. Land Econ 74:230–239

Train KE (2009) Discrete choice methods with simulation, 2nd edn. Cambridge University Press, Cambridge

Wachinger G, Renn O, Begg C, Kuhlicke C (2013) The risk perception paradox—implications for governance and communication of natural hazards. Risk Anal 33:1049–1065

Yamanaka S, Ishiyama N, Senzaki M, Morimoto J, Kitazawa M, Fuke N, Nakamura F (2020) Role of flood-control basins as summer habitat for wetland species - a multiple-taxon approach. Ecol Eng 142:105617

Chapter 24
Assessing Public Preference for Construction of Giant Seawalls Using the Best–Worst Scaling Approach

Takahiro Tsuge, Yasushi Shoji, and Ayumi Onuma

Abstract Giant seawalls have been constructed in areas stricken by the 2011 Great East Japan Earthquake to protect nearby townships from tsunamis, but their construction has caused great controversy. This study conducted a survey using the best–worst scaling (BWS) method in Iwate and Miyagi Prefectures, where the seawalls are constructed to understand what specific factors citizens think should be taken into consideration during construction. The results revealed that the negative impacts on the natural environment and seawall height (safety) were ranked most highly. Additionally, the results of regression analysis with respondents' B-W scores for the negative impacts on the natural environment and seawall height (safety) as the explained variables demonstrated that people who less strongly recognized the need for a seawall were more likely to think the negative impact on the natural environment should be taken into consideration than did people who strongly recognized the need for a seawall; however, people who did not know much about seawalls were more likely to think seawall height should be considered than people who did. The regression analysis also found that people who more strongly felt there was a need to build seawalls were more likely to think seawall height should be a key consideration, whereas people who personally had suffered harm in the tsunami following the Great East Japan Earthquake were more likely to think seawall height should be a key consideration than did those who were not personally affected.

Keywords Seawall · Great East Japan Earthquake · Best–Worst Scaling · Preference

T. Tsuge (✉)
Graduate School of Global Environmental Studies, Sophia University, Tokyo, Japan
e-mail: t-tsuge-8s2@sophia.ac.jp

Y. Shoji
Research Faculty of Agriculture, Hokkaido University, Sapporo, Hokkaido, Japan

A. Onuma
Faculty of Economics, Keio University, Tokyo, Japan

© The Author(s) 2022   401
F. Nakamura (ed.), *Green Infrastructure and Climate Change Adaptation*,
Ecological Research Monographs, https://doi.org/10.1007/978-981-16-6791-6_24

24.1 Introduction

Following the Great East Japan Earthquake, which caused the death or disappearance of 18,425 people (as of March 10, 2021, according to a National Police Agency press release), Japan's Central Disaster Management Council designated high-frequency but low-impact tsunamis that occur every few decades to 100 or so years as Level 1 (L1) Tsunamis, and low-frequency but high-impact giant tsunamis like the Great East Japan Earthquake tsunami that only occur once in several 100 or even 1000 years as Level 2 (L2) Tsunamis (Central Disaster Management Council 2011; Onuma 2015). The council determined to develop coastal protection facilities, including seawalls and seaside protection forests, to protect human life, safeguarding residents' assets, stabilizing the regional economy, and securing efficient production bases if an L1 tsunami occurs. However, because the prevention of an L2 tsunami using artificial structures is unrealistic both from the perspective of the costs required to maintain facilities and the impact on the coastal environments and their usage, the council limited the focus of tsunami countermeasures solely to the protection of residents' lives related to evacuation (Central Disaster Management Council 2011; Onuma 2015).

Iwate, Miyagi, and Fukushima Prefectures were seriously damaged in the Great East Japan Earthquake. As a result of this catastrophic disaster, seawalls of up to 400 km in length have been constructed in approximately 600 locations across these prefectures, costing a total of approximately ¥1.3 trillion. The maximum seawall height is 15.5 m, which was established assuming an L1 tsunami and is not assumed to be able to withstand an L2 tsunami.

The construction of these giant seawalls has caused significant controversy. While some people believe that seawalls of an adequate height are essential to enable residents in coastal areas to live securely, others have expressed concerns about various issues associated with seawall construction including the impact on local industry and coastal landscape, cost of construction, cost for future generations, insufficient consensus building, and negative impact on the natural environment. Although it is impossible to make improvements to the perceived problems on seawalls that have already been built, if a tsunami causes damage in the future and necessitates similar additional seawall construction, it will be important to reflect upon these problem areas and act accordingly.

Given this context, the authors of this study conducted a survey to examine the understanding and feelings of the residents in Miyagi and Iwate Prefectures concerning seawall construction. There are trade-offs in the impacts of seawall construction, for example, between increasing the height of seawalls for safety purpose and preserving coastal landscapes and the natural environment. Therefore, it is necessary to examine citizens' preferences in order to determine what should be prioritized. However, when the importance of each issue is asked in the conventional question format, there is often not much difference in the ratings among the alternatives because respondents are capable of rating any of the alternatives equally. Therefore, the study then analyzes the survey using best–worst scaling (BWS)—a method that has recently become popular in fields such as marketing—in an attempt to capture in quantitative terms the preferences of local residents concerning seawall construction. BWS is a question format in which respondents answer under trade-offs, so there tends to be a clearer difference in evaluation among the alternatives.

24.2   Methods

The study uses BWS to examine citizens' recognition of the problems associated with seawall construction in Miyagi and Iwate Prefectures, where seawalls are being built (Fig. 24.1). BWS is an analytical method introduced by Finn and Louviere (1992) that has become widely utilized in recent years in marketing and other fields (Louviere et al. 2015; Tsuge et al. 2014). To the authors' best knowledge, there are no studies that have used BWS to understand people's preferences for disaster prevention infrastructure such as seawalls.

Using this BWS approach, the respondents were presented with multiple items and asked to choose which one they rated highest ("best") and which one they rated lowest ("worst") based on evaluation criteria such as "utility" and "importance." A

Fig. 24.1 Positions of Iwate Prefecture and Miyagi Prefecture

respondent's preferences were captured by changing the items they were shown and repeating the question (Finn and Louviere 1992; Louviere et al. 2015).

There are three classified types of the BWS method: object case (case 1), profile case (case 2), and multi-profile case (case 3), as described by Louviere et al. (2015). In the object case (case 1), respondents are shown multiple objects such as opinions and products and asked to choose the best and the worst object in the choice set. In the profile case (case 2), respondents are shown a single profile comprising the attribute levels and asked to choose the best and the worst attribute level within the profile. In the multi-profile case (case 3), respondents are presented with multiple profiles comprising the attribute levels and asked to choose the best and the worst profile in the choice set. Since the goal of this study is to capture the relative importance of problems associated with seawall construction, we have selected object case (case 1), which is appropriate to help capture a respondent's relative assessments of multiple items.

Table 24.1 summarizes the typical issues that were highlighted as seawall construction problem areas. A total of seven points with the addition of "seawall height (safety)" were used as the BWS items.

These seven items were combined to create choice sets to be presented to respondents. This study follows the precedent of numerous previous studies in using balanced incomplete block designs (BIBDs) to create choice sets. With BIBDs, each item appears the same number of times in all the choice sets created; moreover, the combinations of each item with other respective items also appear the same number of times (Louviere et al. 2015).

Seven-item BIBDs are used since this study employs seven items. Replacing BIBD numbers one to seven with each of the seven items thus creates the seven choice sets shown in Table 24.2. This table shows how each item appears three

Table 24.1 Main issues in seawall construction (problems that have been highlighted regarding such construction)

1	Impact on local industry	Seawalls divide the town from the sea and are built on sandy beaches, so will they not affect local industries like fishing and tourism?
2	Impact on coastal landscape	Will seawalls not destroy the coastal landscape or make it impossible to see the coastal landscape?
3	Cost of construction	Is the cost of construction not too high with an approximate project cost of ¥1 trillion?
4	Negative impact on the natural environment	Will seawalls not have a negative effect on the natural environment by directly impacting the flora and fauna that exist in the sea and on the beaches, or by altering the flow of sand carried by ocean currents?
5	Building consensus	Were seawalls not built without the opportunity for discussion with local residents?
6	Cost for future generations	While the construction of seawalls is covered by the special reconstruction income tax, will future generations not have to pay large sums of money for their maintenance and renewal?

Table 24.2 Choice sets

Set			
1	Impact on local industry	Building consensus	Cost of construction
2	Seawall height (safety)	Cost of construction	Negative impact on the natural environment
3	Cost of construction	Cost for future generations	Impact on coastal landscape
4	Impact on coastal landscape	Impact on local industry	Seawall height (safety)
5	Cost for future generations	Negative impact on the natural environment	Impact on local industry
6	Building consensus	Seawall height (safety)	Cost for future generations
7	Negative impact on the natural environment	Impact on coastal landscape	Building consensus

Table 24.3 Examples of best–worst scaling (BWS) questions. Please select which of the following three factors you think is the most important to consider and which is the least important to consider

	Most important to consider	Least important to consider
Seawall height (safety)	X	
Cost of construction		X
Negative impact on the natural environment		

times throughout the choice sets, with each item and other items in combination appearing only once.

The BWS questions were posed as shown in the example in Table 24.3. The respondent was asked to assume that the seawall (which had actually already been built) is in the pre-construction stage and was then asked which of the three items they were shown should be considered "most important" and which should be considered "least important." They were asked to answer on the assumption that the seawall was definitely going to be constructed.

Each individual respondent was presented with all seven choice sets to which they provided seven answers. The order in which the choice sets were shown was randomized to minimize to the extent possible the potential impacts that order may have on the way respondents answered.

There are several ways to analyze BWS, which are generally classified as counting analysis and econometric methods (Louviere et al. 2015). An advantage of counting analysis is that it does not require an advanced understanding of statistics and is easy to perform; furthermore, scores calculated using counting analysis have a strong linear relationship with the coefficients estimated through the maximum difference model (Finn and Louviere 1992), which is an econometric method (Marley and Louviere 2005; Tsuge et al. 2014). This study has therefore chosen to utilize counting analysis.

The main score calculation method used in counting analysis is shown in Eqs. (24.1), (24.2), (24.3), (24.4), and (24.5) (Cohen 2009). B_{in} and W_{in} are the number of times item i was chosen as "best" and "worst" in all the questions for respondent n. The B-W score$_{in}$, the score for each individual respondent of each item, was calculated by subtracting the latter from the former (Eq. 24.1). $\Sigma_n B_{in}$ is the number of times item i was chosen as "best" in all respondents' answers. Similarly, $\Sigma_n W_{in}$ is the number of times it was chosen as "worst." The B-W score$_i$ of each item was calculated by subtracting the latter from the former (Eq. 24.2). The *average B-W* score$_i$ was obtained by dividing B-W score$_i$ by the number of survey respondents N and the number of times each item occurs in all choice sets r, which in the case of this study is 3 (Eq. 24.3). Moreover, the value of the most highly ranked item was standardized as 1, and the relative importance$_i$ of each item was found as shown in Eq. 24.5. It should be noted that $sqrt(B/W)_i$ was calculated using Eq. (24.4). Furthermore, max $sqrt(B/W)_i$ represents the maximum value of $sqrt(B/W)_i$. Equations (24.1), (24.2), (24.3), and (24.4) are based on the descriptions from Cohen (2009).

$$B - W \text{ score}_{in} = B_{in} - W_{in} \tag{24.1}$$

$$B - W \text{ score}_i = \sum_n B_{in} - \sum_n W_{in} \tag{24.2}$$

$$\text{average } B - W \text{ score}_i = \frac{B - W \text{ score}_i}{Nr} \tag{24.3}$$

$$sqrt(B/W)_i = \sqrt{\frac{\sum_n B_{in}}{\sum_n W_{in}}} \tag{24.4}$$

$$\text{relative importance}_i = \frac{sqrt(B/W)_i}{\max sqrt(B/W)_i} \tag{24.5}$$

24.3 Results and Discussion

The survey was conducted online between March 19 and 25, 2020. Survey participants were men and women between the ages of 20 and 69 in Iwate and Miyagi Prefectures who were registered contributors with the polling company. Responses were received from a total of 2099 people: 567 in Iwate Prefecture and 1532 in Miyagi Prefecture. Sample collection was coordinated to reflect as closely as possible the population compositions of the two prefectures in terms of gender and age. As a result, the respondents comprised 1053 men and 1039 women. In terms of age, 317 respondents were in their 20s, 402 in their 30s, 476 in their 40s, 443 in their 50s, and 461 in their 60s.

The scores obtained through counting analysis are shown in Table 24.4. Figure 24.2 is a graph of *relative importance$_i$*. Both "negative impact on the natural environment" and "seawall height (safety)" received particularly high scores. The negative impact on the natural environment received the highest score, although only by a small amount. These two items were followed in order by "impact on local industry," "cost for future generations," "building consensus," "cost of construction," and "impact on coastal landscape." It thus follows expectations that seawall height was highly ranked, given that the point of constructing the seawalls is to reduce tsunami damage, so it is noteworthy that the negative impact on the natural environment received an even higher score. This finding shows that people feel strongly that the issue of a seawall's impact on the natural environment should be given very careful consideration during seawall construction.

It is worth analyzing the preferences for the negative impact on the natural environment and seawall height in more detail, given how highly both of these two items were valued. The authors conducted regression analysis to understand the characteristics of the individuals scoring these items highly; in this analysis, the B-W scores of respondents to the two items (Eq. 24.1) were used as the explained variable, while the explanatory variables were each respondent's gender, age, address, knowledge of seawalls, recognition of the need for seawalls, and harm suffered in the tsunami from the Great East Japan Earthquake. A male dummy (d_male), taking the value of 1 for male and 0 for others (females and others, including those who do not wish to disclose their gender), was used for gender as variable. For age, values ranging from 25 to 65 were used as variables for people in their 20s to their 60s, respectively. A Miyagi Prefecture dummy (d_miyagi) was used for the address as a variable, taking the value of 1 for Miyagi Prefecture and 0 for Iwate Prefecture. Dummy variables were set for knowledge of seawalls based on responses to the question "Do you know about seawalls?" with the responses "I know about seawalls and have seen one," "I know about seawalls but have not seen one," "I have heard about seawalls," and "I don't know anything about seawalls" (in order, d_knowledge1, d_knowledge2, d_knowledge3, and d_knowledge4), with the three dummy variables excluding d_knowledge4 used as variables. For recognition of the need for seawalls, respondents were asked the question, "Do you think the seawalls need to be constructed?" with answers on a seven-point Likert scale (Jamieson 2004), with seven being "very necessary" and one "not necessary." Responses choosing one or two—meaning essentially there was no need for a seawall—were set to a dummy variable (d_need12) with a value of 1. Responses choosing three, four, or five—meaning there was no significant need for a seawall—were set to a dummy variable (d_need345) with a value of 1. Responses choosing six or seven—meaning there was a significant need for a seawall—were set to a dummy variable (d_need67) with a value of 1. The dummies d_need12 and d_need67 were used as variables. For harm suffered in the Great East Japan Earthquake tsunami, the authors constructed four dummy variables for responses to the question "Did you suffer harm in the Great East Japan Earthquake tsunami?" The responses

Table 24.4 Calculation results of scores

	Impact on local industry	Building consensus	Cost of construction	Seawall height (safety)	Negative impact on the natural environment	Cost for future generations	Impact on coastal landscape
$\sum_n B_{in}$	2165	1595	1663	3134	2881	1950	1305
$\sum_n W_{in}$	1486	2549	2656	1451	1180	2138	3233
B-W score$_i$	679	−954	−993	1683	1701	−188	−1928
Average B-W score$_i$	0.107829	−0.1515	−0.15769	0.26727	0.270129	−0.02986	−0.30618
Sqrt(B/W)$_i$	1.207034	0.791035	0.791283	1.469656	1.562538	0.955022	0.635334
Relative importance$_i$	0.772483	0.50625	0.506409	0.940557	1	0.611199	0.406604

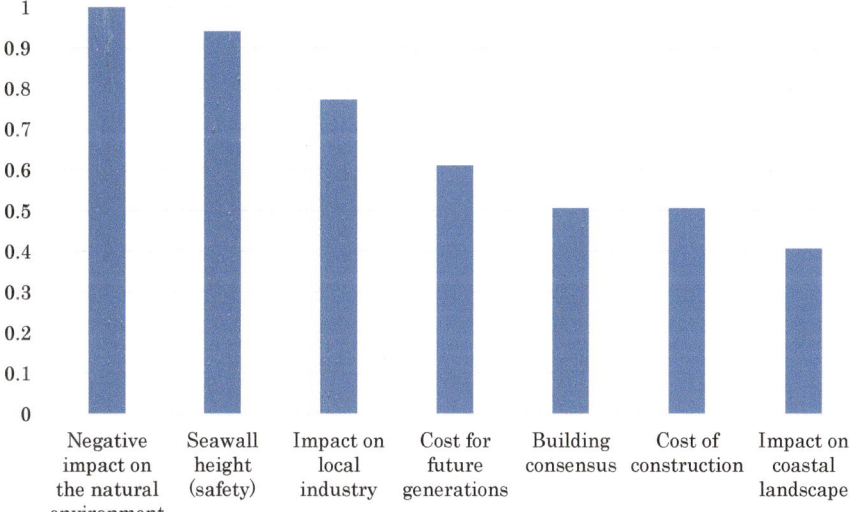

Fig. 24.2 Relative importance

of "I suffered harm," "My family and relatives were harmed," "My friends and acquaintances were harmed," and "No harm was suffered" were assigned respective dummy variables (in order, d_damage1, d_damage2, d_damage3, d_damage4) with a value of 1, and the first three dummy variables (with the exclusion of d_damage4) were used as variables. The descriptive statistics of the variables used in the regression analysis are shown in Table 24.5.

The model where the B-W scores of "negative impact on the natural environment" is taken as the explained variable is referred to as Model 1, and the model where the B-W scores of "seawall height (safety)" is taken as the explained variable is referred to as Model 2. Table 24.6 shows the estimation results through the least-squares method, using the data of the 2099 people.

In Model 1, d_male, age, and d_need67 were significant; d_male was negatively significant, meaning that males had significantly lower B-W scores than others. Age was positively significant, meaning that older respondents had higher B-W scores. On the other hand, d_need67 was negatively significant, meaning that people who think there is a strong need for seawalls had lower B-W scores than did people who think otherwise. It is possible to interpret these various findings to conclude that people who think there is a strong need for seawalls want greater safety and are thus willing to tolerate some negative impact on the natural environment with that end in mind. However, since there are few significant variables and the adjusted coefficient of determination ($adj.R^2$) is relatively low in Model 1, other factors not included in the model may have influenced the explained variable. Further research is needed in this regard.

Table 24.5 Descriptive statistics of the variables used in the regression analysis

		Min	Max	Mean	Std. Dev.
B-W score negative impact on the natural environment	Each respondents' B-W score for negative impact on the natural environment	−3	3	0.8104	1.5859
B-W score seawall height (safety)	Each respondents' B-W score for seawall height (safety)	−3	3	0.8018	1.9796
d_male	=1 for male and = 0 for other	0	1	0.5017	0.5001
Age	=25 for 20s, =35 for 30s, =45 for 40s, =55 for 50s, and = 65 for 60s	25	65	46.5674	13.6437
d_miyagi	=1 for Miyagi prefecture and = 0 for Iwate prefecture	0	1	0.7299	0.4441
d_knowledge1	=1 for the response "I know about seawalls and have seen one" to the question "do you know about seawalls?"	0	1	0.4326	0.4956
d_knowledge2	=1 for the response "I know about seawalls but have not seen one" to the question "do you know about seawalls?"	0	1	0.3492	0.4768
d_knowledge3	=1 for the response "I have heard about seawalls" to the question "do you know about seawalls?"	0	1	0.1258	0.3317
d_need12	=1 for the responses choosing one or two to the question "do you think the seawalls need to be constructed?" with answers on a seven-point Likert scale, with seven being "very necessary" and one "not necessary."	0	1	0.0853	0.2794
d_need67	=1 for the responses choosing six or seven to the question "do you think the seawalls need to be constructed?"	0	1	0.4640	0.4988
d_damage1	=1 for the response "I suffered harm" to the question "did you suffer harm in the Great East Japan Earthquake tsunami?"	0	1	0.1501	0.3572
d_damage2	=1 for the response "my family and relatives were harmed" to the question "did you suffer harm in the Great East Japan Earthquake tsunami?"	0	1	0.2573	0.4372
d_damage3	=1 for the response "my friends and acquaintances were harmed" to the question "did you suffer harm in the Great East Japan Earthquake tsunami?"	0	1	0.4307	0.4953

Table 24.6 Estimation results

Explained variable	Model 1 B-W score negative impact on the natural environment		Model 2 B-W score seawall height (safety)	
	Coefficient	Standard error	Coefficient	Standard error
Constant	0.7095	0.1790***	0.7199	0.1963***
d_male	−0.3257	0.0691***	−0.1137	0.0758
Age	0.0086	0.0026***	−0.0048	0.0028*
d_miyagi	−0.1107	0.0780	0.1010	0.0855
d_knowledge1	0.0843	0.1263	−0.4487	0.1384***
d_knowledge2	0.0298	0.1275	−0.2425	0.1398*
d_knowledge3	−0.1422	0.1491	−0.0508	0.1634
d_need12	0.1441	0.1294	−1.5853	0.1419***
d_need67	−0.2656	0.0724***	1.4917	0.0794***
d_damage1	−0.0865	0.0995	0.2332	0.1091**
d_damage2	0.0094	0.0811	−0.1227	0.0889
d_damage3	0.0850	0.0700	0.0348	0.0767
Adjusted coefficient of determination ($adj.R^2$)	0.02527		0.24797	
Number of respondents	2099		2099	

NB: ***significant at 1% level; **significant at 5% level; *significant at 10% level

In Model 2, age, d_knowledge1, d_knowledge2, d_need12, d_need67, and d_damage1 were significant (noting that age and d_knowlcdgc2 were significant at the 10% level). Age was negatively significant, meaning that B-W scores increased as age decreased. In addition, d_knowledge1 and d_knowledge2 were negatively significant, meaning that people with knowledge of seawalls had lower B-W scores than did people who lacked such knowledge. The fact that the absolute value of the coefficient of d_knowledge1 was greater than the coefficient of d_knowledge2 shows that people who have seen a seawall in person had lower B-W scores. This finding may be interpreted to mean that people with knowledge of seawalls are aware of both their magnitude and their associated problems and therefore do not want the structure to be as high as do people without that direct knowledge. People who have seen seawalls in person immediately have a clear sense of their magnitude and thus have a particularly strong tendency toward this reaction. Additionally, d_need12 was negatively significant, whereas d_need67 was positively significant, meaning that in comparison to people who think there is no significant need for a seawall, those who think there is basically no need for a seawall had lower B-W scores, while people who think there is a strong need for a seawall had higher B-W scores. A reasonable interpretation of these findings suggests that people who think there is basically no need for a seawall have a comparatively lower interest in focusing on safety and people who think there is a strong need for a seawall have a comparatively higher interest in focusing on safety than do people who think there is no significant need for a seawall, respectively; thus, the former are less desirous of high seawalls than are people with other opinions and the latter are more desirous of high seawalls than are people with other opinions, respectively. Finally, d_damage1 was positively significant, meaning that people who suffered harm in the Great East Japan Earthquake tsunami had higher B-W scores than did those who experienced no harm. A reasonable conclusion is that people who personally suffered harm in the Great East Japan Earthquake tsunami now understandably want high seawalls to provide greater safety.

24.4 Conclusions

In this study, the authors used the best–worst scaling method to examine the preferences of citizens in Miyagi and Iwate Prefectures (areas stricken by the Great East Japan Earthquake) regarding the construction of seawalls. The main issues were presented in the form of seven items: impact on local industry, impact on coastal landscape, cost of construction, negative impact on the natural environment, building consensus, cost for future generations, and seawall height (safety). The results demonstrated that citizens believe that the negative impact on the natural environment and seawall height in particular should be taken into consideration during seawall construction.

The authors then conducted regression analysis. Respondents' B-W scores for the negative impact on the natural environment and seawall height (safety) were set as the explained variable, and respondents' gender, age, address, knowledge of seawalls, recognition of the need for seawalls, and harm suffered in the Great East Japan Earthquake tsunami were set as explanatory variables. The results showed that people who did not identify as male were more likely to feel that the negative impact on the natural environment should be taken into consideration more than did those who identified as male; older people were also more likely to feel that the negative impact on the natural environment should be taken into consideration. Additionally, people who felt there was a strong need for seawalls were less likely to think that the negative impact on the natural environment should be taken into consideration than did those who felt otherwise. Furthermore, younger people were more likely to think seawall height should be taken into account, while people who did not know about seawalls were more likely to think seawall height should be taken into account than did people with seawall knowledge. Among people with knowledge of seawalls, those who had not seen a seawall in person were more likely to think seawall height should be a consideration than did people who had seen a seawall in person. People who feel more strongly that there is a need to build seawalls are more likely to think seawall height should be taken into consideration, whereas those who suffered personal harm in the Great East Japan Earthquake tsunami are more likely to think seawall height should be taken into consideration than did people who did not suffer personal harm.

This study has shown that local residents place particular importance on the negative impact on the natural environment and seawall height (safety). However, since there is a trade-off between the two, it is difficult to achieve both with gray infrastructure such as seawalls. Therefore, in order to achieve both, it is necessary to utilize hybrid infrastructures that combine gray and green infrastructures.

As described above, this study successfully clarified what specific issues citizens feel should be thoughtfully addressed during seawall construction. If there is a need to build future seawalls, these findings should be considered and acted upon.

There are, however, unresolved issues in this study. First, the authors chose to use BWS to conduct their analysis, but other analytical methods, including choice experiments (Holmes et al. 2017), are also possible. Comparing the results of this study with the results of analysis by other methods would also help to verify the robustness of these results. Second, this study focused on Iwate and Miyagi Prefectures, but it would be beneficial to conduct similar surveys in other regions of Japan, given that the development of coastal seawalls is planned as part of the National Resilience (Onuma 2015; National Resilience Promotion Office, Cabinet Secretariat, 2020). It would be worthwhile to capture the preferences of people in other regions, rather than just in the areas stricken by the Great East Japan Earthquake, and then use these findings for discussions and consensus building around seawall construction. Third, while this study examined the issues that most matter to citizens during seawall construction, it is also important to investigate which disaster prevention measures they specifically want. In addition to gray and green infrastructure, various hybrids of the two also exist. An important future topic will be to capture citizens' preferences in this regard to assistance with regional consensus building.

Acknowledgments This research was partially supported by Keio Gijuku Academic Development Funds and the Environment Research and Technology Development Fund of the Ministry of the Environment, Japan (Number 4-1805). We would like to express our sincere gratitude. We would also like to thank Mr. Masato Abe for providing the photos used in our survey.

References

Central Disaster Management Council, Committee for Technical Investigation on Countermeasures for Earthquakes and Tsunamis Based on the Lessons Learned from the "2011 off the Pacific coast of Tohoku Earthquake" (2011) Report of the Committee for Technical Investigation on Countermeasures for Earthquakes and Tsunamis Based on the Lessons Learned from the "2011 off the Pacific coast of Tohoku Earthquake"

Cohen E (2009) Applying best-worst scaling to wine marketing. Int J Wine Bus Res 21(1):8–23

Finn A, Louviere JJ (1992) Determining the appropriate response to evidence of public concern: the case of food safety. J Public Policy Mark 11(2):12–25

Holmes TP, Adamowicz WL, Carlsson F (2017) Choice experiments. In: Champ PA, Boyle KJ, Brown TC (eds) A primer on nonmarket valuation. Springer, Dordrecht, pp 133–186

Jamieson S (2004) Likert scales: how to (ab) use them? Med Educ 38(12):1217–1218

Louviere JJ, Flynn TN, Marley AAJ (2015) Best-worst scaling: theory, methods and applications. Cambridge University Press, Cambridge

Marley AAJ, Louviere JJ (2005) Some probabilistic models of best, worst, and best-worst choices. J Math Psychol 49(6):464–480

National Resilience Promotion Office, Cabinet Secretariat (2020) Annual Plan for National Resilience 2020

Onuma A (2015) Sustainable coastal management policy under population decline - disaster management and nature conservation (in Japanese). Rev Environ Econ Policy Stud 8(2):11–17

Tsuge T, Nakamura S, Usio N (2014) Assessing the difficulty of implementing wildlife-friendly farming practices by using the best–worst scaling approach. In: Usio N, Miyashita T (eds) Social-ecological restoration in Paddy-dominated landscapes. Springer, Tokyo, pp 223–236

Chapter 25
Coastal Community Preferences of Gray, Green, and Hybrid Infrastructure Against Tsunamis: A Case Study of Japan

Yui Omori, Koichi Kuriyama, Takahiro Tsuge, Ayumi Onuma, and Yasushi Shoji

Abstract A decade has passed since the 2011 Great East Japan Earthquake and Tsunami struck. Despite increasing awareness that concrete-based coastal infrastructure, such as seawalls, is not sufficient to protect against unfathomable events, engineering structures still play a significant role in fortifying coastal communities. Meanwhile, purely nature-based approaches (i.e., coastal forests) also have limitations against cataclysmic waves, and there remain uncertainties regarding their ecosystem-based disaster risk reduction functions (Eco-DRR). In tackling these issues, hybrid infrastructure, which combines both gray and green components, has received growing interest. However, little research has been conducted to evaluate the economic values of coastal gray, green, and hybrid infrastructures under uncertainties in terms of people's preferences.

Therefore, in this study, we aimed to (1) quantify the economic value of coastal ecosystem services, including species richness, landscape, recreational services, and disaster risk reduction, under uncertainties through choice experiments; (2) clarify the differences in preferences for preparations against long-cycle tsunamis between those who reside in tsunami-prone areas and those who do not, using a conditional logit (CL) model; and (3) discuss the heterogeneities in coastal citizen perceptions by comparing the CL and mixed logit (ML) model. As a result, this study highlights the importance of considering the heterogeneity of preferences.

Y. Omori · K. Kuriyama (✉)
Division of Natural Resource Economics, Graduate School of Agriculture, Kyoto University, Kyoto, Japan
e-mail: omori.yui.27x@st.kyoto-u.ac.jp; kkuri@kais.kyoto-u.ac.jp

T. Tsuge
Sophia University Graduate School of Global Environmental Studies, Tokyo, Japan

A. Onuma
Keio University, Graduate School of Economics, Tokyo, Japan

Y. Shoji
Research Faculty of Agriculture, Hokkaido University, Hokkaido, Japan

© The Author(s) 2022
F. Nakamura (ed.), *Green Infrastructure and Climate Change Adaptation*,
Ecological Research Monographs, https://doi.org/10.1007/978-981-16-6791-6_25

Furthermore, our respondents in the tsunami-prone group (TPG) valued the coastal defense function offered by gray more highly than the non-TPG, demonstrating an especially large gap regarding seawalls against short-cycle tsunamis (willingness-to-pay (WTP) values of 11,233 JPY and 5958 JPY, respectively). However, there was no significance for coastal forests in the TPG, reflecting the importance of disaster prevention function offered by gray infrastructure. In addition, the hybrid landscape (seawalls + coastal forests) received higher positive responses, 71.1% with WTP of 8245 JPY, than the gray landscape (seawalls only) with WTP of -3358 JPY, as estimated by the ML model. These contradictions and heterogeneities in people's preferences may foreshadow the difficulties of applying hybrid approaches; hence developing synthesized both stated preference and other revealed preference methods is indispensable for providing strategic design of gray-green combined coastal defense and bolstering coastal realignment.

Keywords Seawalls · Coastal forests · Hybrid infrastructure · Eco-DRR · Choice experiments

25.1 Introduction

A decade has passed since the 2011 Great East Japan Earthquake and Tsunami occurred, which caused severe damages to both human life and socioeconomic properties both on the coast and in the hinterlands. After these events, artificial coastal defenses, including 10-m-high seawalls, have been constructed on some coastlines in Japan. The reliance on traditional defense methods continues despite increasing awareness that gray-based coastal infrastructure alone does not offer sufficient tsunami protection. Then, discussions regarding the reconstruction of higher and wider seawalls prompted coastal residents to rethink their coastal infrastructure because they both benefit from the ocean but also bear the brunt of natural hazards, such as tsunamis, windstorms, storm surges, and typhoons.

Over centuries, coastal design and planning have become an engineering discipline, initially for economic development, such as harbors and ports. Their spatial advantages have led to the development of human settlements, as coastlines provide resources, trading, and job opportunities, thus the global population benefits directly and indirectly from marine ecosystems (McGranahan et al. 2007). Moreover, population growth in coastal zones is accelerating, and more than half of coastal countries have 80%–100% of their population within 100 km of their coastlines (Martínez et al. 2007). Similarly, the intensity and frequency of natural hazards are increasing, implying that coastal zones are becoming more vulnerable. Thus, minimizing the impact of natural disasters is increasingly imperative. In particular, seawalls are considered to be the last line of defense and essential for maintaining residential livelihood (Reeve et al. 2018), as they have important roles in stabilizing the shoreline and protecting the coastal communities of Japan.

In parallel, *Pinus thunbergii* (black pine trees) has been traditionally afforested along coasts in Japan as a nature-based disaster mitigation method to collect blown sand, mitigate wind speeds, and protect agricultural products and residential buildings. Moreover, coastal forests can be planted to attenuate wave velocity and stop drifts, thereby providing evacuation time when tsunamis occur (Harada and Imamura 2005). However, purely green measures have limitations against catastrophic events. Broadly, artificial coastal barriers (termed gray infrastructure, built infrastructure or hardened structures, encompassing seawalls, levees, culverts, bulkheads) ensure greater protection during extreme weather events than their nature-based counterparts, such as mangroves, coastal forests, dunes, and salt marshes, unless the intensities of natural hazards go beyond the infrastructure's capacity (Onuma and Tsuge 2018). Conversely, coastal armoring can exert negative or unexpected influences on coastal habitats and prevent them from restoring after disturbances (Borsje et al. 2011). Furthermore, the high cost of construction and maintenance poses a financial burden on municipality budgets. To overcome these issues, green infrastructure has been given increasing interest and is recognized as a nature-based approach that can serve disaster mitigation. However, the performance of ecosystem-based disaster risk reduction (Eco-DRR) varies because of coastal morphology and land use configuration (i.e., topography, soil conditions, and vegetation) (Irtem et al. 2009). Furthermore, Gedan et al. (2011) stated that unfathomable events, such as large tsunami waves and storm surges, can overwhelm the attenuation effects provided by vegetation and emphasized the necessity of combining man-made structures with ecological means of coastal protection. As such, hybrid infrastructure, which combines gray and green components, has also received growing interest. Sutton-Grier et al. (2015) summarized the characteristics of coastal gray, green, and hybrid infrastructure, evaluating their strengths and weaknesses. Tanaka (2012) investigated the effectiveness and limitations of coastal pine forests against large tsunami and concluded that their ability to attenuate waves is not better than that of seawalls, which does not necessarily mean their effect is negligible when a large tsunami overtops coastal armoring. In addition, it emphasized that other functions, such as trapping debris, provided by coastal forests should be considered, and gray-green integrated approaches should be applied to future coastal designs. Hence, the hybrid approach of seawalls and coastal forests that has long been utilized in some coastal regions in Japan may be a solution to the current reliance on gray-based coastal defense and the ambiguous effects of green infrastructure on coastal protection (Table 25.1). However, because of the multiple types of gray and green integrations (TCHGG 2018), there is a lack of economic analysis considering hybrid infrastructure that integrates seawalls and coastal forests. Furthermore, the uncertainties of disaster risk reduction (DRR) function provided by green infrastructure (Eco-DRR) hinder policymakers from performing cost–benefit analyses as it is difficult to understand how gray and green combined infrastructure can reduce hazard risks. (Hereafter, we use the term "DRR," which is the abbreviation of disaster risk reduction provided by either gray or green, but again, especially DRR function offered by green is termed Eco-DRR.) This financial challenge has an ongoing issue for green infrastructure (Sutton-Grier et al. 2015), which has also led to limited implementation and data regarding hybrid infrastructure as well.

Table 25.1 Coastal gray and green infrastructure strengths and weaknesses

Infrastructure	Strengths	Weaknesses
Gray (seawalls)	Greater level of protection	Coastal habitat loss
	– Alleviates speed of waves, tsunamis	– Has negative effects on other ecosystem services that coasts and surrounding areas, such as beaches, provide
	– Prevents erosion	Lack of capability and high maintenance cost
	– Withstands storm events soon after seawalls are built and performs stable protection for two to three decades	– Does not adapt to unforeseen events (e.g., tsunamis, sea-level rise)
	Greater understanding of techniques and effects	– Weakens the effects (built-in lifetime)
	– Can perform to cost–benefit analysis	Lack of ocean and community bond
	Significant engineering expertise	– Leads to safety misunderstandings and disaster risks
Green (coastal forests)	Many co-benefits	Ambiguous effects
	– Coastal protection	– Limited understanding regarding protection levels because of topography, vegetation, seasons, and soils
	– Recreational use	Time for mature forests
	– Coastal habitats with many species	– Requires approximately 20 years to mature for sufficient protection
	Lower cost	Pine wilt disease
	Adaptation to unexpected events	– Damaged by diseases and pests
	– Can keep pace with climate change, sea-level rise	Other societal disadvantages (crime, dumping)
		– Requires appropriate maintenance
Hybrid (gray + green: seawalls + forests)	Greater protection with other co-benefits	Little data and limited expertise
	– May require less space than natural approaches alone	– Requires more research regarding potential effects
	Innovative coastal design and planning	– May require more space to introduce both systems
	– Compatible with resilience and authentic value	

Sources: Based on Sutton-Grier et al. (2015) and Harada and Imamura (2005), edited by the authors

From an economic perspective, Costanza et al. (2008) evaluated the value of coastal wetlands for ecosystem protection against hurricanes in the United States and estimated the potential storm protection to be $23.2 billion annually using a regression model. Barbier et al. (2011) evaluated the coastal ecosystem services (ESSs) in different natural environments, including coral reefs, seagrasses, salt marshes, mangroves, sand beaches, and dunes. However, these estimations of nonmarket values do not necessarily consider the trade-offs between ESSs in different coastal planning scenarios. Additionally, information regarding the potential values of coastal forests, especially *Pinus thunbergii*, is lacking. Thus, it is crucial to quantify the monetary values of ecosystem functions provided by concrete structures and natural measures. As there is a significant reliance on concrete structures, such as seawalls, and the uncertainties of DRR functions offered by coastal forests are still unclear, we considered seawalls as coastal gray infrastructure and coastal forests as green infrastructure (Fig. 25.1) in order to (1) evaluate the economic value of coastal ESSs, including species richness, landscape, recreational services, and DRR under uncertainties, (2) describe the preferences of coastal citizens regarding preparations against long-cycle tsunamis using economic models, and (3) discuss better combinations of gray and green infrastructure. In this project, we did not just aim at giving a monetary value of coastal ecosystem services but rather focused on how much coastal people understand the hazard risk and acknowledge the importance of coastal ecosystems; therefore, this study

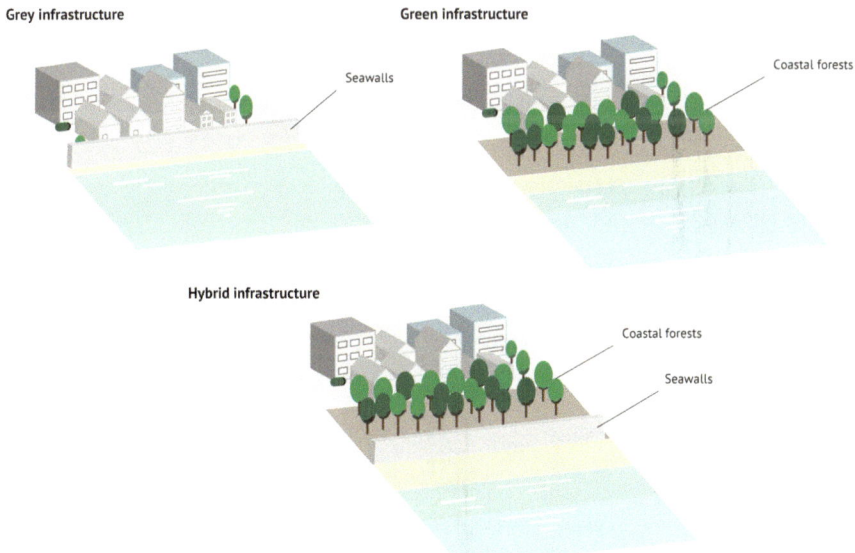

Fig. 25.1 Example of gray, green, and hybrid approaches in Japan's coastal zone. *Source*: Based on Sutton-Grier et al. (2015), edited by the authors

identifies the reasons for which it remains difficult to build a consensus between municipalities and locals and promote hybrid approaches.

25.2 Survey Design

25.2.1 Data Collection

In this project, an online survey intended for people in their 20s to 60s in Japan was administered by Nikkei Research Inc. between March 6 and March 10, 2020. The subjects were randomly selected, and 955 responses were obtained. These data are summarized in Table 25.2. Note that the distances listed in Table 25.2 were calculated with open-source Geographic Information System QGIS 3.10 using

Table 25.2 Data collection

Gender			Distance (from the coastline: km)			Frequency of coastal use		
Female	466	(48.8%)	<5	295	(30.9%)	Almost every day	12	(1.3%)
Male	487	(51%)	5–10	146	(15.3%)	3–5 times/week	4	(0.4%)
Others	2	(0.2%)	10–15	129	(13.5%)	1–2 times/week	13	(1.4%)
			15–20	79	(8.3%)	1–2 times/month	61	(6.4%)
Age			20–30	80	(8.4%)	1–2 times/year	116	(12.1%)
20s	153	(16%)	30–40	51	(5.3%)	Vacation use	112	(11.7%)
30s	176	(18.4%)	40–50	36	(3.8%)	Seldom	345	(36.1%)
40s	242	(25.3%)	>50	45	(4.7%)	None	278	(29.1%)
50s	181	(19%)	Missing data	94	(9.8%)	Others	14	(1.5%)
60s	203	(21.3%)	Minimum value	0.1				
			Maximum value	102.7				
Income (million JPY)			*Occupation*					
<2	101	(10.6%)	Office workers	429	(44.9%)			
2–4	212	(22.2%)	Officials	33	(3.5%)			
4–6	218	(22.8%)	Association staff	22	(2.3%)			
6–8	163	(17.1%)	Self-employed	73	(7.6%)			
8–10	115	(12%)	Housekeepers	140	(14.7%)			
10–12	54	(5.7%)	Part-time jobs	141	(14.8%)			
12–14	29	(3%)	Retired	50	(5.2%)			
14–16	22	(2.3%)	No job	29	(3%)			
16–18	13	(1.4%)	Students	20	(2.1%)			
18–20	7	(0.7%)	Others	18	(1.9%)			
20–22	10	(1%)						
>22	8	(0.8%)						
Missing data	3	(0.3%)						

postal codes. Furthermore, answering postal codes was optional, and responders can type "9999999" or "0000000" in the questionnaires if they did not wish to provide.

25.2.2 Experimental Design

25.2.2.1 Review of Economic Evaluation

To evaluate nonmarket values, such as coastal ESSs, we employed choice experiments. The economic analysis of ESSs is essential for municipalities to conduct cost–benefit analyses for coastal planning. There are several methods to estimate the nonmarket values of coastal ESSs, including travel cost or hedonic methods, which are known as revealed preference. However, these methods are not applicable to coastal ESSs' evaluations because of their far-reaching direct and indirect effects (Börger et al. 2014). Thus, in this study, we applied a conjoint analysis, which is a stated-preference method that enables researchers to understand individual respondents' statements of ESSs and determine their preferences. For instance, Garber-Yonts et al. (2004) used choice experiments to estimate people's willingness to pay (WTP) by changing the levels of biodiversity conservation under different conservation programs in the Oregon Coast. However, the public perceptions of ESSs' benefits that both gray and green infrastructure offer, such as DRR function, richness, recreation, and landscape, are not well understood. To date, choice experiments have been conducted to investigate the preference of coastal citizens regarding seawalls and coastal habitat conservation (Imamura et al. 2016). Based on this, our choice experiments addressed citizen preferences by adding the DRR function attributes offered by either gray or green (or both) approaches. On the other hand, another challenge is that the estimates by stated-preference can be relatively stable over short periods of time, and the results for longer periods are likely to be unstable because of the environmental changes projected for 50–100 years into the future (Börger et al. 2014). Thus, we also included some future settings that coastal infrastructure might be presented with by changing different frequencies of tsunamis. Therefore, we are able to address these issues and reveal the changes in coastal citizens' preference in various scales of tsunamis, and we explain the details of how these settings are generated in the following section.

25.2.2.2 Choice Experiments

A conjoint analysis using choice experiments provides concrete ecological information on a variety of coastal settings and evaluates trade-offs by jointly considering a number of important attributes for different hypothetical situations. It is therefore a useful tool for evaluating people's perceptions of ESSs (Louviere 1988; Louviere et al. 2000). However, it requires careful descriptions of ESSs in the survey design in order to ensure the validity of the ecological data. In this sense, we considered four coastal settings, including the status quo with the six attributes (Fig. 25.2): additional seawall height, coastal forest width, landscape, biodiversity, recreation, and annual tax, owing to ensuring that each attribute could be adequately understood without causing confusion (Bateman et al. 2002) (Table 25.3). The choice experiment was repeated eight times for each respondent, during which the levels of attributes changed. It is noteworthy that the additional seawall height and coastal forest width are associated with gray and green components, respectively, enabling the evaluation of DRR functions. The detailed explanations of the attributes and levels are provided in Sect. 25.2.3.

	Option 1	Option 2	Option 3
Additional seawall height	the preparation for one-in-100 year probability tsunami	storm surge (not well preparation for tsunami)	the preparation for one-in-10 year probability tsunami
Forest width	0 m	500 m	300 m
Landscape	seawall	coastal forest	both
Recreation	No recreational services	Camping & Walking	Camping, Walking & Fishing
Bird species	3 kinds of birds	10 kinds of birds	20 kinds of birds
Annual cost (JPY)	30,000	3,000	10,000
Most preferable	○	○	●

Fig. 25.2 Example of question

Table 25.3 Attributes and levels of coastal settings

Attributes	Levels
Additional seawalls height (security)	±0 (for typhoons and storm surges)/
	+1 m − +2 m (1-in-10 year tsunamis)/
	+2 m − +5 m (1-in-30 year tsunamis)/
	+ over 5 m (1-in-100 year tsunami)
Forest width	0/100 m/200 m/300 m/500 m
Landscape	Coastal forest only/seawall only/both
Coastal recreation	Walking only/camping and walking/fishing only/ camping, walking and fishing/nothing
Coastal biodiversity (birds species)	3/10/20 kinds of birds
Annual tax (JPY)	1000/3000/5000/10,000/30,000

25.2.3 Attributes and Levels

25.2.3.1 Additional Seawall Height

Following the 2011 earthquakes and tsunamis, the MILT (2012) established new tsunami disaster risk management, in which tsunamis are classified into two different levels: Level 1 (L1) tsunamis, which have return periods ranging from approximately several decades to 100 years, and relatively low tsunami inundation depths; and Level 2 (L2) tsunamis, which are likely to occur for longer return periods, such as a few hundred to a few thousand years, and cause widespread damage to both the coast and hinterlands. This Japanese tsunami hazard management strategy further stipulated that the height of a coastal embankment should be designed in preparation of L1 tsunamis, whereas coastal structures are designed in preparation for L2 tsunamis in order to assure more time for evacuation. Then, seawall heights rely on prefectural decisions and computer simulations, causing some seawalls to be higher than 10 m in some coastlines. In our study, coastal gray infrastructure was categorized into four settings (Fig. 25.3), and then we assumed that the seawall with 2 m in height and 2000 m in total length has already existed (Fig. 25.4). First, as

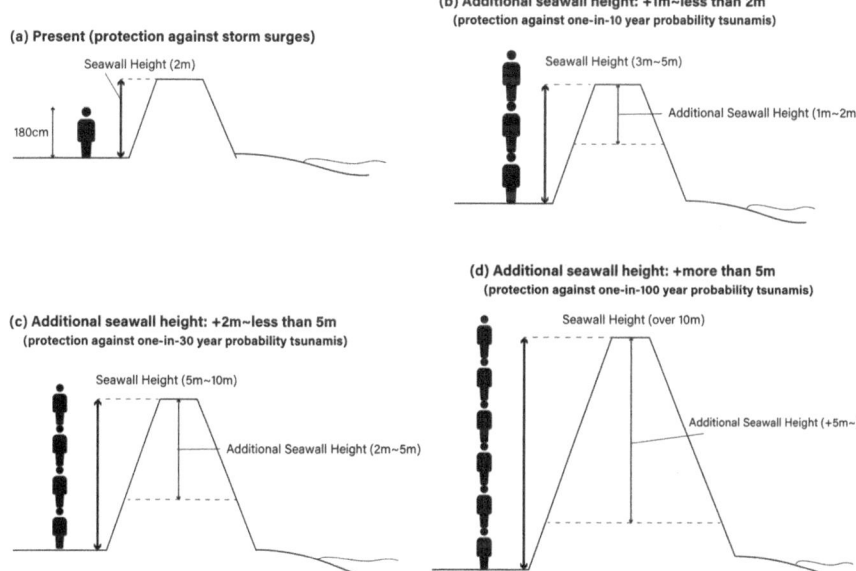

Fig. 25.3 Seawall settings. (**a**) Present situation (storm surge, not well prepared for tsunami); (**b**) +1 m–less than 2 m (the preparation for one-in-10 year probability tsunami), total seawall height: 3–5 m; (**c**) +2 m–less than 5 m (the preparation for a one-in-30 year probability tsunami), total seawall height: 5–10 m; (**d**) + more than 5 m (the preparation for one-in-100 year probability tsunami), total seawall height: more than 10 m. *Note*: The material that was used in our questionnaires is not permitted to use for commercial uses, so we here display the similar image above

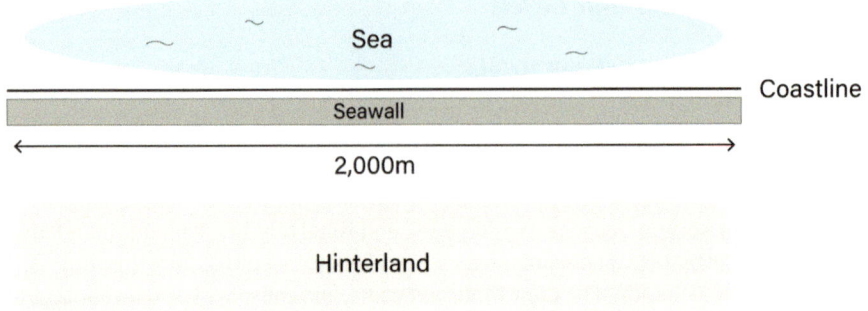

Fig. 25.4 The prerequisite of coastal land use. *Note:* Edited by the authors

shown in Fig. 25.3a (top left), there are no changes in a seawall with a height of approximately 2 m that already exists, providing effective protection from typhoons and storm surges. Second, as shown in Fig. 25.3b (top right), the additional height ranged 1–2 m will be constructed on the existing seawall in order to withstand a tsunami occurrence once a decade. Third, as shown in Fig. 25.3c (bottom left), a 2–5 m high seawall will be added to protect against 30-year probability tsunamis. Finally, a seawall that is over 10 m high will be built to protect against unexpected tsunamis (Fig. 25.3d). To increase responders' comprehension, various tsunami frequencies were assumed, and a total seawall height does not mean the height from the Tokyo Peil (T.P.). The Japanese datum of leveling is commonly used to express a seawall height.

25.2.3.2 Forest Width

Coastal forests, in which *Pinus thunbergii* is predominant, span the interface of beaches and terrestrial environments and are mainly observed on the main island of Japan. In other types of coastal forests, mangrove forests exhibit significant wave attenuation (Bao 2011) and play important roles in DRR, providing habitats for coastal creatures (Barbier et al. 2011; Morris et al. 2018). Although the effects of coastal forests on tsunami mitigation are not well established, Harada and Imamura (2005) examined the effect of coastal forests and observed a reduction in tsunami inundation depth and hydraulic energy behind the forests with increasing forest width, which denotes the cross-shore distance (Fig. 25.5). On the other hand, Tanaka (2012) investigated the tsunami-affected regions of the 2011 earthquakes and tsunamis and found that the forest width that can mitigate tsunamis is at least 200 m. This evidence was then applied to the government disaster management guidelines (MILT 2012). Specifically, if respondents selected coastal forests less than 200 m in width, they may prioritize protection from annual hazards such as wind storms rather than unexpected tsunami and wave attenuation, and if they select wider coastal forests, they desire a stronger protection level in both predicted

(i) Coastal green infrastructure   **(ii) Coastal grey and green infrastructure**

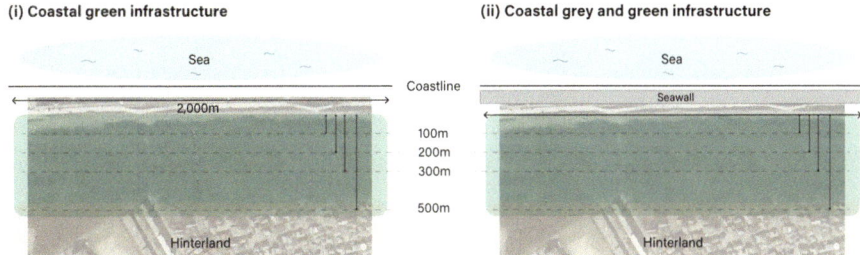

Fig. 25.5 Coastal forest scale. *Source*: Geospatial Information Authority of Japan, edited by the authors

and unpredicted natural disasters. Before beginning the choice experiment, we explained the positive and negative aspects concerning Eco-DRR, including that it has limitations and uncertainties in coastal protection (Onuma and Tsuge 2018). Then, we assume that coastal forests are afforested behind seawalls if they have already been constructed.

25.2.3.3   Landscape

Owing to increasing uncertainties in the face of multiple stressors, such as storm surges, typhoons, and coastal design for preparing against natural hazards, unforeseen events and several future scenarios must be considered. In this regard, coastal landscapes, also called "seascapes," which are associated with land use planning adjacent to the coastline and ocean, that integrate existing structures and natural approaches are gaining increasing attention. As coastal zones provide significant economic, transport, residential, and recreational functions, all of which have foundations in physical characteristics, they should have appealing landscapes, natural resources, and terrestrial biodiversity (European Commission 2000). Perkol-Finkel et al. (2018) used ecologically sensitive designs and concrete technologies to minimize harmful impacts on marine flora and fauna. Concomitant to the growing threats of coastal hazards (Temmerman et al. 2013), anthropogenic changes in tourism and the development of residential districts are accelerating in coastal regions (Martínez et al. 2007). Subsequently, ecological engineering, which is relatively new discipline that combines engineering and ecology, has emerged to address these concerns (Chapman and Underwood 2011), and many studies regarding gray to green regime shifts and gray and green combined approach (hybrid) are available (Andersen et al. 2009; Cheong et al. 2013; Morris et al. 2018; Schoonees et al. 2019; Cooper et al. 2020), thereby augmenting coastal landscape architecture (Bergen et al. 2001; Pioch et al. 2018). Although interdisciplinary gray and green designs have been developed, there is limited information on how people value gray, green, and hybrid landscapes. Hence, to quantify gray and green coastal landscape from an economic point of view, our study placed landscape attributes

Fig. 25.6 Landscape photographs (**a**: gray, **b**: green, and **c**: gray + green). (**a**) Seawall. (**b**) Coastal forest. (**c**) Seawall and coastal forest. *Source*: The photo (**c**) was taken by researcher groups at Tokushima University

into choice experiments to explain the values of gray (seawalls), green (coastal forests), and gray + green (seawalls + coastal forests) landscapes. In addition, three photographs were provided during the choice experiment (Fig. 25.6).

25.2.3.4 Recreation

Recreation is an important ecosystem function and is a key element for evaluating estuarine and coastal ESSs (Barbier et al. 2011). Brenner et al. (2010) assessed the nonmonetary value of ecosystem functions provided by a coastal zone in Spain. They stressed the higher aesthetic and recreation values that contributed to the total coastal ecosystems. In previous studies, the direct uses of recreational benefits have been calculated via travel cost (Blackwell 2007; Ghermandi et al. 2009; Prayaga 2017). However, the indirect monetary value of recreational services is still not estimated, and the comprehensive understanding of recreational services offered by coastal pine forests is limited. Thus, our study classified coastal recreation into five cases: (a) a promenade provided for walking along coastlines, (b) fishing is permitted, (c) a promenade for walking and a space for camping near the ocean, (d) all three aforementioned activities are available, and (e) no recreation (Fig. 25.7).

25.2.3.5 Bird Species

Most marine flora and fauna reside in coastal areas; however, anthropogenic changes to coastlines are responsible for the loss of coastal habitats and associated ESSs

Fig. 25.7 Coastal recreation opportunities. (**a**) Walking. (**b**) Fishing. (**c**) Walking and camping. (**d**) Walking, fishing and camping. *Note*: As the material used in our questionnaires is not permitted for commercial use, we displayed a similar image above. The photos were provided only recreational activities are available for responders' understanding, but no recreation was reflected on the choice experiment (Fig. 25.2)

(Spalding et al. 2007). By accounting for the current ecosystems' values from human dimensions, some species are deemed more important than others. That is to say that the loss of some species at the peak of a food chain might have a very different effect than the loss of less charismatic species further down, which might support an entire ecosystem, called as a "keystone species" (Helm and Hepburn 2012). Thus, evaluating the monetary values of biodiversity is complex and difficult as it requires careful consideration from the perspectives of both human beings and animal conservation. Looking at the onshore environment, Sekercioglu (2006) stressed the ESSs provided by birds and avian ecological functions, such as predation, transportation, and excretion, which play significant roles in seed dispersal, pollination, controlling prey species populations, scavenging, and nutrient cycles. Canterbury et al. (2000) depicted the bird community as a useful tool for monitoring forest conditions. Thus, we regarded birds as keystone species in coastal zones using a scale of 3 to 20 kinds of birds, in accordance with Takagawa et al. (2011), who provided avian data in Japan. The small number of bird species indicated that *Columba livia*, *Larus crassirostris*, and *Corvus* (*Corvus macrorhynchos* or *Corvus corone*), which are generally seen in Japan, can be observed while a wide range of avian species is described as raptors and includes *Pandion haliaetus* and *Accipiter gentilis*, which are associated with higher consumers. These values are based on the annual number of average coastal birds to reduce migratory bird seasonal fluctuations.

25.2.4 Profile Design

To describe efficient settings in the choice experiments, we applied a D-efficiency for optimal profile design using Ngene (ChoiceMetrics 2018), which is a common and comprehensive tool utilized in economic analyses. Unlike an orthogonal design that is statistically uncorrelated in attributes, D-efficiency can consider the inter-attribute correlation and allow analysts to exclude "insignificant" alternatives and to place more weight on practical situations (Hensher et al. 2005).

25.3 Econometric Models

25.3.1 Conditional Logit and Mixed Logit Models

Discrete choice models are derived based on utility maximization, in which the decision-maker selects an alternative that offers the greatest utility. Let U_{ni} denote the utility that respondent n obtains from alternative i in choice set C_n as follows:

$$U_{ni} = V_{ni} + \varepsilon_{ni,} \tag{25.1}$$

where V_{ni} is a deterministic component and ε_{ni} is a random component, which are both assumed to be known to the individual, but unknown to the analyst. The probability that respondent n will choose i from alternative j in choice set C is the probability that U_{ni} is larger than U_{nj} as follows:

$$\Pr(i) = \Pr\left[U_i > U_j\right] = \Pr\left[V_i - V_j > \varepsilon_j - \varepsilon_i\right] \forall j \neq i, \forall j \in C, \tag{25.2}$$

where the index n is omitted for simplification. Note that U_{ni} depends on parameters that are unknown to the researcher and ε_{ni} is the unobservable portion that respondent n chooses alternative i.

This probability is a cumulative distribution as follows:

$$\Pr_i = \Pr\left(\varepsilon_j - \varepsilon_i < V_i - V_j \forall j \neq i\right) = \int I\left(\varepsilon_j - \varepsilon_i < V_i - V_j \forall j \neq i\right) f\left(\varepsilon\right) d\varepsilon, \tag{25.3}$$

where $I(\bullet)$ is an indicator function. Random utility models are obtained from different density specifications, which are then described as follows:

$$U_{ni} = \beta' x_{ni} + \varepsilon_{ni}. \tag{25.4}$$

When the utility is linear in β and x_{ni} is a vector of explanatory variables that are observed by the analysts and encompass the alternative attributes in a choice task, then $V_{ni}(\beta) = \beta' x_{ni}$.

The distribution of the random component (ε_i) is assumed to be a type I extreme value, and the probability that responder n chooses alternative i can be described in the conditional logit (CL) model as follows (McFadden 1974):

$$P_{ni} = \frac{\exp(\beta' x_{ni})}{\sum_{j=1}^{J} \exp(\beta' x_{nj})}, \tag{25.5}$$

where the preference parameter is assumed to be constant for all respondents (Train 2009). When assuming the heterogeneity of preferences that vary across individuals, mixed logit (ML) probabilities are the integrals of standard logit probabilities over density parameters in the following equation:

$$P_{ni} = \int L_{ni}\,(\beta)\, f\,(\beta|\theta)\, d\beta, \tag{25.6}$$

where $L_{ni}(\beta)$ is the logit probability evaluated at β and $f(\beta|\theta)$ is a density function, in which θ refers to the distribution. Thus, the ML probability takes the following form:

$$P_{ni} = \int \left(\frac{e^{\beta' x_{ni}}}{\sum_j e^{\beta' x_{nj}}} \right) f\,(\beta)\, d\beta. \tag{25.7}$$

ML is a mixture of logit functions that evaluate different's with $f(\beta)$ as the mixing distribution, wherein the utility (Eq. 25.4) takes a coefficient form (Train 2009), in which each of the coefficients is given an independent normal distribution with an estimated mean and standard deviation.

25.3.2 Estimation

The log likelihood for the model is described as follows:

$$\text{LL}\,(\theta) = \sum_n \ln P_{ni}\,(\theta). \tag{25.8}$$

Then, Eq. (25.8) is approximated by simulating for any given θ, in which the simulated log likelihood is determined by maximizing LL(θ); thus, Eq. (25.8) can be rewritten as follows:

$$\text{SLL}\,(\theta) = \sum_{n=1}^{N} \sum_{j}^{J} d_{nj} \ln \frac{1}{R} \sum_{r=1}^{R} L_{ni}\,(\beta^r), \tag{25.9}$$

where R is the number of draws and $d_{nj} = 1$ if responder n chooses j plan and $d_{nj} = 0$ if they do not. In the context of discrete choice models, Train (2009) explained the superior coverage of Halton draws and its effectiveness, as compared with a random draw, stating that 100 Halton draws improve model accuracy. Thus, this study utilized the maximum simulated likelihood with 100 Halton draws applying to the ML model.

25.3.3   CL Versus ML Model

Next, we compared CL model with ML model of coastal residents' choices of coastal infrastructure. The limitation of the CL is that it assumes the same parameters for all responders. Thus, a CL facilitates interpretation when assuming that the same categories of responders have the same preferences. Therefore, we divided the data into two categories: a tsunami-prone group for the respondents that reside in the area where it is likely to be inundated by tsunamis (TPG) and others, using ArcGIS 10.3.1 and a tsunami inundation map is provided by Ministry of Land, Infrastructure, and Transport. Consequently, we obtained 105 TPG samples, which accounted for 11% of all responses. Meanwhile, the ML model enables us to consider the heterogeneities and elucidate the distribution of preferences.

25.3.4   WTP

WTP measures are useful for interpreting change in a given attribute and are employed for several reasons. Given the limited budgets in municipalities and the pressure of hydro-meteorological events (i.e., floods and windstorms), WTP can illustrate how much people value goods, properties, and services. Accordingly, our choice experiments assume that a respondent will pay an annual tax to improve the level of a given attribute when they understand the needs or the desirability of coastal designs. WTP is calculated as follows:

$$\mathrm{WTP}_k = -\frac{\beta_k}{\beta_{\mathrm{tax}}}, \tag{25.10}$$

where β_k is the parameter of attribute k and β_{tax} is the parameter of the tax.

25.4   Results

Table 25.4 lists the variables and definitions used in our analysis. Besides, Table 25.5 summarizes the estimation results of the CL model for the TPG and the others. As previously mentioned, the CL model assumes the same parameters for all respondents. In TPG, the attributes, except for forest, bird, and recreation, were found to be statistically significant at the 1% or 5% level. In contrast, for the non-

Table 25.4 Variables and definitions

Variables	Definitions
asc	A dummy variable representing respondents' answers for alternative 4 (choose nothing)
sea10	Whether alternatives including one-in-10 year probability tsunamis were chosen (0–1 dummy)
sea30	Whether alternatives including one-in-30 year probability tsunamis were chosen (0–1 dummy)
sea100	Whether alternatives including one-in-100 year probability tsunamis were chosen (0–1 dummy)
forest	Coastal forest width
bird	Species richness (the number of avian species)
landsc_sea	Whether alternatives including gray (seawalls) landscape were chosen (0–1 dummy)
landsc_both	Whether alternatives including hybrid (seawalls + coastal forests) landscape were chosen (0–1 dummy)
rec_walk	Whether alternatives including walking were chosen (0–1 dummy)
rec_fish	Whether alternatives including fishing were chosen (0–1 dummy)
rec_camp	Whether alternatives including camping were chosen (0–1 dummy)

Table 25.5 Estimation results

	Tsunami-prone sample			Others		
	Coef.	z	p-value	Coef.	z	p-value
sea10	0.4557	2.93	0.00	0.2230	4.1	0.00
sea30	0.5126	2.96	0.00	0.4292	7.14	0.00
sea100	0.4642	4.22	0.00	0.4134	10.7	0.00
forest	0.0020	0.08	0.94	0.0268	3.13	0.00
bird	−0.0061	−0.94	0.35	−0.0054	−2.35	0.02
landsc_sea	−0.2207	−2.02	0.04	−0.1137	−2.97	0.00
landsc_both	0.2273	2.2	0.03	0.2456	6.67	0.00
rec_walk	0.0706	0.53	0.60	0.0249	0.53	0.59
rec_fish	0.1350	1.35	0.18	0.0003	0.01	0.99
rec_camp	−0.0145	−0.11	0.91	0.0114	0.24	0.81
price	−0.0406	−5.42	0.00	−0.0374	−14.45	0.00
asc	−0.5868	−3.13	0.00	−0.4123	−6.3	0.00
Number of obs	3360			27,200		
Log likelihood	−1099.51			−9031.88		
Pseudo R2	0.056			0.042		

TPG group, which has less likelihood that their residences will suffer from tsunami inundation, most of the attributes were statistically significant, with the exception of recreation. Furthermore, Table 25.6 demonstrates the WTP results for both groups based on the findings listed in Table 25.5. Regarding seawalls, one remarkable finding was that the TPG had a higher WTP for short-cycle tsunamis than the non-TPG (11,233 JPY and 5958 JPY, respectively). Moreover, the highest WTP

Table 25.6
Willingness-to-pay (WTP)
estimation (JPY)

	TPG	Other	TPG-other
sea10	11,233	5958	5275
sea30	12,635	11,470	1165
sea100	11,441	11,046	395
forest	48	716	−668
bird	−152	−146	−6
landsc_sea	−5439	−3038	−2401
landsc_both	5602	6564	−962
rec_walk	1741	665	1076
rec_fish	3328	8	3320
rec_camp	−358	305	−663

values were estimated for the mid-term tsunami in both groups, reaching 12,635 JPY and 11,470 JPY for the TPG and others, respectively. In coastal forests, the WTP values/100 m was 716 JPY for the non-TPG, whereas it had no significance in the TPG. As for the number of bird species, both groups showed negative perceptions (−152 JPY in TPG and −146 JPY in others). Likewise, we observed negative WTP values for the gray-based coastal landscape in both groups. Simultaneously, the gray + green landscape had higher WTP values, at 5602 JPY for the TPG and 6564 JPY for the others.

Next, to consider preference heterogeneity, we applied all the data into both the CL and ML models for comparison (Table 25.7). Again, CL assumes the same parameters for all respondents, while ML assumes that individuals have different preferences. Thus, the former tends to overestimate the WTP. The normally distributed coefficients, estimated means, and standard deviations listed in Table 25.8 reflect the distribution of preferences. For example, the distribution of the seawalls with a 30-year-tsunami coefficient had an estimated mean of 0.44 and an estimated standard deviation of 1.19, such that 68.7% of the distribution was above zero and 31.3% was below. This indicates that two-thirds of the respondents view a seawall for a mid-term tsunami as a positive and prefer it, whereas one-third do not prefer it. Similarly, compared with seawalls for three different frequencies, seawalls for long-term tsunamis had an estimated mean of 0.12 and an estimated standard deviation of 1.88, implying 53.2% positive and 46.8% negative attitudes. A similar trend was observed for the short-term tsunamis with an estimated mean of 0.17 and a standard deviation of 1.19, implying 57.2% of respondents supported it and 42.8% did not. Regarding coastal forests, the results revealed that 60.3% of respondents accepted coastal defense by green infrastructure, whereas 39.7% did not. On the one hand, for landscape, gray + green (seawalls + coastal forests) obtained high positive responses (71.1% in positive and 28.9% in negative), while 61.3% of respondents showed negative perceptions toward gray landscape (seawalls only). Looking at bird species, 64.9% of responders had negative preferences regarding an increased number of birds. Conversely, regarding recreation, no significant results were obtained. Nevertheless, in the CL model, the recreation parameters were positive for various activities; however, the mean values of these ones in the ML model were

Table 25.7 Conditional logit (CL) and mixed logit (ML) model results

	CL					
	Coef.	z	p-value			
sea10	0.2485	4.85	0.00			
sea30	0.4380	7.72	0.00			
sea100	0.4188	11.50	0.00			
forest	0.0242	3.00	0.00			
bird	−0.0055	−2.53	0.01			
landsc_sea	−0.1252	−3.47	0.00			
landsc_both	0.2437	7.03	0.00			
rec_walk	0.0304	0.69	0.49			
rec_fish	0.0159	0.47	0.64			
rec_camp	0.0085	0.19	0.85			
price	−0.0378	−15.43	0.00			
asc	−0.4293	−6.95	0.00			
Number of obs.		30,560				
Log likelihood		−10135.66				
	ML					
	Mean			Standard deviation		
	Coef.	z	p-value	Coef.	z	p-value
price	−0.0459	−14.13	0.00			
asc	−0.9752	−11.35	0.00			
sea10	0.1718	2.21	0.03	1.1879	12.62	0.00
sea30	0.4430	5.29	0.00	1.1856	13.03	0.00
sea100	0.1187	1.49	0.14	1.8750	22.81	0.00
forest	0.0439	3.20	0.00	0.2128	11.12	0.00
bird	−0.0319	−6.49	0.00	0.1073	21.29	0.00
landsc_sea	−0.1540	−2.80	0.01	0.6796	8.90	0.00
landsc_both	0.3781	6.86	0.00	0.8967	12.91	0.00
rec_walk	−0.0839	−1.23	0.22	−0.7666	−11.38	0.00
rec_fish	−0.0551	−1.13	0.26	0.1370	1.27	0.21
rec_camp	0.0784	1.28	0.20	0.3985	4.04	0.00
Number of obs.		30,560				
Log likelihood		−8843.73				

contrasting, with the exception of camping. It is worth noting that the ML model can reflect the distribution of respondents' preferences (Fig. 25.8). Finally, the WTP results in each model showed significant differences regarding seawall variables. In particular, in seawalls for long-cycle tsunamis, the WTP calculated by the CL model was 8504 JPY higher than that in the ML model. However, it is not always the case that the CL model overestimates all attributes except coastal forests and landscape (gray + green) variables. Overall, we found that the CL model's overestimation is significant in gray infrastructure and that it is not easy to apply the stated preference method for long periods of time. Moreover, the average WTP/100 m of coastal

Table 25.8 Willingness-to-pay (JPY)

	Conditional logit (CL)	Mixed logit (ML)	CL-ML
sea10	6582	3747	2834
sea30	11,602	9662	1941
sea100	11,094	2590	8504
forest	642	957	−316
bird	−146	−696	550
landsc_sea	−3318	−3358	40
landsc_both	6455	8245	−1791
rec_walk	806	−1830	2636
rec_fish	420	−1202	1622
rec_camp	226	1710	−1484

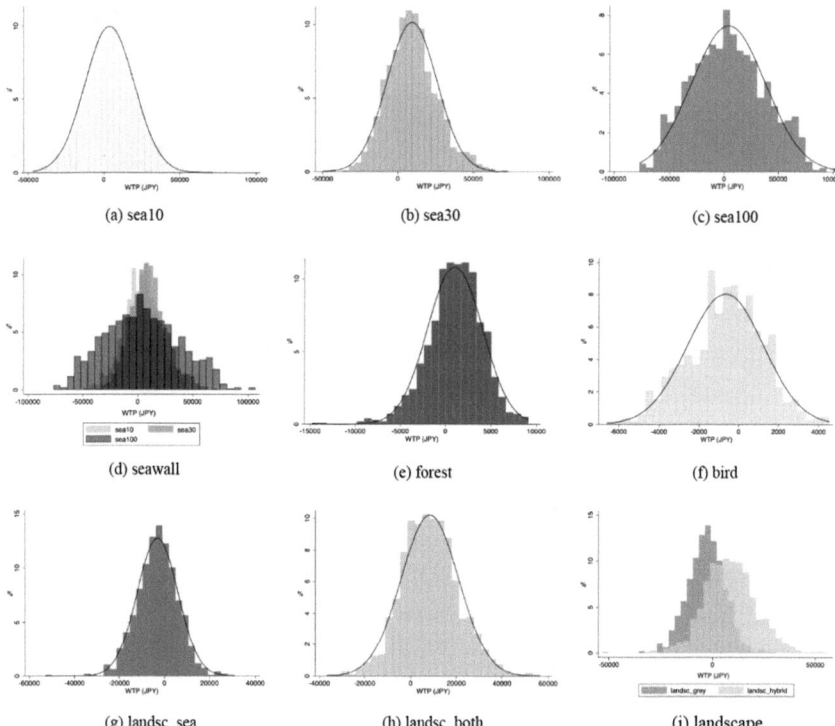

(a) sea10   (b) sea30   (c) sea100

(d) seawall   (e) forest   (f) bird

(g) landsc_sea   (h) landsc_both   (i) landscape

Fig. 25.8 The distribution of willingness-to-pay (WTP). (**a**) Seawalls against one-in-10 year probability tsunamis, (**b**) seawalls against one-in-30 year probability tsunamis, and (**c**) seawalls against one-in-100 year probability tsunamis. (**d**) Graph with integrated (**a**), (**b**), and (**c**). WTP for (**e**) coastal forests, (**f**) species richness, (**g**) gray landscape, and (**h**) hybrid (gray + green) landscape. (**i**) Graph with integrated (**g**) and (**h**)

forests was estimated to be 957 JPY by the ML, while that of the hybrid landscape was 8245 JPY. Conversely, species richness showed a negative value of −696 JPY with an increasing number of bird species.

25.5 Discussion and Conclusion

The analysis presented in this study supports a number of issues, including the trade-off relationship between ESSs, the monetary values of coastal functions, and the preferences for long-term settings using the stated preference method. However, before discussing our results, we must determine the restrictions of this study. To investigate whether the responders live in areas that are susceptible to tsunamis (TPG), we first specified coordinates using zip codes applied for geocoding. However, the points identified by coordinates do not necessarily indicate their precise location. In particular, zip codes apply to large areas, making it likely that there are higher and lower tsunami risks (Fig. 25.9). Moreover, as 213 respondents hesitated to supply their zip codes, we used their answer to the TPG-related question to group them, as shown in Fig. 25.10. Thus, this process depends on the reliability of their answers. Furthermore, the CL model assumes the same parameters for all respondents, implying that the TPG and the non-TPG each have their own

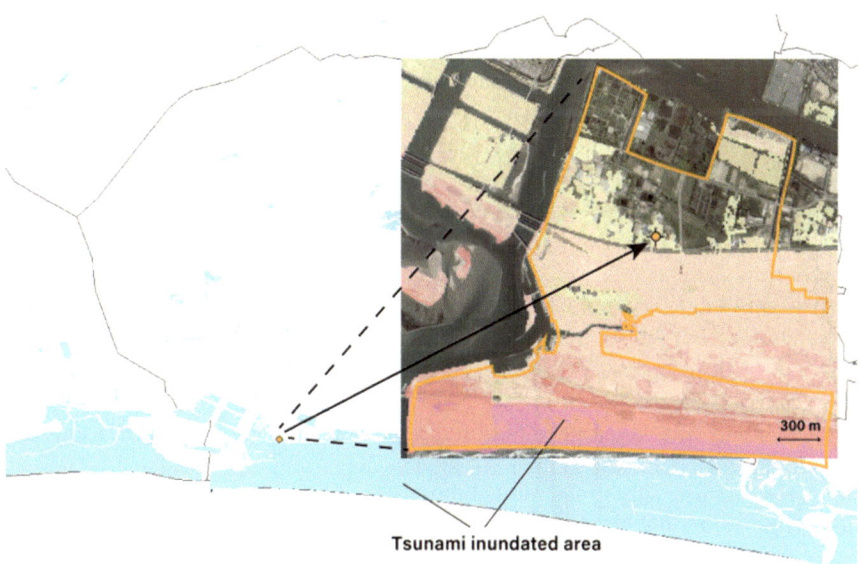

Tsunami inundated area

Fig. 25.9 Precision limitations. *Note*: The polygon represents the typical location identified by the zip codes, and the line depicts a municipal boundary. As the polygon does not identify specific respondents' addresses, it covers both tsunami-prone and lower areas. Hence, there exists a limitation in identifying the respondents in the tsunami-prone areas

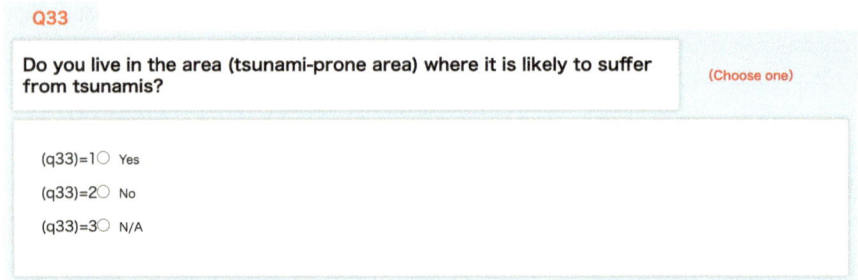

Fig. 25.10 Example question

parameters. Based on the comparisons between the TPG and the non-TPG, the former provided higher WTP values for the DRR function offered by gray than the latter. In particular, the difference in preferences against a short-cycle tsunami was significant (11,233 JPY and 5958 JPY for the TPG and others, respectively). In addition, increasing the scale of coastal forests/100 m in width had no significance in the TPG. This is probably because of the unstable Eco-DRR effects and the fact that expanding coastal forests may overlap with their residences. Then, more birds were not preferable for the TPG. Despite the economic importance of biodiversity (Martínez et al. 2007; TEEB 2010), for which the bird community is a useful tool for monitoring the onshore environment (Canterbury et al. 2000; Sekercioglu 2006), birds may be associated with negative images of bird droppings, and the drawbacks of certain ESSs, known as ecosystem disservices. Thus, evaluating the economic value of biodiversity remains challenging. Also, only animals in the peak of a food chain should not be spotted, so we need to pay attention to the underestimation of other fauna and flora and to consider the way of evaluating coastal biodiversity in both onshore and offshore environments as well. Regarding landscape, both groups had negative attitudes toward gray-based landscape, especially the TPG (−5439 JPY), whereas TPG had a higher WTP for gray + green landscape (5602 JPY). Considering the understanding of the DRR functions of gray, the negative response toward gray landscape reflects the difficulty of coastal disaster management and installation of hybrid infrastructure. Furthermore, the visual image can affect the responders' decision-making; hence, further development of questionnaire design is necessary.

Meanwhile, the heterogeneities in preferences can be examined using the estimation results from the CL and ML models. Regarding additional seawall height, little attention was paid to low-probability tsunamis, in which a WTP of 2590 JPY was estimated by the ML model. This indicates the importance of considering the individual preferences in comparison to the CL model (11,094 JPY). However, the large gap between CL and ML results in long-cycle tsunamis demonstrates the complexity of predicting people's preferences for future settings (Fig. 25.8d). The coastal forest data suggest that respondents expect coastal forests to function as Eco-DRR, despite the presence of uncertainties. Although the WTP for the

DRR function of coastal forests is 957 JPY, which is lower than that for gray, ecosystem multifunctionality, encompassing biodiversity, landscape, and recreation, the approximately 60% of positive answers from respondents must be considered. Overall, a perceived increase in the proportion of coastal disaster reduction in the gray and green infrastructure was shown to attract higher WTP values. As the goods and services resulting from ecosystems are prone to be underestimated (Heal 2000; Beaumont et al. 2008), economic valuation techniques have been developed to capture monetary terms. Moreover, coastal ecosystem functions commensurate with other vegetation (i.e., sand dunes and coral reefs). However, as we have focused on only the onshore environment, only coastal forests were applied in our choice experiments. Chávez et al. (2021) mentioned that coastal green infrastructure encompasses the multi-scale processes of recovery and maintenance, and then it can encourage local people's engagement. Silva et al. (2017) categorized green infrastructure by applying the degree of naturalness ranging from natural to hard engineering and stressed that coastal green infrastructure can be regarded as a series of natural to artificial components. Thus, we need to address how the economic values of coasts in both onshore and offshore environments as a whole can be evaluated. Another thing that needs to be explored is that recreational services were not significant in either model. In order to value coastal recreation and landscape services for cost-benefit analysis, the careful descriptions are required. These economic and ecological constraints have long been a subject of debate. Finally, the ML model demonstrated that the hybrid landscape had over 70% of positive responses, whereas the gray landscape had approximately 60% of negative perceptions. Therefore, our study highlighted the recognition of coastal protection offered by gray and green infrastructure, but the question of whether seawalls prevent people from enjoying ocean views remains. Furthermore, even though the hybrid approach is preferable for coastal citizens, identifying the best locations and practice of coastal designs that optimizes gray, green, and hybrid approaches is inevitable (Conger and Chang 2019).

All in all, the results of the TPG and others clarified that the former valued protection function provided by the gray infrastructure. However, the estimated values in the ML model were lower as the scale of seawalls increased for low-probability tsunamis. In addition, the Eco-DRR of coastal forests was not significant in the TPG, but the ML model showed the necessity of its multifunctionality. Furthermore, hybrid seascapes received positive attention in both groups in both models. These paradoxical views of coastal infrastructure and the heterogeneities of people's preferences may foreshadow the challenges of building a consensus among multiple stakeholders and implementing hybrid approaches. As there is no one-size-fits-all coastal design, it is important to explore long-term resilience planning at regional and local levels from risk reduction socioeconomic and ecological perspectives. In the future, we will examine the risk preferences of people that reside in or will move to hinterlands and higher places where they are less likely to be affected by tsunamis. As their DRR function values may reflect their locations, the hedonic pricing method will be useful. Therefore, the synthesized both stated preference and revealed preference data will help coastal communities provide strategic and long-term coastal planning through the interdisciplinary design stage of gray-green combined coastal defense for DRR while simultaneously enhancing the quality of life.

Acknowledgments This research was partially funded by the Environment Research and Technology Development Fund (Project number: 4-1805). We would like to thank the research group from Tokushima University and Takumi Tsuruma and Makoto Nakata from Niigata University. Furthermore, we received extensive support and advice from Nikkei Research Inc. and would also like to thank the respondents of our online surveys.

Appendix

The questionnaire is provided.

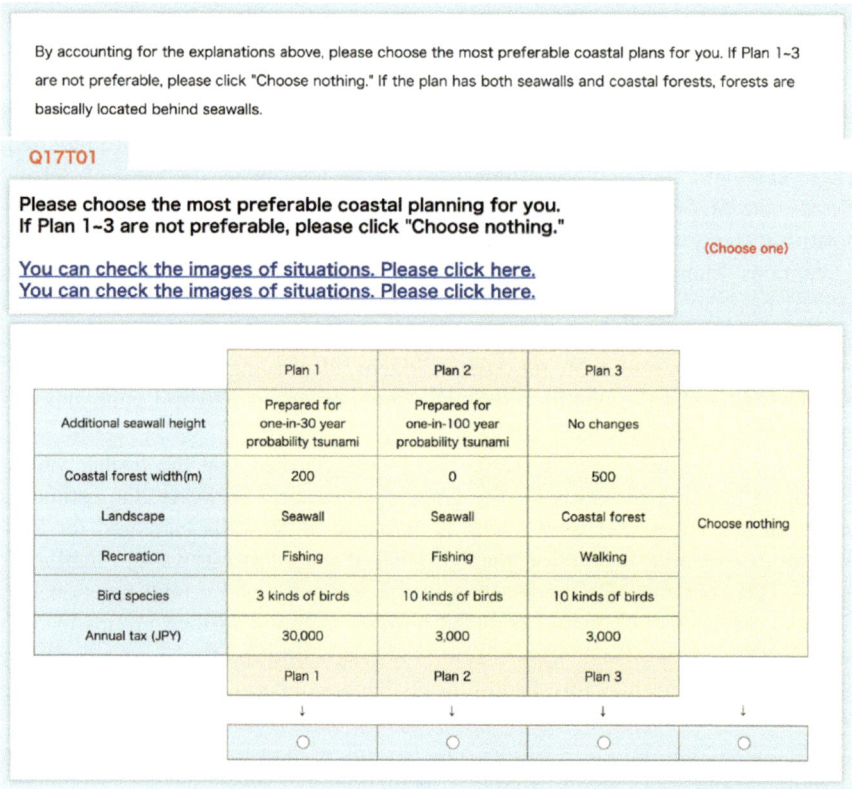

Note: The choice experiments were introduced in Q17T01–Q17T08 in the questionnaire.

References

Andersen T, Carstensen J, Hernandez-Garcia E, Duarte CM (2009) Ecological thresholds and regime shifts: approaches to identification. Trends Ecol Evolut 24(1):49–57

Bao TQ (2011) Effect of mangrove forest structures on wave attenuation in coastal Vietnam. Oceanologia 53(3):807–818

Barbier EB, Hacker SD, Kennedy C, Koch EW, Stier AC, Silliman BR (2011) The value of estuarine and coastal ecosystem services. Ecol Monogr 81(2):169–193

Bateman IJ, Carson RT, Day B, Hanemann WM, Hanley N, Hett T, Jones-Lee M, Loomes G, Mourato S, Özdemiroglu E, Pearce DW, Sugden R, Swanson J (2002) Economic valuation with stated preference techniques. Edward Elgar

Beaumont NJ, Austen MC, Mangi SC, Townsend M (2008) Economic valuation for the conservation of marine biodiversity. Mar Pollut Bull 56(3):386–396

Bergen SD, Bolton SM, Fridley JL (2001) Design principles for ecological engineering. Ecol Eng 18(2):201–210

Blackwell B (2007) The value of a recreational beach visit: an application to Mooloolaba beach and comparisons with other outdoor recreation sites. Econ Analy Policy 37(1):77–98

Börger T, Beaumont NJ, Pendleton L, Boyle KJ, Cooper P, Fletcher S, Haab T, Hanemann M, Hooper TL, Hussain SS, Portela M, Stithou M, Stockill J, Taylor T, Austen MC (2014) Incorporating ecosystem services in marine planning: the role of valuation. Mar Policy 46:161–170

Borsje BW, van Wesenbeeck BK, Dekker F, Paalvast P, Bouma TJ, van Katwijk MM, de Vries MB (2011) How ecological engineering can serve in coastal protection. Ecol Eng 37(2):113–122

Brenner J, Jimenez JA, Sarda R, Garola A (2010) An assessment of the non-market value of the ecosystem services provided by the Catalan coastal zone, Spain. Ocean Coast Manag 53(1):27–38

Canterbury GE, Martin TE, Petit DR, Petit LJ, Bradford DF (2000) Bird communities and habitat as ecological indicators of forest condition in regional monitoring. Conserv Biol 14(2):544–558

Chapman MG, Underwood AJ (2011) Evaluation of ecological engineering of "armoured" shorelines to improve their value as habitat. J Exp Mar Biol Ecol 400(1–2):302–313

Chávez V, Lithgow D, Losada M, Silva-Casarin R (2021) Coastal green infrastructure to mitigate coastal squeeze. J Infrastruct Preservat Resilie 2(1):1–12

Cheong SM, Silliman B, Wong PP, Van Wesenbeeck B, Kim CK, Guannel G (2013) Coastal adaptation with ecological engineering. Nat Clim Chang 3(9):787–791

ChoiceMetrics (2018) Ngene 1.2.0 User Manual & Reference Guide. ChoiceMetrics.http://www.choice-metrics.com/NgeneManual120.pdf

Conger T, Chang SE (2019) Developing indicators to identify coastal green infrastructure potential: the case of the Salish Sea region. Ocean Coast Manag 175:53–69

Cooper JAG, O'Connor MC, McIvor S (2020) Coastal defences versus coastal ecosystems: a regional appraisal. Mar Policy 111:102332

Costanza R, Pérez-Maqueo O, Martinez ML, Sutton P, Anderson SJ, Mulder K (2008) The value of coastal wetlands for hurricane protection. Ambio 37:241–248

European Commission (2000) Communication from the commission to the council and the European parliament on integrated coastal zone management: a strategy for Europe. https://ec.europa.eu/environment/iczm/comm2000.htm

Garber-Yonts B, Kerkvliet J, Johnson R (2004) Public values for biodiversity conservation policies in the Oregon Coast Range. For Sci 50(5):589–602

Gedan KB, Kirwan ML, Wolanski E, Barbier EB, Silliman BR (2011) The present and future role of coastal wetland vegetation in protecting shorelines: answering recent challenges to the paradigm. Clim Chang 106(1):7–29

Ghermandi A, Dias N, Paulo A, Portela R, Nalini R, Teelucksingh, SS (2009) Recreational, cultural and aesthetic services from estuarine and coastal ecosystems. FEEM Working Paper No. 121.2009, Available at http://dx.doi.org/10.2139/ssrn.1532803

Harada K, Imamura F (2005) Effects of coastal forest on tsunami hazard mitigation—a preliminary investigation. In: Satake K (ed) Tsunamis. Springer, Dordrecht, pp 279–292

Heal G (2000) Valuing ecosystem services. Ecosystems:24–30

Helm D, Hepburn C (2012) The economic analysis of biodiversity: an assessment. Oxf Rev Econ Policy 28(1):1–21

Hensher DA, Rose JM, Rose JM, Greene WH (2005) Applied choice analysis: a primer. Cambridge university press

Imamura K, Takano KT, Mori N, Nakashizuka T, Managi S (2016) Attitudes toward disaster-prevention risk in Japanese coastal areas: analysis of civil preference. Nat Hazards 82(1):209–226

Irtem E, Gedik N, Kabdasli MS, Yasa NE (2009) Coastal forest effects on tsunami run-up heights. Ocean Eng 36(3–4):313–320

Louviere JJ (1988) Conjoint analysis modelling of stated preferences: a review of theory, methods, recent developments and external validity. JTEP 22:93–119

Louviere JJ, Hensher DA, Swait DA (2000) Stated choice methods: analysis and applications. Cambridge University Press

Martínez ML, Intralawan A, Vázquez G, Pérez-Maqueo O, Sutton P, Landgrave R (2007) The coasts of our world: ecological, economic and social importance. Ecol Econ 63(2–3):254–272

McFadden D (1974) Conditional logit analysis of qualitative choice behavior. In: Zarembka P (ed) Frontiers in econometrics. Academic Press, pp 105–142

McGranahan G, Balk D, Anderson B (2007) The rising tide: assessing the risks of climate change and human settlements in low elevation coastal zones. Environ Urban 19(1):17–37

Ministry of Land, Infrastructure, Transport and Tourism (2012) Concepts of Comprehensive Tsunami Countermeasures (original Japanese). http://www.mlit.go.jp/common/000146461.pdf

Morris RL, Konlechner TM, Ghisalberti M, Swearer SE (2018) From grey to green: efficacy of eco-engineering solutions for nature-based coastal defence. Glob Chang Biol 24(5):1827–1842

Onuma A, Tsuge T (2018) Comparing green infrastructure as ecosystem-based disaster risk reduction with gray infrastructure in terms of costs and benefits under uncertainty: a theoretical approach. Int J Disast Risk Reduct 32:22–28

Perkol-Finkel S, Hadary T, Rella A, Shirazi R, Sella I (2018) Seascape architecture–incorporating ecological considerations in design of coastal and marine infrastructure. Ecol Eng 120:645–654

Pioch S, Relini G, Souche JC, Stive MJF, De Monbrison D, Nassif S, Simard F, Allemand D, Saussol P, Spieler R, Kilfoyle K (2018) Enhancing eco-engineering of coastal infrastructure with eco-design: moving from mitigation to integration. Ecol Eng 120:574–584

Prayaga P (2017) Estimating the value of beach recreation for locals in the Great Barrier Reef Marine Park, Australia. Econ Anal Polic 53:9–18

Reeve D, Chadwick A, Fleming C (2018) Coastal engineering: processes, theory and design practice. CRC Press

Schoonees T, Mancheño AG, Scheres B, Bouma TJ, Silva R, Schlurmann T, Schüttrumpf H (2019) Hard structures for coastal protection, towards greener designs. Estuar Coasts 42(7):1709–1729

Sekercioglu CH (2006) Increasing awareness of avian ecological function. Trends Ecol Evol 21(8):464–471

Silva R, Lithgow D, Esteves LS, Martínez ML, Moreno-Casasola P, Martell R, Pereira P, Mendoza E, Campos-Cascaredo A, Grez PW, Osorio AF, Osorio-Cano JD, Rivillas GD (2017) Coastal risk mitigation by green infrastructure in Latin America. Proceed Inst Civil Eng Maritime Eng 170(2):39–54

Spalding MD, Fox HE, Allen GR, Davidson N, Ferdaña ZA, Finlayson MAX, Halpern BS, Jorge MA, Lombana A, Lourie SA, Martin KD, McManus E, Molnar J, Recchia CA, Robertson J (2007) Marine ecoregions of the world: a bioregionalization of coastal and shelf areas. Bioscience 57(7):573–583

Sutton-Grier AE, Wowk K, Bamford H (2015) Future of our coasts: the potential for natural and hybrid infrastructure to enhance the resilience of our coastal communities, economies and ecosystems. Environ Sci Pol 51:137–148

Takagawa S, Ueta M, Amamo T, Okahisa Y, Kamioki M, Takagi K, Takahashi M, Hayama S, Hirano T, Mikami O, Mori S, Morimoto G, Yamaura Y (2011) JAVIAN database: a species-level database of life history, ecology and morphology of bird species in Japan. Bird Res (Orig Japan) 7:R9–R12

Tanaka N (2012) Effectiveness and limitations of coastal forest in large tsunami: conditions of Japanese pine trees on coastal sand dunes in tsunami caused by Great East Japan Earthquake. J Japan Soc Civil Eng Ser B1 (Hydraul Eng) 68(4):II_7–II_15

Task Committee for the Hybrid of Green and Gray Infrastructures (2018) Study on The Hybrid of Green and Grey Infrastructures (original Japanese). https://committees.jsce.or.jp/s_research/system/files/%e3%82%b0%e3%83%aa%e3%83%bc%e3%83%b3%e3%82%b0%e3%83%ac%e3%83%bc%e5%a0%b1%e5%91%8a%e6%9b%b8%ef%bc%88%e6%8f%90%e5%87%ba%e7%89%88%ef%bc%89.pdf

TEEB (2010) In: Kumar P (ed) The economics of ecosystems and biodiversity ecological and economic foundations. Earthscan, London

Temmerman S, Meire P, Bouma TJ, Herman PM, Ysebaert T, De Vriend HJ (2013) Ecosystem-based coastal defence in the face of global change. Nature 504(7478):79–83

Train K (2009) Discrete choice methods with simulation. Cambridge University Press

Chapter 26
Forest Green Infrastructure and the Carbon Storage and Substitution Benefits of Harvested Wood Products

Gregory Valatin

Abstract Forest Green Infrastructure (FGI) provides society with a wide range of benefits. Significant climate change mitigation benefits arise outside the forest associated with the use of harvested wood products. These include both carbon storage in wood products and carbon substitution benefits associated with the use of wood instead of more fossil energy-intensive materials such as concrete and steel, or of fossil fuels in energy production. This chapter considers the potential of extending coverage of the UK Woodland Carbon Code to the carbon benefits of wood products associated with woodland creation projects. It builds on previous approaches to including the carbon benefits of harvested wood products under existing carbon market standards. The key recommendations include (1) exploring ways of allocating carbon units between woodland owners and wood users that provide incentives to increase the quality and supply of timber, the carbon storage and substitution benefits per unit of wood, as well as the overall benefit to society; (2) consideration of potential double-counting issues and how these can be minimized; and (3) investigating rebound and leakage effects, which affect by how much fossil fuel use in the economy changes as a result of increased woodfuel use. Depending on the management system and species used, woodland creation projects involving wood harvesting may increase overall carbon benefits once carbon storage and substitution benefits have been accounted for particularly over multiple rotations. Further work would be required to assess whether average and generic values of carbon storage and substitution benefits could be incorporated into the UK Woodland Carbon Code's project-level accounting and impacts on the levels of carbon credits that could then be claimed.

Forest Research is the Research Agency of the Forestry Commission and is the leading UK organization engaged in forestry and tree-related research. The Agency aims to support and enhance forestry and its role in sustainable development by providing innovative, high-quality scientific research, technical support, and consultancy services.

G. Valatin (✉)
The Research Agency of the Forestry Commission, Farnham, UK
e-mail: gregory.valatin@forestresearch.gov.uk

© Crown 2022
F. Nakamura (ed.), *Green Infrastructure and Climate Change Adaptation*,
Ecological Research Monographs, https://doi.org/10.1007/978-981-16-6791-6_26

Keywords Climate change mitigation · Woodland creation · Rebound effect · Double-counting · Incentives

26.1   Introduction

Forest Green Infrastructure (FGI) provides society with a wide range of benefits. Significant climate change mitigation benefits arise outside the forest associated with the use of harvested wood products (HWP). These include both carbon storage in HWP and substituting wood in construction instead of more fossil energy-intensive materials, such as concrete and steel, as well as using wood instead of fossil fuels in energy and heat production. Taking carbon substitution and storage benefits into account can affect comparisons of the benefits for climate change mitigation of different FGI projects. For example, recent analysis indicates that typical carbon benefits to 2100 in the UK from planting native broadleaves as conservation woodlands are 6.2 tons of carbon dioxide (tCO_2)/ha/year, while those for "productive forests" managed for wood production are 7 tCO_2/ha/year once carbon storage and substitution benefits are included (Scottish Forestry 2020).

Carbon standards such as the UK Woodland Carbon Code (Forestry Commission 2014) help underpin market confidence and the claims of project developers about the climate change mitigation benefits associated with undertaking their projects. Accounting for carbon storage and substitution benefits under carbon standards covering woodland creation projects could be desirable for a variety of reasons. First, incentives for carbon sequestration—a process by which carbon dioxide (CO_2) is removed from the atmosphere and held in solid or liquid form—on their own may fail to maximize the overall carbon benefits of woodland creation, and they may potentially provide perverse incentives. (A perverse incentive is an incentive that has an unintended and undesirable result that is contrary to the intended outcome.) This could occur, for example, if incentives reduce wood harvesting and the consequent reduction in carbon storage and substitution is larger than the carbon savings from increased sequestration (Valatin 2012). Second, the wider the coverage of climate benefits of woodland creation, the more comprehensive the estimates and the more attractive forestry becomes as an investment compared with alternative options. Third, if carbon storage and substitution benefits are not covered, there is no incentive for landowners or investors to consider them in their land use and investment decisions, which may lead to woodland creation opportunities being missed. Fourth, product and energy substitution may be more effective long-term climate change mitigation strategies than sequestration (e.g., Niles and Schwarze 2001), although this also depends on the rate at which energy and construction sectors become more efficient in their use of fossil fuels, and how quickly any end-of-pipe carbon sequestration and storage technologies are introduced. Fifth, focusing upon carbon sequestration alone may prove counterproductive if it leads to less harvesting and the use of more fossil energy-intensive products (Miner and Lucier 2004).

In principle, accounting for a broader range of carbon benefits is desirable. However, there are also potential costs associated with taking a more comprehensive approach and significant technical obstacles to this being feasible. The Woodland Carbon Code follows a project-level approach to carbon accounting, and the potential incorporation of average and generic values for carbon storage into this framework would require careful scrutiny.

Quantifying the carbon benefits of HWP is less straightforward than for carbon sequestration as, once harvested, wood is subject to a range of processes and has a wide variety of end uses. The carbon savings of HWP depend not only upon the specific end use, material displaced, efficiency of use, and what recycling or disposal process is used at the end of the product's life, but also upon wider "leakage" and "rebound" effects. Furthermore, there is also potential double-counting to consider.

A review (Valatin 2017) identified four carbon standards that account for the carbon storage benefits of HWP: the American Carbon Registry (ACR), the California Air Resources Board (ARB), the Climate Action Reserve (CAR), and the Verified Carbon Standard (VCS). None of the existing carbon standard protocols for forest projects cover the carbon substitution benefits of HWP, either the use of wood instead of more fossil fuel-intensive materials such as concrete and steel, or of using wood as a source of energy instead of fossil fuels.

To help inform consideration of the potential to extend the coverage of the UK Woodland Carbon Code, this chapter summarizes and develops the findings of the review of approaches to incorporating the carbon benefits of harvested wood products under existing carbon market standards (Valatin 2017). It is structured in sections covering carbon storage in wood products, carbon substitution benefits and wider issues (i.e., potential double-counting and rebound effects), and monitoring and accounting, followed by recommendations.

26.2 Carbon Storage in Wood Products

Total carbon storage in HWP in Great Britain is significant compared with that in British woodlands. For example, one estimate suggests that in 2000 it was around 300 million metric tons of carbon dioxide equivalent ($MtCO_2e$), approximately half the level of the current above-ground forest carbon stock in Great Britain (Forestry Commission 2020).

The findings of the initial review (Valatin 2017) regarding the protocols that cover the carbon storage benefits of HWP under the four carbon standards (i.e., ACR, ARB, CAR, and VCS) were supplemented by also considering newer protocols (ACR 2017, 2018; CAR 2017; VCS 2015, 2016), which also cover these benefits. The following were found:

- A 100-year time frame is used in each case when accounting for carbon storage benefits. Carbon stored for 100 years or longer in HWP is assumed to be stored permanently. This includes wood products in use but also, in some cases, the

Table 26.1 Half-lives for harvested wood products by end use

End use or product category		Half-life (years)
New residential construction	Single family	100
	Multifamily	70
	Mobile homes	12
Residential upkeep and improvement		30
Manufacturing	Furniture	30
	Other products	12
Shipping	Wooden containers	6
	Pallets	6
	Dunnage, etc.	6
Other uses for lumber and panels		12
Solid wood exports		12
Paper		2.6

Source: USDoE (2006, Table D3, p.218)

proportion of carbon stored for 100 years or more in wood products sent to landfills. Carbon stored for less than 100 years is assumed to release the carbon stored immediately or over a fixed period (e.g., 20 years), or according to a fixed decay rate. Table 26.1 presents the half-lives (i.e., the time taken for one-half of the carbon stored to decay and be emitted to the atmosphere) assumed for different categories of HWP under two carbon standards (ARB and CAR). Further information on the half-life recommended for different categories of wood products in various countries can be found in Penman et al. (2003, Table 3a.1.3, p. 3.270); for instance, one study in the Netherlands gives estimates for sawn timber of 18 years for spruce and poplar and 45 years for oak and beech.

- The estimated carbon storage benefits of HWP vary, partly because there are differences in the approaches used to determine the expected net carbon emissions if a project does not go ahead (i.e., "baseline" emissions). Baseline emissions are often assumed to reflect compliance with wider legal requirements such as existing timber harvest plans or specific forest management rules on the diameter of trees harvested. In some cases, historical records, or "common practice" emissions, are used instead. Under one standard, baseline emissions are estimated using economic optimization to determine the legally permissible harvesting scenario that maximizes the net present value of the wood harvested from a perpetual series of rotations. For afforestation and reforestation projects, the wood product component of baseline emissions is often simply assumed to be zero.
- Carbon credits for carbon storage benefits of HWP accrue to the project developer. This is generally the forest owner, even though the carbon storage benefits of HWP depend upon processes that occur outside the forest.

26.3 Carbon Substitution Benefits of Wood

None of the existing carbon standard protocols for forestry projects cover the carbon substitution benefits of HWP, either the use of wood instead of more fossil fuel-intensive materials such as concrete and steel or as a source of energy in place of fossil fuels (Valatin 2017). However, several renewable energy project protocols cover the carbon substitution benefits associated with woodfuel use. The findings of the initial review of protocols under two voluntary carbon standards—the Gold Standard and the VCS—along with the United Nations Clean Development Mechanism (Valatin 2017) were supplemented by considering two more recent protocols (CDM 2017, 2018). The conclusions were the following:

- The impacts of climate change mitigation activities within forests where the biomass originates are seldom taken into consideration when quantifying the carbon benefits of woodfuel use. Only one protocol covers activities within the forest.
- A variety of "emissions factors" that represent emissions per unit of input are used to estimate the carbon benefits of woodfuel use. These include emissions per unit of energy generated and transport emissions per kilometer traveled and fuel type. Differences in emissions factors used partly reflect different project types.
- The forms of leakage (e.g., increased greenhouse gas [GHG] emissions outside the project boundary attributed to the project) accounted for differ, in part reflecting different project types. These include diversion of biomass from other uses, shifts in deforestation, and shifts in other activities.
- CO_2 is the primary focus, but nitrous oxide (N_2O) and methane (CH_4) emissions are also covered in some cases. High uncertainty is associated with some of these. For example, a default of 300% uncertainty is assumed for CH_4 emissions from the combustion of biomass residues under some protocols.
- The project developer running the renewable energy plant, rather than the forest owner, receives the carbon credits.

26.4 Wider Issues

There are several other important considerations. These include whether extending the Woodland Carbon Code to cover carbon storage and the substitution benefits of HWP could give rise to double-counting if the same benefits are accounted for under a different standard in a downstream sector such as construction, potentially undermining the credibility and integrity of climate change mitigation activities. Accounting for potential rebound effects is also important to consider.

Table 26.2 Forms of double-counting

Type		Description
1	Double issuance	More than one carbon unit issued for a single benefit
2	Double certification	A carbon benefit certified under more than one standard
3	Double claiming	A benefit claimed twice toward attaining mitigation pledges
4	Double use	A carbon unit used twice to attain mitigation pledges
5	Double selling	A carbon benefit sold twice to attain different mitigation pledges
6	Double payment	Payments for the same carbon benefit to more than one supplier
7	Double purpose	A carbon unit counted both toward climate change mitigation and to attain another pledge (e.g., international development finance)

26.4.1 Double-Counting

Would extending the Woodland Carbon Code to cover the carbon storage and substitution benefits of HWP fit with wider GHG accounting and carbon standards in downstream sectors, or could it pose intractable issues of double-counting that would risk the integrity of the Code? For example, the Publicly Available Specification (PAS) 2050—a specification for life-cycle assessment of the GHG emissions of goods and services developed in 2008 by the British Standards Institution—takes account of the carbon storage benefits of HWP. Thus, were the Woodland Carbon Code extended to the carbon storage benefits of HWP, there would be a risk that they would be double-counted if the same benefits were also claimed by a construction company under a standard such as PAS2050.

In considering potential double-counting, it is useful to note that definitions vary and a variety of forms can be distinguished (Hood et al. 2014; Schneider et al. 2014; Foucherot et al. 2015). Of the types shown in Table 26.2, the first six are the most relevant when considering the potential for extending the Woodland Carbon Code.

Double-counting can be considered a concern only to the extent that it risks the credibility and integrity of climate change mitigation activities. It is unlikely to be invariably harmful in this respect. Consider, for example, if the same carbon benefit were accounted for under the Woodland Carbon Code and PAS 2050. If each stakeholder and purchaser of any associated carbon units recognizes and accepts the role of others involved in generating the benefit without claiming exclusive ownership, inclusion of the carbon benefit under both standards would be unlikely to undermine the credibility or integrity of either standard.

A direct approach that explicitly addresses the distribution of ownership rights associated with the carbon benefits of HWP between users of wood products and the owners of the woodlands from which they are sourced could avoid double-counting altogether. This could be achieved in cases where a benefit associated with the use of wood in construction is claimed by both a woodland owner and a construction company, for example, by allocating each a share of the associated carbon units. The share of the woodland owner might be issued at the same time as carbon units are issued for carbon sequestration, with the construction company's share kept back until the wood has been harvested, processed, and used, and potentially varied according to the level of carbon saving associated with the specific use selected.

Where the carbon substitution and storage benefits from HWP use are expected to be at least as great (after accounting for permanence issues) as the reduction in net carbon sequestration associated with future wood harvesting, then, compared with a case without harvesting, no reduction in the number of carbon units issued to a woodland owner would appear to be warranted. Focusing on differences in overall carbon benefits would be feasible, providing that carbon substitution and storage benefits from HWP can be reliably tracked, quantified, and verified.

Further exploration of potential double-counting issues can be found in Valatin (2017).

26.4.2 Rebound Effects

Rebound effects are closely related to the concept of "leakage" (emissions that increase elsewhere as a consequence of the project, measure, or policy being introduced). They similarly result inadvertently in increased GHG emissions. Rebound effects occur, for example, where a reduction in unit costs leads to greater use of a product or service or to increased demand for other products or services. In contrast to leakage, increased emissions do not necessarily occur outside the specific (project or geographical) boundary, nor do they always refer to impacts of a project, measure, or policy.

Rebound effects are often considered in relation to energy efficiency. In this context, they not only reduce the energy-efficiency savings anticipated but also sometimes result in a negative overall impact. This result is characterized as a case of "back-fire" or the "Jevons paradox," which refers to William Stanley Jevon's 1865 hypothesis that energy-efficiency improvements increase rather than decrease energy use.[1]

Rebound effects can also result from dynamic feedbacks associated with wider policies and changes in relative prices, aspects that are more relevant to carbon savings associated with HWP. For example, policies encouraging greater use of woodfuel (and other forms of renewable energy) may reduce the demand for fossil fuels in energy generation, consequently leading to a reduction in fossil fuel prices, thus stimulating their greater use in other activities. Similarly, policies to encourage the use of HWP and other low-carbon materials in construction may reduce the prices of fossil fuel-intensive materials such as concrete and steel, stimulating their greater use elsewhere in the economy (e.g., transport infrastructure). Ultimately, policies encouraging greater HWP use and the development of a low carbon

---

[1] Noting that the reduction in coal used (per ton of iron produced) to less than a third of the previous level had been followed by a tenfold increase in coal consumption in the Scottish iron industry, Jevons argued that "it is wholly a confusion of ideas to suppose that the economical use of fuel is equivalent to a diminished consumption. The very contrary is the truth .... Every improvement of the engine when effected will only accelerate anew the consumption of coal" (cited in Sorrell 2009, p. 138).

economy may stimulate innovations leading to economic growth, which in turn may lead to increased emissions. (For wider discussion of innovation feedbacks, see Fölster and Nyström 2010).

Table 26.3 lists various rebound effects for a range of contexts, where each one is classified according to its association with energy-efficiency or substitution measures, time-saving, taxes, consumption, or input choices.

Table 26.3 Types of rebound effect by relevance to quantifying harvested wood product (HWP) carbon savings

Category	Type	Description
E	*Price*	*Increased use of "low fossil energy-intensive" products* (e.g., HWP) *reduces* (demand for and) *prices of fossil fuels, stimulating* demand for and *greater use of fossil fuels* in the wider economy and/or other countries
E	*Structural*	*Increased use of "low fossil energy-intensive" products* (e.g., HWP) *reduces* the (demand for and) *prices of fossil energy-intensive goods and services, increasing* demand for the latter and associated *energy use*
E	*Infrastructure*	*Increased demand for "low fossil energy-intensive" products* (e.g., woodfuel) *necessitates new infrastructure* (e.g., local wood transport and storage facilities), thus *increasing energy use*
E	*Installation*	*Adoption of renewable energy* (e.g., woodfuel use) or energy efficiency measures *requires energy for the manufacture, transport, and installation* of new equipment (e.g., new boilers to use woodfuel), thus *increasing energy use*
E	*Norm*	*Adoption of renewable energy* (e.g., biomass boilers) or energy efficiency *measures provides a pretext to neglect wider social norms on limiting emissions, leading to higher emissions in other areas* (e.g., flights)
E	*Transport*	*Increased demand for "low fossil energy-intensive" products leads to* economies of scale and *reduced transportation costs* (e.g., for shipping woodfuel), *stimulating longer distance trade in these products and increasing associated energy use*
E	*Usage*	*Increased energy efficiency or use of "low fossil energy-intensive" products leads to less* attention to *switching off appliances* when not in use, *increasing energy usage*
E	*Income*	*Increased energy efficiency of using a good* (e.g., of woodfuel due to more efficient boilers) *makes it cheaper to use, thus stimulating increased use*
E	*Production*	*Reduction in unit production costs leads* producers to lower prices and raise *output, increasing energy use* (i.e., lower prices stimulate consumer demand)
E	*Substitution*	*Cost savings* (e.g., switching to woodfuel where less expensive) *lead to increased spending on other goods and services, increasing energy use*
T	Activity	Reduced time required for a specific economic activity (e.g., installing a new boiler) increases the time available for and energy use in other activities

(continued)

Table 26.3 (continued)

Category	Type	Description
I	Capital	Substitution of manufactured inputs (e.g., insulation) for the use of fossil fuel (e.g., central heating) increases energy use in manufacturing
C	Consumption	Reduced consumption of goods and services by some leads to price reductions, increasing demand by others, and associated energy use
E	Downstream	Increased energy efficiency in producing final goods reduces unit costs, leading to a reduction in sales prices and increased demand, and creates additional demand for inputs, increasing the energy use associated with their production and transport
E	Growth	Increased energy efficiency raises productivity and stimulates economic growth, increasing demand for goods and services and their associated energy use in the wider economy
I	Labor	Measures involving greater use of human power instead of fossil fuels (e.g., cycling rather than travelling by car) may lead to an increase in associated expenditure (e.g., on bicycles) and energy used in their production
E	Multiplier	Shifts to higher-priced "low fossil energy-intensive" products and services (e.g., rail travel) from lower-priced "high fossil energy-intensive" products and services (e.g., air travel) may increase total profits and payments to staff and shareholders of supply companies, increasing their consequent demand and energy use
R	Tax	An environmental (e.g., carbon) tax increasing government receipts and expenditure increases demand for goods and services in the wider economy and associated energy use
T	Time	Reduced time required to use a specific service (e.g., train travel between cities) stimulates demand from users and increases energy use
E	Upstream	Increased energy efficiency in manufacturing reduces unit costs, leading to a reduction in the sales price and increased demand, as well as increased output and demand for producer goods, with each leading to higher energy use

C consumption, *E* energy-efficiency or substitution measures, *I* input choices, *R* taxes, *T* time-saving

The two key types of rebound effect (price and structural) to consider in quantifying the carbon savings of HWP are given in the first two rows. Others that are particularly relevant for quantifying the carbon benefits of HWP—infrastructure, installation, norm, transport, and usage—are listed in the next five rows. The next most directly relevant rebound effects, namely, income, production, and substitution, follow in the next three rows.

Some of these rebound effects (e.g., the installation rebound) are accounted for in traditional life-cycle assessments (LCAs)—for an example of approaches to LCA, refer to Matthews et al. (2018)—while others (e.g., the production rebound) may be limited by wider regulations (e.g., GHG emission limits for different sectors). There has been little work to date on rebound effects in relation to HWP, and none of the current protocols attempt to account for them. However, Grafton et al. (2012) identify a potential "green paradox" in regard to subsidies for renewable energy use, whereby the direct effect of a reduction in demand for fossil fuels on the extraction of fossil fuels is outweighed by the indirect effect of a reduction in fossil fuel prices. This helps to illustrate the potential importance of a price rebound (see the first row of Table 26.3).

26.5 Monitoring and Accounting

The inclusion of carbon storage benefits associated with HWP may appear far from straightforward, given their range of potential uses. Also, the further the wood products move through the value chain, the more uncertain the carbon storage estimates become (Mensink 2007). These benefits depend on manufacturing, transportation, and end use, as well as end-of-life recycling and disposal processes. Monitoring based upon periodic sampling of carbon storage in wood products is, in general, far more difficult and expensive than for forest carbon pools. To allow for variations in the carbon storage benefits of HWP, uncertainty discounts (Ingerson 2011) can be used, or buffers that involve withholding a proportion of carbon units to cover the risk that some potential benefits will not arise.

Relatively simple approaches to the inclusion of the carbon storage benefits associated with HWP exist based upon applying fixed decay rates to different categories of wood products. The costs of implementing such approaches are expected to be modest as they avoid the necessity for long-term monitoring. Although the proportions of different HWP categories produced in the UK differ from those in the USA, a similar approach would be simple to apply where the UK Woodland Carbon Code extended. Country averages, possibly adjusted for different species, could be used, along with fixed decay rates, such as those used for national-level GHG accounting.

However, the extent to which such simple approaches offer robust metrics is unclear. Uncertainties exist, not only concerning the proportion of wood harvested from any given woodland that will be used in the future for different types of HWP and the level of associated wood processing emissions, but also regarding wood product decay rates, as well as in quantifying baseline emissions. (Uncertainty about baseline emissions is pervasive because the baseline is a hypothetical construct, although this is true more widely in relation to quantifying carbon benefits and does not only affect carbon storage and the substitution benefits of HWP.) Where material is processed and used outside the region in which the wood has been grown, emissions associated with the transport of HWP can be significant. In the USA, for example, carbon emissions from processing and transportation may approach the levels of long-term carbon storage in HWP in some cases (Ingerson 2011), although there still may be carbon substitution savings compared with using alternatives such as concrete and steel, which are associated with relatively high emissions.

Providing a single recommendation detailing the best method to account for the carbon benefits of HWP is beyond the scope of this chapter. However, it is overly conservative to assume that all the carbon in wood products with a lifetime of less than 100 years is emitted immediately. Assuming this in the baseline can inflate the credits issued because it implies that projects could receive credits for emission reductions that may only occur in 99 years' time (Pearson et al. 2012). Instead, it is preferable to use an average based on modeling retirement and emissions from wood products over many cycles and the associated carbon stored in the HWP pool over the long term. Alternatively, a radiative forcing approach can be adopted, based upon estimating the atmospheric impact of keeping carbon out of the atmosphere over a product's lifetime (Pearson et al. 2012).

Permanence and equivalence issues between carbon sequestration and the carbon substitution and storage benefits of HWP need to be considered when developing an approach to accounting for both. The carbon sequestration benefits of woodland creation projects are currently computed under the Woodland Carbon Code over up to 200 years, with those involving cycles of clear-cutting and restocking based upon a long-term average that is typically between 30% and 50% of the cumulative total carbon sequestered over one rotation (West 2018). In contrast to carbon sequestration, the average carbon substitution and storage benefits of HWP associated with a woodland creation project tend to increase over time (due to more wood being harvested). Nonetheless, a long-term average could potentially also be used to take account of the carbon substitution and storage benefits of HWP, although detailed consideration of the best approach to this is beyond the scope of this chapter.

The failure of existing carbon standards to account for rebound effects may be because they are complex and expensive to estimate. However, this represents a significant potential weakness in quantifying the carbon substitution benefits of HWP, and particularly in regard to reductions in fossil fuel use. Increased HWP use could potentially influence fossil fuel supply and demand in the wider economy in ways that stimulate greater use of fossil fuels in other activities.

Whether extending carbon standards for woodland creation projects to cover the benefits of HWP is worthwhile depends partly on the cost of quantifying and certifying these benefits. While forestry options deliver a range of ecosystem services in addition to climate change mitigation—such as the absorption of other pollutants like ammonia and nitrates—there can also be potential disbenefits. The emission of particulates—matter in the form of minute separate particles—is associated with the use of (especially) damp woodfuel in domestic fires. This is a cause of concern: for instance, the UK Committee on Climate Change advised the UK Government not to support the use of biomass for heat in urban areas (Committee on Climate Change 2018). Potentially adverse impacts on forest carbon stocks (e.g., Matthews et al. 2018) should also be considered if the extension of carbon standards to cover projects involving wood production through forest management of existing "underutilized" woodlands is to be contemplated.

26.6   Recommendations

This chapter provides a technical contribution to discussions about whether it is feasible to extend the UK Woodland Carbon Code to the carbon storage and substitution benefits of HWP. Further work will be needed to assess the practical feasibility and whether it can be done in a robust way that underpins market confidence and maintains the integrity of the Woodland Carbon Code. In terms of exploring the technical potential, it is recommended to do the following:

- Consider adopting a system of units for carbon storage that takes account of the expected lifespan of different product types (e.g., sawn softwood, sawn hardwood).
- Consider adopting a simple approach that accounts for carbon storage benefits over a fixed time horizon (e.g., the longest lifespan of the different product types).
- Consider applying a simple decay function to the carbon stored for each product category.
- Consider how transport emissions can best be included in estimating net carbon savings and the extent to which their inclusion would provide incentives for local processing and HWP use to help increase overall carbon benefits to society.
- Explore potential mechanisms to allocate units for carbon storage and substitution between woodland owners and wood users that would provide incentives to increase domestic timber supply and quality, the carbon storage and substitution benefits per unit of wood, and overall net benefits to society.
- Explore the costs and benefits of empirical monitoring of carbon storage and carbon substitution.
- Consider potential double-counting issues further and how these can be minimized.
- Investigate how carbon storage and substitution benefits, taking rebound and leakage effects into account, can be quantified by drawing on international trade and inter-sectoral models.
- Consider further whether covering carbon storage and substitution benefits of HWP on the basis of national average wood use, product assortment, and generic half-lives would fit with project-level projections of carbon sequestration on which the Woodland Carbon Code is based.

References

ACR (2017) Methodology for the quantification, monitoring, reporting and verification of greenhouse gas emission reductions and removals from afforestation and reforestation of degraded land. Version 1.2. American Carbon Registry, Arlington, VA

ACR (2018) Improved forest management methodology for quantifying GHG removals and emission reductions through increased forest carbon sequestration on non-federal U.S. forestlands. Version 1.3. American Carbon Registry, Arlington, VA

CAR (2017) Forest project protocol. Version 4.0. Climate Action Reserve, Los Angeles, CA

CDM (2017) ACM0018. Large-scale consolidated methodology: electricity generation from biomass residues in power-only plants. Version 4.0. Clean Development Mechanism. United Nations Framework Convention on Climate Change, Bonn

CDM (2018) ACM0006. Approved consolidated baseline and monitoring methodology: consolidated methodology for electricity and heat generation from biomass. Version 14. Clean Development Mechanism. United Nations Framework Convention on Climate Change, Bonn

Committee on Climate Change (2018) Biomass in a low-carbon economy, London

Fölster S, Nyström J (2010) Climate policy to defeat the green paradox. Ambio 39:223–235

Forestry Commission (2014) The woodland carbon code. Version 1.3. Forestry Commission, Edinburgh

Forestry Commission (2020) Forestry statistics 2020. Forestry Research, Edinburgh

Foucherot C, Grimault J, Morel R (2015) Contribution from I4CE on how to address double counting within voluntary projects in Annex B countries. Note. Institute for Climate Economics, Paris

Grafton RQ, Kompas T, Van Long N (2012) Substitution between bio-fuels and fossil fuels: is there a green paradox? J Enviro Econ Manag 64(3):328–341

Hood C, Briner G, Rocha M (2014) GHG or not GHG: accounting for diverse mitigation contributions in the post-2020 climate framework. Paper 2014(2). Climate Change Expert Group, OECD

Ingerson A (2011) Carbon storage potential of harvested wood: summary and policy implications. Mitigat Adaptat Strat Global Chang 16:307–323

Matthews R, Hogan G, Mackie E (2018) Carbon impacts of biomass consumed in the EU: supplementary analysis and interpretation for the European Climate Foundation. Project report for ECF. Forest Research, Farnham

Mensink M (2007) Framework for the development of carbon footprints for paper and board products. Confederation of European paper industries, Brussels, Belgium

Miner R, Lucier A (2004) A value chain assessment of climate change and energy issues affecting the global forest-based industry. A report to the WBCSD Sustainable Forest Products Industry Working Group. NCASI. www.greenbiz.com/sites/default/files/document/CustomO16C45F60545.pdf. Accessed 21 April 2021

Niles J, Schwarze R (2001) Editorial. Clim Chang 49:371–376

Pearson T, Swails E, Brown S (2012) Wood product accounting and climate change mitigation on projects involving tropical timber. Report TMT-PA 007/11 Rev.1 (1). International Tropical Timber Organization/Winrock International, Washington DC

Penman J, Gytarsky M, Hiraishi T, Krug T, Kruger D, Pipatti R, Buendia L, Miwa K, Ngara T, Tanabe K, Wagner F (eds) (2003) Good practice guidance for land use, land-use change and forestry. Intergovernmental Panel on Climate Change (IPCC) National Inventories Programme Technical Support Unit, Institute for Global Environmental Strategies. Hayama

Schneider L, Kollmuss A, Lazurus M (2014) Addressing the risk of double counting emission reductions under the UNFCCC. Working Paper 2014-02. Stockholm Environmental Institute, Sweden

Scottish Forestry (2020) Climate mitigation: woodland creation and management. Information Note, Edinburgh

Sorrell S (2009) Economy-wide rebound effects. In: Sorrell S, Herring H (eds) Energy efficiency and sustainable consumption: the rebound effect. Palgrave Macmillan, Basingstoke, pp 136–164

US Department of Energy (2006) Technical guidelines for voluntary reporting of greenhouse gas program. Part I: Appendix. Office of Policy and International Affairs, US Department of Energy, Washington, DC

Valatin G (2012) Additionality and climate change mitigation by the UK forest sector. Forestry 85(4):445–462

Valatin G (2017) Harvested wood products and carbon substitution: approaches to incorporating them in market standards. Research report. Forest Research, Farnham

VCS (2015) VM0034. British Columbia Forest Carbon Offset Methodology. Version 1.0. Verified Carbon Standard. Verra, Washington, DC

VCS (2016) VM00010. Methodology for improved forest management: conversion from logged to protected forest: approved VCS methodology. Version 1.3. Verra, Washington, DC

West V (2018) Using the WCC carbon calculation spreadsheet version 2.0. UK Woodland Carbon Code. https://woodlandcarboncode.org.uk/images/PDFs/WCC_CarbonCalculation_Guidance_V2.0_March2018.pdf. Accessed 21 Apr 2021

Part VIII
Governance

Chapter 27
Social System in Collaborative Activities for Conserving Coastal Pine Forest in Karatsu City, Kyusyu, Japan

Fumika Asanami and Mahito Kamada

Abstract We studied the social system to maintain collaborative activities for restoring and conserving coastal pine forests in Karatsu City, Saga Prefecture, Kyushu, Japan. Governance has been structured and functioned in most conservation processes; administrative offices of the state, prefecture, and city, business sector, and Civil Society Organization have acted interdependently, and network members have continuous interaction under the management of an NPO group known as KANNE. KANNE has filled in a structural hole in the social network. KANNE plays a special role as the hub. Social ties, however, are very weak, and hence the system continuity is fragile. If KANNE stops its work, the network structure will collapse. As an internal circumstance of KANNE, most management works have been governed by the secretary-general, and any person who can act as a substitute is now absent. Installing a system to develop human resources within KANNE is important to retain conservation activities.

Keywords Local governance · NPO · Niji-no-Matsubara · Social network · Underuse of natural resources

27.1 Introduction

Pine forests along sandy beaches are a familiar landscape for Japanese people (Fig. 27.1). Historically, coastal pine forests have been planted and maintained by local people. In the seventeenth century, under a promotion of the government of the era, people who lived along sandy beaches started to plant pines, mainly *Pinus thunbergii,* to protect residential and farming areas from strong coastal wind and

F. Asanami (✉) · M. Kamada (✉)
Research Center for Management of Disaster and Environment, Tokushima University, Tokushima, Japan

Department of Civil and Environmental Engineering, Graduate School of Technology, Industrial and Social Sciences, Tokushima University, Tokushima, Japan
e-mail: asanami@tokushima-u.ac.jp; kamada@ce.tokushima-u.ac.jp

© The Author(s) 2022
F. Nakamura (ed.), *Green Infrastructure and Climate Change Adaptation,*
Ecological Research Monographs, https://doi.org/10.1007/978-981-16-6791-6_27

Fig. 27.1 Pine forest along a sandy beach, a typical landscape in Japan

wind-blown sand (Oda 2003; Ohta 2015). Such coastal pine forests have protected residential areas from storm surge and tsunami in addition to extreme wind because the pine is a saline tolerant species. People used to gather fallen pine needles as fuel for cooking and mushrooms as food from mature pine forests. The coastal pine forest has provided shade to people coming to swim in summer, and the landscape composed of green forest with the white beach has been beautiful scenery for both visitors and local residents. Thus, the coastal pine forest is an important green infrastructure, which provides several ecosystem services (Asanami et al. 2020), and most coastal pine forests have been designated as forest reserves by the Forest Act for keeping a function of disaster risk reduction (hereafter DRR). In addition, several forests have been preserved as scenic beauty reserves by the Act on Protection of Cultural Properties for retaining the scenic value of the area.

The situation of coastal pine forests in Japan, however, has changed and declined due to natural succession caused by the underuse of forest resources and pine-wilt disease (Yoshizaki 2012; Ohta 2015). Because both a sandy surface and sufficient sunlight are necessary for regeneration of pine trees, it is necessary to remove grass and pine needles from the forest floor in addition to shrubs colonizing the forest (Kamada et al. 1991; Kamada 2018). People used to collect pine needles and shrubs for daily fuel and sometimes cut pine trees for timber before the 1960s in Japan. In that social situation, pine forests could be continuously maintained. With the change of household fuel to petroleum and electricity due to rapid economic growth and globalization from the 1960s, people stopped using biomass, and thus forest succession has progressed (Kamada et al. 1991; Kamada 2018). In some areas, forest composition has completely changed from pine to evergreen-broad leaved forest by succession (Morisada et al. 2020). The outbreak of pine-wilt disease, which is caused by pinewood nematode (*Bursaphelenchus xylophilus*), has caused serious damage to pine forests in all regions of Japan (Kamada et al. 1991, Mamiya 1988, Yoshizaki 2012, Kamada 2018).

In order to sustain coastal pine forests, national and local governments have taken measures to protect pine trees from nematode attacks in the preserved areas. Insecticides have been sprayed to kill the longicorn beetle (*Monochamus alternatus*), which is the vector of pinewood nematode. In addition, dead trees containing larva of the beetles have been removed from the forests. As for the compositional decline due to succession, however, it is difficult to take measures by the government alone. It is necessary to establish a public consensus for the way of management because the image of forest composition to be conserved is different for each person. For example, people disagree with cutting shrubs. Hence, public involvement is essential to restore and conserve coastal pine forests, and governance is required to progress such works (Rhodes 1997).

We studied collaborative activities for restoring and conserving a coastal pine forest in Karatsu City, Saga Prefecture, Kyushu, Japan, and evaluated a social system from local governance perspective. The coastal pine forest is called "Niji-no-Matsubara."

27.2 Study Area

Activities for restoring and conserving Niji-no-Matsubara were studied (Fig. 27.2). Niji-no-Matsubara is in the area of Genkai Quasi-National Park, which extends 4 km in length and 700 m in width (Fig. 27.3). It is owned and governed by the National Forestry Agency of Japan and designated as a Protected Forest to maintain DRR function and Recreation Forest for the general public. This coastal pine forest

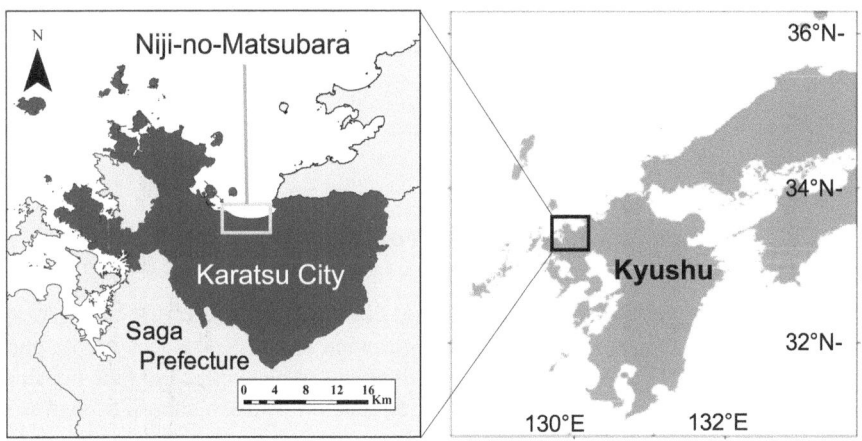

Fig. 27.2 Map of the study site

Fig. 27.3 Studied coastal pine forest, *Niji-no-Matsubara*

has also been designated as a Place of Scenic Beauty by the National Agency for Cultural Affairs since 1955.

Many collaborative activities for restoring and conserving "Niji-no-Matsubara" have been carried out through management by the NPO KANNE (https://www.facebook.com/npokanne).

27.3 Methods

In order to clarify the rolls of stakeholders, motivation and incentive, type of activities, and social system to support activities, literature documentation, focus group interview, and participant observation were conducted referring to methods of Patton (2014) and Flick (2018). Informants were (a) secretary-general and a staff of the NPO KANNE, (b) staff of Saga District Forest Office, (c) high school students, (d) junior high school students, and (e) volunteers.

27.4 Results

27.4.1 Background of Restoration and Conservation Works

Niji-no-Matsubara was governed by a feudal clan until the end of the Edo period (1603–1868) and then by the Japanese government after 1869. Local people had been forbidden to fell pine trees by the governments, while collecting pine needles had been permitted for use as daily fuel. The coastal pine forest had been maintained by the usage of local people. The pine-wilt disease occurred and caused damage to aged pine trees in 1958, and part of the Site of Scenic Beauty was invalidated in 1959 (Watanabe et al. 2006).

In this situation, the "Council on Protective Measure for Niji-no-Matsubara (CPM)" was established in 1966. The CPM has been composed of the enterprise sector of tourism, Saga District Forest Office, Saga Prefecture, Saga Prefectural Board of Education, Saga Prefectural Police Office, Karatsu City, and Karatsu Municipal Board of Education. The mayor of Karatu City is the representative, and Tourism Division of Karatsu City plays the role of the executive office.

Vegetation succession has proceeded following the outbreak of pine-wilt disease, and dense shrubs grew in the forest around the year 2000 due to changes of fuel material and lifestyle. In order to take measures against forest decline, Saga District Forest Office made the "Basic Plan for Restoration and Conservation of Niji-no-Matsubara" in 2007 (Saga District Forest Office, Japan Forest Technology Association 2007). The "Project Team for Completing the Basic Plan for Restoration and Conservation of Niji-no-Matsubara (PT)" was formed in the CPM, and the "Execution Plan of Restoration and Conservation of Niji-no-Matsubara" was established in 2008. PT included Civil Society Organization (CSO; NPOs, citizen groups, association of parents and teachers, women's association, etc.), Saga prefectural and Karatsu municipal governments, and Saga District Forest Office (Council on Protective Measure for Niji-no-Matsubara 2014).

Because several stakeholders from different sectors are concerned with decision-making. People make various images of forest structures to be established and conserved for different purposes, such as disaster prevention, cultural properties, tourism, education, daily use for health of local people, etc. Hence, a process of public involvement is necessitated. The CPM has acted as a platform for the process, and the PT has been installed to obtain wider opinions from the CSO.

According to the Execution Plan, Saga District Forest Office, Saga Prefectural Office, and Karatsu City made a "Memorandum on the Preservation and Conservation of Niji-no-Matsubara" in 2008 to establish a collaborative framework to take measures against the pine-wilt disease.

Saga District Forest Office also concluded an Agreement on Forest Improvement for Recreational Use with CPM in 2008. Roles of the CPM are to (1) assist governmental work for controlling pine-wilt disease, (2) promote works for cleaning and beautifying Niji-no-Matsubara, (3) cooperate on the works for enlightening nature conservation, (4) carry out fund-raising activities, and (5) support and advice on measures to control pine-wilt disease. The CPM entrusted the management of those works to NPO KANNE. The CPM pays a trust cost of 8.1 million Japanese yen (JPY) every year, and KANNE has employed two permanent and two part-time staffs to carry out the mission.

Tasks of the KANNE for Niji-no-Matsubara conservation are to (1) manage and operate litter cleanup activities, (2) do legal proceedings to carry out activities in the area preserved by laws, (3) encourage organizations/enterprises to participate in activities, (4) call the general public to join cleanup events, and (5) support activities of CSO.

In parallel to those measures, students of Karatsu-Minami High School established "Shoro Project Team" in 2004 to revive the Shoro mushroom in the coastal pine forest. Shoro, *Rhizopogon roseolus*, is truffle-like mushroom that occurs in the coastal pine forest with a clear floor. It used to be very familiar for old people in the region. People who have experiences to collect pine leaves for fuel have also experiences collecting and eating Shoro. However, due to the colonization of shrubs into the forest floor, Shoro has disappeared, and old people have missed the mushroom.

27.4.2   Activities Managed by NPO KANNE

27.4.2.1   Conservation Activities in the Pine Forest

Litter Cleanup Activities Through an "Adoption Program"

Maintenance of "Niji-no-Matsubara" has been carried out in two ways. The first one, called "adoption program," is that the area to be managed is divided into small lots, and each lot is entrusted to a volunteer group and cleaned by its members.

In order to recruit participants, the secretary-general of KANNE visited enter-prises individually and asked them to collaborate. According to the secretary-general, some enterprises were just seeking a chance to contribute to environmental activities in Karatsu City and/or Niji-no-Matsubara, and thus it was easy to obtain an agreement with them. However, it was difficult to gain collaboration from the other enterprises without any incentives. KANNE has provided an award program and introduced voluntary activities of enterprises through the web to make incentives for them.

As a result of the continuous effort of the secretary-general, the numbers of volunteer groups and persons increased from 129 and 5013 in 2011 to 219 and 7140 in 2017, respectively (Fig. 27.4). And accordingly, the cleaned area also increased from 48 ha to 56 ha. Each adopted group can do cleaning activities whenever members want. Members gather and remove pine branches and needles fell on the

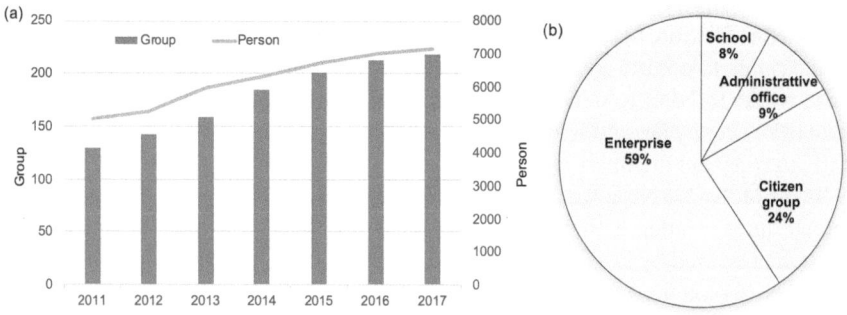

Fig. 27.4 Transition of the number of volunteer groups and persons (**a**) and the ratio of group/organization number in different sectors in 2019 (**b**)

forest floor in the lot. As a total in 2018, 245 times of activities were conducted, and 6752 volunteers participated (NPO KANNE 2019).

Litter Cleanup Event, "Keep Pine Project"

KANNE calls on the general public to participate in events to collect and remove fallen pine needles. The events, called "Keep Pine Project (KPP)," have been held four times a year, and 300–400 volunteers have participated every time. According to the participant observation on 17 December 2017, ca. 300 volunteers cleaned the forest floor of 2 ha in 1.5 h; the event started at 9:00 and ended at 10:30 in the morning, and five truckloads of pine matter was collected in total; the maximum load of the truck is about 4 m^3 or 750 kg (Fig. 27.5).

For managing the event, about 30 students of Karatsu-Minami High School worked in addition to staff of the KANNE. Students said that the "Keep Pine Project" was named by former students at the high school, expecting that Niji-no-Matsubara would come to be known and loved by anyone and to be conserved by collaborative activities of many people. Students of Karatsu-Minami High School visited Karatsu Commercial High School, Karatsu Higashi High School, and Higashi Karatsu Elementary School in order to explain the necessity of conservation activities and to ask them to join. Students of Junior-high and high schools joined the events as well as general citizens. The students of Karatsu Daiichi Junior High School said, "it is fun for me to come here," "here is a comfortable place," and "I will continue participating in this activity as long as I live here."

Fig. 27.5   Litter cleanup event, "Keep Pine Project"

As a result of those litter cleanup activities, Shoro has revived in the Niji-no-Matsubara.

27.4.2.2 Efforts to Increase Participants in Conservation Activity

In order to increase the number of participants in activities for Niji-no-Matsubara conservation, KANNE has provided several opportunities for the public to know about the ecological and social situation of Niji-no-Matsubara and to become interested in conservation activities.

Publicity

KANNE has produced a mascot character, named "Nijimatsu Mamoru," and made it to appear in events held in Karatsu City (Fig. 27.6). Nijimatsu Mamoru comes from the name of the coastal pine forest, Niji-no-Matsubara, and conservation activity; "Mamoru" is a Japanese word meaning conservation. A promotional song and video have also been produced. KANNE asked a singer–song writer to make a song, and the song has been sung at events and broadcast on radio programs.

KANNE has continuously published a newsletter several times a year and distributed it via the website as well as a printed version available at some public places. KANNE also posts an invitation before and report after every event on SNS such as Twitter, Instagram, and Facebook.

Fig. 27.6 Mascot character, "Nijimatsu Mamoru" (https://www.facebook.com/npokanne/)

Environmental Education

KANNE has provided nature observation events for the public several times a year, such as bird watching and mushroom gathering in Niji-no-Matsubar. Picture books of plants in the pine forest and beach have been made, and anyone can download them from the website (https://npokanne.com/資料ダウンロード/). KANNE also has held workshops for crafts using pinecones at a community center.

The secretary-general of the KANNE has been asked by kindergartens and schools to give a lecture on Niji-no-Matsubara. She has visited elementary, junior-high, and high schools several times a year. Coloring books have been made for children and used at the lectures in kindergarten (https://npokanne.com/資料ダウンロード/).

Some schools, visiting Karatsu City from other districts as a school excursion, have a request to learn about the culture and social activities in the region. KANNE provides an opportunity for students at such schools to experience a litter cleanup activity.

Fund-Raising

KANNE obtains 8.1 million JPY a year from the CPM as a contract and uses about 7.5 million yen to employ staff. There is a little money left over, and an additional fund must be obtained to carry out all tasks. Hence, the staff of KANNE always make efforts to obtain subsidies and donations.

KANNE has received subsidies from several foundations such as Seven-Eleven Foundation, Japan Fund for Global Environment, and Suntory Fund for Bird Conservation, and from governments of state, Saga Prefecture, and Karatsu City.

A light truck was donated by the Karatsu Corporation Association for the use of removing pine needles collected during litter cleanup activities. Also, 360 sets of cotton work gloves for activities were donated by Suehiro Ltd. Asahi Breweries Co., Ltd. donated 1.5 million JPY, and Itoen Co., Ltd. provided bottles of tea to volunteers after events such as KPP.

27.5 Discussion

Stakeholders and their roles in activities for the conservation of Niji-no-Matsubara are summarized in Fig. 27.7. Saga District Forest Office, Saga Prefecture, and Karatsu City take measures under the Forest Act to protect pine trees in the Protected forest from pine-wilt disease. In this case, the purpose and goal of the work are obvious, and it is not difficult to obtain public consensus.

The CPM has acted as a platform for the process of public involvement and decision-making, and the PT has been included to obtain wider opinions from the CSO. KANNE has participated in PT as one of the NPOs and input opinions

Fig. 27.7 Role of Council on Protective Measures for Niji-no-Matsubara (CPM) and NPO KANNE

regarding the way of Niji-no-Matsubara conservation. Also, KANNE has accepted the management office and acted as the hub to implement the execution plan, which was established by the PT and authorized by the CPM. In order to achieve this aim, the secretary-general of KANNE and the staff of Saga District Forest Office contact each other and exchange information frequently.

KANNE is always required to obtain funding from outside the network by way of its own effort. The incentive for the KANNE to make a continuous effort even during poor economic situations. This is both a social responsibility and pride of an NPO. KANNE was established in 2006 as an NPO working in Karatsu City for the purpose of supporting (1) community development, (2) environmental conservation activity, (3) disaster rescue operations, and (4) activity for regional security (https://npokanne.com/wp-content/uploads/2019/10/ teikan.pdf). Working as a management office of the activities is well matched to KANNE's purpose.

Rhodes (1997) summarized the characteristics of governance as follows: (1) interdependence between organizations as boundaries between public, private, and voluntary sectors became shifting and opaque; (2) continuous interactions between network members, caused by the need to exchange resources and negotiate shared purposes; (3) game-like interactions, rooted in trust and regulated by rules of the game negotiated and agreed by network participants; and (4) a significant degree of autonomy from the state. Networks are not accountable to the state as they are self-organizing. Although the state does not occupy a sovereign position, it can indirectly and imperfectly steer networks.

According to Rhodes's criteria, it seems that governance is structured and functioned in almost all conservation processes. (1) Administrative offices of the state, prefecture, and city, business sector, and CSO have acted interdependently. (2) Network members have continuous interaction under the management of KANNE. (3) Staff of KANNE and Saga District Forest Office frequently and freely discuss and make decisions on managing conservation activities. In litter cleanup events such as KPP, high school students and KANNE's staff also have game-like interactions. The CPM, however, acts only as a platform for the exchange of official or governmental information and does not have game-like interactions among the members. (4) Activities of volunteers are independent of the state, although a framework for the treatment of the area preserved by laws is set by the Forestry Agency of Japan and National Agency for Cultural Affairs.

From a viewpoint of social networks, KANNE has filled a structural hole in the social network (Burt 1992). KANNE plays a special role as the hub, but social ties are very weak and hence system continuity is fragile. If KANNE ceases its work, the network structure will collapse. As an internal circumstance of KANNE, most management works have been carried out by the secretary-general, and a person who can act as a substitute is now lacking. Creating a system to develop human resources within KANNE is important to maintain conservation activities.

Another flaw in the governance is little concern and less responsibility of the CPM to the management process, and little opportunities for KANNE to advocate new policy/plan for completing the PDCA cycle. In order to build a more sustainable governance system, participation of the CPM in the PDCA process is strongly recommended as well as setting the KANNE to even relationship with CPM, not as a subcontractor.

NPOs play important roles in governance for sustaining regional ecosystems as common property in Japan (Kamada 2018). External and internal situations of the NPOs are probably similar to KANNE, or worse. Recognition and support of local government and society to develop stable NPOs is required in order to establish a mature and reliable governance system for maintaining natural and social capital.

Acknowledgments We would like to express our gratitude to Ms. Wakako Fujita, the secretary-general of NPO KANNE, and Mr. Hitoshi Hida, a staff of Saga District Forest Office, for their cooperation on interview survey and for help at participation observation. We also thank students of Saga Prefectural Karatsu-Minami High School, Karatsu Municipal Daiichi Junior High School, and Karatsu Municipal Hamatama Junior High School for their cooperation in the litter cleanup event. This study was supported by the Environment Research and Technology Development Fund (JPMEER20184005).

References

Asanami F, Ito K, Kamada M (2020) Self-managerial activity by local people for keeping coastal pine forest and complemental policy of Fukutsu City, Fukuoka prefecture, Japan. Landsc Ecol Manag 25:53–68. (in Japanese with English Abstract)

Burt RS (1992) Structural holes: the social structure of competition. Harvard University Press, Cambridge

Council on Protective Measure for Niji-no-Matsubara (2014) Execution Plan of Restoration and Conservation of Niji-no-Matsubara, 1st revised Edition. (in Japanese)

Flick U (2018) An introduction to qualitative research, 6th edn. SAGE Publications Ltd., Los Angeles, p 696

Kamada M (2018) *Satoyama* landscape of Japan –past, present, and future. In: Hong S-K, Nakagoshi N (eds) Landscape ecology for sustainable society. Springer, Cham, pp 87–109. https://doi.org/10.1007/978-3-319-74328-8_6

Kamada M, Nakagoshi N, Nehira K (1991) Pine forest ecology and landscape management: a comparative study in Japan and Korea. In: Nakagoshi N, Golley FB (eds) Coniferous Forest ecology from an international perspective. SPB Academic Publishing, The Hague, pp 43–62

Morisada S, Nozaki T, Ogawa M, Kamada M (2020) Succession from pine forest to evergreen broad-leaved forest at Ohki coastal beach in Kochi prefecture, Shikoku, Japan. Landsc Ecol Manag 25:75–86. (in Japanese with English Abstract)

NPO KANNE. 2019. Business report on the 2018 fiscal year. (in Japanese)

Oda T (2003) The people who made the coastal forest (Kaigan-rin o tsukutta hitobito). Hokuto Publisher, Tokyo, p 254. (in Japanese)

Ohta T (2015) Nowadays and future for coastal forests. J Japanese Soc Revegetat Technol 41:332–333. (in Japanese)

Patton MQ (2014) Qualitative evaluation and research methods, 4th edn. SAGE Publications Ltd., London, p 832

Rhodes RAW (1997) Understanding governance - policy networks, governance, reflexivity and accountability. Open University Press, Maidenhead, p 235

Saga District Forest Office, Japan Forest Technology Association (2007) Survey report on measures for conservation and restoration of Niji-no-Matsubara. Saga District Forest Office, p 68. (in Japanese)

Watanabe T, Yokouchi K, Okada T, Mitsumizo H (2006) A study on the landscape management strategy in "Niji-no-Matsubara": relationship between the management and the landscape value of pine forest. J Architect Infrastruct Environ 1:107–114. (in Japanese)

Yoshizaki S (2012) Contemporary problems relating to the coastal forest in Japan. Water Sci 56(3):14–27. (in Japanese)

Chapter 28
Governance for Realizing Multifunctional Floodplain: Flood Control, Agriculture, and Biodiversity in Yolo Bypass Wildlife Area, California, USA

Mahito Kamada, Jun Nishihiro, and Futoshi Nakamura

Abstract Yolo Bypass (YB) is an engineered floodplain bypass for flood control in Sacrament River Valley in California, USA. Although the main purpose of the YB is to prevent the capital city of Sacramento and nearby riverside communities, the bypass provides wildlife habitats for various organisms and a good farming ground for multiple seasonal crops. Thus, we can refer to this facility as green infrastructure. We conducted an interview with key persons involving in the establishment and management of Yolo Bypass Wildlife Area (YBWA) in 2015 and reviewed the history of collaboration of various sectors with literature mining. Finally, we clarified the schematic relationship of stakeholders in the establishment and management of YBWA in which federal, state, and local entities and individuals take years of meetings, discussions, negotiation, and trust-building to reach consensus for restoration and management of the wildlife area.

Keywords Governance · Nature-based solution · Green infrastructure · Restoration · Wetland

M. Kamada (✉)
Research Center for Management of Disaster and Environment, Tokushima University, Tokushima, Japan

Department of Civil and Environmental Engineering, Graduate School of Technology, Industrial and Social Sciences, Tokushima, Japan
e-mail: kamada@ce.tokushima-u.ac.jp

J. Nishihiro
Center for Climate Change Adaptation, National Institute for Environmental Studies, Tsukukba, Japan

F. Nakamura
Laboratory of Ecosystem Management, Graduate School of Agriculture, Hokkaido University, Sapporo, Japan

© The Author(s) 2022
F. Nakamura (ed.), *Green Infrastructure and Climate Change Adaptation*,
Ecological Research Monographs, https://doi.org/10.1007/978-981-16-6791-6_28

28.1   Introduction

Nature-based solutions (NbS) in flood risk reduction, which are "actions (1) inspired by, (2) supported by or (3) copied from nature (European Commission 2015)," has been required as an adaptation to climate change (Jongman 2018; Hartmann et al. 2019). The NbS bring multiple benefits to people and social systems, such as ecological issues, agriculture, and tourism (Hartmann et al. 2019), and incorporating green infrastructure is an effective way to reduce disaster risk in a changing climate. While for realizing the NbS as well as installation and management of the green infrastructure, involvement with more diverse stakeholders, including different administrations, businesses, NGOs, landowners, and citizens, is required than in the traditional engineering approach, and governance strategies are necessitated (Driessen et al. 2018).

Yolo Bypass (YB) is an engineered floodplain setting in Sacrament River Valley in California, USA, for flood control (Figs. 28.1 and 28.2). Several studies clarified that YB has an important role not only for flood control but also for agricultural use and biodiversity (Sommer et al. 2001; Garnche and Howitt 2011; Greco and Larsen 2014; Suddeth and Lund 2016). In addition, Wildlife Area has been established in the YB through the involvement with many stakeholders under the facilitation of a nonprofit organization (NPO), Yolo Basin Foundation (YBF) (Brice 2015). Yolo Bypass Wildlife Area (YBWA) provides excellent examples for designing multifunctional floodplain (Garnche and Howitt 2011; Greco and Larsen 2014) and for governance structure for realizing NbS (Salciso 2012; Brice 2015).

https://upload.wikimedia.org/wikipedia/commons/0/0
4/Sacramento_River_basin_map.png

Fig. 28.1 Map of Sacrament River Valley. Square area is shown in Fig. 28.2

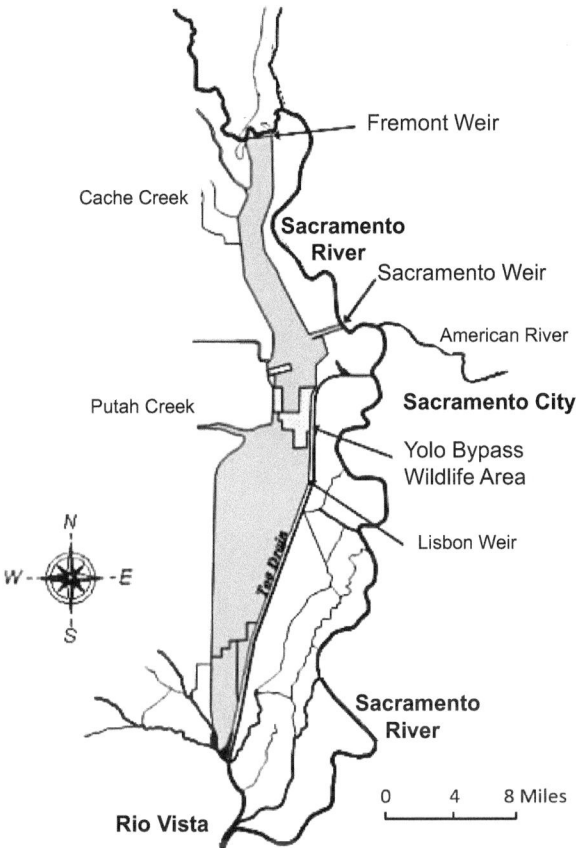

Fig. 28.2 Location of Yolo Bypass and Yolo Bypass Wildlife Area

In this report, we review a process of establishment and management of YBWA based on an interview with key persons as well as literature mining.

28.2 Interview and Inspection

The interview was conducted on December 1, 2015, at the headquarter and field in the YBWA. Ms. Robin Kulakow, Mr. Jack DeWit, and Mr. Jeff Stoddard explained to us the history, structure, and management system of the YBWA (Fig. 28.3). Kulakow was a Founder and Executive Director of the YBF, and DeWit was a Director of the YBF and a farmer who kept rice fields in the YBWA. Stoddard was a manager of the YBWA, worked at the California Department of Fish and Wildlife (CDFW).

Fig. 28.3 Interview with the key persons was conducted at the headquarter and field in the YBWA on December 1, 2015

28.3 Background of Yolo Bypass

The Sacrament River flows the south in the Sacrament Valley and connects with Feather and American Rivers just above Sacrament City (Figs. 28.1 and 28.2). Before reclamation, the Sacramento River flooded almost yearly in response to winter rains and spring snowmelt (Suddeth and Lund 2016). The Yolo Basin along the Sacramento River, filled with water from three rivers for most of the winter months; the basin could be inundated for more than 100 days (Christian-Smith 2010). Flooding provided seasonal marshy habitat for tule elk, fishes, and waterfowl; as the valley was located along the Pacific Flyway, millions of waterfowl migrated for wintering in the early nineteenth century (Brice 2015).

From the middle of the nineteenth century, after California Gold Rush, reclamation for agriculture began (Brice 2015), and now only 6% of the historical wetlands and 11–13% of the riparian vegetation remain (Katibah 1984; Suddeth and Lund 2016). In accordance with land development, social demand for flood control was increased. In 1911, the Flood Control Act was established by the state and construction of two bypasses, the Yolo and Sutter, were started. In 1917, the Sacramento River Control Project as a federal flood control act was established (Brice 2015).

Based on the concept to construct a broad system mimicking the Sacrament River's natural floodplain function, which was originally proposed by Will Green in 1860s, the US Army Corps of Engineers (Corps) developed a network of weirs and bypasses to protect Sacrament City and other communities. The bypass systems were completed by 1930, and a series of dams were completed around 1943 (Sommer et al. 2001; Greco and Larsen 2014).

The YB is 24,000 ha leveed floodplain, 64 km long and 2–4 km wide. The maximum design flow for the Sacrament River channel below the Sacrament metropolitan area is 3100 m^3/sec, and the YB is engineered to convey 14,200 m^3/sec for 100-year protection (Sommer et al. 2001; Christian-Smith 2010; Greco and Larsen 2014; Brice 2015). The main flood flows into the bypass from Fremont Weir (Fig. 28.2) when the water level of Sacrament River exceeds 10 m (Suddeth and Lund 2016). The flood frequency is approximately 1.6 years, varying in duration from 3 to 83 days in the period of 1980–2010 (Greco and Larsen 2014).

According to Garnche and Howitt (2011) and Salciso (2012), about 75% of the YB is privately owned, except for the YBWA. Two-thirds of the floodplain are used for farming and grazing in the spring and summer, while the rest is mostly wetlands. Farmers have water rights and are subject to flood easement. The State Reclamation Board has the right to inundate the land with floodwaters. They prevent landowners from building structures and berms and need not provide compensation for losses due to flooding.

28.4 Yolo Bypass Wildlife Area and its Management

The creation of the YBWA has resulted in restoration and management of wetland, riparian, and grassland communities that provide habitat for a diverse assortment of plant, wildlife, and fish species and the creation of educational and interpretive programs and partnerships to serve the public. The notable traditions of agriculture have also been maintained throughout the YBWA, employing innovative wildlife-friendly management strategies to achieve multiple resource objectives (CDFG 2008).

In 1991, California Wildlife Conservation Board (WCB) purchased the land of 1275 ha in YB for $4.75 million and 237 ha in 1994 (WCB 2001). The US House of Representatives approved $1.6 million for the Yolo Basin Wetland Project implemented by the Corps as the federal sponsor (Brice 2015). The NPO Ducks Unlimited cooperated on the project in designing ponds and contracting bulldozers, backhoes, and tree-planters in the project (Hayes 1999). The first stage of wetland restoration was completed, and the Corps turned over the 1512 ha YBWA to the California Department of Fish and Game (CDFG, forerunner of CDFW) to manage on November 12, 1997 (Hayes 1999; Brice 2015).

The YBWA was expanded to 6782 ha through land acquisitions by WCB in 2001 (5182 ha), 2002 (40 ha), and 2004 (48 ha) (WCB 2001, 2002, 2004). Ducks Unlimited and California Waterfowl Association each received $4 million from the state and implemented restoration project (Brice 2015). According to Kulakow, Yolo County agreed with the wetland restoration only at the area unsuitable for farming. Finally, 3240 ha in the YBWA have been restored to seasonal and permanent wetland from fallow farmland (Christian-Smith 2010).

Stoddard said that CDFW has been responsible for managing 6776 ha habitat and 6 ha headquarters in YBWA. About a half of the habitat area has been leased for farming and grazing and managed by farmers, and the rest has been managed by the CDFW directly. According to YBF (2015), new five-year leases were negotiated in 2015, and about 810 ha for wild and white rice and about 2400 ha for grazing cattle were agreed with two farmers, Jack DeWit and Tom Schene, respectively. Both lessees also agree to grow food plots for wildlife, such as milo, safflower, sunflower, Japanese millet, and water grass. Dixon Resource Conservation District has handled the lease and the funds generated by these leases (Brice 2015; YBF 2015). The leasing charge is set cheaper than the area outside YBWA to be an incentive for farmers, according to Stoddard, and it brings around $650,000 a year and supports operations on the YBWA (YBF 2015).

Figure 28.4 shows one of the restored wetlands and facilities for pouring water. According to Kulakow and Stoddard, wetlands were designed by the Ducks Unlimited and the California Waterfowl Association. The land in YBWA is not

Fig. 28.4 Restored wetlands (upper photo) and facilities for pouring water (lower photos) (December 1, 2015)

Fig. 28.5   A line of riparian willow trees has been remained and used as a resting site by raptors; great horned owl (L) and sharp-shinned hawk (R) (December 1, 2015)

always inundated, and therefore it is necessary to pump water into the area. The operation device is set at 7 m above the ground to prevent inundation during flooding.

Riparian woods have been seriously reduced and should be conserved and restored in Sacramento Valley. According to Kulakow, the federal office that is responsible for flood control allowed to maintain riparian trees along a direction of flooding flow, not to obstruct the flow. And thus, a line of willow trees has been remained and used as a resting site by raptors (Fig. 28.5).

Figure 28.6 shows a part of the area where DeWit has leased for rice farming. Rice cultivation is a practical way to create seasonal wetlands and stay vegetation in an early successional stage; rice is planted from April and harvested by October, and hence the field can be used for waterfowl in the winter season (CDFG 2008). Flooding rice fields after harvest helps decompose the rice straw, and the flooded fields provide an important food source with the rice that is left after harvest (YBF 2012). In addition, DeWit fallows about 94 ha of his 810 ha rice fields and then floods them around 10 cm deep for the shorebirds in the summer, usually 47 ha in

Fig. 28.6 Rice field, where Jack DeWit has rent and farmed, has been used as a wintering site for waterfowl (December 1, 2015)

July and 47 ha in August (Brice 2015). According to DeWit, he grows white rice and wild rice and exports white-waxy rice to Japan. The yield of wild rice is lower than the white rice, but the seeds become good feed for shorebirds and waterfowl. He can continue agricultural activities with additional efforts for birds because of the low leasing charge.

Under the land management plan (CDFG 2008), cattle grazing occurs on an extensive portion in the southern end of the YBWA (Fig. 28.7). According to Stoddard, cattle has been used as an alternative to elk, which has been already extinct in the area, for vegetation management through grazing. Cattles remove alien plants and keep vegetation height low, and it saves indigenous plant species adapting to the conditions with frequent flood disturbances and elk grazing before the reclamation of the floodplain. The moderate density of cattle grazing also contributes to the maintenance of wet grasslands called vernal pools, which provide habitat for rare animals such as the Mexican tiger salamander.

In the grassland area, "Umbrella Barn" has been kept as a historical resource; it was built around 1913 for providing refuge for the livestock during high water (CDFG 2008). Raccoon and short-eared owl were using when we visited (Fig. 28.8).

Fig. 28.7   Grassland area managed by cattle grazing (December 1, 2015)

Fig. 28.8   "Umbrella Barn" has been remained in the grassland area as a historical monument. It was occupied by raccoon (L) and short-eared owl (R) (December 1, 2015)

Duck hunting is another significant use of the YBWA. The rise of duck clubs was in the first half of the twentieth century, and it was the beginning of waterfowl protection (Brice 2015). Various duck hunting clubs operate in the YBWA. During the season, waterfowl, coots, moorhens, snipe, pheasant, and doves are hunted (Brice 2015). Hunting license ($160/year) and hunting ($26/day) fees are important sources of income to help with the management. A variety of other activities, such as fishing, wildlife viewing, and environmental education, occur throughout the year (CDFG 2008).

28.5   Significant Role of Yolo Basin Foundation in Governance Structure

Figure 28.9 summarizes the schematic relationship of stakeholders in the establishment and management of YBWA. According to the CDFG (2008), "planning for the YBWA encompasses issues that cross regional, local, and project area boundaries. This section identified the federal, state, county, and local agencies and other planning influences that effect the function and management planning of the YBWA." YBF took a significant role in the process. The YBF is an NPO created in 1990 as a community-based organization to facilitate the creation of the YBWA (Salciso 2012).

Brice (2015) described the starting point of the actions forward the YBWA as follows: Ted Beedy and his fellow birders knew that the YB area was a significant place for birds when flooded, and "a year-round managed wetland" became the dream. They drew a conceptual map and later developed a plan for a refuge. Beedy took US Congressman Vic Fazio to the proposed site, and the plan was favored by him. The plan gained momentum with support from the Yolo County Board Supervisors, the US Fish and Wildlife Service, and California Waterfowl Association. Then, it brought the decision-making of WCB to purchase the land in the YB for the wildlife area in 1991.

Kulalow, a founder of the YBF, joined the core group with the idea of forming an NPO to support the proposal. That group was to become the YBF. In the interview, Kulakow said, "the idea was born in the kitchen, when I was talking and drinking tea with friends,, and wrote, "in 1989 Susan and I had filed the incorporation papers for Putah Creek Council. One year later we were it again for a new nonprofit named Yolo Basin Foundation. We wrote newsletter articles, membership appeals, thank you letters, and press releases while our sons played on the floor (YBF 2016)."

YBF recognized from the beginning that there is a complex web of policies and other influences that need to be reconciled (CDFG 2008). The ground-level work to make consensus among federal, state, and local entities and individuals took years of meetings, discussions, negotiation, and trust-building (Salciso 2012). The YBF, as an NPO outside of the formal structure of the legal management of the area, had an important role in analyzing and coordinating exiting legal obligations and policy objectives with restoration plans (Salciso 2012).

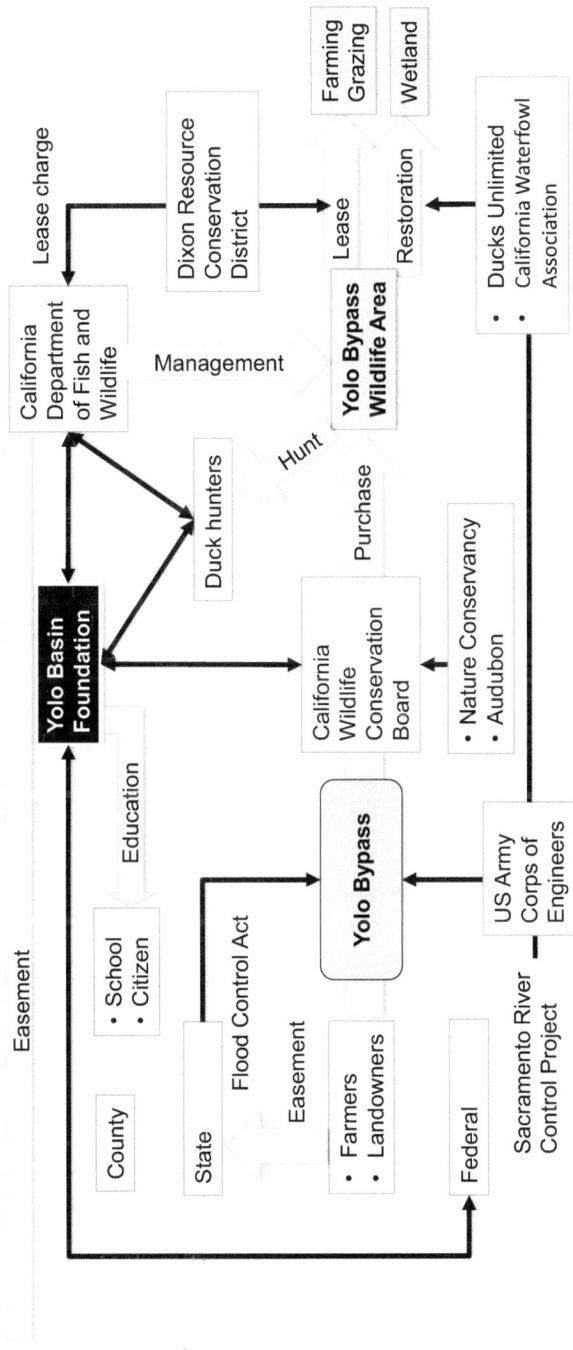

Fig. 28.9 Schematic relationship of stakeholders in establishment and management of YBWA

In 1998, the YBF initiated the Yolo Bypass Working Group (YBWG) under a CALFED Ecosystem Restoration Program Grant (CDFG 2008). The YBWG meetings were served as a forum to educate and inform all parties interested in the Yolo Bypass, approximately every 2–4 months. Information on bypass-related land use, flood management, resource policy, proposed projects, economics, and ecological issues is presented and openly discussed by members (CDFG 2008). More than 30 people representing a wide range of stakeholders regularly attended these meetings (Table 28.1).

Before the YBWG was formed, many landowners and other stakeholders were not informed about issues and decision-making processes. These meetings gave local stakeholders the chance to provide direct input, helping to protect their interests and guide projects proposed by others (CDFG 2008). After frequent meetings, the YBWG produced "A Framework for the Future: Yolo Bypass Management Strategy." Two general land use categories, agriculture with integrated enhancement and habitat enhancement as the primary land use, were recommended in the framework. The YBWG said that the YBWA could be a place where realistic goals and objectives could be achieved, resulting in benefits for all parties involved (YBF 2001; Salciso 2012). The YBWG meetings were continuously held, a total of 37 meetings until 2006, and then finally "Yolo Bypass Wildlife Area Land Management Plan (LMP)" was publicized in 2008 as the fundamental guideline (CDFG 2008).

Table 28.1 Participants in Yolo Bypass Working Group, after CDFG (2008)

• Landowners and their tenants (farmers, ranchers, duck hunters)
• California Department of Fish and Game
• California Department of Water Resources
• State Reclamation Board
• US Fish and Wildlife Service
• California Department of Food and Agriculture
• Natural Resources Conservation Service
• Sacramento-Yolo Mosquito and Vector Control District
• Dixon and Yolo Resource Conservation Districts
• Sacramento Area Flood Control Agency
• Yolo County
• Cities of West Sacramento
• Woodland and Davis
• California Waterfowl Association
• Ducks Unlimited
• National Oceanic and Atmospheric Administration
• National Marine Fisheries Service
• National Weather Service
• Port of Sacramento

28.6   Environmental Education by Yolo Basin Foundation

YBF's activities on managing YBWG are based on its mission, "expanding public appreciation and stewardship of wetland in the Yolo Basin through education and innovative partnerships (https://www.yolobasin.org/about/)." Another and rather their main activity is the facilitation of environmental education in the YBWA, carried out by a 20-member board of directors, a small staff, and over 100 volunteers (Brice 2015). Through the Discover the Flyway school program in the 2016–2017 school year, the YBF provided an educational program to 3656 students and 172 teachers of 181 classes from 58 schools (YBF 2017). The YBF is also the sponsoring NPO for California Duck Days, publishes the Yolo Flyway Newsletter, introduces the public to natural places in the community through public field trips, and hosts the popular Flyway Nights speaker series (CDFG 2008; Brice 2015).

Kulalow said in the interview that "because the area surrounding YBWA, such as West Sacramento, is now under urban development (Fig. 28.10), many people move into the area from outside. People living in the urban area, newcomers in particular, are surprised by explosive sound roaring which is happened to drive off birds from farmyards and loathe occurrence of mosquitoes from wetlands. Environmental education for the people is important to avoid or reduce the conflict."

Funding for YBF's programs is provided through individual and business memberships as well as by a wide variety of private-sector sponsors. Bucks for Ducks, a dinner and auction, is the YBF's fundraising event held every October since 1991 (CDFG 2008).

Fig. 28.10 Urban area developing at the area adjacent to wetlands for wildlife (December 1, 2015)

M. Kamada et al.

28.7 Concluding Remarks

The YBWA has provided multiple functions for various people with different interests, and we can learn many things about the process of getting NbS. Ground-level work to make consensus among federal, state, and local entities and individuals is essential, and a community-based organization takes a central role in forming a platform for meetings, discussions, negotiation, and trust-building. Kulalow, a founder and former Executive Director of the YBF, wrote that "no one worked with me because of the money. Everyone did it because we had shared values (YBF 2016)," and it must be a key for governance.

Another feature of the Yolo bypass is the funding mechanism. Most of the expenses required to manage YBWA are covered by paddy rental fees and hunter license fees. By incorporating these ecosystem services into the economic mechanism, management that balances flood control and biodiversity conservation has been achieved.

Acknowledgments We are grateful to Ms. Robin Kulakow, Mr. Jack DeWit, and Mr. Jeff Stoddard for their cooperation on the interview and to Mr. Kenji Seki, Mr. Ken-ichi Yoshiya, and Ms. Reiko Niwano of Ecosystem Conservation Society Japan for conducting and facilitating the survey on the Yolo Bypass Wildlife Area.

References

Brice A (2015) The yolo bypass wildlife area: history, management and significance for birds. CVBC Bulletin/Winter 2015. http://www.cvbirds.org/wp-content/uploads/2016/09/The-Yolo-Bypass-Wildlife-Area-History-Management-and-Significance-for-Birds.pdf

CDFG (California Department of Fish and Game) (2008) Yolo Bypass Wildlife Area Land Management Plan. 370pp. https://nrm.dfg.ca.gov/FileHandler.ashx?DocumentID=84924&inline

Christian-Smith J (2010) Managing for multiple benefits: farming, flood protection, and habitat restoration in the yolo bypass wildlife area. In: Christian-Smith J et al (eds) California farm water success stories. Pacific Institute, Oakland, pp 17–24. https://pacinst.org/wp-content/uploads/2010/03/success_stories3.pdf

Driessen PPJ, Hegger DLT, Kundzewicz ZW, van Rijswick HFMW, Crabbé A, Larrue C, Matczak P, Pettersson M, Priest S, Suykens C, Raadgever GT, Wierug M (2018) Governance strategies for improving flood resilience in the face of climate change. Water 10(1):1595. https://doi.org/10.3390/w10111595

European Commission (2015) Towards an EU research and innovation policy agenda for Nature-Based Solutions & Re-Naturing Cities. In: Final report of the Horizon 2020 expert group on nature-based solutions and re-Naturing cities. European Commission, Brussels. https://doi.org/10.2777/765301

Garnche C, Howitt RE (2011) Species conservation on a working landscape: the joint production of wildlife and crops in the Yolo Bypass floodplain. https://ageconsearch.umn.edu/record/103973/files/AAEAv4.pdf

Greco S, Larsen EW (2014) Ecological design of multifunctional open channels for flood control and conservation planning. Landsc Urban Plan 131:14–26

Hartmann T, Slavíkoá L, McCarthy S (2019) Nature-based solutions in flood risk management. In: Hartmann T, Slavíkoá L, McCarthy S (eds) Nature-based food risk management on private land – disciplinary perspectives on a multidisciplinary challenge. Springer Open, New York, pp 3–8

Hayes PJ (1999) Yolo bypass wildlife area: birth of a wintering waterfowl wildland. Outdoor California, January–February 1999: 16–17

Jongman B (2018) Effective adaptation to rising flood risk. Nat Commun 9:1986. https://doi.org/10.1038/s41467-018-04396-1

Katibah EF (1984) A brief history of riparian forests in the Central Valley of California. In: Warner RE, Hnedrix KM (eds) California riparian ecosystems: ecology, conservation and productive management. University of California Press, Berkeley, pp 23–19

Salciso RE (2012) The success and continued challenges of the yolo bypass wildlife area: a grassroots restoration. Ecol Law Q 39(4):1085–1134

Sommer T, Harrell B, Nobriga M, Brown R, Moyle P, Kimmerer W, Schemel L (2001) California's yolo bypass – evidence that flood control can be compatible with fisheries, wetlands, wildlife, and agriculture. Fisheries 26(8):6–16

Suddeth GR, Lund JR (2016) Multi-purpose optimization for reconciliation ecology in an engineered floodplain: yolo bypass, California. San Francisco Estuary and Watershed Science 14(1):1–23. https://doi.org/10.15447/sfews.2016v14iss1art5

WCB (California Wildlife Conservation Board) (2001) State of California The Resource Agency Department of Fish and Games Wildlife Conservation Board Minutes August 30, 2001, 130pp

WCB (California Wildlife Conservation Board) (2002) State of California The Resource Agency Department of Fish and Games Wildlife Conservation Board Minutes May 23, 2002, 55pp

WCB (California Wildlife Conservation Board) (2004) State of California The Resource Agency Department of Fish and Games Wildlife Conservation Board Minutes February 19, 2004, 67pp

YBF (Yolo Basin Foundation) (2001) A Framework for the Future: Yolo Bypass Management Strategy. Jones & Stokes (J&S 99079), Sacramento, CA

YBF (Yolo Basin Foundation, Satter H) (2012) Jack DeWit, rice farmer at the Wildlife Area. Yolo Flyway 21(3):3. https://secureservercdn.net/72.167.242.48/h9k.1b4.myftpupload.com/wp-content/uploads/PDF/yfnl/ybf_flyway_2012_fall.pdf

YBF (Yolo Basin Foundation) (2015) Farm and ranch leases at the Yolo Bypass Wild Area. Yolo Flyway 24(3):1–2. http://yolobasin.org/wp-content/uploads/2015/10/17077-YBF-Flyway-Fall-2015.pdf

YBF (Yolo Basin Foundation) (2016) A letter from Robin . . . . Yolo Flyway 25(1):1–2. https://secureservercdn.net/72.167.242.48/h9k.1b4.myftpupload.com/wp-content/uploads/2016/02/17428-YBF-Flyway-Winter-2016.pdf

YBF (Yolo Basin Foundation) (2017) Wildlife area news. Yolo Flyway 26(3):1–2. https://www.yolobasin.org/wp-content/uploads/2017/10/YBF-Flyway-Fall-2017.pdf

Chapter 29
Analysis of the Description of the Multifunctionality of Farmland in the Administrative Plans of Local Municipalities

Yosuke Masuda, Takashi Oka, Erika Yoshinari, Takaaki Nishida, and Tadashi Ikeda

Abstract Farmland has various beneficial functions, such as flood control, water purification, and habitat provision, in addition to food production. These functions are highly compatible with green infrastructure, and the use of farmland as green infrastructure has been discussed in recent years. In order to utilize these functions of farmland, it is preferable to include their usefulness and utilization measures in administrative plans and link them to actual projects. In this research, we collected eight types of administrative plans from local governments across Japan that could be related to the multifunctionality of farmland and reviewed the extent to which they contain descriptions of the multifunctionality of farmland as basic information for promoting the utilization of the multifunctionality of farmland. As a result, we discovered that farmland's multifunctionality was incorporated into the plans of many municipalities. Municipalities with a certain population size and a high financial strength index, in particular, tended to mention the multifunctional role of farmland in their plans more frequently. In addition, we found that some of the functions were mentioned less frequently in the plans. While descriptions of "conservation of natural environment" and "landscape/culture formation and recreation" were common in many plans, descriptions of "disaster mitigation and response" and "water and food supply" in times of disaster were less common.

Y. Masuda (✉)
MISE Solutions Co., Ltd., Tokyo, Japan
e-mail: y.masuda@mises.co.jp

T. Oka · E. Yoshinari ·
Mitsubishi UFJ Research and Consulting Co. Ltd., Osaka, Japan

T. Nishida
Mitsubishi UFJ Research and Consulting Co. Ltd., Osaka, Japan

Kyoto Sangyo University, Kyoto, Japan

T. Ikeda
CS Tower, Tokyo, Japan

© The Author(s) 2022   487
F. Nakamura (ed.), *Green Infrastructure and Climate Change Adaptation*,
Ecological Research Monographs, https://doi.org/10.1007/978-981-16-6791-6_29

Finally, we drew some recommendations that can be used as a reference for future planning and project promotion, including dissemination of knowledge and information of farmland's multifunctionality to government and citizens.

Keywords Multifunctionality · Farmland · Green infrastructure · Policy analysis · Administrative plan

29.1 Background and Goal of the Study

With the increasing risk of disasters due to climate change and decreasing opportunities for contact with nature due to urbanization, the multifunctionality of farmland is attracting attention. For example, farmlands can be utilized as evacuation sites and routes in the event of a disaster (Hara et al. 2016), and paddy fields can store water during floods while providing habitats for a variety of species (Kamda 2019; Osawa 2017). Farmlands also provide scenic, cultural, and recreational values through the formation of rural landscapes, providing opportunities to touch plants and soil for urban residents (see, e.g., Abler (2004), OECD (2001), Pretty (2004), and Van Huylenbroeck et al. (2007) for details of multifunctionality of farmland). This multifunctionality of farmlands has a high affinity with green infrastructure, which has become a growing concern in recent years in Japan. Moreover, the effectiveness of utilizing these functions of farmlands has been argued in a variety of settings (Ichinose 2015; Nishihiro et al. 2020).

While these diverse functions are highly beneficial to the public, they have not been sufficiently promoted due to the difficulty of converting them to market values. Therefore, to utilize these functions, support from the government is necessary. On that note, as a first step, it is important to mention the multifunctionality of farmland in administrative plans. This is because when it is included in the administrative plan, it becomes the basis for project proposals and budget requests, which will lead to the materialization of projects that will demonstrate the multiple functions of farmlands. For example, Yoshikawa et al. (2011) claimed that public support, including institutional development, is necessary for paddy fields to fulfill their flood mitigation function.

Some administrative plans of local governments already mentioned the multifunctionality of farmland, and it is meaningful to understand how it is referred to in such administrative plans to utilize these functions as green infrastructure. As for the studies on the description of the multifunctionality of farmland in local administrative plans, Aragane et al. (2017) reviewed the reference to the disaster prevention functions of urban "greenery" (not limited to farmland) in the Green Basic Plan, and the Ministry of Land, Infrastructure, Transport and Tourism (2018) also reviewed the description of green infrastructure in related administrative plans. However, there is no study focusing on the description of the multifunctionality of "farmland" in administrative plans of municipalities.

To obtain the basic information to promote the utilization of multifunctionality of farmland as green infrastructure, this study reviews eight kinds of administrative plans in the fields of the natural environment, urban planning, land use, disaster prevention, and agriculture in 179 municipalities nationwide to ascertain which functions of farmland are mentioned in what plans. In addition, the tendency of the description is analyzed according to the characteristics of each municipality, which include population size, public finance, scale of agriculture, and flood risk.

29.2   Method

29.2.1   Content Analysis

In this study, analysis of textual information of administrative plans is attempted using the method of content analysis, which is a research method that systematically organizes and quantitatively analyzes textual information (see, e.g., Neuendorf (2001) for details of content analysis). This method has also been recently used in the analysis of environmental policies (e.g., Masuda 2017). Although the content analysis is intended to be as objective as possible, it should be noted that a certain degree of subjective judgment is included in this study since extracting references to the multifunctionality of farmland completely mechanically is difficult due to its complexity.

In the following, we outline the target municipalities and the administrative plans, the classification of the multiple functions of farmland, the criteria for determining the presence or absence of descriptions, and the viewpoint of analysis.

29.2.2   Municipalities for Analysis

To compare and analyze the information on the multifunctionality of farmland in administrative plans depending on the characteristics of each municipality, 179 municipalities in Japan were selected for this study. Since it is often difficult for small municipalities to formulate their own plans due to staffing and financial constraints, this study focused on municipalities with a certain population size: government-designated cities (20), special wards (21), core cities (58), and municipalities that do not belong to any of the above categories with a total population of 150,000 or more (80).

29.2.3   Administrative Plans for Analysis

To compare and analyze the mentioning of the multifunctionality of farmland among different kinds of administrative plans, this study reviewed eight kinds of municipal administrative plans that are most likely to mention the multifunctionality of farmland from the perspectives of land use, agricultural promotion, environment, and disaster prevention. The plans were collected from the websites of each municipality. In cases where the main body of the plan was not published, the summary version of the plan was reviewed (the latest versions of plans were reviewed as of December 2018). In addition, the survey was focused on the main content of the plan such as measures, basic policies, and goals because those described in other sections such as the appendix information of the plan are less likely to be linked to the implementation of projects (the areas to be analyzed for each plan are shown in Table 29.1).

Table 29.1 Analytical administrative plans

No	Plan's name[a]	Foundational law[a]	Formulation ratio[b]	Reviewed contents
1	Green Master Plan (*Midori no Kihon Keikaku*)	Urban Green Space Conservation Act (*Toshi Ryokuchi Hou*)	91.1%	Basic concept/future vision, measures/programs
2	Basic Environment Plan (*Kankyo Kihon Keikaku*)	The Basic Environment Law (*Kankyo Kihon Hou*)	98.9%	Future vision, measures
3	Urban Master Plan (*Toshi Keikaku Masuta Puran*)[c]	City Planning Act (*Toshi Keikaku Hou*)	97.2%	Basic concept, basic policy
4	Regional Biodiversity Strategy (*Seibutsu Tayousei Chiiki Senryaku*)	Basic Act on Biodiversity (*Seibutsu Tayousei Kihon Hou*)	31.8%	Goals/philosophy, basic policy, measures
5	National Land Use Plan (*Kokudo Riyou Keikaku*)	National Land Use Planning Act (*Kokudo Riyou Keikaku Hou*)	10.1%	Basic policy/future vision, measures
6	Fundamental Plans for Regional Resilience (*Kokudo Kyojinka Chiiki Keikaku*)	Basic Act for National Resilience (*Kokudo Kyoujinka Kihon Hou*)	19.6%	Basic concept, goals, required measures
7	Basic Plan for Urban Agriculture Promotion (*Toshi Nougyou Shinkou Kihon Keikaku*)	Basic Law on the Promotion of Urban Agriculture (*Toshi Nougyou Shinkou Kihon Hou*)	14.0%	Vulnerability assessment, basic policy, measures/programs
8	Agricultural Development Plan (*Nougyou Shinkou Keikaku*)	–	40.8%	Basic concept/policy, measures/programs

[a]Italic letters indicate plans and laws in Japanese
[b]Formulation ratio: Percentage of the 179 municipalities that have formulated the plan
[c]Only Urban Master Plan is mandatory

29.2.4 Organizing and Classifying the Multiple Functions of Farmland

To review what functions of farmland are referred to in the administrative plans of municipalities, the multiple functions of farmland are first classified. Regarding the classification, Article 3 of the Food, Agriculture and Rural Areas Basic Act explains that the "multiple roles that agriculture plays through stable production in rural areas include conservation of national land, water resources, natural environment, formation of good landscape, and respect for the cultural tradition in addition to its conventional role as a primary food supplier." Based on the description in this law, this study classified the functions into "national land conservation and disaster prevention," "conservation of natural environment," "landscape/culture formation and recreation," and "water sources conservation." More detailed functions included in these categories were then classified based on the classification of the Science Council of Japan (2001).

Furthermore, from the perspective of farmland as infrastructure, functions were categorized into "functions in normal times" and "functions in times of disaster." The details of each function and the examples of related words and terms are listed in Table 29.2.

29.2.5 Establishment of Criteria for the Reference to the Multifunctionality of Farmland in Administrative Plans

Descriptions related to the multifunctionality of farmland (including fields, rice paddies, valleys, and productive green lands) were extracted in administrative plans and were reviewed whether or not they are mentioned in each plan. The criteria for the presence of the descriptions are explained below.

Although it is ideal to have completely objective criteria, it is difficult to judge the presence of the descriptions by simply searching particular words or phrases. For example, with regard to the flood control function of farmland, simply extracting the word "flood control" does not allow us to determine whether it belongs to farmland. Also, "flood control" can be termed in various ways, such as "flood prevention," "rainwater storage," "storing water," and "reducing the amount of water flowing into rivers." Therefore, this study decided not to extract specific words or phrases but to read the texts based on certain rules and make qualitative judgments on whether there are references to the multifunctionality of farmland. The rules are as follows:

1. Only functions clearly stated as functions that the "farmland" (including paddy fields, fields, valleys, rice paddies, and production green spaces) performs are counted (descriptions about functions of "green spaces" or "open spaces" that include farmland but have a broader meaning are not counted).

Table 29.2 Categorization of the functions of farmlands

Categories		Functions	Examples of related phrases
Normal times	National land conservation and disaster prevention	• Flood control • Soil erosion prevention • Mitigation of heat island phenomenon	• Flood prevention • Rainwater storage
	Conservation of natural environment	• Provision of habitats • Ecological network formation • Water purification • Organic waste decomposition • Air purification • Prevention of resource over-accumulation and deprivation • Carbon sequestration	• Sustainable farming • Biodiversity conservation
	Landscape/culture formation and recreation	• Education and learning • Amenity • Landscape formation • Culture formation • Community building	• Farming activity • Rural landscapes • Contact with greenery and soil • Agricultural tourism • Biological survey for education
	Water resources conservation	• Water resources conservation	• Water source protection • Groundwater conservation
In times of disaster	Disaster mitigation and response	• Flood control • Evacuation site provision • Fire spread prevention	• Flood plain • Securing evacuation routes
	Water and food supply	Food and water provision (for fire suppression and daily use)	• Fire suppression water

2. Regarding the disaster prevention function of farmland, descriptions related to preventing damage to the farmland itself are not counted.
3. General descriptions such as "multiple functions" that do not refer to specific functions (such as flood prevention) are not counted.

29.2.6 Analysis of Descriptions According to Municipal Characteristics

To analyze the differences in the reference to the multifunctionality of farmland according to municipal characteristics (socioeconomic status and natural environment status), the correlation coefficients between the reference to the multifunctionality in each plan and municipal characteristics were calculated. The

Table 29.3 List of attribute information of municipalities used in the analysis

Attribute info	Data	Data source
Population size	Population size	National census, Ministry of Internal Affairs and Communications (2015)
Public finance	Financial strength index	List of major financial indicators for local governments, Ministry of Internal Affairs and Communications (2016)
Scale of agriculture	Farmland acreage	Census of Agriculture and Forestry, Ministry of Agriculture, Forestry and Fisheries (2015)
	Farmland acreage (ratio)	Census of Agriculture and Forestry, Ministry of Agriculture, Forestry and Fisheries (2015) and Area survey of prefectural cities, towns, and villages nationwide, Institute of Land and Geography (2016)
	Number of agricultural enterprises	
	Farmland abandonment rate	
Flood risk	Flood affected area of residential area and others	Statistical survey on flood damage, Ministry of Land, Infrastructure, Transport and Tourism (aggregated the data from 2006 to 2016)
	Flood affected area of farmland	
	Total flood affected area	

Table 29.4 Categories according to population size and farmland acreage

• Municipality type I: Population size and farmland acreage are above the average of the target municipalities
• Municipality type II: Only the population size is above the average of the target municipalities
• Municipality type III: Only the acreage of farmland is above the average of the target municipalities
• Municipality type IV: Population size and farmland acreage are less than the average of the target municipalities

objective variable was the number of plans referring to the multifunctionality of farmland, and the explanatory variables were the population size and financial strength index as the socioeconomic conditions and the area of farmland, farmland ratio, abandoned farmland ratio, the number of agricultural enterprises, and acreage of the flooded area as the natural environmental conditions (Table 29.3).

A total of 179 municipalities were classified into the four categories (Table 29.4) according to population size and farmland acreage, and the number of plans with descriptions of multifunctionality for each category was counted to clarify the differences among the categories.

29.3 Results and Discussion

29.3.1 Description of the Multifunctionality of Farmland in each Administrative Plan

The result of the review process revealed that the description of the multifunctionality of farmland varied depending on the plan (Table 29.5). The three plans with a relatively high formulation ratio had a high percentage of descriptions of the categories of natural environment conservation, landscape/culture formation, and recreational functions of farmland, while the Fundamental Plans for Regional Resilience and the Basic Plan for Urban Agriculture Promotion, which exhibit low formulation ratios, had a high percentage of descriptions of all categories of multiple functions including disaster prevention. The following is a summary of the status of the description by each plan.

• In terms of Green Basic Plan, the Basic Environment Plan, and the Regional Biodiversity Strategy, many of them included descriptions of natural environment conservation and conservation of landscape and culture during normal times (more than 60%). On the other hand, only a few (less than 30%) mentioned disaster mitigation and response in times of disaster.
• Regarding Urban Master Plans, a high percentage of them mentioned natural environment conservation (58%) and landscape/culture formation and recreation

Table 29.5 Percentage of each kind of plan that refers to each category of farmland's functions

Name of plans	Formulation ratio	Normal times				In times of disaster	
		National land conservation and disaster prevention	Conservation of natural environment	Landscape/ culture formation and recreation	Water resources conservation	Disaster mitigation and response	Water and food supply
Green Master Plan	91%	48%	64%	79%	15%	28%	0%
Basic Environment Plan	99%	34%	76%	73%	34%	5%	0%
Urban Master Plan	97%	44%	58%	91%	19%	50%	0%
Regional Biodiversity Strategy	32%	32%	95%	74%	18%	11%	0%
National Land Use Plan	10%	83%	83%	83%	61%	17%	0%
Fundamental Plans for Regional Resilience	20%	43%	17%	17%	14%	29%	20%
Urban Agricultural Promotion Basic Plan	14%	96%	100%	100%	36%	92%	12%
Agricultural Development Plan	41%	74%	93%	96%	66%	26%	4%

(91%) during normal times. Disaster mitigation and response in times of disaster were mentioned by 50%.

- In respect of National Land Use Plan, the percentage of descriptions of national land conservation and disaster prevention, natural environment conservation, landscape and culture formation and recreation, and water resource conservation during normal times was generally high (more than 60%). On the contrary, the percentage of descriptions of disaster mitigation and response in times of disaster was limited (17%).

- In the Fundamental Plans for Regional Resilience, the description of the multifunctionality of farmland was limited in both normal times and in times of disaster, but among them, disaster prevention and land conservation functions in normal times (43%) and disaster mitigation and response in times of disaster (29%) were relatively high.

- In terms of the Basic Plan for Urban Agriculture Promotion, every plan has descriptions of national land conservation and disaster prevention, conservation of natural environment, and landscape/culture formulation and recreation during normal times. In addition, a high percentage (92%) of the plans described disaster mitigation and response in times of disaster.

- In the Agricultural Development Plan, the percentage of descriptions of land conservation and disaster prevention, natural environment conservation, landscape/culture formulation and recreation, and water resource conservation during normal times was generally high (more than 60%).

Results showed that only 2% of the municipalities covered all the functions, and disaster mitigation and response and water and food supply functions in times of disaster, in particular, tended not to be mentioned.

The rate of municipalities that did not mention the disaster prevention function of farmland in any of their plans for disaster prevention was 19.8% and 36.8% in normal times and in times of disaster, respectively. This indicates that at least some municipalities may not even recognize farmland's disaster prevention and mitigation functions.

29.3.2   Tendency of the Description of the Multifunctionality of Farmland According to the Characteristics of Municipalities

The number of plans that refer to the multifunctionality of farmland had a positive correlation with the population, the financial strength index, and the number of agricultural enterprises. Meanwhile, the number of plans had a negative correlation with the farmland abandonment rate (Table 29.6). The absolute values of the correlation coefficients for Green Master Plans, Environmental Master Plans, and

Table 29.6 Correlation coefficients between the number of plans describing the multifunctionality of farmland and municipal attribute values (eight plans)

Name of plans	Normal times				In times of disaster		Total
	National land conservation and disaster prevention	Conservation of natural environment	Landscape/ culture formation and recreation	Water resources conservation	Disaster mitigation and response	Water and food supply	
Population size	0.314	0.347	0.270	0.192	0.260	0.172	0.354
Financial index	0.267	0.265	0.259	0.109	0.330	0.054	0.309
Farmland acreage	0.118	0.106	0.086	0.288	-0.240	0.122	0.097
Farmland acreage (ratio)	0.080	0.114	0.062	0.207	-0.118	0.052	0.089
Number of agricultural enterprises	-0.220	-0.287	-0.329	-0.144	-0.235	-0.114	-0.312
Farmland abandonment rate	0.274	0.234	0.219	0.402	-0.196	0.038	0.237
Flood affected area of residential area and others	0.246	0.114	0.112	0.065	0.009	0.030	0.143
Flood affected area of farmland	0.119	0.067	0.065	0.305	-0.061	-0.051	0.113
Total flood affected area	0.249	0.124	0.121	0.250	-0.035	-0.014	0.174

*Orange cell indicates a correlation coefficient of 0.2 or higher, light blue cell indicates a correlation coefficient of -0.2 or lower, red font indicates a negative correlation coefficient

Urban Master Plans were similar (Tables 29.7, 29.8, and 29.9, respectively), although the absolute value of the correlation coefficients was smaller.

The results suggest that population size, financial strength, and farmland status are indicators of how well an administrative plan is written and formulated. On the other hand, the scale of floods has little impact on the description of farmland multifunctionality in the administrative plan, suggesting that there are some municipalities with many disasters but limited efforts in promoting the utilization of farmland as disaster prevention and mitigation measures.

29.3.3 Number of Plans with Descriptions with Respect to Population Size and Farmland Acreage

Figure 29.1 shows the differences in descriptions of the multifunctionality of farmlands by municipality type. Municipality type I, where both population size and farmland acreage are above average, had the highest number of plans mentioning all functions. In Municipality type II, where only the population size was above average, the natural environment and landscape/culture formation and recreation were described most frequently. In Municipality type III, which is above average

Table 29.7 Correlation coefficients between the number of plans describing the multifunctionality of farmland and municipal attribute values (Green Master Plan)

Name of plans	Normal times				In times of disaster		Total
	National land conservation and disaster prevention	Conservation of natural environment	Landscape/ culture formation and recreation	Water resources conservation	Disaster mitigation and response	Water and food supply	
Population size	0.142	0.228	0.182	0.209	0.032	-	0.220
Financial index	0.261	0.236	0.196	0.123	0.211	-	0.294
Farmland acreage	-0.075	-0.027	-0.001	-0.065	-0.112	-	-0.078
Farmland acreage (ratio)	0.015	0.129	0.030	-0.019	0.025	-	0.056
Number of agricultural enterprises	-0.179	-0.266	-0.337	-0.095	-0.155	-	-0.296
Farmland abandonment rate	0.048	0.010	0.068	-0.033	-0.085	-	0.007
Flood affected area of residential area and others	0.115	0.048	-0.003	-0.070	-0.037	-	0.025
Flood affected area of farmland	-0.050	0.001	-0.023	0.010	-0.056	-	-0.035
Total flood affected area	0.045	0.034	-0.018	-0.041	-0.063	-	-0.007

*Orange cell indicates a correlation coefficient of 0.2 or higher, light blue cell indicates a correlation coefficient of -0.2 or lower, red font indicates a negative correlation coefficient

Table 29.8 Correlation coefficients between the number of plans describing the multifunctionality of farmland and municipal attribute values (Basic Environmental Plan)

Name of plans	Normal times				In times of disaster		Total
	National land conservation and disaster prevention	Conservation of natural environment	Landscape/ culture formation and recreation	Water resources conservation	Disaster mitigation and response	Water and food supply	
Population size	0.057	-0.004	0.011	-0.029	-0.044	-	0.005
Financial index	0.165	0.129	0.183	0.042	0.121	-	0.196
Farmland acreage	0.021	0.154	0.062	0.186	-0.030	-	0.141
Farmland acreage (ratio)	0.054	0.167	0.160	0.155	0.085	-	0.198
Number of agricultural enterprises	-0.069	-0.167	-0.276	-0.019	-0.053	-	-0.186
Farmland abandonment rate	0.168	0.316	0.159	0.352	0.024	-	0.343
Flood affected area of residential area and others	0.020	0.116	0.047	0.161	-0.001	-	0.117
Flood affected area of farmland	0.079	0.121	0.034	0.175	-0.041	-	0.133
Total flood affected area	0.067	0.161	0.055	0.228	-0.029	-	0.170

*Orange cell indicates a correlation coefficient of 0.2 or higher, light blue cell indicates a correlation coefficient of -0.2 or lower, red font indicates a negative correlation coefficient

Table 29.9 Correlation coefficients between the number of plans describing the multifunctionality of farmland and municipal attribute values (Urban Master Plan)

Name of plans	Normal times				In times of disaster		Total
	National land conservation and disaster prevention	Conservation of natural environment	Landscape/ culture formation and recreation	Water resources conservation	Disaster mitigation and response	Water and food supply	
Population size	-0.009	0.146	0.012	0.180	0.071	-	0.128
Financial index	0.053	0.051	0.074	0.019	0.161	-	0.118
Farmland acreage	0.167	0.042	0.120	0.216	-0.236	-	0.078
Farmland acreage (ratio)	0.127	0.052	0.115	0.110	-0.203	-	0.048
Number of agricultural enterprises	-0.125	-0.145	-0.290	-0.131	-0.190	-	-0.269
Farmland abandonment rate	0.243	0.020	0.201	0.216	-0.223	-	0.120
Flood affected area of residential area and others	0.132	0.078	0.040	-0.024	-0.091	-	0.045
Flood affected area of farmland	0.057	-0.063	-0.026	0.224	0.003	-	0.056
Total flood affected area	0.129	0.012	0.011	0.134	-0.060	-	0.069

*Orange cell indicates a correlation coefficient of 0.2 or higher, light blue cell indicates a correlation coefficient of -0.2 or lower, red font indicates a negative correlation coefficient

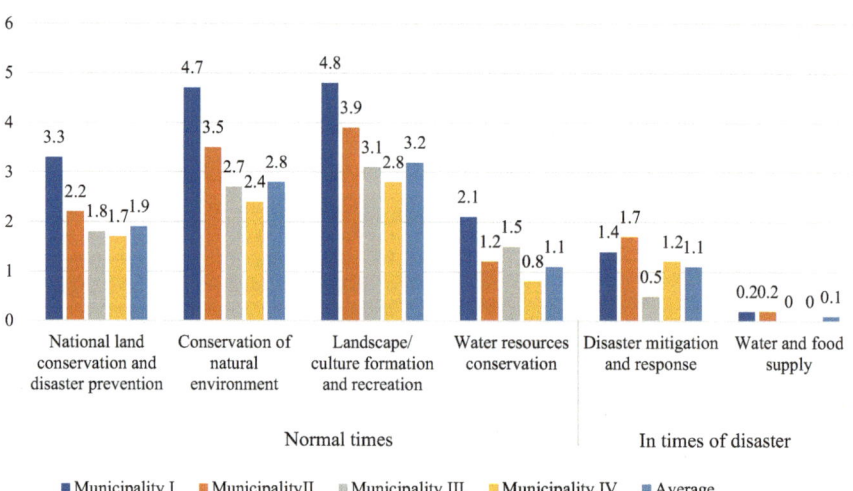

Fig. 29.1 Number of plans that describe the multifunctionality of farmland (by category, based on population size and farmland area)

only in terms of farmland acreage, the number of plans with descriptions of multifunctionality was larger than in Municipality II in terms of water resource conservation, but lower than in Municipality II in all other functions. Municipality type IV, where both population size and farmland acreage are below average, surpassed Municipality type III in disaster mitigation and response in times of disaster but was below the other municipality types in the other functions. In terms of the functions during a disaster, all municipalities had limited descriptions compared to normal times.

29.4   Conclusion: Issues and Future Directions of Green Infrastructure Policies for Farmland

Results of this study reveal that the multifunctionality of farmland has been mentioned in the plans of many municipalities in the fields of natural environment, urban planning, and agriculture. In particular, municipalities with a certain population size and financial strength tended to mention the multifunctionality of farmland more in their plans. However, some functions tended not to be referred to in the plans, especially in the area of "national land conservation and disaster prevention."

The results of this study indicate that the following four approaches may be effective in promoting measures to utilize farmland as green infrastructure.

1. With regard to the Basic Environment Plan, the Green Master Plan, and the Urban Master Plan, the rate of plan formulation is high, but there is currently little mention of the national land conservation and disaster prevention functions of farmland. Therefore, in the process of reviewing and revising these plans, it would be effective to inform local governments of the importance of the national land conservation and disaster prevention functions of farmland and to include this point in the plans, which will consequently lead to the implementation of the measures.

2. Although only a few municipalities have formulated the National Land Use Plan, the Basic Plan for Urban Agriculture Promotion, and the Agricultural Development Plan, these plans tend to refer to the multifunctionality of farmland. The formulation of these plans could lead to projects that promote the utilization of the multifunctionality of farmland since the formulation of the plans is often with the description of multifunctionality, which could be a basis to promote the activities. Therefore, it may be effective to promote the formulation of the plan, and it is desirable to promote appropriate support from the national and prefectural governments for municipalities with limited human resources and budgets.

3. As for municipalities at high risk of flooding, the percentage of those that currently include the use of farmland for disaster reduction in their plans is not higher than other municipalities, but the potential need for disaster reduction is considered to be high. Therefore, it would be effective to make the effectiveness

of the disaster reduction function of farmland known to the local government and public so that it can be written in the plan and specific measures can be promoted.

4. For municipalities with a large farmland size and a small population (Category III), there were few plans that mentioned the multifunctionality of farmland despite its high potential for utilization. For these areas, it would be effective to specifically consider how to utilize the multifunctional functions of farmland and describe them in the plans to materialize them.

This study comprehensively reviewed the descriptions of the multifunctionality of farmland in the administrative plans of local governments and derived suggestions for utilizing the multifunctionality of farmland in the future.

In this analysis, only the presence or absence of mentions was identified based on certain rules, and there was no research on how the multifunctionality of farmland was described or what kind of descriptions will more likely lead to effective measures. In addition, while this study investigated the status of the description of the multifunctionality of farmland in administrative plans, it did not delve deeply into why the description in the plans has not progressed in the first place or why the formulation of plans that facilitate the description itself has not progressed. Furthermore, while this research focused on the multifunctionality of farmland, it is important to understand and analyze descriptions of the multifunctionality of other land uses and ecosystems, such as forests and parks. We believe that further research on these issues will lead to the formulation of more effective administrative plans for utilizing the multifunctionality of farmland and other green spaces.

References

Abler D (2004) Multifunctionality, agricultural policy, and environmental policy. Agric Resour Econ Rev 33(1):8–17

Aragane K, Nishimura A, Funakubo S (2017) A study on the role of the green master plan in disaster prevention. J Japanese Ins Landscape Architect 80(5):673–676. (in Japanese)

Hara Y, Yoshii T, Tsujimura K, Sanpei Y (2016) Land use and land ownership characteristics of designated disaster evacuation farmland in Sakai City, Japan. In: Memories of the center for research and education of disaster reduction, vol 2. Wakayama University, Wakayama, pp 31–37. (in Japanese)

Ichinose T (2015) Disaster risk reduction based on green infrastructure in rural landscapes of Japan amid population decline. J Rural Plan 34(3):353–356. (in Japanese)

Kamda M (2019) Utilization of Paddy field as green infrastructure for eco-DRR. J Rural Plan 37(4):358–361. (in Japanese)

Masuda Y (2017) Quantitative analysis of chronological changes of the wetland mitigation policy in the United States. J Environ Assess 15(2):47–59. (in Japanese)

Ministry of Land, Infrastructure, Transport and Tourism (2018) Status survey on the description of green infrastructure in major administrative plans. (in Japanese)

Neuendorf KA (2001) The content analysis guidebook. SAGE Publications, Newbury Park

Nishihiro J, Ohtsuki K, Kohzu A, Kato H, Ogasawara S, Satake Y, Shoji T, Hasegawa M, Kondoh A (2020) "SATOYAMA green-infrastructure", a measure for climate change adaption: potential for multipurpose usage of abandoned paddy on small valley bottoms in the Lake Inbanuma watershed, Chiba. Japan Ecol Civil Eng 22(2):175–185. (in Japanese)

Organisation for Economic Co-operation and Development (OECD) (2001) Towards an analytical framework. OECD, Paris

Osawa T (2017) Agricultural land use policy in Japan in an era of declining population. Wildlife Human Soc 8:43–45. (in Japanese)

Pretty JN (2004) Agri-environmental stewardship schemes and "multifunctionality". Rev Agric Econ 26(2):220–237

Science Council of Japan (2001) Multifunctionality of Agriculture and Forest regarding Global Environment and Human Life (report). (in Japanese)

Van Huylenbroeck G, Vandermeulen V, Mettepenningen E, Verspecht A (2007) Multifunctionality of agriculture: a review of definitions, evidence and instruments. Living Rev Landscape Res 1:3

Yoshikawa N, Arita H, Misawa S, Miyazu S (2011) Evaluation of social function of paddy field dam and its technical prospects. J Japan Soc Hydrol Water Resour 24(5):271–279. (in Japanese)

Index